Proceedings of the Institute of Industrial Engineers Asian Conference 2013

Yi-Kuei Lin · Yu-Chung Tsao
Shi-Woei Lin
Editors

Proceedings of the Institute of Industrial Engineers Asian Conference 2013

Volume II

Springer

Editors
Yi-Kuei Lin
Yu-Chung Tsao
Shi-Woei Lin
Department of Industrial Management
National Taiwan University of Science and Technology
Taipei
Taiwan, R.O.C.

ISBN 978-981-4451-97-0   ISBN 978-981-4451-98-7 (eBook)
DOI 10.1007/978-981-4451-98-7
Springer Singapore Heidelberg New York Dordrecht London

Library of Congress Control Number: 2013943711

© Springer Science+Business Media Singapore 2013
This work is subject to copyright. All rights are reserved by the Publisher, whether the whole or part of the material is concerned, specifically the rights of translation, reprinting, reuse of illustrations, recitation, broadcasting, reproduction on microfilms or in any other physical way, and transmission or information storage and retrieval, electronic adaptation, computer software, or by similar or dissimilar methodology now known or hereafter developed. Exempted from this legal reservation are brief excerpts in connection with reviews or scholarly analysis or material supplied specifically for the purpose of being entered and executed on a computer system, for exclusive use by the purchaser of the work. Duplication of this publication or parts thereof is permitted only under the provisions of the Copyright Law of the Publisher's location, in its current version, and permission for use must always be obtained from Springer. Permissions for use may be obtained through RightsLink at the Copyright Clearance Center. Violations are liable to prosecution under the respective Copyright Law.
The use of general descriptive names, registered names, trademarks, service marks, etc. in this publication does not imply, even in the absence of a specific statement, that such names are exempt from the relevant protective laws and regulations and therefore free for general use.
While the advice and information in this book are believed to be true and accurate at the date of publication, neither the authors nor the editors nor the publisher can accept any legal responsibility for any errors or omissions that may be made. The publisher makes no warranty, express or implied, with respect to the material contained herein.

Printed on acid-free paper

Springer is part of Springer Science+Business Media (www.springer.com)

Preface

The organizers of the IIE Asian Conference 2013, held July 19–22, 2013 in Taipei, Taiwan, are pleased to present the proceedings of the conference, which includes about 180 papers accepted for presentation at the conference. The papers in the proceedings were selected from more than 270 paper submissions. Based on input from peer reviews, the committee selected these papers for their perceived quality, originality, and appropriateness to the theme of the IIE Asian Conference. We would like to thank all contributors who submitted papers. We would also like to thank the track chairs and colleagues who provided reviews of the submitted papers.

The main objective of IIE Asian Conference 2013 is to provide a forum for exchange of ideas on the latest developments in the field of industrial engineering and management. This series of conferences were started and supported by the Institute of Industrial Engineers (IIE) in 2011 as a meeting of organizations striving to establish a common focus for research and development in the field of industrial engineering. Now after just 2 years of the first event, the IIE Asian Conference has become recognized as an international focal point annually attended by researchers and developers from dozens of countries around the Asian.

Whether the papers included in the proceedings are work-in-progress or finished products, the conference and proceedings offer all authors opportunities to disseminate the results of their research and receive timely feedback from colleagues, without long wait associated with publication in peer-reviewed journals. We hope that the proceedings will represent a worthwhile contribution to the theoretical and applied body of knowledge in industrial engineering and management. We also hope that it will attract some other researchers and practitioners to join future IIE Asian Conferences.

Finally, the organizers are indebted to a number of people who gave freely their time to make the conference a reality. We would like to acknowledge Dr. Ching-Jong Liao, the president of the National Taiwan University of Science and Technology (Taiwan Tech), and all faculties and staffs in the Department of Industrial

Management at the Taiwan Tech, who have contributed tremendous amount of time and efforts to this conference. The Program Committee would also like to express its appreciation to all who participate to make the IIE Asian Conference 2013 a successful and enriching experience.

Program Committee of IIE Asian Conference 2013

Contents

An Optimal Ordering Policy of the Retailers Under Partial Trade Credit Financing and Restricted Cycle Time in Supply Chain   1
Shin'ichi Yoshikawa

An Immunized Ant Colony System Algorithm to Solve Unequal Area Facility Layout Problems Using Flexible Bay Structure   9
Mei-Shiang Chang and Hsin-Yi Lin

Characterizing the Trade-Off Between Queue Time and Utilization of a Manufacturing System 19
Kan Wu

Using Latent Variable to Estimate Parameters of Inverse Gaussian Distribution Based on Time-Censored Wiener Degradation Data 29
Ming-Yung Lee and Cheng-Hung Hu

Interpolation Approximations for the Performance of Two Single Servers in Series 37
Kan Wu

On Reliability Evaluation of Flow Networks with Time-Variant Quality Components 45
Shin-Guang Chen

Defect Detection of Solar Cells Using EL Imaging and Fourier Image Reconstruction 53
Ya-Hui Tsai, Du-Ming Tsai, Wei-Chen Li and Shih-Chieh Wu

Teaching Industrial Engineering: Developing a Conjoint Support System Catered for Non-Majors 63
Yoshiki Nakamura

An Automatic Image Enhancement Framework for Industrial Products Inspection................................ 73
Chien-Cheng Chu, Chien-Chih Wang and Bernard C. Jiang

Ant Colony Optimization Algorithms for Unrelated Parallel Machine Scheduling with Controllable Processing Times and Eligibility Constraints................................ 79
Chinyao Low, Rong-Kwei Li and Guan-He Wu

General Model for Cross-Docking Distribution Planning Problem with Time Window Constraints............................. 89
Parida Jewpanya and Voratas Kachitvichyanukul

A New Solution Representation for Solving Location Routing Problem via Particle Swarm Optimization..................... 103
Jie Liu and Voratas Kachitvichyanukul

An Efficient Multiple Object Tracking Method with Mobile RFID Readers................................ 111
Chieh-Yuan Tsai and Chen-Yi Huang

A New Bounded Intensity Function for Repairable Systems........ 119
Fu-Kwun Wang and Yi-Chen Lu

Value Creation Through 3PL for Automotive Logistical Excellence................................ 127
Chin Lin Wen, Schnell Jeng, Danang Kisworo, Paul K. P. Wee and H. M. Wee

Dynamics of Food and Beverage Subsector Industry in East Java Province: The Effect of Investment on Total Labor Absorption................................ 133
Putri Amelia, Budisantoso Wirjodirdjo and Niniet Indah Arvitrida

Solving Two-Sided Assembly Line Balancing Problems Using an Integrated Evolution and Swarm Intelligence................ 141
Hindriyanto Dwi Purnomo, Hui-Ming Wee and Yugowati Praharsi

Genetic Algorithm Approach for Multi-Objective Optimization of Closed-Loop Supply Chain Network..................... 149
Li-Chih Wang, Tzu-Li Chen, Yin-Yann Chen, Hsin-Yuan Miao, Sheng-Chieh Lin and Shuo-Tsung Chen

Replacement Policies with a Random Threshold Number of Faults ... 157
Xufeng Zhao, Mingchih Chen, Kazunori Iwata, Syouji Nakamura
and Toshio Nakagawa

A Multi-Agent Model of Consumer Behavior Considering Social Networks: Simulations for an Effective Movie Advertising Strategy ... 165
Yudai Arai, Tomoko Kajiyama and Noritomo Ouchi

Government Subsidy Impacts on a Decentralized Reverse Supply Chain Using a Multitiered Network Equilibrium Model ... 173
Pin-Chun Chen and I-Hsuan Hong

A Capacity Planning Method for the Demand-to-Supply Management in the Pharmaceutical Industry ... 181
Nobuaki Ishii and Tsunehiro Togashi

Storage Assignment Methods Based on Dependence of Items ... 189
Po-Hsun Kuo and Che-Wei Kuo

Selection of Approximation Model on Total Perceived Discomfort Function for the Upper Limb Based on Joint Moment ... 197
Takanori Chihara, Taiki Izumi and Akihiko Seo

Waiting as a Signal of Quality When Multiple Extrinsic Cues are Presented ... 205
Shi-Woei Lin and Hao-Yuan Chan

Effect of Relationship Types on the Behaviors of Health Care Professionals ... 211
Shi-Woei Lin and Yi-Tseng Lin

A Simulation with PSO Approach for Semiconductor Back-End Assembly ... 219
James T. Lin, Chien-Ming Chen and Chun-Chih Chiu

Effect of Grasp Conditions on Upper Limb Load During Visual Inspection of Objects in One Hand ... 229
Takuya Hida and Akihiko Seo

A Process-Oriented Mechanism Combining Fuzzy Decision Analysis for Supplier Selection in New Product Development ... 239
Jiun-Shiung Lin, Jen-Huei Chang and Min-Che Kao

Reliability-Based Performance Evaluation for a Stochastic Project Network Under Time and Budget Thresholds.............. 249
Yi-Kuei Lin, Ping-Chen Chang and Shin-Ying Li

System Reliability and Decision Making for a Production System with Intersectional Lines............................. 257
Yi-Kuei Lin, Ping-Chen Chang and Kai-Jen Hsueh

Customer Perceptions of Bowing with Different Trunk Flexions..... 265
Yi-Lang Chen, Chiao-Ying Yu, Lan-Shin Huang, Ling-Wei Peng and Liang-Jie Shi

A Pilot Study Determining Optimal Protruding Node Length of Bicycle Seats Using Subjective Ratings...................... 271
Yi-Lang Chen, Yi-Nan Liu and Che-Feng Cheng

Variable Neighborhood Search with Path-Relinking for the Capacitated Location Routing Problem................... 279
Meilinda F. N. Maghfiroh, A. A. N. Perwira Redi and Vincent F. Yu

Improving Optimization of Tool Path Planning in 5-Axis Flank Milling by Integrating Statistical Techniques.............. 287
Chih-Hsing Chu and Chi-Lung Kuo

A Multiple Objectives Based DEA Model to Explore the Efficiency of Airports in Asia–Pacific.................................. 295
James J. H. Liou, Hsin-Yi Lee and Wen-Chein Yeh

A Distributed Constraint Satisfaction Approach for Supply Chain Capable-to-Promise Coordination...................... 303
Yeh-Chun Juan and Jyun-Rong Syu

Design and Selection of Plant Layout by Mean of Analytic Hierarchy Process: A Case Study of Medical Device Producer...... 311
Arthit Chaklang, Arnon Srisom and Chirakiat Saithong

Using Taguchi Method for Coffee Cup Sleeve Design............. 319
Yiyo Kuo, Hsin-Yu Lin, Ying Chen Wu, Po-Hsi Kuo, Zhi-He Liang and Si Yong Wen

Utilizing QFD and TRIZ Techniques to Design a Helmet Combined with the Wireless Camcorder...................... 327
Shu-Jen Hu, Ling-Huey Su and Jhih-Hao Laio

**South East Asia Work Measurement Practices Challenges
and Case Study**... 335
Thong Sze Yee, Zuraidah Mohd Zain and Bhuvenesh Rajamony

**Decision Support System: Real-Time Dispatch
of Manufacturing Processes**................................. 345
Chung-Wei Kan and An-Pin Chen

**The Application of MFCA Analysis in Process Improvement:
A Case Study of Plastics Packaging Factory in Thailand**...... 353
Chompoonoot Kasemset, Suchon Sasiopars and Sugun Suwiphat

Discussion of Water Footprint in Industrial Applications..... 363
Chung Chia Chiu, Wei-Jung Shiang and Chiuhsiang Joe Lin

**Mitigating Uncertainty Risks Through Inventory Management:
A Case Study for an Automobile Company**..................... 371
Amy Chen, H. M. Wee, Chih-Ying Hsieh and Paul Wee

Service Quality for the YouBike System in Taipei............ 381
Jung-Wei Chang, Xin-Yi Jiang, Xiu-Ru Chen,
Chia-Chen Lin and Shih-Che Lo

Replenishment Strategies for the YouBike System in Taipei.... 389
Chia-Chen Lin, Xiu-Ru Chen, Jung-Wei Chang,
Xin-Yi Jiang and Shih-Che Lo

**A Tagging Mechanism for Solving the Capacitated
Vehicle Routing Problem**.................................... 397
Calvin K. Yu and Tsung-Chun Hsu

**Two-Stage Multi-Project Scheduling with Minimum
Makespan Under Limited Resource**............................ 405
Calvin K. Yu and Ching-Chin Liao

**A Weighting Approach for Scheduling Multi-Product
Assembly Line with Multiple Objectives**..................... 415
Calvin K. Yu and Pei-Fang Lee

Exploring Technology Feature with Patent Analysis........... 423
Ping Yu Hsu, Ming Shien Cheng, Kuo Yen Lu and Chen Yao Chung

Making the MOST® Out of Economical Key-Tabbing Automation... 433
P. A. Brenda Yap, S. L. Serene Choo and Thong Sze Yee

Integer Program Modeling of Portfolio Optimization with Mental Accounts Using Simulated Tail Distribution 441
Kuo-Hwa Chang, Yi Shou Shu and Michael Nayat Young

Simulated Annealing Algorithm for Berth Allocation Problems...... 449
Shih-Wei Line and Ching-Jung Ting

Using Hyperbolic Tangent Function for Nonlinear Profile Monitoring... 457
Shu-Kai S. Fan and Tzu-Yi Lee

Full Fault Detection for Semiconductor Processes Using Independent Component Analysis....................... 465
Shu-Kai S. Fan and Shih-Han Huang

Multi-Objective Optimal Placement of Automatic Line Switches in Power Distribution Networks...................... 471
Diego Orlando Logrono, Wen-Fang Wu and Yi-An Lu

The Design of Combing Hair Assistive Device to Increase the Upper Limb Activities for Female Hemiplegia 479
Jo-Han Chang

Surgical Suites Scheduling with Integrating Upstream and Downstream Operations.............................. 487
Huang Kwei-Long, Lin Yu-Chien and Chen Hao-Huai

Research on Culture Supply Chain Intension and Its Operation Models 497
Xiaojing Li and Qian Zhang

Power System by Variable Scaling Hybrid Differential Evolution 505
Ji-Pyng Chiou, Chong-Wei Lo and Chung-Fu Chang

Investigating the Replenishment Policy for Retail Industries Under VMI Strategic Alliance Using Simulation................. 513
Ping-Yu Chang

Applying RFID in Picker's Positioning in a Warehouse 521
Kai Ying Chen, Mei Xiu Wu and Shih Min Chen

An Innovation Planning Approach Based on Combining Technology Progress Trends and Market Price Trends..................... 531
Wen-Chieh Chuang and Guan-Ling Lin

Contents

Preemptive Two-Agent Scheduling in Open Shops Subject to Machine Availability and Eligibility Constraints 539
Ming-Chih Hsiao and Ling-Huey Su

Supply Risk Management via Social Capital Theory and Its Impact on Buyer's Performance Improvement and Innovation 549
Yugowati Praharsi, Maffie Linda Araos Dioquino and Hui-Ming Wee

Variable Time Windows-Based Three-Phase Combined Algorithm for On-Line Batch Processing Machine Scheduling with Limited Waiting Time Constraints 559
Dongwei Yang, Wenyou Jia, Zhibin Jiang and You Li

Optimal Organic Rankine Cycle Installation Planning for Factory Waste Heat Recovery 569
Yu-Lin Chen and Chun-Wei Lin

Evaluation of Risky Driving Performance in Lighting Transition Zones Near Tunnel Portals 577
Ying-Yin Huang and Marino Menozzi

Application of Maple on Solving Some Differential Problems 585
Chii-Huei Yu

Six Sigma Approach Applied to LCD Photolithography Process Improvement 593
Yung-Tsan Jou and Yih-Chuan Wu

A Study of the Integrals of Trigonometric Functions with Maple 603
Chii-Huei Yu

A Study of Optimization on Mainland Tourist Souvenir Shops Service Reliability 611
Kang-Hung Yang, Li-Peng Fang and Z-John Liu

The Effects of Background Music Style on Study Performance 619
An-Che Chen and Chen-Shun Wen

Using Maple to Study the Multiple Improper Integral Problem 625
Chii-Huei Yu

On Reformulation of a Berth Allocation Model 633
Yun-Chia Liang, Angela Hsiang-Ling Chen
and Horacio Yamil Lovo Gutierrezmil

Forecast of Development Trends in Cloud Computing Industry 641
Wei-Hsiu Weng, Woo-Tsong Lin and Wei-Tai Weng

Self-Organizing Maps with Support Vector Regression for Sales Forecasting: A Case Study in Fresh Food Data 649
Annisa Uswatun Khasanah, Wan-Hsien Lin and Ren-Jieh Kuo

State of Charge Estimation for Lithium-Ion Batteries Using a Temperature-Based Equivalent Circuit Model................. 657
Yinjiao Xing and Kwok-Leung Tsui

Linking Individual Investors' Preferences to a Portfolio Optimization Model.. 665
Angela Hsiang-Ling Chen, Yun-Chia Liang and Chieh Chiang

Models and Partial Re-Optimization Heuristics for Dynamic Hub-and-Spoke Transferring Route Problems 673
Ming-Der May

Hazards and Risks Associated with Warehouse Workers: A Field Study 681
Ren-Liu Jang and An-Che Chen

Green Supply Chain Management (GSCM) in an Industrial Estate: A Case Study of Karawang Industrial Estate, Indonesia........... 687
Katlea Fitriani

Limits The Insured Amount to Reduce Loss?: Use the Group Accident Insurance as an Example........................... 695
Hsu-Hua Lee, Ming-Yuan Hsu and Chen-Ying Lee

Manipulation Errors in Blindfold Pointing Operation for Visual Acuity Screenings 705
Ying-Yin Huang and Marino Menozzi

3-Rainbow Domination Number in Graphs..................... 713
Kung-Jui Pai and Wei-Jai Chiu

A Semi-Fuzzy AHP Approach to Weigh the Customer Requirements in QFD for Customer-Oriented Product Design 721
Jiangming Zhou and Nan Tu

An Optimization Approach to Integrated Aircraft and Passenger Recovery 729
F. T. S. Chan, S. H. Chung, J. C. L. Chow and C. S. Wong

Minimizing Setup Time from Mold-Lifting Crane in Mold Maintenance Schedule 739
C. S. Wong, F. T. S. Chan, S. H. Chung and B. Niu

Differential Evolution Algorithm for Generalized Multi-Depot Vehicle Routing Problem with Pickup and Delivery Requests 749
Siwaporn Kunnapapdeelert and Voratas Kachitvichyanukul

A Robust Policy for the Integrated Single-Vendor Single-Buyer Inventory System in a Supply Chain 757
Jia-Shian Hu, Pei-Fang Tsai and Ming-Feng Yang

Cost-Based Design of a Heat Sink Using SVR, Taguchi Quality Loss, and ACO 765
Chih-Ming Hsu

Particle Swarm Optimization Based Nurses' Shift Scheduling 775
Shiou-Ching Gao and Chun-Wei Lin

Applying KANO Model to Exploit Service Quality for the Real Estate Brokering Industry 783
Pao-Tiao Chuang and Yi-Ping Chen

Automated Plastic Cap Defect Inspection Using Machine Vision 793
Fang-Chin Tien, Jhih-Syuan Dai, Shih-Ting Wang and Fang-Cheng Tien

Coordination of Long-Term, Short-Term Supply Contract and Capacity Investment Strategy 801
Chiao Fu and Cheng-Hung Wu

An Analysis of Energy Prices and Economic Indicators Under the Uncertainties: Evidence from South East Asian Markets 809
Shunsuke Sato, Deddy P. Koesrindartoto and Shunsuke Mori

Effects of Cooling and Sex on the Relationship Between
Estimation and Actual Grip Strength.................................... 819
Chih-Chan Cheng, Yuh-Chuan Shih and Chia-Fen Chi

Data Clustering on Taiwan Crop Sales Under Hadoop Platform..... 827
Chao-Lung Yang and Mohammad Riza Nurtam

Control with Hand Gestures in Home Environment: A Review....... 837
Sheau-Farn Max Liang

An Integrated Method for Customer-Oriented Product Design...... 845
Jiangming Zhou, Nan Tu, Bin Lu, Yanchao Li and Yixiao Yuan

Discrete Particle Swarm Optimization with Path-Relinking
for Solving the Open Vehicle Routing Problem
with Time Windows 853
A. A. N. Perwira Redi, Meilinda F. N. Maghfiroh and Vincent F. Yu

Application of Economic Order Quantity on Production
Scheduling and Control System for a Small Company 861
Kuo En Fu and Pitchanan Apichotwasurat

CUSUM Residual Charts for Monitoring Enterovirus Infections..... 871
Huifen Chen and Yu Chen

A Study on the Operation Model of the R&D Center
for the Man-Made Fiber Processing Industry Headquarter......... 879
Ming-Kuen Chen, Shiue-Lung Yang and Tsu-Yi Hung

Planning Logistics by Algorithm with VRPTWBD for Rice
Distribution: A Case BULOG Agency in the Nganjuk
District Indonesia 889
Kung-Jeng Wang, Farikhah Farkhani and I. Nyoman Pujawan

A Systematic and Innovative Approach to Universal Design
Based on TRIZ Theories 899
Chun-Ming Yang, Ching-Han Kao, Thu-Hua Liu,
Hsin-Chun Pei and Yan-Lin Lee

Wireless LAN Access Point Location Planning..................... 907
Sung-Lien Kang, Gary Yu-Hsin Chen and Jamie Rogers

Contents

The Parametric Design of Adhesive Dispensing Process with Multiple Quality Characteristics.......................... 915
Carlo Palacios, Osman Gradiz and Chien-Yi Huang

The Shortage Study for the EOQ Model with Imperfect Items....... 925
Chiang-Sheng Lee, Shiaau-Er Huarng, Hsine-Jen Tsai and Bau-Ding Lee

Power, Relationship Commitment and Supplier Integration in Taiwan....................................... 935
Jen-Ying Shih and Sheng-Jie Lu

A Search Mechanism for Geographic Information Processing System... 945
Hsine-Jen Tsai, Chiang-Sheng Lee and Les Miller

Maximum Acceptable Weight Limit on Carrying a Food Tray....... 953
Ren-Liu Jang

Fatigue Life and Reliability Analysis of Electronic Packages Under Thermal Cycling and Moisture Conditions................. 957
Yao Hsu, Wen-Fang Wu and Chih-Min Hsu

Clustering-Locating-Routing Algorithm for Vehicle Routing Problem: An Application in Medical Equipment Maintenance....... 965
Kanokwan Supakdee, Natthapong Nanthasamroeng and Rapeepan Pitakaso

Whole-Body Vibration Exposure in Urban Motorcycle Riders....... 975
Hsieh-Ching Chen and Yi-Tsong Pan

Analysis of Sales Strategy with Lead-Time Sensitive Demand........ 985
Chi-Yang Tsai, Wei-Fan Chu and Cheng-Yu Tu

Order and Pricing Decisions with Return and Buyback Policies..... 993
Chi-Yang Tsai, Pei-Yu Pai and Qiao-Kai Huang

Investigation of Safety Compliance and Safety Participation as Well as Cultural Influences Using Selenginsk Pulp and Cardboard Mill in Russia as an Example.................. 1001
Ekaterina Nomokonova, Shu-Chiang Lin and Guanhuah Chen

Identifying Process Status Changes via Integration of Independent Component Analysis and Support Vector Machine... 1009
Chuen-Sheng Cheng and Kuo-Ko Huang

A Naïve Bayes Based Machine Learning Approach and Application
Tools Comparison Based on Telephone Conversations 1017
Shu-Chiang Lin, Murman Dwi Prasetio, Satria Fadil Persada
and Reny Nadlifatin

Evaluating the Development of the Renewable Energy Industry 1025
Hung-Yu Huang, Chung-Shou Liao and Amy J. C. Trappey

On-Line Quality Inspection System for Automotive
Component Manufacturing Process 1031
Chun-Tai Yen, Hung-An Kao, Shih-Ming Wang and Wen-Bin Wang

Using Six Sigma to Improve Design Quality: A Case of Mechanical
Development of the Notebook PC in Wistron 1039
Kun-Shan Lee and Kung-Jeng Wang

Estimating Product Development Project Duration
for the Concurrent Execution of Multiple Activities 1047
Gyesik Oh and Yoo S. Hong

Modeling of Community-Based Mangrove Cultivation Policy
in Sidoarjo Mudflow Area by Implementing
Green Economy Concept 1055
Diesta Iva Maftuhah, Budisantoso Wirjodirdjo and Erwin Widodo

Three Approaches to Find Optimal Production Run Time
of an Imperfect Production System 1065
Jin Ai, Ririn Diar Astanti, Agustinus Gatot Bintoro
and Thomas Indarto Wibowo

Rice Fulfillment Analysis in System Dynamics Framework
(Study Case: East Java, Indonesia) 1071
Nieko Haryo Pradhito, Shuo-Yan Chou, Anindhita Dewabharata
and Budisantoso Wirdjodirdjo

Activity Modeling Using Semantic-Based Reasoning to Provide
Meaningful Context in Human Activity Recognizing 1081
AnisRahmawati Amnal, Anindhita Dewabharata, Shou-Yan Chou
and Mahendrawathi Erawan

An Integrated Systems Approach to Long-Term Energy
Security Planning 1091
Ying Wang and Kim Leng Poh

An EPQ with Shortage Backorders Model on Imperfect Production System Subject to Two Key Production Systems....... 1101
Baju Bawono, The Jin Ai, Ririn Diar Astanti
and Thomas Indarto Wibowo

Reducing Medication Dispensing Process Time in a Multi-Hospital Health System.......................... 1109
Jun-Ing Ker, Yichuan Wang and Cappi W. Ker

A Pareto-Based Differential Evolution Algorithm for Multi-Objective Job Shop Scheduling Problems 1117
Warisa Wisittipanich and Voratas Kachitvichyanukul

Smart Grid and Emergency Power Supply on Systems with Renewable Energy and Batteries: An Recovery Planning for EAST JAPAN Disaster Area 1127
Takuya Taguchi and Kenji Tanaka

Establishing Interaction Specifications for Online-to-Offline (O2O) Service Systems.. 1137
Cheng-Jhe Lin, Tsai-Ting Lee, Chiuhsiang Lin, Yu-Chieh Huang
and Jing-Ming Chiu

Investigation of Learning Remission in Manual Work Given that Similar Work is Performed During the Work Contract Break.. 1147
Josefa Angelie D. Revilla and Iris Ann G. Martinez

A Hidden Markov Model for Tool Wear Management........... 1157
Chen-Ju Lin and Chun-Hung Chien

Energy Management Using Storage Batteries in Large Commercial Facilities Based on Projection of Power Demand....... 1165
Kentaro Kaji, Jing Zhang and Kenji Tanaka

The Optimal Parameters Design of Multiple Quality Characteristics for the Welding Thick Plate of Aerospace Aluminum Alloy 1173
Jhy-Ping Jhang

Synergizing Both Universal Design Principles and Su-Field Analysis to an Innovative Product Design Process 1183
Chun-Ming Yang, Ching-Han Kao, Thu-Hua Liu, Ting Lin
and Yi-Wun Chen

The Joint Determination of Optimum Process Mean, Economic Order Quantity, and Production Run Length 1191
Chung-Ho Chen

Developing Customer Information System Using Fuzzy Query and Cluster Analysis................................ 1199
Chui-Yu Chiu, Ho-Chun Ku, I-Ting Kuo and Po-Chou Shih

Automatic Clustering Combining Differential Evolution Algorithmand k-Means Algorithm 1207
R. J. Kuo, Erma Suryani and Achmad Yasid

Application of Two-Stage Clustering on the Attitude and Behavioral of the Nursing Staff: A Case Study of Medical Center in Taiwan ... 1217
Farn-Shing Chen, Shih-Wei Hsu, Chia-An Tu and Wen-Tsann Lin

The Effects of Music Training on the Cognitive Ability and Auditory Memory.................................. 1225
Min-Sheng Chen, Chan-Ming Hsu and Tien-Ju Chiang

Control Scheme for the Service Quality 1233
Ling Yang

Particle Swam Optimization for Multi-Level Location Allocation Problem Under Supplier Evaluation 1237
Anurak Chaiwichian and Rapeepan Pitakaso

Evaluation Model for Residual Performance of Lithium-Ion Battery.. 1251
Takuya Shimamoto, Ryuta Tanaka and Kenji Tanaka

A Simulated Annealing Heuristic for the Green Vehicle Routing Problem................................. 1261
Moch Yasin and Vincent F. Yu

Designing an Urban Sustainable Water Supply System Using System Dynamics.................................. 1271
S. Zhao, J. Liu and X. Liu

An Evaluation of LED Ceiling Lighting Design with Bi-CCT Layouts 1279
Chinmei Chou, Jui-Feng Lin, Tsu-Yu Chen, Li-Chen Chen and YaHui Chiang

Contents

Postponement Strategies in a Supply Chain Under the MTO Production Environment 1289
Hsin Rau and Ching-Kuo Liu

Consumer Value Assessment with Consideration of Environmental Impact. 1297
Hsin Rau, Sing-Ni Siang and Yi-Tse Fang

A Study of Bi-Criteria Flexible Flow Lines Scheduling Problems with Queue Time Constraints 1307
Chun-Lung Chen

Modeling the Dual-Domain Performance of a Large Infrastructure Project: The Case of Desalination 1315
Vivek Sakhrani, Adnan AlSaati and Olivier de Weck

Flexibility in Natural Resource Recovery Systems: A Practical Approach to the "Tragedy of the Commons" 1325
S. B. von Helfenstein

The Workload Assessment and Learning Effective Associated with Truck Driving Training Courses 1335
Yuh-Chuan Shih, I-Sheng Sun and Chia-Fen Chi

Prognostics Based Design for Reliability Technique for Electronic Product Design 1343
Yingche Chien, Yu-Xiu Huang and James Yu-Che Wang

A Case Study on Optimal Maintenance Interval and Spare Part Inventory Based on Reliability. 1353
Nani Kurniati, Ruey-Huei Yeh and Haridinuto

Developing Decision Models with Varying Machine Ratios in a Semiconductor Company 1361
Rex Aurelius C. Robielos

New/Advanced Industrial Engineering Perspective: Leading Growth Through Customer Centricity 1371
Suresh Kumar Babbar

Scheduling a Hybrid Flow-Shop Problem via Artificial Immune System. 1377
Tsui-Ping Chung and Ching-Jong Liao

Modeling and Simulation on a Resilient Water Supply System Under Disruptions 1385
X. Liu, J. Liu, S. Zhao and Loon Ching Tang

A Hybrid ANP-DEA Approach for Vulnerability Assessment in Water Supply System 1395
C. Zhang and X. Liu

An Integrated BOM Evaluation and Supplier Selection Model for a Design for Supply Chain System 1405
Yuan-Jye Tseng, Li-Jong Su, Yi-Shiuan Chen and Yi-Ju Liao

Estimation Biases in Construction Projects: Further Evidence 1413
Budi Hartono, Sinta R. Sulistyo and Nezar Alfian

Exploring Management Issues in Spare Parts Forecast 1421
Kuo-Hsing Wu, Hsin Rau and Ying-Che Chien

Artificial Particle Swarm Optimization with Heuristic Procedure to Solve Multi-Line Facility Layout Problem 1431
Chao Ou-Yang, Budi Santosa and Achmad Mustakim

Applying a Hybrid Data Preprocessing Methods in Stroke Prediction 1441
Chao Ou-Yang, Muhammad Rieza, Han-Cheng Wang, Yeh-Chun Juan and Cheng-Tao Huang

Applying a Hybrid Data Mining Approach to Develop Carotid Artery Prediction Models 1451
Chao Ou-Yang, Inggi Rengganing Herani, Han-Cheng Wang, Yeh-Chun Juan, Erma Suryani and Cheng-Tao Huang

Comparing Two Methods of Analysis and Design Modelling Techniques: Unified Modelling Language and Agent Modelling Language. Study Case: A Virtual Bubble Tea Vending Machine System Development 1461
Immah Inayati, Shu-Chiang Lin and Widya Dwi Aryani

Persuasive Technology on User Interface Energy Display: Case Study on Intelligent Bathroom 1471
Widya Dwi Aryani, Shu-Chiang Lin and Immah Inayati

Investigating the Relationship Between Electronic Image of Online
Business on Smartphone and Users' Purchase Intention 1479
Chorng-Guang Wu and Yu-Han Kao

Forecast of Development Trends in Big Data Industry 1487
Wei-Hsiu Weng and Wei-Tai Weng

Reliability Analysis of Smartphones Based on the Field
Return Data 1495
Fu-Kwun Wang, Chen-I Huang and Tao-Peng Chu

The Impact of Commercial Banking Performance
on Economic Growth..................................... 1503
Xiaofeng Hui and Suvita Jha

Data and Information Fusion for Bio-Medical Design
and Bio-Manufacturing Systems 1513
Yuan-Shin Lee, Xiaofeng Qin, Peter Prim and Yi Cai

Evaluating the Profit Efficiency of Commercial Banks:
Empirical Evidence from Nepal............................ 1521
Suvita Jha, Xiaofeng Hui and Baiqing Sun

Explore the Inventory Problem in a System Point of View:
A Lot Sizing Policy 1529
Tsung-shin Hsu and Yu-Lun Su

Global Industrial Teamwork Dynamics in China
and Southeast Asia: Influence on Production Tact Time
and Management Cumulative Effect to Teamwork Awareness-1/2 ... 1539
Masa-Hiro Nowatari

Global Industrial Teamwork Dynamics in Malaysia—Effects
of Social Culture and Corporate Culture to Teamwork
Awareness—2/2... 1551
Masa-Hiro Nowatari

Relaxed Flexible Bay Structure in the Unequal Area Facility
Layout Problem 1563
Sadan Kulturel-Konak

Comparisons of Different Mutation and Recombination Processes of the DEA for SALB-1 1571
Rapeepan Pitakaso, Panupan Parawech and Ganokgarn Jirasirierd

An Exploration of GA, DE and PSO in Assignment Problems...... 1581
Tassin Srivarapongse and Rapeepan Pitakaso

Author Index 1589

Particle Swarm Optimization Based Nurses' Shift Scheduling

Shiou-Ching Gao and Chun-Wei Lin

Abstract The nurse scheduling is a multifaceted problem with the extensive number of constraints requires. In the past, many researchers tried to find the high-quality nursing schedule by analyzing their different aspects and targets, including the lowest cost and the highest efficiency. However, the study lacks of the consideration for nurses' happiness. The nurses with bad mood will feel distracted in their working time, and even quit their jobs in the end. This thesis not only contains the scheduling constraints by the Administrative Regulations and the Hospital Regulations to construct the mathematical models, but also includes the consideration of nurses' happiness. The main algorithm of this research is the Particle Swarm Optimization (PSO). We used PSO to look for the most suitable schedule for nursing staffs to maximize their working happiness.

Keywords Nurses · Shift scheduling · Happiness · PSO

1 Introduction

Hospital is an all year-round, the 24 h operation of the service units, in addition to the outpatient nurses to care for the sick as the main core work, that necessarily take a substantial period of time to get along with patients, their performance is more important, so the nurses the scheduling core operations of the hospital medical management, improve staff morale and productivity has considerable

---

S.-C. Gao (✉) · C.-W. Lin
Industrial Engineering and Management, National Yunlin University of Science and Technology, 123 University Road, Section 3, Douliou, Yunlin 64002, Taiwan, Republic of China
e-mail: m10021011@yuntech.edu.tw

C.-W. Lin
e-mail: lincwr@yuntech.edu.tw

influence (Wen 2008). How to properly manage the nursing manpower arrangement is a major issue. For an adequate supply of services during working hours manpower, the current shift system as the main arrangement, they should make every certain time interval repeating in a new shift table. The a suitable schedules for whether or not the new nurses, on the quality of medical care and other medical services have considerable influence (Lee 2005). Especially nursing staff is not just a job, but a burden of relating to other people's lives and safety of nurses not only is the physical exertion, but also the psychological pressure (Liu 2011).When the accumulation of physical and mental unhappiness and pressure, but can not be properly relieve to cause nurses can not concentrate on work, or even leave. Therefore, the scheduling of the nurses in addition to emphasis on fairness and feasibility, as far as possible to meet the nurses' degree of happiness, construct nurses the highest degree of happiness schedules.

Nurse Scheduling in general hospitals in Taiwan in 24 h shifts a day divided into three classes eight hours, respectively, for the day shift, evening shift and night shift, some differences in different classes of the nature of work, manpowerdemand is not the same (Huang et al. 2009). Must be arranged in accordance with the demand for each shift, to avoid the nurses work excessive burden is too heavy, will have a negative impact on physiological conditions and job preferences (Berrada et al. 1996). Therefore the Nurse Scheduling need to be considered seven factors as fair, reasonable, flexible, humane considerations, etc. (Chang and Lee 1992). And through the unit scheduling, each unit head nurses considering the nurses' personal wishes to schedule. It can improve not only the lack of understanding of the centralized scheduling for nurses demand conditions, and help reduce the self-scheduling to provide high autonomy, easy to increase the difficulty of communication and coordination between the nursing staff, it is both time-consuming and expensive effort.

2 Happiness

The word "happiness" of the Chinese people, including the meaning of happy, pleasant, joy, and it symbolizes a person's heart was filled with joyand happy. In the Analects of Confucius, the word "happiness" expresses personal inner pleasant feelings and spiritual enrichment. It's a self-inner feelings (Bai 2008). Homer and Herodotus believes that happiness is the human body and mind feel exhilarated. And Allard et al. believes that happiness is an emotional joy and happiness, when forward personal emotions than negative emotions will feel happy (Shin and Johnson 1978).

Happiness is a psychological feeling of self-inner, according to the level of consciousness can be divided into three levels: 1. Competitive happiness: The lowest level of happiness, compared competing with others to get happy, the competitive process is not necessarily happy, even in the final happiness is limited; if the lost, the result is not only unhappy, but pain. It is the worst kind. 2. Conditional happiness:The most common and most easily achieve happiness. People tend to be happy with some of the specific conditions tied together, people will be happy when the conditions are met. 3. Unconditional happiness: True happiness does not come from their own people and things, is derived from the self-bliss, not subject to change due to the interference and influence of external things (Ye 2009). Everyone in the pursuit of happiness, in other words the person's life is the target of numerous stacked or chained together (Ye 2009). Each person's definition of happiness is different; in this study for the definition of happiness is a subjective feeling of pleasant personal feelings on their own lives. Freud once said: make up the not enough make people happiness and pleasure (Ma 2005). When a person to get what he wants, needs, or goals to achieve satisfactory level, he has sufficient reason that he is happiness (Simpson 1975). On the contrary, when the demand cannot be met will feel pain (Ma 2005).

Argyle also pointed out that the degree of happiness can be understood as the extent of personal life satisfaction or positive emotional intensity of feelings (Argyle et al. 1997). Diener also believes that the degree of happiness is a personal assessment of the cognitive experience of own lives (Bai 2008). So when the needs and wants of a person's life as much as possible to meet, there will be a happy life (Shin and Johnson 1978). If a person can do what he wants to do, to get what he needs, he will be glad. Therefore, it is a pursuit of happiness behavior, happiness can be obtained through certain events (Dilman 1982).

Happiness is a feeling. People like a bottle filled with Sense and Sensibility, when put in more rational, emotional lacking something more sensual, so that we have the courage to pursue their full measure of happiness. The ratio to adjust the standard is that our own "three benefits": Benefit our feeling, benefit our existence, and benefit our lives (Ye 2009). When beneficial conditions exist, we can feel happy and get happiness. We can through the conditions which would be benefit people are able to do what he wants to do to get what he needs, so that he can be happiness.

3 Model Construction

Nurse Scheduling addition to consider the degree of happiness, must also consider the existing restrictions. Nurse Scheduling limits can be classified as administrative law and hospitals, and regulations of the Association, as well as meet particular target or nurses wishes (Berrada et al. 1996). This study refer to the collation Nurse Scheduling restrictions and limitations, and review the limitations of the study. Summarized as follows:.

1. In accordance with Article 30 of the Labor Standards Law, the labor daily work can not be more than eight hours, and the bi-weekly work hours should not be more than eighty-four hours.
2. In accordance with Article 34 of the Labor Standards Law, If a rotation system, the work shift should be rotate at least once a week. After the shift has been replaced, should be given appropriate time to rest.

3. In accordance with Article 36 of the Labor Standards Law, Workers should have at least a day as official holiday in every seven days.
4. The minimum limit demand, the number of nurse required to meet the demand of the daily each shift.
5. Continuous scheduling restrictions, each nurse only can arrange one shift, to avoid the night shift followed by the evening shift.
6. Each nurses cannot work continuously for more than six days.
7. The combination of each nurse for three days continuously cannot shift off day, works, off day.
8. Each nurse duty at the night shift cannot take every other day the day shift or evening shift.
9. Each nurse duty at the evening shift cannot take every other day the day shift.
10. Each nurse cannot off day continuously for more than five days.

The purpose of this study is that the maximum degree of happiness pursuit of nurses schedules. Mathematical functions in addition to the inclusion in the above constraints, and joined the nurses measure of the degree of happiness, in the limited solution space to find the optimal solution allows nurses to obtain the maximum degree of happiness.

3.1 Parameters

i	Index for nurses; $\forall i = 1, 2, \ldots, I$
j	index for days within the scheduling horizon; $\forall j = 1, 2, \ldots, J$
k	Shift type; $\forall k = 0, 1, 2, 3$ (off day, day, evening, and night shifts, respectively)
d_{jk}	Demand for nurses of shift type k on day j
h_{ij}	Happiness for nurse i on day j

3.2 Variables

x_{ijk} nurse i on day j for available shift type l to be assigned

$$x_{ijl} = \begin{cases} 1, \text{if nurse } i \text{ on day } j \text{ for available shift type } k \text{ to be assigned}; \forall k = 0, 1, 2, 3 \\ 0, \text{otherwise} \end{cases}$$

A_{ij} $A_{ij} = k$, when $x_{ijk} = 1$ $\forall i = 1, 2, \ldots, I$; $\forall j = 1, 2, \ldots, J$;
$$\forall k = 0, 1, 2, 3$$

B_{ij} The shift type which nurse i have on the table with happiness on day j; $B_{ij} \in 0, 1, 2, 3$ (off day, day, evening, and night shifts, respectively)

L_{ij} $L_{ij} = A_{ij} - B_{ij}$

U_{ij} $U_{ij} = \begin{cases} 1, L_{ij} = 0 \\ 0, L_{ij} \neq 0 \end{cases}$

3.3 Mathematical Model

The objective of this study is to obtain the greatest happiness of the nurses. The objective function is the sum of all shifts consistent happiness for nurse i on day j. That is:

Maximize H

$$H = \sum_{i=1}^{I} \sum_{j=1}^{J} (U_{ij} \times h_{ij}) \quad (1)$$

s.t

$$\sum_{k=0}^{3} x_{ijk} = 1 \quad \forall i = 1, 2, \ldots, I; \forall j = 1, 2, \ldots, J \quad (2)$$

$$\sum_{i=1}^{I} \sum_{k=1}^{3} x_{ijk} \geq \sum_{k=1}^{3} D_{jk} \quad \forall j = 1, 2, \ldots, J \quad (3)$$

$$D_{jk} = \sum_{i=1}^{I} x_{ijk} \quad \forall j = 1, 2, \ldots, J; \forall k = 1, 2, 3 \quad (4)$$

$$\sum_{j}^{j+6} \sum_{k=1}^{3} x_{ijk} \leq 6 \quad \forall i = 1, 2, \ldots, I; \forall j = 1, 8, 15, 22 \quad (5)$$

$$\sum_{j}^{j+13} \sum_{k=1}^{3} x_{ijk} \leq 10 \quad \forall i = 1, 2, \ldots, I; \forall j = 1, 8, 15 \quad (6)$$

$$\sum_{j}^{j+5} x_{ij0} \leq 5 \quad \forall i = 1, 2, \ldots, I; \forall j = 1, 2, \ldots, J-5 \quad (7)$$

$$\sum_{j}^{j+6} \sum_{k=1}^{3} x_{ijk} \leq 6 \quad \forall i = 1, 2, \ldots, I; \forall j = 1, 2, \ldots, J-6 \quad (8)$$

$$x_{ij0} + \sum_{t=1}^{3} x_{i(j+1)k} + x_{i(j+2)0} \leq 2 \quad \forall i = 1, 2, \ldots, I; \forall j = 1, 2, \ldots, J-2 \quad (9)$$

$$x_{ij3} + x_{i(j+1)1} \leq 1 \quad \forall i = 1, 2, \ldots, I; \forall j = 1, 2, \ldots, J-1 \quad (10)$$

$$x_{ij3} + x_{i(j+1)2} \leq \quad \forall i = 1, 2, \ldots, I; \forall j = 1, 2, \ldots, J-1 \quad (11)$$

$$x_{ij2} + x_{i(j+1)1} \leq 1 \quad \forall i = 1, 2, \ldots, I; \forall j = 1, 2, \ldots, J-1 \quad (12)$$

$$x_{ijk}, A_{ij}, B_{ij}, U_{ij} \geq 0; \ L_{ij} \in Z; \ x_{ijk}, U_{ij}, \in \{0, 1\}; \ A_{ij}, B_{ij} \in 0, 1, 2, 3 \quad (13)$$

(1) The objective function. (2)–(13) The constraints. (2) The nurses i only arrange a shifts type k on day j. (3) The minimum limit demand, the sum of the nurses arrange shifts type k on day j should greater than or equal to the demand. (4) The demand on duty, the sum of the nurses arrange not off day shifts type k on day j should greater than or equal to the demand. (5) Minimum weekly days off limit, workers should have at least a day as official holiday in every seven days (6) Minimum bi-weekly days off limit, every two weeks working days must be less than or equal to ten days. (7) Number of consecutive off days shall not exceed five. (8) Consecutive days limit, number of consecutive days shall not exceed six. (9) The combination of each nurse for three days continuously cannot shift off day, works, off day. (10) , (11) each nurse duty at the night shift cannot take every other day the day shift or evening shift. (12) Each nurse duty at the evening shift cannot take every other day the day shift. (13) Description of all variables limit.

4 Instance Validation

In this study mathematical functions use the particle swarm algorithm (PSO) to solve the optimal solution. PSO was published by Kennedy and Eberhart in 1995. By observing nature birds foraging search behavior, they developed the optimized search algorithms. It was applied in various fields for more than ten years. Its biggest advantage is that it only requires a few parameters to adjust, can be used to solve most of the optimization problem. In this algorithm, every possible solution is called particle. Each particle to move in the search space dimension D, and note the move once the optimal solution here, to convey the message of the particles. Including the location of the particles x now, acceleration v and the particles adaptation values, For each generation, each particle updates the acceleration and a new location by using the following equation:

$$v_i(t+1) = w \cdot v_i(t) + c_1 \cdot rand() \cdot [p_i - x_i(t)] + c_2 \cdot rand() \cdot [p_g - x_i(t)] \quad (14)$$

Rand() is a random number between 0 and 1, i means the i particle, t means the number of iterations, w means the inertia weight, p_i means the particle once moved best local solution, p_g means the particle once moved best global solution, c_1 and c_2 are the learning factors. Calculate the acceleration of a particle, and then update the new location of the particles according to the following formula, and find the optimal solution:

$$x_i(t+1) = x_i(t) + v_i(t+1) \quad (15)$$

PSO step-by-step instructions are as follows:

Step 1: Set parameters.

Particle Swarm Optimization Based Nurses' Shift Scheduling

Table 1 Happiness table (part)

No.	1	2	3	4	5	6	7	8	9	10	11
1				0	0						
				100	100						
2										0	0
										90	100
3											

Table 2 Shift table (part)

NO	1	2	3	4	5	6	7	8	9	10	11
1	0	1	1	0	0	1	1	1	3	3	0
2	3	0	2	2	1	1	1	3	3	0	0
3	0	2	2	0	1	1	1	0	0	2	2

shifts :0: off day, 1: day, 2: evening, 3: night

Step 2: Establish initial solution and randomly generated initialization particle swarm position and speed.
Step 3: Based on the objective function to calculate the fitness values.
Step 4: Compare particles fitness values, if relatively good, updat the location.
Step 5: According to Eqs. (14), (15), adjustment and update the position and velocity of the particles.
Step 6: Back to the Step3 and calculate the new particle fitness value. Stop searching when the iterations reaches the maximum number of searches.

The goal of this study is to obtain the scheduling with the maximum degree of happiness. Happiness comes from people's own feelings, Through to meet the demand, you can get pleasure. So through the investigation of nurses demand for the next month, learning how the scheduling arrangement enables nurses to obtain the maximum degree of happiness. For example, the date 4, 5, and 10, 11 for the holidays, No. 2 nurse hope can arrange shift 0 on date 10 and 11. If come true, it can get 90 and 100 degree of happiness.

Using Table 1 and parameters set into the mathematical functions, and use PSO to find the optimal solution as the following table, nurses can get at least 390 degree of happiness: (Table 2).

5 Conclusion

This study constructs a mathematical model to maximize the nurse happiness, and solve it through the PSO. Use computer program to help nurses arrange schedules, solving the maximum degree of happiness schedules. That not only can reduce the time-consuming manual scheduling, but also try to meet the needs of each nurse, to obtain the maximum degree of happiness schedules.

Acknowledgments The authors are grateful to nurses and friends for careful guidance and help to overcome all the difficulties and trials. Finally thanks the reviewers members of this seminar with the relevant staff.

References

Argyle M, Shi J-B, Lu L (1997). The psychology of happiness. CHULIU, Taipei
Bai S-H (2008) Positive life events on happiness, happiness, self-esteem. Dissertation, Fo Guang University
Berrada I, Ferland JA, Michelon P (1996) A multi-objective approach to nurse scheduling with both hard and soft constraints. Socio-Economic Plann Sci 30(3):183–193
Chang C-Y, Lee S-S (1992) Nurse scheduling system of experimental design. Hosp Comp 8:65–70
Dilman I (1982) Happiness. J Med Ethics 8(4):199–202. doi:10.2307/27716097
Huang Y-C, Jian S-H, Kang J-R (2009) Nurses rostering under the consideration of staff preferences. J Health Sci 11(1):57–69
Lee J-D (2005) Constraint programming models for 7 × 24 manpower scheduling problem: a case of nurse scheduling application. Dissertation, National Chiao Tung University
Liu Z-Y (2011) The research of the relationship between employee assistance programs and employee's turnover intention- taking employee's job satisfaction as a mediator. Dissertation, National Central University
Ma Z-F (2005) Turn the key to happiness: Oscar Wilde, "The happy prince analysis". Elementary Educ J 52(1):50–57
Shin DC, Johnson DM (1978) Avowed happiness as an overall assessment of the quality of life. Soc Indic Res 5(1):475–492
Simpson RW (1975) Happiness. American Philos Q 12(2):169–176. doi:10.2307/20009571
Wen J-B (2008) A study of nurse scheduling factors in a regional hospital. Dissertation, Southern Taiwan University of Science and Technology
Ye Z (2009). Peking university professor given 28 Happy rule, 1st edn. Taiwan: by the Court Press (Knowledge and Innovation Agency), Taiwan

Applying KANO Model to Exploit Service Quality for the Real Estate Brokering Industry

Pao-Tiao Chuang and Yi-Ping Chen

Abstract This research applies Kano model to exploit service quality for the REB industry in Taiwan. The study, firstly, collected data about the importance degree of each REB quality requirements through a questionnaire survey. Meanwhile, the Kano questions and evaluation criteria were designed in the functional and dysfunctional questions of the questionnaire to collect data for further classifications of the two-dimensional quality model of REB industry. Then, a factor analysis was used to group the quality requirements into quality dimensions. Finally, a satisfaction-dissatisfaction matrix analysis was performed to confirm what quality attribute that each quality dimension of REB belongs to in the Kano's two dimensional model. Results show that quality dimensions of security and dependability are confirmed as the "must-be" attributes of Kano model. Reliability is the "one-dimensional" attribute. Profession and information belong to the "attractive" attributes. Whereas, tangibility, communication, and empathy are identified as the "indifference" attributes of Kano model. Managerial suggestions are also provided.

Keywords Service quality · KANO model · Satisfaction coefficient · Dissatisfaction coefficient · Satisfaction-dissatisfaction matrix

---

P.-T. Chuang (✉)
Department of Asia–Pacific Industrial and Business Management, National University of Kaohsiung, Kaohsiung, Taiwan
e-mail: ptchuang@nuk.edu.tw

Y.-P. Chen
Catcher Technology Co., Ltd, Tainan, Taiwan
e-mail: zeus925854@yahoo.com.tw

1 Introduction

According to the Directorate General of Budget, Accounting and Statistics (DGBAS) of Executive Yuan, Taiwan, the percentage of households that self-owned the house or apartment in Taiwan is 84.58 % in year of 2011. In most European nations and the United States, relatively, the percentage is around 60 %. From this statistics, we can realize that most Taiwanese would prefer living in self-owned house or apartment. Therefore, finding a suitable and comfortable self-owned house or apartment becomes an important issue to most Taiwanese family. Nevertheless, it always spends lots of time and energy to find an ideal house, and needs to consider many factors such as location, traffic condition, as well as living functions of the vicinity. Thus, most people would consign this job to the real estate brokers, who are more professional in searching an appropriate house for the clients and saving their time and energy.

Real estate brokering (REB) industry can provide service to satisfy customer's requirements on searching a house or apartment. Like other service industries, REB may either bring good service to satisfy customers or disregard customer's needs that results in unsatisfied customers. That is, when a customer who needs a brokering service for either buying or renting a house/apartment, he or she would have some quality requirements (expectations) from the service. Throughout the entire process, the customer would perceive how well (perceptions) the service is provided and evaluate the REB service quality. Literatures have shown that the higher the service quality is, the higher the satisfaction and the loyalty of customers will be; and this would bring more profitability to the service company (Heskett et al. 1994, Heskett and Sasser Jr 2010).

Service quality is defined as the measure of how well the service level delivered matches the customer expectations (Grönroos 1982, Parasuraman et al. 1985, 1988). Delivering a quality service means conforming to customers' expectations on a consistent basis. Perceived service quality can be evaluated as the degree and direction of discrepancy between consumers' perceptions and their expectations about the particular service provided by the service company. Following this definition, the service quality can then be evaluated, compared to its counterpart of manufacturing product quality, from the perspective of customer. After this debut, a lot of literatures regarding service quality has been evaluated and analyzed, both practical and academic applications, for either a specific service industry or a particular service company, (e.g., Berry et al. 1994, 1997, Harrison-Walker 2002, Taylor 2004, Peiró et al. 2005, Pakdil and Aydin 2007, Lin 2010). Regarding the REB service quality, Zeng (1990), Mao (1995), and Wang (2007) adopted the SERVQUAL instrument, which proposed by Parasuraman et al. (1988), to empirically studied the service quality of the real estate brokering industry in Taipei city, Taichung city, and Taipei county of Taiwan, respectively.

Most of the existing research about REB service quality adopted a one-dimensional viewpoint to evaluate each of the quality requirements. It means the higher the perceived service quality, the higher the customer's satisfaction, and vice versa. In fact, nevertheless, not all the quality requirements have linear contributions to the overall customer satisfaction. Some of them may have curvilinear relationship between the perception of quality requirement and the extent of customer satisfaction. In this aspect, Kano et al. (1984) proposed a two-dimensional quality model, the so-called Kano model, which distinguished five categories of quality attributes (must-be, one-dimensional, attractive, indifference, and Reverse) that influence customer satisfaction on different ways when met (Matzler and Hinterhuber 1998).

This research applies Kano model to exploit service quality for the REB industry in Taiwan. The study, firstly, collected data about the importance degree of each REB quality requirements through a questionnaire survey. Meanwhile, the Kano questions and evaluation criteria were designed in the functional and dysfunctional questions of the questionnaire to collect data for further classifications of the two-dimensional quality model of REB industry. Then, a factor analysis was used to group the quality requirements into quality dimensions, which identify latent dimensions that direct analyses may not and reduce number of variables that consist of quality requirements. Finally, a satisfaction-dissatisfaction matrix analysis was performed to confirm what quality attribute that each quality dimension of REB belongs to in the Kano's two dimensional model. Based on this analysis, corporate resource allocation strategy for improving service quality is suggested for the REB companies.

2 Kano Two-Dimensional Model

Unlike conventional one-dimensional viewpoint on quality attributes, the two dimensional model addresses the non-linear relationship between quality attribute performance and overall customer satisfaction. Kano et al. (1984) were the first to propose that quality attributes can be classified into five categories based on the level of impact of individual attributes on overall customer satisfaction. The five categories are must-be (basic), one-dimensional (performance), attractive (excitement), indifference, and reverse quality. The definition of those five quality attributes is also shown below.

(1) Must-be (Basic) quality: The "must-be" attributes are basic criteria of a product/service. If these attributes are not fulfilled, the customer will be extremely dissatisfied. On the other hand, as the customer takes these attributes for granted, their fulfillment will not increase his satisfaction.
(2) One-dimensional (Performance) quality: For "one-dimensional" attribute, customer satisfaction is proportional to the level of fulfillment. The higher the level of fulfillment is, the higher the customer's satisfaction is; and vice versa.
(3) Attractive (Excitement) quality: The relation curve for the "attractive" attribute is upward with a steeper slope in the presence of the attribute, implying

that a lack of this attribute has little impact on customer satisfaction but the presence of the attribute provides great satisfaction.
(4) Indifference quality: With regards to the "indifference" attribute, customer satisfaction will not be affected no matter whether this quality attribute is provided or not. That is, providing indifference quality attributes do not bring advantages to customer satisfaction.
(5) Reverse quality: The influencing direction of "reverse" attribute on customer satisfaction is opposite to that of "one-dimensional". Customers will be dissatisfied if this quality attribute is provided; otherwise, they will be satisfied.

3 Factor Analysis for Quality Dimensions of REB Service

To construct the Kano questionnaire, this research firstly listed 34 items of service quality requirements of REB by explorative investigation and synthesizing those literatures of Zeng (1990), Mao (1995) and Wang (2007). These quality requirements are shown in Table 1. The importance degree for each quality requirements were investigated by a questionnaire survey, in which the data regarding to customer responses for those Kano questions were also collected. The Likert-five-scale was used for the importance degree. A total of 300 questionnaires were distributed to those respondents who had bought or rent houses by way of the REB companies. Among those 153 questionnaires were collected and 127 were determined as valid.

Based on the importance degree, the factor analysis was used to group the quality requirements into 8 dimensions, by deleting 5 items of quality requirements that have factor loadings less than 0.5. This leaves a total of 29 quality requirements that spread in 8 dimensions.

Table 1 Kano's functional and dysfunctional questions in the questionnaire

Kano question	Respondent's answer
Functional form of the question (e.g., If the service is conforming to the facts, how do you feel?)	I like it that way
	It must be that way
	I am neutral
	I can live with it that way
	I dislike it that way
Dysfunctional form of the question (e.g., If the service is not conforming to the facts, how do you feel?)	I like it that way
	It must be that way
	I am neutral
	I can live with it that way
	I dislike it that way

4 Kano Questionnaire Design and Evaluation Criteria

In the questionnaire survey, the Kano questions were designed to collect data for further classification of quality attributes in the two-dimensional quality model. The paper adopted the types of question that were designed and proposed by Kano et al. (1984). The Kano questions are shown as Table 1. In the questionnaire, questions regarding to the extent of customer satisfaction when individual quality requirements are presented and the extent of customer dissatisfaction when individual quality requirements are absent were answered by the respondents. For each quality requirement a pair of questions is formulated to which the customer can answer in one of five different ways when he or she faces the functional or dysfunctional situations, respectively. That is, the first question concerns the reaction of the customer if the REB service has that quality requirement; and the second question concerns the reaction of the customer if the REB service does not have that quality requirement.

For each individual questionnaire, the category for each of the quality requirements is determined by the Kano evaluation table, shown as Table 2. If the customer answers, for example, "I like it that way" for the functional question and answer "I am neutral" for the dysfunctional question, the corresponding quality requirement is determined as "Attractive" quality attribute in the Kano classification.

5 Satisfaction-Dissatisfaction Matrix Analysis

To confirm which quality attribute of Kano classification that each quality dimension belongs to, the paper proposed a matrix analysis that displays the relative location of each pair of standardized dissatisfaction coefficient with the dysfunctional quality dimension (Zx) and standardized satisfaction coefficient with the functional quality dimension (Zy).

The dissatisfaction coefficient (X), when a particular quality dimension is dysfunctional, is computed as Eq. (1). The Satisfaction coefficient (Y), when a particular quality dimension, is functional is computed as Eq. (2). The

Table 2 Kano evaluation table

		Dysfunctional form of the question					
		Like	Must-be	Neutral	Live with	Dislike
Functional form of the question	Like	Q	A	A	A	O
	Must-be	R	I	I	I	M
	Neutral	R	I	I	I	M
	Live with	R	I	I	I	M
	Dislike	R	R	R	R	Q

Table 3 Computation results for dissatisfaction and satisfaction coefficients

Factor (Quality Dimension)	Quality require-ment	A	O	M	I	R	Q	Dissat. Coeff.	Sat. Coeff.	Zx (Dissat. With Dysfunc.)	Zy (Satis. With Func.)
Security	Q1	11	43	52	21	0	0	0.459	0.791	1.958	−0.206
	Q2	9	51	54	13	0	0				
	Q3	10	42	58	13	3	1				
	Q4	19	41	49	15	2	1				
	Q5	13	49	57	7	1	0				
Reliability	Q7	15	37	39	33	3	0	0.482	0.595	0.550	0.143
	Q8	29	31	33	32	2	0				
	Q9	13	44	42	24	3	1				
	Q10	35	36	34	21	1	0				
Information	Q12	29	49	21	27	1	0	0.533	0.503	−0.110	0.945
	Q13	22	41	24	37	1	2				
	Q14	42	24	25	33	3	0				
	Q15	18	32	60	17	0	0				
	Q16	43	34	30	18	1	1				
	Q17	49	18	21	39	0	0				
Communication	Q18	30	38	39	19	1	0	0.434	0.466	−0.378	−0.601
	Q19	27	26	23	44	7	0				
	Q20	26	27	25	47	2	0				
	Q21	13	28	25	58	3	0				
Profession	Q22	19	48	33	26	1	0	0.591	0.484	−0.246	1.860
	Q23	53	29	12	32	1	0				
Empathy	Q27	28	30	28	41	0	0	0.466	0.391	−0.915	−0.093
	Q28.	28	32	9	57	0	1				
Tangibility	Q29.	32	11	6	78	0	0	0.392	0.340	−1.279	−1.266
	Q30	29	13	26	59	0	0				
	Q31	31	21	16	58	0	1				
	Q32	13	48	31	34	1	0				
Depend-ability	Q33	12	42	46	26	1	0	0.422	0.577	0.422	−0.784
	Q34	23	29	28	45	1	1				

standardized dissatisfaction coefficient with the dysfunctional quality dimension (Zx) is computed as Eq. (3) and the standardized satisfaction coefficient with the functional quality dimension (Zy) is computed as Eq. (4). Results of these coefficients are shown in Table 3.

$$X_i = \sum_{i=1}^{k_i} \left(\frac{A+O}{A+O+M+I} \right) \quad (1)$$

$$Y_i = \sum_{i=1}^{k_i} \left(\frac{O+M}{A+O+M+I} \right) \quad (2)$$

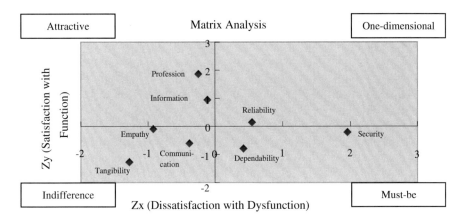

Fig. 1 Satisfaction-dissatisfaction matrix analysis

$$Z_{X_i} = \frac{X_i - \overline{X}}{s_X} = \frac{X_i - \frac{\sum_{i=1}^{8} X_i}{8}}{\sqrt{\frac{\sum (X_i - \overline{X})^2}{7}}} \quad (3)$$

$$Z_{Y_i} = \frac{Y_i - \overline{Y}}{s_Y} = \frac{Y_i - \frac{\sum_{i=1}^{8} Y_i}{8}}{\sqrt{\frac{\sum (Y_i - \overline{Y})^2}{7}}} \quad (4)$$

where, A is the frequency of respondents whose answer is classified as "attractive" quality attribute; O is the frequency for "one-dimensional"; M for "must-be; I for "Indifference"; R for "reverse"; and Q for questionable result. k_i represent the number of quality requirements that are grouped into the ith quality dimension by factor analysis.

Results of the satisfaction-dissatisfaction Matrix is shown in Fig. 1. From the figure, quality dimensions of security and dependability are confirmed as the "must-be" attributes of Kano model. Thus, REB companies should ensure those quality dimensions work functionally well to prevent from refusal of customers. Reliability is the "one-dimensional" attribute. That is, the REBs need to continue increase the function of reliability to increase the customer's satisfaction and decrease the dissatisfaction. Profession and information belong to the "attractive" attributes. It tells the REBs that providing good functions of profession and information could be sources of increasing customer loyalty. Whereas, tangibility, communication, and empathy are identified as the "indifference" attributes of Kano model. Therefore, some of resources that were previously put in those quality dimensions can be adjusted to strengthen other attributes of quality dimensions.

6 Conclusion

Existing research about REB service quality always adopted a one-dimensional viewpoint to evaluate each of the quality requirements. In fact, nevertheless, not all the quality requirements have linear contributions to the overall customer satisfaction. Some of them may have curvilinear relationship between the perception of quality requirement and the extent of customer satisfaction. This research applies Kano model to exploit service quality for the REB industry in Taiwan. In the research, a factor analysis was used to group the quality requirements into eight quality dimensions. The Kano questions and evaluation criteria were designed in the functional and dysfunctional questions of the questionnaire to collect data for further classifications of the two-dimensional quality model of REB industry. A satisfaction-dissatisfaction matrix analysis was performed to confirm what quality attribute that each quality dimension of REB belongs to in the Kano's two dimensional model. Results show that quality dimensions of security and dependability are confirmed as the "must-be" attributes of Kano model. Reliability is the "one-dimensional" attribute. Profession and information belong to the "attractive" attributes. Whereas, tangibility, communication, and empathy are identified as the "indifference" attributes of Kano model. Based on this analysis, corporate resource allocation strategy for improving service quality is suggested for the REB companies.

Acknowledgments This work is partially supported by the National Science Council of Taiwan under the grant (NSC 101-2221-E-390-012). The authors also gratefully acknowledge the helpful comments and suggestions of the reviewers, which have improved the presentation.

References

Berry LL, Parasuraman A, Zeithaml VA (1994) Improving service quality in America: lessons learned. Acad Manage Executive 8:32–45
Berry LL, Parasuraman A (1997) Listening to the customer: the concept of a service-quality information system. Sloan Manage Rev 38:65–66
Grönroos C (1982) Strategic management and marketing in the service sector. Swedish Sch Econ Bus Adm, Helsingfors, Sweden
Harrison-Walker LJ (2002) Examination of the factorial structure of service quality: a multi-firm analysis. Serv Ind J 22:59–72
Heskett JL, Jones TO, Loveman GW, Sasser WE Jr, Schlesinger LA (1994) Putting the service-profit chain to work. Harv Bus Re 72:164–174
Heskett JL, Sasser WE Jr (2010) The service profit chain. In: Maglio PP et al (eds) Handbook of service science. Springer Science+Business Media, LLC, pp 19–29
Kano N, Seraku N, Takahashi F, Tsuji S (1984) Attractive quality and must-be quality. Hinshitsu: J. Japanese Soc for Quality Control, 39–48
Lin HT (2010) Fuzzy application in service quality analysis: an empirical study. Exp Sys Appl 37:517–526
Mao CK (1995) A study of service quality of real estate agent in Taichung city. Unpublished Master Thesis. Dept. of Land Management, Feng Chia University

Matzler K, Hinterhuber HH (1998) How to make product development projects more successful by integrating Kano's model of customer satisfaction into quality function deployment. Technovation 18:25–38

Pakdil F, Aydin Ö (2007) Expectations and perceptions in airline services: an analysis uing weighted SERVQUAL scores. J Air Transp M 13:229–237

Parasuraman A, Zeithaml VA, Berry LL (1985) A conceptual model of service quality and its implications for future research. J Marketing 49:41–50

Parasuraman A, Zeithaml VA, Berry LL (1988) SERVQUAL: a multiple item scale for measuring consumer perceptions of service quality. J Retail 64:12–40

Peiró JM, Martínes-Tur V, Ramos J (2005) Employees' overestimation of functional and relational service quality: a gap analysis. Serv Ind J 25:773–788

Taylor A (2004) A journey to the truth: achieving top box customer satisfaction at enterprise. Executive Speeches 19:12–18

Wang HY (2007) The real estate brokerage service quality research of Taipei county. Unpublished Master Thesis. Graduate School of Business and Management, Vanung University

Zeng YM (1990) A study of service quality research: a study case of rehouse industry in Taipei. Unpublished Master Thesis. Institute of Management Science, Tamkang University

Automated Plastic Cap Defect Inspection Using Machine Vision

Fang-Chin Tien, Jhih-Syuan Dai, Shih-Ting Wang and Fang-Cheng Tien

Abstract Plastic caps are the most commonly seen bottle caps used in beverage and food containers. They are widely used to seal freshness of beverage or liquids in bottles. Threads are usually grooved inside the caps for easy twist-off caps and sealing rings prevent the liquids from bacterial infection. Companies print logos or pictures on the top surface of plastic cap, such that the quality of printing also indirectly affects the customers purchase. Inspection of plastic caps, including the surface printing, thread, and sealing ring, is a great issue during the caps production currently. The objective of this study is to use machine vision to inspect the defect of the sealing area and the printing surface of a plastic cap. An automated inspection system, which includes two CCD camera, lighting source, sensors, and a cap transporter, is constructed, and a digital image processing software is designed to learn good caps and screen out the defective ones. The experimental results show that the proposed inspection system can self-learn the features of a good surface printing, and effectively detect the defective caps under very few parameters setting, while the major defects in the sealing ring and thread area such as malformation, contamination, overfill, incomplete, scratches, can be successfully identified under the rate of 1,200 piece per minute.

Keywords Machine vision inspection · Plastic cap · Cap print inspection · Cap sealing ring inspection

---

F.-C. Tien (✉) · J.-S. Dai · S.-T. Wang
Department of Industrial Engineering, Taipei Tech, 1, Chung-Hsia East Road Sec. 3, Taipei, Taiwan, Republic of China
e-mail: fctien@ntut.edu.tw

F.-C. Tien
Department of Applied Statistics, Chung Hua University, 707, Sec.2, WuFu Rd, Hsinchu, Taiwan, Republic of China30012,

1 Introduction

Plastic cap is the most commonly used item to seal the PET bottles for most of the drinks. In plastic caps, threads are usually grooved inside the caps for easily twist-off, while sealing rings are designed to prevent the liquids from bacterial infection. Besides, producers print logos or pictures on the top surface of plastic caps, such that the quality of printing also indirectly affects the customers purchase. In this study, the size of cap with printed logos and sealing ring is ranged between 28 and 30 mm. The automated detection technology (Chin and Harlow 1982; Newman and Jain 1995) is an integrated technology, including optical technology, mechanical design, image processing, computer-aided processing and control. A machine vision system is composed with hardware and software, which using a variety of algorithms and computing capabilities to develop different applications. The hardware is composed of light sources, lens, CCD, video capture card, computer and I/O devices. The software usually includes the technology of pattern recognition (Belongie et al. 2002), feature extraction (Torralba et al. 2007), spatial data capture, etc.

The objective of this study is to use machine vision to inspect the defect of the sealing area and the printing surface of a plastic cap. An automated inspection system, which includes two CCD camera, lighting source, sensors, and a cap transporter, is constructed, and a digital image processing software is designed to learn good caps and screen out the defective ones.

2 Proposed Method

The configuration of proposed system is shown in Fig. 1. This study sets up a set of lighting device and two CCD cameras on a conveyor, which transport caps to a proper position; the CCD cameras are triggered by sensors to acquire the cap top and bottom images; and to use appropriate image processing method (Qing hua et al. 2010) to determine the cap defects. Then, the inspected cap is transported to screener to screen out the defective caps. In particular, this study develops a learning process which acquires cap patterns to find inspection parameters automatically for surface inspection, while the sealing ring inspection uses connected-component labeling (Haralick and Shapiro 1992) to get image parameters. The details of inspection process are described in the following sections.

3 Inspection of Sealing Ring

This study adopts high speed black-and-white cameras to acquire images with 900×900 pixels on the fly for sealing ring inspection. As shown in Fig. 2a, the cap and its background are highly different, so can be easily differentiated through

Automated Plastic Cap Defect Inspection Using Machine Vision

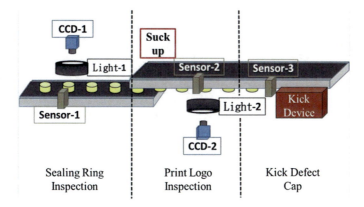

Fig. 1 Configuration of proposed inspection system

a simple threshold. After that, the center of the cap and its radius is calculated by blob analysis such that the image of cap is segmented into five circular regions, so called cap edge, cap base, bottle top, cap sealing ring, and cap center, which are denoted by R0, R1, R2, R3, R4 respectively as shown in Fig. 2.

After segmentation, the gray vale of each region can be calculated. The maximum and minimum gray values are set as upper and lower limits for double thresholding processing. A set of normal caps are collected to learn the average of

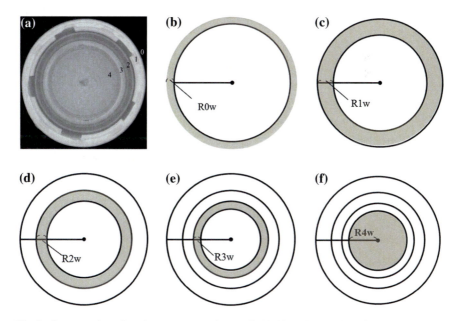

Fig. 2 Segmentation of cap image. a A cap image divided into five sections. b Cap edge. c Cap base. d Bottle top. e Cap sealing ring. f Cap center. *Riw* represents the radius of ith region

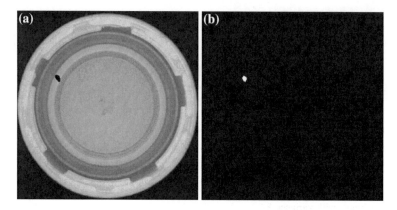

Fig. 3 Defect detection process. **a** A cap with defect in region 3. **b** After defect detection

Table 1 Gray value limits of each region

Region	Min gray value	Max gray value
R0	0	37
R1	103	228
R2	102	155
R3	118	195
R4	114	184

upper and lower limits, and then use these two limits to identify any abnormal gray value in these region. An example is shown in Fig. 3 and Table 1. The upper and lower limits of trained caps are first calculated, and then a cap with a defect in region 3 can be easily detected by its lower limit of gray value.

4 Inspection of Printing Logo

Most of caps have distinctive features such as changeable print logos and color patterns. This study uses a color CCD camera to acquire images for print logos on top surface with the same image size, and then a two-step inspection process, learning and inspecting, are developed.

4.1 Learning Process

Each cap owns different printing logos and color patterns. In this study, we first convert the color image into three color channels (RGB) of image, and calculate the image variance of each image. Instead of processing three images, we select

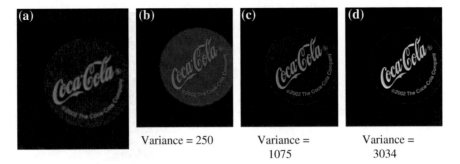

Fig. 4 Color image convert to three grayscale image and variance of cap image. **a** Color image. **b** *Red grayscale* image with variance 250. **c** *Green grayscale* image with variance 1,075. **d** *Blue grayscale* image with variance 3,034

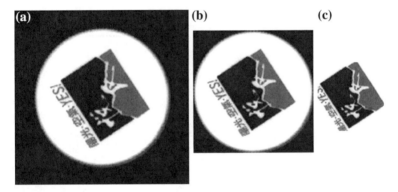

Fig. 5 Selection of ROI. **a** Normal cap image. **b** ROI selection. **c** Remove background by image pre-processing

the image with the maximum variance as the target image to conduct the following inspection procedure. As show in Fig. 4, the original image is converted into R, G, and B three channels of monochrome images. Based on the calculated variance, the Blue channel image is selected as the target image. After that, an ROI, which defines the normal cap pattern, is selected and its background is removed through an image pre-processing as shown in Fig. 5a–c.

4.2 Inspecting Stage

The inspection process of print logo cap is described as below:

(1) Image Matching with learned pattern:By conducting a pyramid matching process, we derive the information of the target cap including: the similarity of

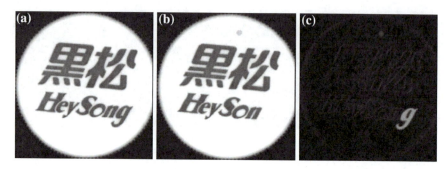

Fig. 6 Image comparing. **a** Normal cap image. **b** Defective cap image. **c** Difference image

inspected cap (matching score), the pattern angle (matching angle), and the center position of cap. Matching score value shows matching level which can filter some large defect. Matching angle is a direction gap between object cap and learned normal cap. Matching center position is the location of the target cap.

(2) Image rotating: It is important preprocessing of image compare. The target cap is rotated so that its direction is same as the learned cap.

Fig. 7 Ten different testing caps

(3) Image comparing: By subtracting the gray value in the corresponding coordinates of good cap (Fig. 6a) and the target cap (Fig. 6b), we obtain the difference image (Fig. 6c), which shows the cap with some contamination and an incomplete printing.
(4) Image thresholding and morphology: There are many noises in difference-image due to system noise and printing deviation of every cap. Image thresholding is used to segment the defective region, and then the morphology, including: opening and closing, are used to filter out the noise and retain the defects.

In this study, we validate our method with ten caps with different print logos and color patterns. After testing with our proposed method, the proposed method successfully detects the defects in the case study company (Fig. 7).

5 Conclusion

The proposed inspection method is applicable to a variety of caps with sealing rings and printing logos. A normal cap is first learned in this system and parameters for inspecting are learned to reduce the complexity of inspection. The proposed method segments bottom cap image into five key regions and derive the parameters of each region automatically. Then, based on the derived parameters of regions, the defect of the cap can be detected. For the top of cap, the standard printing pattern is learned first, and then the defect and abnormal printing (missing, extra or blur printing) can be detected by matching and comparing the inspecting cap with the standard printing pattern. Ten different printing logos and color patterns cap are used to validate the print inspection process. The experimental results show that the proposed inspection system can self-learn the features of a good surface printing, and effectively detect the defective caps under very few parameters setting, while the major defects in the sealing ring and thread area such as malformation, contamination, overfill, incomplete, scratches, can be successfully identified under the rate of 1,200 piece per minute.

References

Belongie S, Malik J, Puzicha J (2002) Shape matching and object recognition using shape contexts IEEE trans. Pattern Anal Mach Intell 24:557–573
Chin RT, Harlow CA (1982) Automated visual inspection: a survey. IEEE Trans Pattern Anal Mach Intell 4:557–573
Haralick RM, Shapiro LG (1992) Computer and robot vision volume I. Addison-Wesley Publishing Company Inc, United States of America
Newman TS, Jain AK (1995) A survey of automated visual inspection. Comput Vis Image Underst 61:231–262

Qing hua W, Xunzhi L, Zhen Z, Tao H (2010) Defects inspecting system for tapered roller bearings based on machine vision. Paper presented at International Conference on Electrical and Control Engineering 25–27 June 2010

Torralba A, Murphy KP, Freeman WT (2007) Sharing visual features for multicalss and multiview object detection. IEEE Trans Pattern Anal Mach Intell 29:854–868

Coordination of Long-Term, Short-Term Supply Contract and Capacity Investment Strategy

Chiao Fu and Cheng-Hung Wu

Abstract This research studies contract design and capacity investment problem in a two-echelon supply chain consisting of a supplier and a downstream retailer who has in-house capacity. After building in-house capacity, the retailer would use his own capacity first. Under such situation, the risk of the variance of capacity utilization would be transferred to suppliers. The objective of this research is to protect the suppliers' profit by exploring the coordination of supply contract (combining long-term and short-term contract) and capacity investment strategies. At the beginning of each period, the demand uncertainty would be realized, and then the supplier would offer both long-term and short-term contracts. In long-term contract, the retailer makes a reservation for the next two successive periods; in short-term contract, the retailer orders products to fulfill the reserved deficiency. Additionally, both parties would make capacity investment decision in every period. The supplier has higher market power, making the capacity investment decision first and deciding the contracts. To solve the problem, we build a mathematical model, using game theory to decide the short-term decisions and exercising the dynamic programming to obtain the optimal policy in long-term.

Keywords Supply chain management · Supply chain contract · Capacity investment

---

C. Fu (✉) · C.-H. Wu
Institute of Industrial Engineering, National Taiwan University, Taipei, Taiwan, Republic of China
e-mail: r00546018@ntu.edu.tw

C.-H. Wu
e-mail: wuchn@ntu.edu.tw

1 Introduction

In this research, we study contract design and capacity investment problem under demand uncertainty, multi-period, and the situation that a retailer begins to build in-house capacity. We consider the two-echelon supply chain including an upstream supplier (who provides finished product to the downstream firm) and a downstream retailer (who buys product from the supplier or produces by himself, and sells the product to the market).

The downstream retailer's capacity would cause enormous impact on the performances of upstream firms. If business environment stays prosperous, both upstream and downstream firms' capacity could be utilized well; while if a recession occurs, the retailers would use in-house capacity, and the suppliers' capacity would be idle. Thus suppliers might have to face not only the dropped order but also the risk of the variance of capacity utilization which is transferred by those retailers.

Taking a point of suppliers' view, we want to help the upstream firms to protect their profit. To do this, we propose the long-term contract and short-term contract in order to reallocate the risk in such supply chain. Additionally, both of them could make capacity investment at each period. Each party's goal is to maximize their own expected profit. After defining the problem, we would build a mathematical model to determine the optimal contracts structure and capacity investment policies for each decision maker. Then, we would examine the interaction of capacity investment among such supply chain contracts, and discuss the comprehensive decision making scheme of combined contracts.

2 Research Problem

We consider a two-echelon supply chain, in which a supplier sells products to a retailer who has in-house capacity. Both of the two parties' production plans are constrained by their own capacity level, and they could do capacity investment decisions at each period. In our model, we propose a long-term contract and a short-term contract. In long-term contract, the retailer makes a reservation for the next two successive periods; in short-term contract, the retailer orders products to fulfill the reserved deficiency. The retailer faces demand uncertainty. The ε is used to described the demand scenarios, which are adapted to stand for possible business environment (high, medium and low).

$$d_A = a - bp_A + \varepsilon \quad (1)$$

To simplify the problem we want to solve, we make assumptions as following:

- It is a complete information market.
- The business environment transition follows Markov process.

- The result of capacity investment has no uncertainty, and the invested capacity would be available at the next period.
- Both parties produce products only for the current period, they can't sell inventory.

3 Model Description

We break down the 2 parties' decisions into two parts: the short-term decisions and the long-term decisions. In our model, the short-term decisions would cause effects only on the current profit; while the long-term decisions are those could influence the future state. As we mentioned previously that the supplier is a leader and the retailer is a follower, and so the decision process could be modeled as a Stackelberg game. We would use dynamic programing method to find the solution. The objective of each decision maker is to obtain capacity expansion policy that maximizes their respective expected profits.

3.1 Short-term Decisions

For the supplier, the short-term decision is the short-term contract price; while for the retailer, the short-term decisions include the short-term contract order quantity, the product quantity, and the market price.

According to the contract price announced by the supplier and the retailer's produce cost per product, the retailer would decide the priority of using his in-house capacity or making a short-term order to fulfill his reserved deficiency.

Thus, when the reserved quantity (the total ordered quantity of long-term contract in the last two periods) is deficient, the short-term contract price would be separated to two intervals: (1) the contract price per product is higher than the retailer's produce cost per product. (2) the contract price per product is lower than the retailer's produce cost per product. First, the supplier would choose contract price in these respective intervals, and then choose the price which could make the maximum current profit for him.

3.2 Long-term Decisions

To solve the long-term decisions, we use the result of the short-term decisions, and conduct dynamic programming model to determine the optimal policy of long-term contract and capacity investment. For supplier, his long-term decisions are the long-term contract price, and the capacity expansion amount; while for retailer, it includes the long-term contract order, and the capacity expansion amount.

In this supply chain, each party's goal is to maximize their own expected value. By implementing backward induction and value iteration, the optimal policy for each firm can be obtained. Since the supplier is a leader, the retailer would make response to leader's action.

4 Numerical Study and Discussion

The following parameters were used in our numerical study and discussion. In Table 1 we illustrate the parameter settings.

4.1 Numerical Result

A. Optimal Capacity Expansion

Due to the high cost of capacity investment, both of the two firms would expand their capacity at early period as Fig. 1 shows (the business environment is medium, the supplier's capacity level is 5, the retailer's capacity is 0 and there is no reserved quantity). Thus, they could have sufficient time to earn profit to cover the cost.

B. Long-term Contract

The result of long-term contract would be affected by different time period. We find that at earlier period, the contract price is lower than the price at later period, taking Fig. 2 for example (the business environment is medium, the supplier's capacity level is 10, the retailer's capacity is 3 and there is no reserved quantity).

Table 1 Parameter settings

Time horizon		Constraints of capacity quantity	
T	15	Supplier's UB	10
Market states		Retailer's UB	5
Market size	32	Cost of capacity investment	
Price sensitivity of product	1	Supplier	60
Fluctuation rate	2	Retailer	100
Market transition		Cost of production per product	
$\quad\quad$ H $\;$ M $\;$ L		Supplier	6
H $\begin{bmatrix} 0.6 & 0.3 & 0.1 \\ 0.1 & 0.8 & 0.1 \\ 0.1 & 0.3 & 0.6 \end{bmatrix}$ M L		Retailer	10
		Initial setting	
		Business environment	Medium
		Supplier's capacity level	5
		Retailer's capacity	0

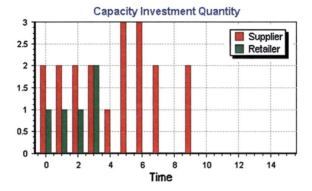

Fig. 1 The capacity investment quantity

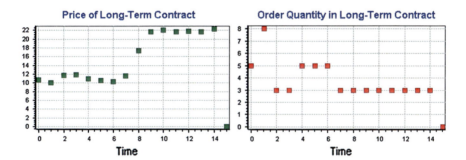

Fig. 2 The optimal long-term decision

The reason is that at early period, the high long-term price would make the retailer be unwilling to reserve but invest in-house capacity. Once the retailer expands capacity, the supplier's expected profit would drop dramatically as Fig. 3 presents (the order quantity and the expansion amount of the retailer are at right-axis; the 2 parties expected profits are at left-axis).

5 Conclusions

In this research, we combine long-term and short-term contracts to provide a risk-sharing mechanism that encourages the supplier to expand capacity and to raise the flexibility of capacity utilization. Given these contracts, we investigate the interaction between the two parties' capacity investment behaviors.

We have shown that the retailer's bargaining power is stronger at early period or with high in-house capacity level. Once the retailer has in-house capacity, supplier's profit would dramatically drop. Therefore, we advise the supplier to

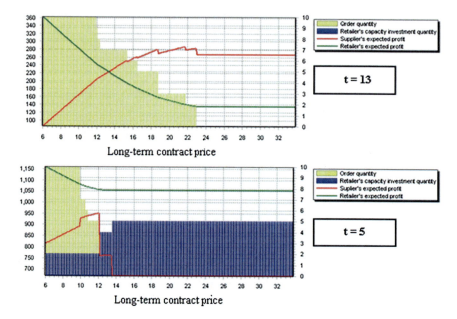

Fig. 3 The long-term price decision process when $t = 13$ and $t = 5$

announce low long-term contract price at early period in order to avoid the retailer to build in-house capacity.

We also explained how each party's capacity level influence their expected profit. Surprisingly, an increase in supplier's capacity level can cause a reduction in supplier's expect profit. We conclude that when the supplier's capacity level is over threshold, he has to reduce contract price to maximum short-term profit, which would eat into long-term profit.

Our results come with several limitations. There are many possible extensions to this research. For example, our model has only one retailer. In real world, the retailer could have more than one supplier and have other competitors. In future study, we could add more downstream retailers to study the consequence of competition between retailers.

References

Cachon GP (2003) Supply chain coordination with contracts. Handbooks Oper Res Manag Sci 11:229–340

Erkoc M, Wu SD (2005) Managing high-tech capacity expansion via reservation contracts. Prod Oper Manag 14(2):232

Serel DA, Dada M, Moskowitz H (2001) Sourcing decisions with capacity reservation contracts. Eur J Oper Res 131(3):635–648

Sethi SP, Yan H, Zhang H (2004) Quantity flexibility contracts: optimal decisions with information updates. Decision Sci 35(4):691–712

Spengler JJ (1950) Vertical integration and antitrust policy. J Polit Econ 58(4):347–352

Spinler S, Huchzermeier A (2006) The valuation of options on capacity with cost and demand uncertainty. Eur J Oper Res 171(3):915–934

Tomlin B (2003) Capacity investments in supply chains: Sharing the gain rather than sharing the pain. Manuf Serv Oper Manage 5(4):317–333

Tsay AA (1999) The quantity flexibility contract and supplier-customer incentives. Manage Sci 45(10):1339–1358

Wang TM (2012) Coordination of supply contract design and long term capacity strategy. Dissertation, National Taiwan University

Wu SD, Murat Erkoc M, Karabuk S (2005) Managing capacity in the high-tech industry: a review of the literature. Eng Econ 50:125–158

An Analysis of Energy Prices and Economic Indicators Under the Uncertainties: Evidence from South East Asian Markets

Shunsuke Sato, Deddy P. Koesrindartoto and Shunsuke Mori

Abstract It is well known that the worldwide financial crisis has caused various international issues. For instance, it is pointed out that the Euro-crisis has spread the financial instability to the other markets e.g. equity, oil, gold, etc. In addition, the energy and the commodity market are known as major factors which influence the decision making of investors in currency market or stock market in both long and short term. Thus it is an important thing for the decision making for investors to find out the relationships among various markets. There are two scenarios in this paper. In the first one, we analyze the future forecasts by applying Vector Error Collection Model (VECM) to economic indicators which have influential power all over the world, and then, we get the relationship among these markets via Granger Causality test. On the other hand, it is also important to predict which factors would be market driven in the future. Then in the second one, since South East Asian market is known as the potential markets for driving the market in the future, we add the variables which belong to South East Asian markets to the

---

S. Sato (✉)
Shunsuke Sato, Master of Engineering in Industrial Administration, Graduate School of Science and Technology, Tokyo University of Science, Yamazaki 2641, Noda, Chiba Prefecture, Japan
e-mail: syunnsuke0130@gmail.com

S. Sato
Master of Science in Management, Graduate School of Business and Management, Bandung Institute of Technology, Jl. Ganesha 10, Bandung, West Java, Indonesia

D. P. Koesrindartoto
Assistant Professor at School of Business and Management, Bandung Institute of Technology, Jl. Ganesha 10, Bandung, West Java, Indonesia
e-mail: deddypri@sbm-itb.ac.id

S. Mori
Department of Industrial Administration, Faculty of Science and Technology, Tokyo University id Science, Yamazaki 2641, Noda, Chiba Prefecture, Japan
e-mail: mori@ia.noda.tus.ac.jp

variables in the first scenario. This outcome will give some opportunity to get the interest to the investors.

Keywords Time series analysis · Vector error collection model (VECM) · Oil price · Index · Currency · South East Asia

1 Introduction

The factor of energy price volatility has been a controversial issue for the policy makers and researchers who have various positions depending on the country conditions as seen in the international conferences. As one of the claims, there is a world energy model where the price is determined by the relationship between supply and demand. For instance, Matsui et al. explored the future trends of the global warming problem mainly due to carbon dioxide by using a world energy supply and demand model (Matsui et al. 1995). In addition, International Energy Agency (IEA) reported about the effects of pushing down real GDP due to higher oil prices by using the large-scale computer simulation model which seems based on "World Energy Model" (IEA 2004).

However, the movement of energy prices in recent years might not simply be a behavior caused by the supply and demand equilibrium in energy prices. In 2008, Saudi Arabia's Oil Minister Ali al-Nuaimi said that "Clearly Something other than supply and demand fundamentals is at work here, and a simplistic focus on supply expansion is therefore unlikely to tame the current price behavior".

There are a large number of papers suggesting that energy prices are moved by the temporal market principles. These are generally analyzed by using time-series analysis such like Generalized Autoregressive Conditional Heteroscedasticity model (GARCH) and Vector Error Collection Model (VECM). Mohamed et al. found dynamic linkages between the European stock market and crude oil price by GARCH. Our Results show strong significant linkages between oil price changes and stock markets for most European sectors (Hedi Arouri and Khuong Nguyen 2010). As well, Kofi et al. showed that crude oil prices affect the stock market by VECM. While currency prices significantly explain Israeli stock returns, crude oil futures prices relate significantly to the Egyptian and Saudi Arabian stock exchanges (Amoateng and Kargar 2004). Therefore, energy market may also be affected from another market such like stock market, currency market and etc. This is referred as market integration in the world market.

1.1 Southeast Asian Market

As an example of the market is being integrated, there is a Association of Southeast Asian Nations (ASEAN). The growth of the Southeast Asian market in recent years, there are remarkable. In particular, Indonesia, has been reached 6.5 % GDP growth in 2012, is attracting attention as a country to lead the Southeast Asian market. Goldman Sachs estimated that GDP of Indonesia will be greater than Japan in 2050 (Wilson and Stupnytska 2007). As well, Indonesia is counted as core countries, is called as P4, in the TPP which is one of the significant international frameworks. Developed countries such as Japan and the United States are also included as TPP negotiations participating countries, and in the future deep cooperation is expected. However, papers exploring dynamic linkages like this are small.

1.2 Statistical Analyzing Method

As a model for analyzing the dynamic linkage between markets, Vector Error Collection Model (VECM) is widely known. VECM is one of the regression models extending Vector Auto Regressive Model (VAR) by using the method of cointegration. It is possible to find the dynamic coordination using Granger Causality Test. However, it must be noted that the statistical time-series analysis is not necessarily a panacea for every events. Lucas criticized, known as Lucas critique, that the macro-econometric model does not have micro natures (Lucas 1976). Thus, I would like to note in advance that current statistical analysis cannot solve this bias that arises from empirical results.

1.3 Research Objectives

By the flow of market integration, it becomes more complex for policy makers and investors to identify the significant factors. In this study, the authors show a method for the decision making by examining the following 2 scenarios:

Scenario 1: Analyze the dynamic linkages in global economy by using the significant indicators which have a power across the world, and

Scenario 2: Consider the current and future trends and linkages in Southeast Asian market by addition the variables of Southeastern Asian market including Singapore and Indonesia to scenario 1 variables.

2 Data and Methodology

2.1 Datasets in Scenario 1

In this paper, the authors deal with a world model in Scenario 1, and the key indicators in each region are selected as variables as follows:

North America: Canadian Dollar (CAD/USD), NASDAQ (GSPC)

Europe: Germany Mark (DEM/USD), U.K. Pound (GBP/USD), Swiss Franc (CHF/USD), FTSE (FTSE)

Asia: Japanese Yen (JPY/USD), Nikkei 225 (N225), Hong Kong Hang Seng Index (HSI)

Commodity markets: West Texas Intermediate as crude oil (XCT/USD), Gold (XAU/USD)

Term: Jan 1987–Dec 2012

Quantity: 312

Frequency: Monthly

We chose Germany Mark before 1999 and then converted them into EURO after that by using exchange coefficient. The authors don't adopt the Chinese Yuan which is currently one of the most important currencies, because of the fixed exchange rate until 2005 As well as the HK dollars which also adopts a dollar peg.

2.2 Datasets in Scenario 2

In scenario 2, in addition to variables in Scenario 1, the indicators of Singapore and Indonesia are adopted. However, after the Asian financial crisis, Indonesia had adopted the dollar peg until July 1997. Therefore, it is difficult to add this period until July 1997. In Scenario 2, the period set from August 1997, on which Indonesian rupiah has already shifted to a floating exchange rate system, to December 2012.

Southeast Asia: Singapore Dollar (SGD/USD) Indonesia Rupiah (IDR/USD) Straits times index (STI) Jakarta composite index (JCI)

Term: Aug 1997–Dec 2012

Quantity: 185

Frequency: Monthly

2.3 Time Series Analysis Method

In the time series analysis, stationary of the data is the most important factor, since the shock stays permanently and the spurious correlation becomes apparent when non-stationary is involved. Therefore, it is necessary to check the data condition first.

As one of the methods measure the stationary of datasets, there are some well established procedures. As the unit root tests, Augmented Dickey-Fuller Test (ADF), Phillips-Perron Test (PP) is well known. In this paper, the authors find the absence of unit root in 10 % level except Indonesian Rupiah by ADF. The table of unit root test in Scenario 2 is following (see Table 1).

An Analysis of Energy Prices and Economic Indicators 813

Table 1 Unit root test in scenario 2

Variables	T-value	P value	Variables	T-value	P-value
CAD/USD	−0.625	0.861	GSPC	−2.186	0.212
DEM/USD	−1.360	0.601	FTSE	−2.000	0.288
GBP/USD	−2.010	0.282	N225	−2.187	0.212
CHF/USD	−0.721	0.838	HSI	−1.453	0.555
JPY/USD	−1.345	0.608	STI	−0.812	0.813
SGD/USD	−0.302	0.921	JCI	−0.349	0.914
IDR/USD	−6.169*	0.000	XAU/USD	0.125	0.967
			XCT/USD	−1.285	0.636

* the variable is rejected in 1 % level

Before we start the analysis, it is necessary to check which model is most suitable. In the time-series analysis, if the dataset has both the cointegration and unit root, the most suitable model for analyzing the dynamic linkage is VECM.

The cointegration is one of the indicators showing relationships between datasets. When the presence of cointegration can be confirmed, the objective variable might has a relationship with explanatory variables. As cointegration test, Johansen cointegration test is famous. In this paper, authors can find at most 2 cointegration in Scenario 1 and at most 11 cointegration in Scenario 2 by Johansen cointegration test. There are cointegration and unit root. Thus, in this paper VECM is adopted.

When the presence of cointegration is accepted in datasets, we can know the causal relationship. Granger devised a Granger Causality Test to be able to get Granger Causality by F test of the coefficient (Granger 1969).

3 Analysis

3.1 Analysis in Scenario 1

In the Scenario 1, the authors adopted lag-2 based on Akaike Information Criterion (AIC) and Schwartz Criterion (SBIC) (see Table 2).

As a results of Granger Causality Test Using VECM(2), it is shown that Swiss franc, FTSE and Nikkei 225 have no self-correlation. This may indicate that these markets be strongly influenced from other markets by dynamic linkages.

In currency market, it shows that Canadian dollar has an impact on many indicators. Canadian dollar has a relevance to the currency except JPY and affects stock markets such like FTSE and HSI (see Table 3). However, it is known that Canadian dollar has a strong relationship with the U.S. Dollar. Hence, it is expected that U.S. dollar has had a significant impact on other markets.

Table 2 Lag decision

Lag		2	3	4	5	6
Scenario 1	Akaike information criterion	1.717143	1.966077	2.24305	2.450815	2.760428
	Schwarz criterion	5.3055	7.028356	8.786278	10.48208	12.28687
Scenario 2	Akaike information criterion	26.64379	26.94945	26.14073	24.8449	23.23195
	Schwarz criterion	40.63931	44.97413	48.22535	51.02064	53.53039

In commodity markets, each indicator has autocorrelation, but they seem to have not been explained so much from other markets. There are significant linkages between crude oil and Japanese Yen and NASDAQ.

In addition, as a result of prediction, NASDAQ, FTSE, HSI and England Pound have upward trend, and Swiss franc, Japanese yen, Nikkei 225, Gold and Crude oil have downward trend. In particular, there is a depreciation tendency of currency against the U.S. dollar, therefore it is expected that U.S. dollar will be high. In commodity market including gold and oil prices, substantial decline is expected. However, since these predictions cannot reject Lucas criticism, it is difficult to consider the macro factors such as policy changes. Thus, authors note that these prediction is a simple result calculated by dynamic linkages among markets.

3.2 Analysis in Scenario 2

In this Scenario, authors adopted lag 2 by using Schwartz Criterion (SBIC) (see Table 1).

The result shows that U.K. Pond, Swiss franc, FTSE and Crude oil have no self-correlation. We must note that Swiss franc and FTSE have no self-correlation in scenario 1 as well as scenario 2.

In currency market, Western currencies are influenced by themselves and Asian currencies are also influenced by themselves (see Table 4). This result may evidence that the dynamic linkage advances in each regions, i.e. Western region and Asian region. For instance, Canadian Dollar influences all of Western currencies including Germany Mark, U.K. Pound and Swiss Franc as well as Scenario 1. U.K. Pound has no self-correlation, but it also influenced all of Western currencies. On the other hand, Japanese Yen doesn't have a strong power for driving Western Currencies, but it effects all of Asian currencies including Japanese Yen, Singapore Dollar and Indonesia Rupiah.

In stock market, Western index including NASDAQ and FTSE has a strong power driving all of currencies. Influence of Nikkei 225 may be week comparing these indexes, but although Western indexes have no influence to Asian stock market, Nikkei 225 widely influences to Asian stock markets. Thus Japanese indicators have a strong power to Asian market instead of no-influence to Western markets.

Table 3 Coefficients about Canadian dollar in scenario 1

Coefficients	Currency market				
	D(CAD_USD)	D(DEM_USD)	D(GBP_USD)	D(CHF_USD)	D(JPY_USD)
D(CAD_USD(−1))	1.94595**	1.81652**	2.99301*	1.84624**	1.11808
D(CAD_USD(−2))	0.56007	0.61602	0.71009	0.11267	0.5431
	Stock market				Commodity market
	D(GSPC)	D(FTSE)	D(N225)	D(HSI)	D(XAU_USD)	D(XCT_USD)
D(CAD_USD(−1))	1.45608	2.32937*	1.23001	1.69456**	0.55377	0.08257
D(CAD_USD(−2))	0.80748	1.17325	2.12117*	0.13342	0.18808	1.20079

* the relation is significant in 10 % level
** the relation is significant in 5 %

Table 4 Currency relationship in developed countries

Coefficients		Currency market							
		Western countries				Asian countries			
		D(CAD_USD)	D(DEM_USD)	D(GBP_USD)	D(CHF_USD)	D(JPY_USD)	D(SGD_USD)	D(IDR_USD)
Western Countries	D(CAD USD(-1))	1.604	1.85749**	1.71003**	1.58013	0.31841	0.42701	0.83649
	D(CAD_USD(-2))	0.22033	0.33298	2.09139*	1.07777	1.51883	0.45485	0.57555
	D(DEM_USD(-1))	0.17326	2.4996*	2.00276*	2.1224*	0.16516	1.49074	0.58674
	D(DEM_USD(-2))	0.29923	1.38026	0.1284	0.2889	1.4493	0.07803	0.37544
	D(GBP USD(-1))	1.60545	2.7704*	0.77473	2.93051*	0.02092	1.81371**	1.09204
	D(GBP_USD(-2))	0.32117	1.06288	0.57231	1.86925**	0.64428	0.94582	1.06056
Asian Countries	D(JPY_USD(-1))	0.49689	1.17259	0.43759	1.40506	3.25121*	1.34817	1.76999**
	D(JPY_USD(-2))	0.71088	0.59841	0.35027	0.77469	1.81256**	1.9446**	1.9981*

* the relation is significant in 10 % level
** the relation is significant in 5 %

Table 5 Dynamic linkage in Southeastern Asian market

Coefficients		Currency market								
		D(CAD_USD)	D(DEM_USD)	D(GBP USD)	D(CHF USD)	D(JPY USD)	D(SGD_USD)	D(IDR USD)		
Currency market	D(SGD USD(−1))	0.33394	0.5994	0.47211	1.17719	0.33757	0.57367	0.40256		
	D(SGD USD(−2))	1.23997	1.7973**	1.67387**	1.36971	0.69604	1.50104	1.32803		
	D(IDR USD(−1))	0.45931	1.42981	0.19642	2.4726*	1.96192*	0.01666	2.77496*		
	D(IDR USD(−2))	1.2771	0.15384	0.80634	0.07286	2.58164*	1.38145	0.50468		
Stock market	D(STI(−1))	1.16183	0.1872	1.14439	0.26166	2.46663*	1.49838	1.45135		
	D(STI(−2))	1.04279	1.0826	1.12691	1.2431	0.19052	1.08728	2.01226*		
	D(JCI(−1))	0.68477	0.15252	0.59573	0.29101	0.21774	0.29027	0.00295		
		Stock market						Commodity market		
		D(GSPC)	D(FTSE)	D(N225)	D(HS)	D(ST)	D(JCI)	D(XAU USD)	D(XCT_USD)
Currency market	D(SGD USD(−1))	1.11315	1.36677	0.50726	0.40816	0.7832	0.09321	1.33825	1.33404
	D(SGD USD(−2))	1.16515	0.78825	0.43618	0.93921	0.36029	0.47975	1.25581	1.39715
	D(IDR USD(−1))	0.07297	0.75452	1.61562	0.42727	0.13287	0.01689	0.52226	0.42479
	D(IDR USD(−2))	1.49869	0.96814	1.03901	0.19672	0.9079	0.83366	0.61211	1.3531
Stock market	D(STI(−1))	1.05932	1.7912B**	0.20288	0.18819	0.44737	0.76281	1.74602**	1.20981
	D(STI(−2))	1.90932**	1.81751**	1.67301**	0.41118	2.02272*	0.43676	0.15946	0.93647
	D(JCI(−1))	1.58068	1.69791**	1.32701	1.87813**	0.79594	0.66737	0.37451	1.10452

* the relation is significant in 10 % level
** the relation is significant in 5 %

In commodity market, Gold is effected from Crude oil and FTSE, and Crude oil is effected from Japanese Yen and NASDAQ. In Scenario 2 the variables which explain crude oil same with Scenario 1.

From the view of Southeastern Asian market, currency and index influence stock market and currency market respectively (see Table 5). This result also shows that there is no strong market integration which overcomes each commodity sector like dynamic linkage between stock market and currency market. In addition, the number of variables which explain STI and JCI is small. Hence, these indexes contribute portfolio-diversification for investment.

4 Conclusions

In this paper, the authors analyzed the dynamic linkage among various market indicators assuming two different scenarios employing the time series analysis. The results would give the investors and policy makers some new perspectives for decision making.

The authors evidenced that there is a dynamic linkage over each commodity sector in Western indicators. For instance Western indexes influence Western currency market while the results did not show that Southeastern Asian markets have strong linkages with developed countries. It is however suggested that the dynamic linkage in Asia including Japan, China and Southeastern Asia may have been advancing.

Since the analysis in this paper fails to contain Chinese currency which is one of important indicators as well as Indian and Brazilian indicators, further the challenges will be needed to investigate the dynamic linkages focusing on developing countries expected the strong future growth.

References

Amoateng KA, Kargar J (2004) Oil and currency factors in middle east equity returns. Manag Financ 30:3–16

Granger CWJ (1969) Investigating causal relations by econometric models and cross-spectral methods. Econometrica 37:424–438

Hedi Arouri ME, Khuong Nguyen D (2010) Oil orices, stock markets and portfolio investment: Evidence from sector analysis in Europe over the last decade. Energ Policy 38: 4528–4539

IEA (2004) Analysis of the impact of high oil prices on the global economy. Leonardo Energy Organization. http://www.leonardo-energy.org/sites/leonardo-energy/files/root/Documents/2009/high_oil_prices.pdf. Accessed 11 May

Lucas Robert (1976) Econometric policy evaluation: a critique. Carnegie-Rochester Conf Ser Pub Policy 1:19–46

Matsui K, Ito H, Yamada A (1995) Simulation analysis by a model projecting world energy supply and demand over the very long term—a case study on China. INSS J 2:77–103

Wilson D, Stupnytska A (2007) The N-11: more than an acronym. Goldman Sacks Global Economics Paper 153

Effects of Cooling and Sex on the Relationship Between Estimation and Actual Grip Strength

Chih-Chan Cheng, Yuh-Chuan Shih and Chia-Fen Chi

Abstract Handgrip strength is essential in manual operations and activities of daily life, but the influence of cold on estimation of handgrip strength is not well documented. Since direct measurement of force is often somewhat difficult, estimations are frequently applied, and these estimations are sometimes used as a criterion for employee selection and screening. Therefore, the aim of the present study is to investigate the relationship between estimated and actual handgrip strength at various target force levels (TFLs, in percentage of MVC) for both sexes under hand was cooled or not. A cold pressor test in a 14 °C-water bath was used to lower the hand skin temperature, and this served as the cooled condition. The uncooled condition, without cold immersion, was the control condition. Ten males and 10 females were recruited. The results indicated that cooling the hand could result in lighter estimation, which could increase the risk of musculoskeletal disorders. Furthermore, females tended to be less reliable than males in the estimation, and greater absolute deviations occurred in the middle range of TFLs for both sexes.

Keywords Cold immersion · Hand manipulation · Handgrip strength estimation

---

C.-C. Cheng (✉) · C.-F. Chi
Department of Industrial Management, National Taiwan University of Science and Technology, Taipei, Taiwan, Republic of China
e-mail: damn0623@yahoo.com.tw

C.-F. Chi
e-mail: chris@mail.ntust.edu.tw

Y.-C. Shih
Department of Logistics Management, National Defense University, Taipei, Taiwan, China
e-mail: river.amy@msa.hinet.net

1 Introduction

The hands offer the most effective means of accomplishing complex work, given their ability to perform specialized tasks that require dexterity, manipulability, and tactile sensitivity. Handgrip force is one of the most important forces required in both daily life and the workplace. Additionally, overexertion has been shown to be one of the critical factors causing musculoskeletal disorders in the workplace, especially in cold environments. Since direct measurement of force is often somewhat difficult, estimations are frequently applied, and these estimations are sometimes used as a criterion for employee selection and screening. Recently, numerous studies have investigated the relationship between estimation and actual force. Unfortunately, the influence of cooling the hands on the estimation of handgrip force is still not well documented, even though working in a cold environment is unavoidable. Therefore, the main goal of the present paper is to explore whether the relationship between the estimated and actual hand grip strength is different between cooled and uncooled hands for both sexes.

2 Literature Survey

Handgrip force is one of the most essential forces for manual operation. Besides poor postures and repetitive motions, force demands have been consistently considered as main risk factors associated with work-related musculoskeletal disorders (Silverstein et al. 1987). In addition, several epidemiologic studies have shown that cold may be a risk factor for the occurrence or aggravation of musculoskeletal disorders, such as in the fish-processing industry (Chiang et al. 1993; Nordander et al. 1999) and meat-processing factories (Kurppa et al. 1991; Piedrahita et al. 2004). Wiggen and colleagues indicated that even petroleum workers must often be exposed to harsh and extreme environments while performing not only heavy lifting tasks but also tasks demanding grip strength and dexterity, for which such workers have to remove their gloves (Wiggen et al. 2011). Therefore, it is unavoidable that bare hands will be exposed in a cold environment. A report by the European Agency for Safety and Health at Work also noted that the risk of musculoskeletal disorders increases with work in cold environments (Schneider et al. 2010).

Exposure in a cold environment reduces the skin temperature of uncovered parts, especially the forearm/hand. Hand/finger skin temperature is considered the vital factor in the reduction in tactile sensitivity (Enander 1984), hand dexterity (Enander and Hygge 1990; Heus et al. 1995; Riley and Cochran 1984; Schiefer et al. 1984), tracking performance (Goonetilleke and Hoffmann 2009), and handgrip strength (Brajkovic and Ducharme 2003; Chen et al. 2010; Chi et al. 2012; Enander 1984; Enander and Hygge 1990; Schiefer et al. 1984).

The differences in findings about selected models could be due to different postures, exertion intensities, muscles involved, and/or experimental protocols (e.g. training or not, MVC measured prior or post to exertion estimated). Due to the deficiency for cooling hand on the relationship between estimated and actual grip strength, therefore, the aim of the present study is to try to determine the relationship between estimation and actual exertion.

3 Methods

3.1 Participants

A convenience sample of 20 volunteers, including 10 males and 10 females, was recruited for this study. All were right-handed, healthy and free of musculoskeletal disorders in the upper extremities. The means (standard deviation, SD) for age, weight, and height for males and females were 28.7 (5.5) and 24.2 years (3.6); 68.8 (6.9) and 53.9 kg (4.7); 172.8 (2.1) and 162.7 cm (4.6), respectively. During the experiments, each participant was dressed on a short-sleeve T-shirt, short pants, and sports shoes.

3.2 Apparatus and Materials

The cold pressor test was employed. The apparatus and materials used in this study included a water bath, a submersible cooler, a digital thermometer and hygrometer, a digital 4-channel thermometer, and a grip gauge with a load cell. They were the same as those used in Chen et al. (2010) and Chi et al. (2012). The handles of the grip gauge, which had a 5-cm grip span, were wrapped in bandages to prevent slippage during exertion.

3.3 Experimental Procedures and Data Acquisition

First, the experimental procedure was explained, and all participants signed an informed consent. Additionally, participants' demographic datexposed and non-cold exposed workers.a were collected. Two sessions associated with two HSTs (cooled (14 °C) and uncooled) were scheduled randomly on two different days, and each session contained three stages: grip MVC measurement, training, and the main experiment. Cooled HST meant the participant had to immerse the dominant hand into a 14°C-water bath up to the elbow joint, and uncooledRiley MW, Cochran DJ (1984) HST meant the forearm was uncooled by cold water.

Prior to formal measurement, the probes of the thermometer were attached by sponge tape on the dorsal side of the middle phalanx of the middle finger (FST), on the middle of the third metacarpal of the dorsal side of the hand (HST), and on the muscles of the extensor digitorum (ED) and flexor digitorum superficialis (FDS) of the forearm (named FAST-E and FAST-F, respectively). When thermometer probes were attached properly and the cooled condition was selected, participants were first asked to immerse their dominant hands into the 14° C-water bath up to the elbow joint for 30 min, and then the grip MVC was measured. After 30-min immersion, the mean HST (SD) was 14.4 °C (0.2) for both sexes, and the mean FSTs (SD) were 13.7 °C (0.3) and 13.8 °C (0.2) for males and females, respectively. Additionally, the corresponding mean FAST-Es (SD) for males and females were 19.3 °C (0.4) and 19.0 °C (0.9), and the mean FAST-Fs (SD) were 19.6 °C (0.6) and 19.4 °C (0.6), respectively.

For handgrip MVC measurement, each participant sat erect in a chair with the elbow at 90° flexion and the upper arm parallel to the trunk. Handgrip MVC was replicated three times, and a 2-min rest was given between successive trials to avoid muscular fatigue. The maximal value of each contraction was recorded, and these three maximal values were averaged to serve as the personal MVC.

Next was the training stage, in which 30, 60, and 90 %MVC, calculated according to each participant's MVC, served as the target force levels (TFLs). For a given TFL, a horizontal marker line appeared and remained in the same central position of the monitor for visual guidance to maintain the required TFL. The 30 %MVC TFL was first practiced, in which participants were informed how much TFL, in terms of percentage of MVC, they were to exert. During exertion, they were asked to reach the TFL for 3 s and to try to perceive it. The same procedure was applied then for 60 %MVC, and finally for 90 %MVC. Here also, 2-min rests were given between successive trials.

In the main experiment, the procedures were the same as those in the training stage, but participants had no idea about how high the TFLs (15, 30, 45, 60, 75, 90 %MVC) were. They were just instructed to exert to match the TFL line demonstrated on the same central position of the PC monitor, and to try to perceive and estimate how much the TFL was in terms of a percentage of their own maximal strength. Then they reported the value to the instructor immediately, which was recorded and denoted by F_{Est}. All six TFLs were arranged randomly, and 2-min rests were given between successive trials.

In the uncooled condition, participants performed the trials at the initial skin temperature. The mean HST (SD, min–max) was 31.5 °C (1.2, 30.3–33.7) and 31.9 °C (1.3, 30.1–33.6) for males and females, respectively.

3.4 Experimental Design and Data Analysis

A combined design of nested-factorial and split-plot was employed. Independent variables, all fixed, included HST (uncooled and cooled (14 °C), the whole-plot), TFL (15, 30, 45, 60, 75, and 90 %MVC, the sub-plot), sex, and participant (nested within sex and serving as a block effect). To precisely examine the main effects and interactions, the variations associated with each participant were extracted and the highest interaction of each participant served as the error term. Furthermore, the responses were force estimations reported by participants (F_{Est}), and the absolute value of Er_{Est} ($F_{Est} - TFL$), denoted by $Abs(Er_{Est})$.

In addition, both Pearson product-moment correlation analysis and intra-class correlation (ICC) were used to test the agreement of grip force estimation among participants. The software Statistica 8.0 was used for data analysis, and a post hoc Bonferroni test was used to test paired differences for significant main effects and interactions. The level of significance (α) was 0.05.

Results indicated that the uncooled hand generated greater MVC (40.9 kgw) than the cooled HST (36.5 kgw, with 11 % MVC reduction); male MVC was greater than female MVC (47.1 vs. 30.3 kgw); and female MVC was about 64 % of male MVC. This reduction in grip strength caused by cooling is consistent with past studies (Brajkovic and Ducharme 2003; Chen et al. 2010).

4 Results and Discussion

For the estimation reliability, the agreement among participants in using %MVC to evaluate the strength they perceived needed to be examined. To do so, the correlation analysis and intra-class correlations (ICCs) were employed. The correlations between actual force (TFL) and estimated force (F_{Est}) were also calculated. Under uncooled HST, the correlation coefficients were 0.98–0.99 ($p < 0.05$) for males and 0.91–0.99 ($p < 0.05$) for females. Under cooled HST (14 °C), they were 0.96–0.99 ($p < 0.05$) for males and 0.92–0.99 ($p < 0.05$) for females. Importantly, the ICCs for males and females under both HSTs were close to 1. That is, there was consistent agreement between participants when they assessed their perceived effort.

As to validity, a paired t test was used to test the F_{Est} s against the TFLs for both sexes at different HSTs. The results showed that not all were significantly different; only that of females under uncooled HST at 45 %MVC was significantly overestimated (mean = 51.7 %MVC; 95 % C.I. = 45.4–58.0 %MVC). The trend of F_{Est} association with TFLs and descriptive statistics are presented in Table 1.

Table 1 Descriptive statistics and paired t-test between actual and estimated forces

Actual (%MVC)	Estimation							
	Uncooled				Cooled			
	Male		Female		Male		Female	
	Mean	SD	Mean	SD	Mean	SD	Mean	SD
15	16.1	4.6	17.7	6.9	15.0	4.7	15.6	4.1
30	29.1	5.0	34.8	11.0	27.8	7.3	29.5	7.8
45	46.5	6.3	51.7*	8.8	43.0	10.1	46.1	15.5
60	61.2	7.6	62.3	14.0	53.7	9.7	60.2	13.3
75	77.0	9.8	78.3	8.0	75.0	6.2	70.7	11.9
90	93.1	5.4	90.0	10.5	89.0	7.7	87.0	7.2

*$p < 0.05$

Fig. 1 a The TFL effect on estimation for HST **b** The TFL effect on estimation for sex

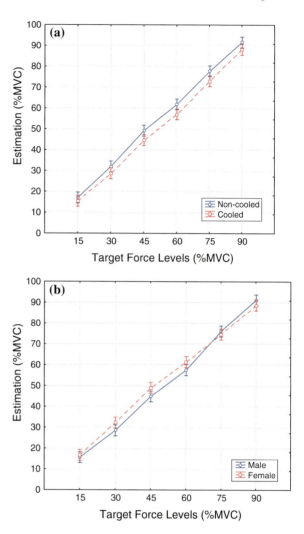

Figure 1a and b illustrate the TFL effect for different sexes and HSTs. Both the photos demonstrate that the increased force estimation followed the increase of TFL (actual exertion). Additionally, Figure 1a and Table 1 indicate that participants tended to overestimate in the uncooled condition and to underestimate in the cooled condition. On the other hand, the absolute error of estimation ($Abs(Er_{Est})$) was considered. Figure 2 displays the TFL effect on $Abs(Er_{Est})$ for both sexes. Most importantly, both sexes had greater $Abs(Er_{Est})$ in the middle range of TFL than at both ends of TFLs. Figure 2 further showed that females had a larger $Abs(Er_{Est})$ than males at all TFLs.

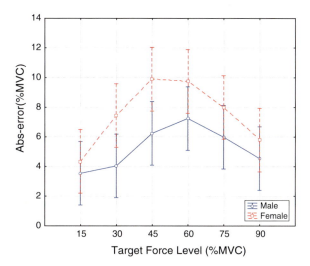

Fig. 2 The TFL effect on absolute error for both sexes

5 Conclusions

For estimation of grip strength, cooling the hand could result in lighter estimation and also cause underestimation. This underestimation could increase the risk of musculoskeletal disorders in the cold. Unfortunately, working in a cold environment is still unavoidable; therefore, how to keep the hands warm and maintain tactile sensitivity is important in avoiding underestimations of hand strength. On the other hand, females tended to be less reliable than males in the estimation. Of interest is that the middle range of TFLs in terms of percentage of MVC produced greater deviations from the TFLs for both sexes.

Acknowledgments This paper presents the results from Project No. NSC97-2221-E-606-026, sponsored by the National Science Council of Taiwan, the Republic of China.

References

Brajkovic D, Ducharme MB (2003) Finger dexterity, skin temperature, and blood flow during auxiliary heating in the cold. J Appl Physiol 95(2):758–770
Chen WL, Shih YC, Chi CF (2010) Hand and finger dexterity as a function of skin temperature, EMG, and ambient condition. Hum Factors J Hum Factors Ergon Soc 52(3):426–440
Chi CF, Shih YC, Chen WL (2012) Effect of cold immersion on grip force, EMG, and thermal discomfort. Int J Ind Ergon 42(1):113–121
Chiang HC, Ko YC, Chen SS, Yu HS, Wu TN, Chang PY (1993) Prevalence of shoulder and upper-limb disorders among workers in the fish-processing industry. Scand J Work Environ Health 126–131
Enander A (1984) Performance and sensory aspects of work in cold environments: a review. Ergonomics 27(4):365–378

Enander AE, Hygge S (1990) Thermal stress and human performance. Scand J Work Environ Health 44–50

Goonetilleke RS, Hoffmann ER (2009) Hand-skin temperature and tracking performance. Int J Ind Ergon 39(4):590–595

Heus R, Daanen HAM, Havenith G (1995) Physiological criteria for functioning of hands in the cold: a review. Appl Ergon 26(1):5–13

Kurppa K, Viikari-Juntura E, Kuosma E, Huuskonen M, Kivi P (1991) Incidence of tenosynovitis or peritendinitis and epicondylitis in a meat-processing factory. Scand J Work Environ Health 17(1):32–37

Nordander C, Ohlsson K, Balogh I, Rylander L, Pålsson B, Skerfving S (1999) Fish processing work: the impact of two sex dependent exposure profiles on musculoskeletal health. Occup Environ Med 56(4):256–264

Piedrahíta H, Punnett L, Shahnavaz H (2004) Musculoskeletal symptoms in cold exposed and non-cold exposed workers. Int J Ind Ergon 34(4):271–278

Riley MW, Cochran DJ (1984) Dexterity performance and reduced ambient temperature. Hum Factors J Hum Factors Ergon Soc 26(2):207–214

Schiefer R, Kok R, Lewis M, Meese G (1984) Finger skin temperature and manual dexterity—some inter-group differences. Appl Ergon 15(2):135–141

Schneider E, Irastorza X, Copsey S, Verjans M, Eeckelaert L, Broeck V (2010) OSH in figures: work-related musculoskeletal disorders in the EU—facts and figures. Luxembourg Eur Agency Saf Health Work

Silverstein BA, Fine LJ, Armstrong TJ (1987) Occupational factors and carpal tunnel syndrome. Am J Ind Med 11:343–358

Wiggen ØN, Heen S, Frevik H, Reinertsen RE (2011) Effect of cold conditions on manual performance wearing petroleum industry protective clothing. Ind Health 49:443–451

Data Clustering on Taiwan Crop Sales Under Hadoop Platform

Chao-Lung Yang and Mohammad Riza Nurtam

Abstract Hadoop is one of the most promising cloud computing platforms to execute a Big Data analytics task which is a process of discovering hidden patterns, unknown correlations, and other valuable information from an extremely large distributed dataset. In this paper, a data clustering was implemented under Hadoop platform to study a large crop sales dataset collected distributedly in Taiwan. Hadoop infrastructure was built to give access of the distributed data centers. An online clustering algorithm utilizing Mahout, a scalable machine learning library, was performed to analyze crop price and yield data from the distributed datasets. This clustering analysis is usually exhausting and time consuming if a single machine is in charge of the whole process. Therefore, in this research, the clustering jobs will be handled under an experimental distributed Hadoop environment. The result can be used to help decision making of crop planning by forecasting or detecting demand changes in the market as early as possible.

Keywords Big data analytics · Hadoop · Mahout · Clustering · Distributed computing

---

C.-L. Yang (✉) · M. R. Nurtam
Department of Industrial Management, National Taiwan University of Science and Technology, No. 43, Sec. 4, Keelung Rd, Da'an District, Taipei 10607, Taiwan, Republic of China
e-mail: clyang@mail.ntust.edu.tw

M. R. Nurtam
e-mail: muhammadriza@gmail.com

1 Introduction

Nowadays, more and more data is collected and stored every day in the world (Gopalkrishnan et al. 2012) and the trend of data size growth has been closer to Moore's Law (Fisher et al. 2012). That means that the volume of collected data will be almost doubled every year. How to analyze the collected huge dataset and create values from it has catch a lot of attentions. Big data is a term coined by data scientists to name this huge dataset. However, the definition of big data is varied. The simple definition to describe big data is a dataset that is too large to fit in a single drive, so it has to be stored in distributed storage (Fisher et al. 2012). Moreover, IBM defines that big data have 3 characteristics called V3 (Volume, Variety, and Velocity). These characteristics simply state that we have data that are so big in size, comes in structured or unstructured form and gets bigger over time with speed (IBM, Zikopoulos et al. 2011). To extract valuable information from big data, a special tool sets for analyzing big data is needed to handle the relatively large data repository by utilizing fast data computation resource. Big data analytics is an emerging research area to perform the process of examining large amounts of data of a variety of types to uncover hidden patterns, unknown correlations and other useful information (Rouse 2012).

In Taiwan, the crop sales data including vegetable and fruit price and sale volume is collected daily in the distributed crop sales markets geographically (AFA 2013). In each market, there are huge amounts of crop sales transactions are processed manually. Regarding the variety of crops, understanding the crop sales pattern and further predicting the price of crop in the coming year or season are important for farmers to conduct the crop cultivation plan. In this research, we utilize the public database to study the crop sales data. A Hadoop platform was built to analyze Taiwan crop sales data using clustering algorithm in Mahout to perform data clustering analysis on crop sales data to study and discover hidden pattern in the data.

The reminder of this paper is organized as follows. Section 2 introduces Hadoop platform and the basic operation of MapReduce. Section 3 describes how the Mahout under Hadoop platform was utilized to perform a data mining analysis on Taiwan crop data and the experimental results. Finally, summary and managerial implications are concluded in Sect. 4.

2 Literature Review

2.1 Hadoop

Hadoop is a distributed computing platform developed by open source community to work with large dataset. Hadoop enables data scientist to store and analyze data by using multiple computing machines as a cluster. The development of Hadoop was initially inspired by a paper presented in 2004 by Google about MapReduce (Dean and Ghemawat 2004, 2008). That contributive research inspired open source community to implement the map-reduce paradigm that exists in functional programming (Hughes 1989) into open source project, and Hadoop is the result (Harris 2013).

Hadoop consists of several core components: Hadoop common, Hadoop YARN, Hadoop Distributed File System (HDFS), and Hadoop MapReduce (Hadoop 2013). Hadoop common is a set of utilities for working with Hadoop platform, while YARN is a framework for computational job scheduling and cluster management. HDFS is a data management service to handle the distributed storage, and Hadoop MapReduce is a divide-conquer system to perform large-scale data computation. The typical architecture for Hadoop cluster is shown in Fig. 1.

A cluster of Hadoop consists of several machines and services. At least one node of Hadoop cluster should have HDFS service and MapReduce service. In HDFS, a name node is a service which handles the task management, data assignment, and scheduling. Usually, the secondary name node is also established in case the primary name node fails to work properly. In the same manner, each node is able to take over other node when a failure occurs. For security reason, the data copied onto HDFS will be duplicated to multiple data nodes to increase the reliability. This replication process also allows the ability of retrieving data from the nearest node (Shvachko et al. 2010).

Hadoop uses map-reduce paradigm to provide the distributed and parallel processing of large data set. This programming model consists of the *Map* function that performs filtering and sorting on the dataset. On the other hand, the *Reduce* function performs summary routines to aggregate the data from distributed data note. The output of map function is an intermediate <key,value> pairs and this pairs will be the input of reduce function.

An example of the MapReduce operation shown in Fig. 2 which is a process of counting average demand of vegetables (cabbage, broccoli and carrot) from HDFS data source (Dean and Ghemawat 2008; Leu et al. 2010; Espinosa et al. 2012). The data stored in HDFS has to split into several partitions and assigned to mapping

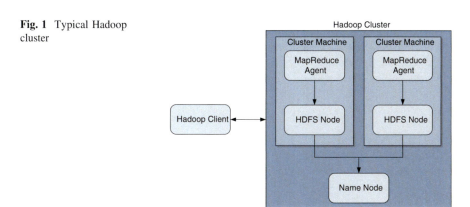

Fig. 1 Typical Hadoop cluster

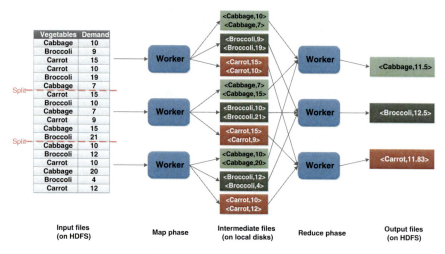

Fig. 2 MapReduce operation

workers. Mapping worker nodes processes the input data and map it into <key,-value> intermediate pairs. Then, the pairs are shuffled and stored into local files, where each file holds data with one specific key. After the mapping process finished, reducing worker retrieve the mapping files from remote machine and start the reducing process and finally store the result to output files.

2.2 Mahout

Mahout is a scalable data mining and machine learning library that can be run on Hadoop distributed platform or the local system (Esteves and Chunming 2011; Esteves et al. 2011). Mahout can be used to process large data set with many data mining algorithms that are already implemented by Java language. Mahout supports four different data mining tasks currently: classification, clustering, recommendation mining, and frequent itemset mining. Mahout is developed based on the multicore MapReduce algorithm (Chu et al. 2006).

The input and result file of the data mining process is saved into sequence file format. To be able to read this result, we need to convert the result file with utility programs called *clusterdump* and *seqdumper* that are provided with Mahout, to dump the result file to readable files.

3 Experiment and Result

Taiwan crop sales data are provided by Taiwan Agriculture and Food Agency at http://amis.afa.gov.tw. This web-based database has multiple query pages and the data has several attributes such as city, crop category, weather, sales price, and sales volumes, and so on. Taiwan crop sales data are collected from major Taiwan markets every day, from 1st January 1996 (85-01-01 in Minguo calendar format) and from 2,767 crop commodities from three categories: vegetables, fruits, and flowers. By using the automatic data retrieval program, we retrieve the data from public database and store the data in a MariaDB database. From this database, a text file in vector format can be created by using SQL query, and sent to HDFS for analyzing. This collected data was used for data analysis and Mahout experiments which will be addressed in the following sections.

3.1 Preliminary Data Analysis

To have better understanding about crop sales data, a preliminary data analysis was conducted. For this analysis, we selected a particular crop, persimmon tomato, with crop ID 'FJ1' to demonstrate the preliminary analysis. Persimmon tomato is a variant of tomato, a big size tomato with some lines on the body of the fruit.

In Fig. 3, two datasets are plotted together. The red curve indicates time series data of the sales price of persimmon tomato in 1999; the blue dashed-line curve indicates another time series data of sales volume of persimmon tomato in the

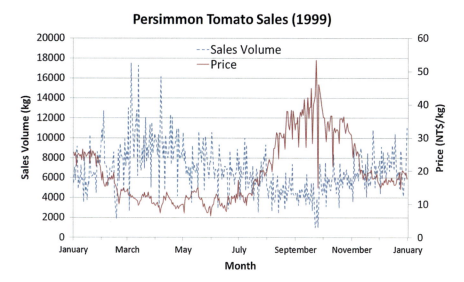

Fig. 3 Sample data of persimmon tomato sales

same year. As can be seen, the sales price for this tomato is relatively low from February to July. The sales price tends to increase in the middle of summer and reach the top price around October (at the beginning of winter in Taiwan). On the other hand, sales volume data has different type of trend and is not that steep as price data. The Sales tend to increase in the beginning of the year but with very larger fluctuation day by day. The volume slightly decreases through a year until October. After passing October, the volume starts to increase. Obviously, the seasonal effect is clear on sales price and demand of persimmon tomato. If looking at the data carefully, it can be found that the trend of sales price data is the opposite of sales volume. It means the demand decreases in higher price market, while demand increases if the price is low. To summarize, persimmon tomato are available in large stock and traded in large volume during the beginning of spring and the price are cheaper when compared with the price in summer (Fig. 4).

3.2 Data Clustering on Crop Sales Data

The seasonal pattern is discussed in Chap. 98, but this analysis heavily relies on domain expert's judgment and sometime it is very difficult to separate the data, especially in large scale dataset like Taiwan crops sales data. By using Mahout k-means algorithm under Hadoop (3 nodes in this case), the clustering analysis was performed 5 times. The average of Sum of Squared Error (SSE) of each run with different number of cluster, K, was computed and the results are shown in Fig. 4. As shown in the plot, the SSE is dropping from K = 2 to K = 10. However, the dropping of SSE when K = 3 seems largest. In order to visual data clustering, the scatter plots of K = 3 are shown in Fig. 4.

In Fig. 5, different colors and symbols are used to indicate 3 different clusters. As can be seen, three clusters is revealed by price–volume combination. The cluster 1 is high-volume–low-price; cluster 2 is middle-volume-middle-price. Interestingly, cluster 3 is a low-volume group across all price range. These grouping of price–volume data actually provide another aspect of tomato sales beyond the time series data. The clustering results can easily indicate the sales patterns where might be correlated with other factors such as the weather or market. The further data analysis is needed to find the influential factors which can causes these clustering results (Fig. 6).

Fig. 4 SSE on various number of K in Mahout K-means clustering

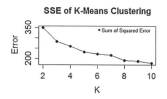

Fig. 5 Bimonthly sales data

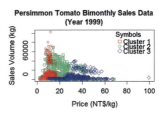

Fig. 6 1,999 crops sales data in 3 clusters

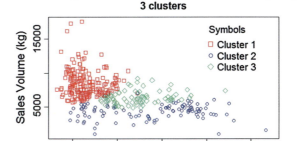

The k-means method is performed with different number of computational node as shown in Fig. 7. Obviously, the more computational node we use, the faster the algorithm is. The number of K for K-means algorithm seems not influential on running time. Although it is intuitive, this scalable structure in fact is the advantage of Hadoop platform because once the more computational effort is needed, the more nodes can be assigned to the job for shortening the processing time.

In this research, we focus on applying the Hadoop platform to demonstrate the data mining capability on Taiwan crop sales data. Mahout that is utilizing MapReduce framework can be applied to perform data mining work across different

Fig. 7 Hadoop performance

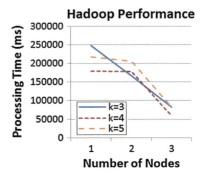

computational notes. Once the data is very big, the Hadoop platform is useful to empower the data analysis simultaneously which has been demonstrate by this experiment.

4 Conclusions

Nowadays, Hadoop is a prominent platform for data scientists that work with big data. Hadoop implements map-reduce programming paradigm and provides the distributed and parallel processing of big data. To be able to analyze the data, an analysis program should be provided and run under MapReduce framework. Mahout is one of promising tool of machine learning and data mining on big data. By running upon Hadoop framework, Mahout can utilize the distributed and parallel processing to enhance analysis performance.

In this research, we used Taiwan crops sales data as a sample of big data, which can be difficult to analyze in a single computer because of its big size. The Mahout under Hadoop was applied to analyze one experimental example, persimmon tomato data, from Taiwan crop sales dataset which collected from public agriculture database. The k-means clustering was used to perform clustering analysis on sales price and volume. The result of the analysis shows that Taiwan crop sales data have seasonal effect in sales price and volume, and the sales can be grouped into multiple clusters in which deferent patterns can be revealed.

Acknowledgments This study was conducted under the "Project Digital Convergence Service Open Platform" of the Institute for Information Industry which is subsidized by the Ministry of Economy Affairs of the Republic of China.

References

AFA (2013) Agriculture market information system. Retrieved July 2013. http://amis.afa.gov.tw

Chu C, Kim S, Lin Y, Yu Y, Bradski G, Ng A, Olukotun K (2006) Map-reduce for machine learning on multicore. NIPS, MIT Press

Dean J, Ghemawat S (2004) MapReduce: simplified data processing on large clusters. In: Proceedings of the 6th conference on symposium on operating systems design and implementation, vol 6. San Francisco, CA, USENIX Association: 10-10

Dean J, Ghemawat S (2008) MapReduce: simplified data processing on large clusters. Commun ACM 51(1):107–113

Espinosa A, Hernandez P, Moure JC, Protasio J, Ripoll A (2012) Analysis and improvement of map-reduce data distribution in read mapping applications. J Supercomputing 62(3):1305–1317

Esteves RM, Chunming R (2011) Using Mahout for clustering Wikipedia's latest articles: a comparison between K-means and fuzzy C-means in the cloud. In: 2011 IEEE third international conference on cloud computing technology and science (CloudCom)

Esteves RM, Pais R, Chunming R (2011) K-means clustering in the cloud: a Mahout test. In: 2011 IEEE workshops of international conference on advanced information networking and applications (WAINA)

Fisher D, DeLine R, Czerwinski M, Drucker S (2012) Interactions with big data analytics. Interactions 19(3):50–59

Gopalkrishnan V, Steier D, Lewis H, Guszcza J (2012) Big data, big business: bridging the gap. In: 1st international workshop on big data, streams and heterogeneous source mining: algorithms, systems, programming models and applications, BigMine-12: held in conjunction with SIGKDD conference, August 12, 2012: Beijing, China, Association for Computing Machinery

Hadoop (2013) Welcome to Apache™ Hadoop®! Retrieved 15 May 2013. http://hadoop.apache.org/

Harris D (2013) The history of Hadoop: From 4 nodes to the future of data. Retrieved 15 May 2013. http://gigaom.com/2013/03/04/the-history-of-hadoop-from-4-nodes-to-the-future-of-data/

Hughes J (1989) Why functional programming matters. Comput J 32(2):98–107

IBM, Zikopoulos P, Eaton C, Deutsch T, Lapis G (2011) Understanding big data: analytics for enterprise class Hadoop and streaming data, McGraw-Hill Education

Leu J-S, Yee Y-S, Chen W-L (2010) Comparison of map-reduce and SQL on large-scale data processing. In: 2010 international symposium on parallel and distributed processing with applications (ISPA)

Rouse M (2012) Definition; big data analytics. http://searchbusinessanalytics.techtarget.com/definition/big-data-analytics

Shvachko K, Hairong K, Radia S, Chansler R (2010) The Hadoop distributed file system. In: 2010 IEEE 26th symposium on mass storage systems and technologies (MSST)

Control with Hand Gestures in Home Environment: A Review

Sheau-Farn Max Liang

Abstract With many advances made in the area of automatic gesture recognition, gestural control has gradually gained its acceptance and popularity. However, the research to date has tended to focus on recognition technologies rather than human behaviors. With an emphasis on users, this paper reviews recent research literature on hand gestural control in home environment. The aim is to summarize and analyze current development processes of gesture vocabularies for commanding home appliances. A semiotic dual triadic model is proposed for the review of the control commands for the appliances, the types of hand gestures, the derivation processes of designer-defined and user-defined gestures. A typical derivation process for a user-defined gesture set was first collecting raw data by inquiring potential users to perform the most suitable gestures that they thought for triggering the given commands or functions, and then applying algorithms for the selection of final gesture vocabularies. A brief comparison among the research results of user-defined freehand gestures for TV control commands was provided as an example to show a need of research in this area. Further research direction includes the exploration of broader user population and the refinement of current gesture selection algorithms.

Keywords Gestural user interface · Human computer interaction · Human centered design · Gesture set

1 Introduction

Many commercialized gestural user interfaces can be found in current home environment, such as smart TV and video game consoles. With the advances in automatic recognition technology, control with hand gestures has gradually gained

S.-F. M. Liang (✉)
National Taipei University of Technology, 1, Sec. 3, Zhong-Xiao E. Rd, Taipei, Taiwan, Republic of China
e-mail: maxliang@ntut.edu.tw

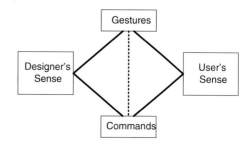

Fig. 1 Dual triadic model for gestural control (modified from Liang 2006)

its acceptance and popularity. However, the research to date has tended to be more technology-driven rather than human-centered. Hence, this paper reviews recent research literature on gestural control for home appliances with a focus on human-centered design. A semiotic dual triadic model is proposed as the framework of the review.

In a semiotic sense, a gesture is a form to represent its referent and to let the interpreter make sense on this representation. The suitability of the representation depends on the sense made of the gesture, its referent, and their relationship, known as semiotic triangle (Chandler 2002). A dual triadic model has been proposed for icon design (Liang 2006), and is also applicable to understand the gestural control. As shown in Fig. 1, the interpreters can be either designers or users. The referents in home environment are usually the control commands to home appliances. Based on their knowledge, experience, and mental models (Norman 1983) on possible control gestures and commands (solid lines in Fig. 1), designers or users define the representations, that is, the associations between each pair of gestures and commands (the dotted line in Fig. 1).

The remainder of this paper is divided into five sections. Section 2 describes the home appliances and associated control commands that have been included in the literature. Section 3 presents different types of hand gestures that have been studied. Section 4 explains the derivation procedures of gesture sets from both designers and users. Section 5 briefly summarizes and compares the research results of user-defined freehand gesture-command mappings for TV control found. Finally, Sect. 6 concludes the paper with discussions.

2 Home Appliances and Control Commands

Television was the most popular home appliance studied in literature among other appliances such as lamps/lights, blinds, curtains, air conditioners, video recorders, answering machines, radios, music players, doors, windows and even robots (Choi et al. 2012; Guesgen and Kessell 2012; Kühnel et al. 2011; Obaid et al. 2012). General control commands found in literature were turn-on/off, increase/decrease (volume or temperature), change to previous/next (channel, station or song), open/close (blinds, curtains, doors, or windows). Specific commands for video or music playback were play, stop, pause, mute, and forward/backward (Ishikawa et al. 2005; Löcken et al. 2012; Vatavu 2012; Wu and Wang 2012, 2013). Some commands for menu selection and electronic program guide (EPG) on television, such as commands for directions and numbering, have also been studied (Aoki et al. 2011b, c; Bailly et al. 2011a, b; Bobeth et al. 2012; Jeong et al. 2012; Shimada et al. 2013; Tahir et al. 2007; Wu et al. 2012). Other relevant gestural control applications include web on TV (Morris 2012) and photo sharing on TV (Freeman et al. 2012).

3 Freehand, Handheld or Hand-worn Gestures

Control with hand gestures for home appliances can be categorized into three main types: freehand, handheld, and hand-worn. While freehand gestures could be one-handed or two-handed, handheld and hand-worn gestures were often one-handed. Freehand gestures are often the gestures made in air. An exception is to use a palm as a control panel and an index finger as a pointer (Dezfuli et al. 2012). For handheld gestures, smart phones were the common devices held by hand. Users can hold the phone and make gestures in air like making freehand gestures (Kühnel et al. 2011), or they can make gestures on the screen of the phone (Aoki et al. 2011b, c). Gesture on surface is a prevalent research topic, but it is beyond the scope of this paper. Other devices held by hand were a Wiimote controller (Bailly et al. 2011a), a laser pen (Aoki et al. 2012), or some tailor-made devices (Kela et al. 2006; Pan et al. 2010; Tahir et al. 2007). Deformable devices have also been proposed with two-handed operation (e.g., Lee et al. 2011). For hand-worn gestures, rings seems the most popular devices that have been applied (Aoki et al. 2011a; Jing et al. 2011; De Miranda et al. 2010, 2011a), though other hand-wear, such as gloves and thimbles, have also been considered (De Miranda et al. 2011b).

Any gesture can be analyzed by its initial state, transition state, and end state. For example, a gesture for turning on a TV might be moving an index finger forward as if pointing a button. The initial state is a posture with the index finger pointing forward. The transition state is to keep the initial posture still and move the hand forward. The end state in this case is the same posture as in the initial state. Therefore, to define a gesture is to identify the form of hand(s), and its spatial and temporal characteristics in each state.

4 User or Designer Defined Gestures

The mappings between gestures and commands were either defined by users or designers. Designer-defined gestures were mainly influenced by the applied gesture recognition technologies (e.g., Aoki et al. 2011a, 2012; Bailly et al. 2011a; Ishikawa et al. 2005; Jeong et al. 2012; Jing et al. 2011; Pan et al. 2010;

Shimada et al. 2013), some of the studies have further conducted usability evaluations by potential users (e.g., De Miranda et al. 2010; Guesgen and Kessell 2012; Lee et al. 2011).

For user-defined gestures, a set of commands was first identified. Participants were then recruited to propose gestures that they thought best for representing the corresponding commands. If the number of participants is N, and the number of commands is M. Initially there should be N × M gestures derived from the participants. However, it is common that many proposed gestures are similar or identical, so they are considered as repeats of each other if they are proposed to the same command. Eventually, each command is associated with several gestures that have different numbers of repetition, and the command "win" the associated gesture with the highest number of repetition (Wobbrock et al. 2005). An extended approach has been proposed to include psychological and physiological measures and to treat the derivation of a gesture set as a mathematical optimization process (Stern et al. 2008).

5 User-Defined Freehand Gesture Sets

Several studies have been done on the user derivation of freehand gestures for TV commands. A brief comparison of the results was made among three studies as shown in Table 1. While agreement can be found for some gesture-command

Table 1 Results of user-defined freehand gestures for TV commands in previous research

Reference (participant)	Choi et al. 2012 (30 Koreans)	Vatavu 2012 (12 Romanians)	Wu and Wang 2012 (12 Chinese)
TV command		Suggested gesture	
Turn on	Make a square shape and increase the size with two index fingers and thumbs	Move index finger forward (air click)	Palm faces forward, fingers point up, and move the hand forward
Turn off	(1) Make a square shape and decrease the size with two index fingers and thumbs (2) Palm faces up, and move thumb to the palm	Draw an X with index finger	Wave one hand (wave goodbye)
Increase the volume	Pinch with the thumb and other fingers and then open them (as if open a mouth)	Left hand serves as the reference while move right hand up or to the right	Palm faces forward, fingers point up, and move the hand up
Decrease the volume	(1) Open the thumb and other fingers and then pinch with them (as if close a mouth) (2) Palm faces down, fingers point forward, and move the hand down	Left hand serves as the reference while move right hand down or to the left	Palm faces forward, fingers point up, and move the hand down
Select previous channel	Move right hand to the right	Move right hand to the right	Move right hand to the right
Select next channel	Move right hand to the left	Move right hand to the left	Move right hand to the left

mappings, such as gestures for selecting previous or next channel, conflicts over the mappings are common even for the basic turn-on/off commands. More empirical data seem necessary for being able to draw a conclusion on what the most appropriate gestures are for certain commands.

6 Conclusion

A review of the literature indicates that the research on hand-gestural control in home environment is still at an early stage in finding appropriate gesture vocabularies for corresponding commands. With current sparse and diverse research results, additional research in this area should prove quite beneficial. One possibility is to continue to collect data about the gesture-command mappings. In addition, issues about how individual or cultural differences effect user's selection of gestures seem worth to be further explored. Finally, the procedures of establishing a set of gestures can be refined by considering user's mental model and physical capabilities and limitations, as well as the competence of gesture recognition technology.

Acknowledgments This work is funded by the National Science Council of Taiwan (NSC 101-2221-E-027-005-MY3).

References

Aoki R, Karatsu Y, Ihara M, Maeda A, Kobayashi M, Kagami S (2011a) Gesture identification based on zone entry and axis crossing. In: Paper presented at the 14th international conference on human-computer interaction. Orlando, USA, 9–14 July 2011 LNCS 6762-2: 194–203

Aoki R, Karatsu Y, Ihara M, Maeda A, Kobayashi M, Kagami S (2011b) Unicursal gesture interface for TV remote with touch screens. In: Paper presented at the 2011 IEEE international conference on consumer electronics (ICCE). Las Vegas, USA, 9–12 Jan 2011, pp 99–100

Aoki R, Karatsu Y, Ihara M, Maeda A, Kobayashi M, Kagami S (2011c) Expanding kinds of gestures for hierarchical menu selection by unicursal gesture interface. IEEE T Consum Electr 57-2:731–737

Aoki R, Ihara M, Kobayashi T, Kobayashi M, Chan B, Kagami S (2012) A gesture recognition algorithm for vision-based unicursal gesture interfaces. In: Paper presented at the 10th European conference on interactive TV and video (EuroITV'12). Berlin, Germany, 4–6 July 2012, pp 53–56

Bailly G, Vo D-B, Lecolinet E, Guiard Y (2011a) Gesture-aware remote controls: Guidelines and interaction techniques. In: Paper presented at the 13th international conference on multimodal interfaces (ICMI'11). Alicante, Spain, 14–18 Nov 2011, pp 263–270

Bailly G, Walter R, Müller J, Ning T, Lecolinet E (2011b) Comparing free hand menu techniques for distant displays using linear, marking and finger-count menus. In: Paper presented at the 13th IFIP TC 13 international conference on human-computer interaction (INTERACT'11). Lisbon, Portugal, 5–9 Sept 2011 LNCS 6947-2: 248–262

Bobeth J, Schmehl S, Kruijff E, Deutsch S, Tscheligi M (2012) Evaluating performance and acceptance of older adults using freehand gestures for TV menu control. In: Paper presented at the 10th European conference on interactive TV and video (EuroITV'12). Berlin, Germany, 4–6 July 2012, pp 35–44

Chandler D (2002) Semiotics: the basics. Routledge, New York

Choi E, Kwon S, Lee D, Lee H, Chung MK (2012) Can user-defined gesture be considered as the best gesture for a command? Focusing on the commands for smart home system. In: Paper presented at the human factors and ergonomics society 56th annual meeting. Boston, USA, 22–26 Oct 2012, pp 1253–1257

De Miranda LC, Hornung HH, Baranauskas C (2010) Adjustable interactive rings for iDTV. IEEE T Consum Electr 56(3):1988–1996

De Miranda LC, Hornung HH, Baranauskas C (2011a) Adjustable interactive rings for iDTV: first results of an experiment with end-users. In: Paper presented at the 14th international conference on human-computer interaction. Orlando, USA, 9–14 July 2011 LNCS 6776: 262–271

De Miranda LC, Hornung HH, Baranauskas C (2011b) Prospecting a new physical artifact of interaction for iDTV: results of participatory practices. In: Paper presented at the 14th international conference on human-computer interaction. Orlando, USA, 9–14 July 2011 LNCS 6770: 167–176

Dezfuli N, Khalilbeigi M, Huber J, Müller FB, Mühlhäuser M (2012) PalmRC: imaginary palm-based remote control for eyes-free television interaction. In: Paper presented at the 10th European conference on interactive TV and video (EuroITV'12). Berlin, Germany, 4–6 July 2012, pp 27–34

Freeman D, Vennelakanti R, Madhvanath S (2012) Freehand pose-based gestural interaction: Studies and implications for interface design. In: Paper presented at the 4th international conference on intelligent human computer interaction. IIT Kharagpur, India, 27–29 Dec 2012, pp 1–6

Guesgen H, Kessell D (2012) Gestural control of household appliances for the physically impaired. In: Paper presented at the 25th international Florida artificial intelligence research society conference. Macro Island, USA, 23–25 May 2012, pp 353–358

Ishikawa T, Horry Y, Hoshino T (2005) Touchless input device and gesture commands. In: Paper presented at the IEEE international conference on consumer electronics. Las Vegas, USA, 8–12 Jan 2005, pp 205–206

Jeong S, Jin J, Song T, Kwon K, Jeon JW (2012) Single-camera dedicated television control system using gesture drawing. IEEE T Consum Electr 58–4:1129–1137

Jing L, Yamagishi K, Wang J, Zhou Y, Huang T, Cheng Z (2011) A unified method for multiple home appliances control through static finger gestures. In: Paper presented at the IEEE/IPSJ 11th international symposium on applications and the internet. Munich, Germany, 18–21 July 2011, pp 82–90

Kela J, Korpipää P, Mäntyjärvi J, Kallio S, Savino G, Jozzo L, Di Marca S (2006) Accelerometer-based gesture control for a design environment. Pers Ubiquit Comput 10–5:285–299

Kühnel C, Westermann T, Hemmert F, Kratz S, Müller A, Möller S (2011) I'm home: defining and evaluating a gesture set for smart-home control. Int J Hum Comput St 69:693–704

Lee S-S, Maeng S, Kim D, Lee K-P, Lee W, Kim S, Jung S (2011) FlexRemote: exploring the effectiveness of deformable user interface as an input device for TV. In: Paper presented at the 14th international conference on human-computer interaction. Orlando, USA, 9–14 July 2011 CCIS 174: 62–65

Liang SFM (2006) Analyzing icon design by the axiomatic method: a case study of alarm icons in process control displays. Asian J Ergon, 7(1–2):11–28

Löcken A, Hesselmann T, Pielot M, Henze N, Boll S (2012) User-centred process for the definition of free-hand gestures applied to controlling music playback. Multimedia Syst 18:15–31

Morris MR (2012) Web on the wall: insights from a multimodal interaction elicitation study. In: Paper presented at the ACM international conference on interactive tabletops and surfaces. Cambridge, USA, 11–14 Nov 2012, pp 95–104

Norman DA (1983) Some observations on mental models. In: Gentner D, Stevens AL (eds) Mental models. Erlbaum, Hillsdale, pp 7–14

Obaid M, Häring M, Kistler F, Bühling R, André E (2012) User-defined body gestures for navigational control of a humanoid robot. In: Paper presented at the international conference on social robotics. Chengdu, China, 29–31 Oct 2012 LNAI 7621: 367–377

Pan G, Wu J, Zhang D, Wu Z, Yang Y, Li S (2010) GeeAir: a universal multimodal remote control device for home appliances. Pers Ubiquit Comput 14:723–735

Shimada A, Yamashita T, Taniguchi R (2013) Hand gesture based TV control system. In: Paper presented at the 19th Korea-Japan joint workshop on frontiers of computer vision. Incheon, Korea, 30 Jan–01 Feb 2013, pp 121–126

Stern HI, Wachs JP, Edan Y (2008) Designing hand gesture vocabularies for natural interaction by combining psycho-physiological and recognition factors. Int J Semantic Comput 2(1):137–160

Tahir M, Bailly G, Lecolinet E (2007) ARemote: A tangible interface for selecting TV channels. In: Paper presented at the 17th international conference on artificial reality and telexistence. Esbjerg, Denmark, 28–30 Nov 2007, pp 298–299

Vatavu R-D (2012) User-defined gestures for free-hand TV control. In: Paper presented at the 10th European conference on interactive TV and video (EuroITV'12). Berlin, Germany, 4–6 July 2012, pp 45–48

Wobbrock JO, Aung HH, Rothrock B, Myers BA (2005) Maximizing the guessability of symbolic input. In: Paper presented at the conference on human factors in computing (CHI). Oregon, USA, 2–7 April 2005 1869-1872

Wu H, Wang J (2012) User-defined body gestures for TV-based applications. In: Paper presented at the 4th international conference on digital home. Guangzhou, China, 23–25 Nov 2012, pp 415–420

Wu H, Wang J (2013) Understanding user preferences for freehand gestures in the TV viewing environment. AISS 5–4:709–717

Wu H, Chen X, Li G (2012) Simultaneous tracking and recognition of dynamic digit gestures for smart TV systems. In: Paper presented at the 4th international conference on digital home. Guangzhou, China, 23–25 Nov 2012, pp 351–356

An Integrated Method for Customer-Oriented Product Design

Jiangming Zhou, Nan Tu, Bin Lu, Yanchao Li and Yixiao Yuan

Abstract This paper introduces a customer-oriented design method in new product development (NPD). This method comprises of persona creation, analytic hierarchy process (AHP), quality function deployment (QFD) and usability engineering. Persona creation based on fuzzy cluster analysis is first employed to identify the typical behaviors and motivations of a broader range of customers. The customer requirements (CRs) are extracted from the persona profiles and then prioritized with a semi-fuzzy AHP approach. Using the QFD method, the prioritized CRs are translated into measurable design requirements (DRs) to set the design targets. After product prototyping, scenario-based usability testing techniques are employed to access and make recommendations to improve usability in product redesign. The above process may be repeated with several iterations until a product that more closely matches the customer needs with high usability can be designed. A new sports earphones design is given as an example to illustrate the implementation of this method.

Keywords Customer-oriented product design · Persona creation · Analytic hierarchy process · Quality function deployment · Usability engineering

1 Introduction

Throughout the new product development (NPD) process, there exist difficulties in accommodating the value elements, which leads to substantial failure rates and high risks. Extensive studies have been carried out to search for the factors that

J. Zhou (✉) · N. Tu · B. Lu · Y. Li · Y. Yuan
Research Center for Modern Logistics, Graduate School at Shenzhen, Tsinghua University, Shenzhen 518055, People's Republic of China
e-mail: jamin.zjm@gmail.com

N. Tu
e-mail: dr.nan.tu@gmail.com

separates winners from losers. Karkkainen and Elfvengren (2002) concluded that the two success factors strongly pointed out in previous studies are early-phase product innovation and careful assessment of customer needs.

Customer-oriented product design aims to design products that better meet the customer requirements (CRs) to secure the competitive advantages and success of a NPD. The designer needs to define the CRs clearly and produce an appropriate design solution with a crucial connection between CRs and products (Aoussat et al. 2000).

Quality function deployment (QFD) is a commonly used tool to incorporate CRs into product design (Akao 1990), using a matrix called house of quality (HOQ) to prioritize the design requirements (DRs) based on the CRs (Hauser and Clausing 1988). As a support method in solving multiple criteria decision making (MCDM) problems (Ho 2008), the analytic hierarchy process (AHP) is employed to determine the relative importance weights of CRs in QFD for precision improvements. The combined AHP-QFD method has been applied in product design selection situations. For example, Myint (2003) proposed the approach to aid the product design where four matrices were constructed to link the CRs, assembly characteristics, main assembly components, sub-components and attributes.

Primal data for QFD is collected from interviews, claims information and even experiences of the product design team (Akao 1997). Such tools as affinity charts (K-J diagrams) and tree diagram account for most industry applications in classifying CRs in HOQ (Griffin and Hauser 1993). These qualitative methods have met some criticisms as being human subjective and imprecise. An interactive design tool called persona can be used to help designers focus on the typical behaviors and motivations of customers. Tu et al. (2010) employed quantitative fuzzy cluster analysis to address the shortcomings of subjective matters. However, we have yet found a method that integrates persona creation with QFD in NPD.

Most researches on NPD are limited to product prototyping, while product design is an iteration process and feedbacks are essential to diagnose and improve existing products. Commonly used in user interface design, scenario-based usability testing can be employed in NPD to access the customer satisfaction. Then the assessment results can be fed back to the designers for use in product redesign.

In this article, the techniques of persona creation, AHP, QFD and scenario-based usability testing are integrated to design a customer-oriented product so that customers' needs can be met with better product usability. Finally a new sports earphones design is employed as an illustrative case study to this approach.

2 Theoretical Background

The overall procedures are divided into the following six steps (Fig. 1):

Step 1: Create persona based on fuzzy cluster analysis.

An Integrated Method for Customer-Oriented Product Design

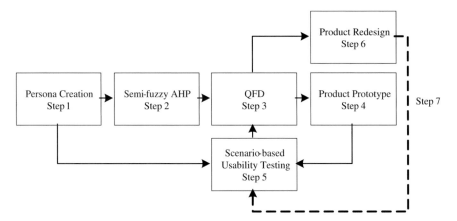

Fig. 1 Overall procedure of the integrated method

Step 2: Extract the CRs and prioritize using the semi-fuzzy AHP technique.
Step 3: Translate the prioritized CRs into DRs using QFD and set the targets.
Step 4: Create a product prototype.
Step 5: Conduct scenario-based usability testing of product prototype.
Step 6: Collect feedbacks and redesign product.
Step 7: Return to step 5 if necessary.

2.1 Persona Creation

In persona creation, questionnaires are filled out help the design team select the dimensions in classifying prototypes and perform fuzzy cluster analysis based on the fuzzy c-mean (FCM) algorithm. Respectively, typical customers from the classified groups are invited for an in-depth interview to create personas.

The target function of the FCM algorithm is to minimize the distance sum of squares between samples and clustering prototypes.

$$J = \sum_{k=1}^{n} \sum_{i=1}^{c} (\mu_{ik})^m (d_{ik})^2 \tag{1}$$

Where μ is the membership degree, m is the weighting exponent and d is the Euclidean Distance between the kth sample and the ith clustering prototype.

The constraint is that $\sum_{i=1}^{c} \mu_{ik} = 1$.

To initialize, set the number of cluster classifications c and the threshold value of iteration ε. Calculate the clustering prototypes using the density function.

In iterative operation, first update the membership degree.

$$\mu_{jk} = \begin{cases} 1 \Big/ \sum_{i=1}^{c} \left(\frac{d_{jk}}{d_{ik}}\right)^{\frac{2}{m-1}} & \forall i, k, d_{ik} > 0 \\ 1, j = i & \exists k, d_{ik} = 0 \\ 0, j \neq i & \end{cases} \quad (2)$$

Second, update the clustering prototype.

$$p_i^{(a+1)} = \frac{\sum_{k=1}^{n} (\mu_{ik})^m \cdot x_k}{\sum_{k=1}^{n} (\mu_{ik})^m}, \quad i = 1, \ldots, c \quad (3)$$

If $\|p^{(a)} - p^{(a+1)}\| < \varepsilon$, the iteration ends, otherwise set $a = a + 1$ and repeat.

2.2 Semi-Fuzzy AHP Approach

The first step is to structure the CRs into different hierarchy levels. After that, comparisons are carried out on a pairwise basis using triangular fuzzy numbers.

$$\tilde{a}_{ij}^{\alpha_{ij}} = [a_{ijl}^{\alpha_{ij}}, a_{iju}^{\alpha_{ij}}] = \begin{cases} [a_{ij}, a_{ij} + 2 - 2\alpha_{ij}], & a_{ij} = 1 \\ [a_{ij} - 2 + 2\alpha_{ij}, a_{ij} + 2 - 2\alpha_{ij}], & a_{ij} = 3, 5, 7 \\ [a_{ij} - 2 + 2\alpha_{ij}, a_{ij}], & a_{ij} = 9 \end{cases} \quad (5)$$

The level of uncertainty is indicated by different index α.

When selecting a value from the given interval, agreements have to be reached. In this semi-fuzzy AHP approach, fuzzy numbers are used only when the preference is significant. The index of optimism μ is employed.

$$\hat{a}_{ij}^{\alpha_{ij}} = \begin{cases} \mu a_{ijl}^{\alpha_{ij}} + (1-\mu) a_{iju}^{\alpha_{ij}}, & (a_{ijl} + a_{iju})/2 \geq 1 \\ 1/\hat{a}_{ji}^{\alpha_{ij}}, & (a_{ijl} + a_{iju})/2 < 1 \end{cases} \quad (6)$$

The pairwise comparison matrix is constructed using the selected value.

$$\hat{A} = \begin{pmatrix} 1 & \hat{a}_{12}^{\alpha_{12}} & \cdots & \hat{a}_{1n}^{\alpha_{1n}} \\ \hat{a}_{21}^{\alpha_{21}} & 1 & \cdots & \hat{a}_{2n}^{\alpha_{2n}} \\ \vdots & \vdots & \ddots & \vdots \\ \hat{a}_{n1}^{\alpha_{n1}} & \hat{a}_{n2}^{\alpha_{n2}} & \cdots & 1 \end{pmatrix} \quad (7)$$

The method of calculating the principle eigenvector is used. The priority vector needs to be normalized and synthesized to get the overall weights at each level.

The final step is to verify the consistency. Calculate the consistency index (CI) as $CI = (\lambda_m - n)/(n-1)$ where λ_m is the principal eigenvalue and n is the

Table 1 Random index of AHP

n	2	3	4	5	6	7
RI	0	0.58	0.90	1.12	1.24	1.32

number of elements. Compare it to the random index (RI). It is acceptable if the ratio $CI/RI < 0.1$, otherwise the comparisons should be reviewed and revised.

The reference values of RI for different n are shown below (Winston 1994) (Table 1).

2.3 QFD

The basic idea of QFD is to translate CRs into DRs, and subsequently into parts characteristics, process plans and production requirements (Park and Kim 1998). The HOQ matrix used consists of a few components: (1) CRs in rows and DRs in columns; (2) relationship matrix between CRs and DRs; (3) weights of CRs on the right and correlations of DRs at the top; (4) weights of DRs at the bottom (Fig. 2).

The absolute importance weights of DRs are calculated as $AI_j = \sum_{i=1}^{n} x_i R_{ij}$ and normalized as $RI_j = AI_j \bigg/ \sum_{j=1}^{m} AI_j$. R_{ik} denotes the relationship rating between the

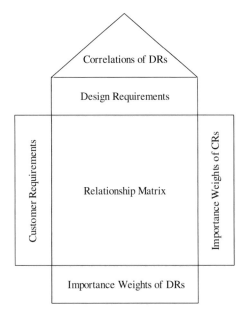

Fig. 2 HOQ matrix model

ith CR and the kth DR. x_i is the importance weight of the ith CR, given by the semi-fuzzy AHP.

2.4 Scenario-Based Usability Testing

After translating CRs into DRs, product prototyping is performed by the design team (step 4). In scenario-based usability testing, the scenarios are generated from the persona profiles. Evaluation forms and interview questionnaires are designed. Two or three internal staffs are asked to participate in the pre-testing to access and help modify the scenarios and tasks. After that, typical customers are invited again to try out the product prototype under the observation of usability engineers. In-depth interviews are carried out to collect customers' feedbacks.

3 A Case Study

An example of designing new sports earphones for joggers and hikers is given to illustrate this integrated method.

3.1 Persona Creation

Consulted with the persona usage toolkit by Olson (2004), questionnaires containing 33 multiple choice questions were invented for specific groups of people who are fond of jogging and hiking. 69 questionnaires were retrieved at last.

The clustering prototype matrix is calculated using the fuzzy cluster analysis. Based on their membership degree to the clustering prototypes, the 69 samples are divided into three groups. The persona (Fig. 3) of the enjoyment group is used.

3.2 Semi-Fuzzy AHP and QFD

The CRs were extracted and structured into a three-level hierarchy. Questionnaires were used to gather preferences with separate indexes of certainty. Applying the semi-fuzzy approach, the final comparison matrix is constructed and the eigenvectors are calculated using MATLAB. The eigenvectors were normalized and synthesized to prioritize the overall weights of CRs.

Specific DRs are discussed by designers and engineers concerning the company's existing technologies. The QFD matrix (Fig. 4) is constructed and the normalized importance weights of DRs are calculated.

An Integrated Method for Customer-Oriented Product Design 851

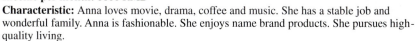

Basic information
Name: Anna
Gender: female
Age: 30
Education background: bachelor
Occupation: mid-level management
Income per month: 8000 RMB
Characteristic: Anna loves movie, drama, coffee and music. She has a stable job and wonderful family. Anna is fashionable. She enjoys name brand products. She pursues high-quality living.
Outdoor experience: Anna has been to outdoor for a long time. She owns a complete set of branded outdoor gears. Although Anna is busy with job, she will go out once every two weeks at least.
Use's goal for listening to music in outdoor activities:
1. Music has magical power. I can inspire myself when feel tired.
2. I love music. Sport and music are indispensable parts of my life.

Fig. 3 Primary persona profile

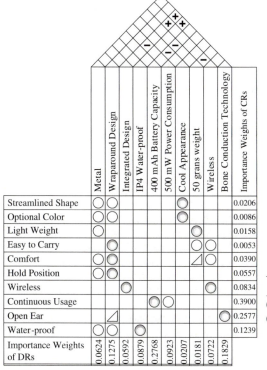

Fig. 4 QFD matrix

3.3 Product Prototype and Scenario-Based Usability Testing

Based on the importance weights of DRs and the correlations, the top five DRs (400 mAh battery capacity, bone conduction technology, wraparound design, 500 mV Power Consumption and IP4 water-proof) are selected as main instructions for the design team. The product prototype is designed.

To access the usability and user experience of the product prototype, scenarios as hiking in groups and jogging alone are generated from the persona profiles. 7 typical customers were invited to participate in the usability testing.

The feedbacks were later translated into DRs using the QFD technique for further product redesign. Through necessary iterations, the improved sports earphones are designed. Featuring major improvements are: (1) 30 % smaller, lighter in-line controller, (2) Clearer ON/OFF switch and micro USB charging port, (3) Patent pending dual suspension bone conduction with improved quality, (4) Reflective safety strip on back of wraparound headband for added safety.

Acknowledgments This work is partially supported by teams from Vtech Inc. The authors also gratefully acknowledge the helpful comments and suggestions of the reviewers, which have improved the presentation.

References

Akao Y (1990) Quality function deployment: integrating customer requirements into product design. Productivity Press, Cambridge
Akao Y (1997) QFD: past, present and future. In: Proceedings of the International Symposium on QFD'97
Aoussat A, Christofol H, Coq ML (2000) The new product design: a transverse approach. Eng Des 11(4):399–417
Griffin K, Hauser JR (1993) The voice of the customer. Mark Sci 12(1):1–27
Hauser JR, Clausing D (1988) The house of quality. Harvard Bus Rev 66:63–73
Ho W (2008) Integrated analytic hierarchy process and its applications: a literature review. European Journal of Operations Research 186:211–228
Karkkainen H, Elfvengren K (2002) Role of careful customer need assessment in product innovation management-empirical analysis. Int J Prod Econ 80(1):85–103
Myint S (2003) A framework of an intelligent quality function deployment (IQFD) for discrete assembly environment. Comput Ind Eng 45(2):269–283
Olsen G (2004) Persona Creation and Usage Toolkit. http://www.interactionbydesign.com/presentations/olsen_persona_toolkit.pdf
Park T, Kim KJ (1998) Determination of an optimal set of design requirements using house of quality. J Oper Manag 16:569–581
Tu N et al (2010) Combine Qualitative and quantitative methods to create persona. In: Proceedings of the 3rd International Conference on Information Management, Innovation Management and Industrial Engineering 3:597–603
Winston WL (1994) The analytic hierarchy process. Operations research: applications and algorithms, Wadsworth, Belmont, CA, pp 798–806

Discrete Particle Swarm Optimization with Path-Relinking for Solving the Open Vehicle Routing Problem with Time Windows

A. A. N. Perwira Redi, Meilinda F. N. Maghfiroh and Vincent F. Yu

Abstract This paper presents a discrete version of the particle swarm optimization with additional path-relinking procedure for solving the open vehicle routing problem with time windows (OVRPTW). In OVRPTW, a vehicle does not return to the depot after servicing the last costumer on its route. Each customer's service is required to start within a fixed time window. To deal with the time window constraints, this paper proposed a route refinement procedure. The result of computational study shows that the proposed algorithm effectively solves the OVRPTW.

Keywords Open vehicle routing problem · Time windows · Discrete particle swarm optimization · Path-relinking

1 Introduction

The goods distribution constitutes an important part of the overall operational costs of companies. The problem intensifies from a practical perspective when the companies do not own a fleet of vehicles (Sariklis and Powell 2000). In this case, outsourcing the distribution functions to third party logistic (3PL) providers is a beneficial business practice. Since the vehicles belong to the 3PL providers, they do no need to turn to the depot when they finish the deliveries. The above

A. A. N. Perwira Redi (✉) · M. F. N. Maghfiroh · V. F. Yu
Department of Industrial Management, National Taiwan University of Science and Technology, Taipei, Taiwan, Republic of China
e-mail: wira.redi@gmail.com

M. F. N. Maghfiroh
e-mail: meilinda.maghfiroh@gmail.com

V. F. Yu
e-mail: vincent@mail.ntust.edu.tw

described distribution model is referred to as the open vehicle routing problem (OVRP). Other typical real-life OVRP applications include home delivery of packages and newspapers, and school buses. The goal of the OVRP is to design a set of Hamiltonian paths (open routes) to satisfy customer demands in a way that minimizes the number of the vehicles, and for a given number of vehicles, minimizes the total distance (or time) traveled by the vehicles.

In some real-life situations, arrival time at a costumer will affect the customer's satisfaction level or sales to the customer. If the vehicle's service to the customer starts later than the time required by the customer, the company may lose sales to the customer. Therefore, the company should start serve every customers within the customer's desired time windows while minimizing the total length of the vehicle routes. School bus routing is another example. In the morning, a school bus arrives at the first pick-up point where its route starts, and travels along a predetermined route, picking up students at several points and taking them to school, within predefined time windows. This type of problem can be modeled as the open vehicle routing problem with time windows (OVRPTW). The objective of this problem reflects the trade-off between the vehicle's fixed cost and the variable travel cost while considering the time windows constraints. As a variant of OVRP and vehicle routing problem with time windows (VRPTW) which are proven to be NP-hard problems, OVRPTW is also NP-hard.

The OVRPTW has been addressed only once in the literature by Repoussis et al. (2006). They developed a heuristic to solve the OVRPTW by utilizing greedy 'look-ahead' solution framework, customer selection and route-insertion criteria. They introduced a new criterion for next-customer insertion, which explicitly accounted for vehicle's total waiting time along the partial constructed routes, and enhanced different types of strategies to determine the 'seed' customers. The proposed algorithm performed very well on test datasets taken from literature, providing high-quality solutions with respect to the number of active vehicles compared to other heuristics approaches. Although optimal solutions can be obtained using exact methods, the computational time required to solve adequately large problem instances is still prohibitive (Repoussis et al. 2010). For this reason, this paper attempt to develop a new efficient metaheuristic, called discrete particle swarm optimization with path-relinking (PSO-PR), to solve OVRPTW.

The rest of the paper is organized as follows. Section 2 defines the problem. Section 3 describes the proposed PSO-PR algorithm. Section 4 discusses the computational experiment and results. Finally, Sect. 5 draws conclusions and points out future research directions.

2 Problem Definition

The OVRPTW can be described as follows. Given a complete graph $G = (V, A)$, where $V = \{0,1,\ldots,n\}$ is the set of nodes and A is the set of arcs. Node 0 represents the depot while other nodes represent the customers. Each customer i has a demand

d_i, a service time s_i and a service time window $[e_i, l_i]$ within which its service must start. There is a non-negative cost c_{ij}, a travel time t_{ij} and a distance d_{ij} associated with the travel from i to j. Let $K = \{1,2,\ldots,k\}$ be the set of vehicles. Each vehicle $k \in K$ has a maximum capacity C and a maximum route length L that limits the maximum distance it can travel. Activation of a vehicle $k \in K$ incurs a fixed cost w_k. Each activated vehicle travels exactly one route.

All routes must satisfy both the capacity and the time window constraints. Time window constraints state that a vehicle cannot start servicing a customer i before the earliest time e_i or after the latest time l_i. The routes must be designed such that each customer is visited only once by exactly one vehicle. Let NV denote the number of vehicles used to service all customers in a feasible manner. The objective of the OVRPTW is first to find the minimum NV required and second to determine the sequence of customers visited by each vehicle such that the total distance is minimized.

The basic assumptions of OVRPTW are summarized as follows:

a. All vehicles depart from a depot, have only one travel route and will not return to the depot after finishing its services;
b. Each customer's demand is known and each customer can only be served once by a vehicle;
c. Every vehicle has the same capacity;
d. Vehicles must not violate customer's time windows;
e. If a vehicle arrives before a customer's earliest time, the vehicle must wait until the customer's earliest time to start the service. On the other hand, if a vehicle arrives after a customer's latest time, the service must be assigned to another vehicle.

3 Methodology

In PSO algorithm, each particle moves from its initial position to promising positions in the search space based on its velocity, its personal best position, and the particle swarm's global best position. Since Eberhart and Kennedy (1995) introduced the algorithm, PSO has been implemented to solve many problems. Since particles are encoded using real variables in the original PSO, the algorithm is not suitable for combinatorial optimization (Xiaohui et al. 2003). Therefore, this study proposes the PSO-PR algorithm for OVRPTW.

The proposed PSO-PR follows three basic assumptions of PSO: particles move based on the updated velocity; particles moves toward its personal best position; particles are attracted to the global best particle. At each iteration of PSO-PR, a particle first move based on a local search procedure. Then the particle's movement is determined by a path-relinking procedure with its personal best as the target. Lastly, the particle moves based on another path-relinking procedure with global best as its target. The detail of PSO-PR is presented in Fig. 1.

> **Algorithm1. Pseudo code of the PSO-PR algorithm**
> 1: Set parameters
> 2: **for** $i = 0$ to Nparticles **do**
> 3:   initialize solution X^i
> 4: $G^i_{best} \leftarrow P^i_{best} \leftarrow X^i$
> 5: **end for**
> 6: **while** iter < Iteration **do**
> 7:   **for** $i = 0$ to Nparticles **do**
> 8: $X_i \leftarrow$ localSearch(X^i)
> 9: $X_i \leftarrow$ pathRelinking(X^i; P^i_{best})
> 10: $X_i \leftarrow$ pathRelinking(X^i; G^i_{best})
> 11: $P^i_{best} \leftarrow$ best among X_i and P^i_{best}
> 12: $G^i_{best} \leftarrow$ best among P^i_{best} and G^i_{best}
> 13:   **end for**
> 14:   localSearch(G^i_{best})
> 15: **end while**
> 16: **return** Best G^i_{best} of all particles

Fig. 1 Pseudo code of PSO-PR algorithm

3.1 Initial Solution

Time windows constraint has a great impact in assigning customer to serve vehicles. A key determinant factor for obtaining good solutions for routing problems with time window constraints is the effective utilization of time window constraints themselves in the solution approach (Ioannou et al. 2001). Therefore, the proposed PSO-PR algorithm clearly accounts for all possible time windows related information in order to expand the set of customers that are feasible for route insertion at each stage of route construction. We designed a mechanism to ensure the customers with earlier time windows will be served first. The initial solution is generated as follows.

1. Sort the list of customers in ascending order of the earliest times of their time windows.
2. Based on the sorted list of customers, customers are assigned to service vehicles. Two feasibility checks are performed for each customer.
   (a) Check vehicle capacity. If inserting a customer into a route violates vehicle capacity, the route is terminated and the customer is assigned to a new vehicle.
   (b) Check the time windows constraint. If the vehicle arrives later than the latest time of a customer, the route is terminated and the customer is assigned to a new vehicle.
3. Terminate when all customers are assigned to a vehicle.

3.2 Local Search

The current solution s has a neighborhood $N(s)$ which is a subset of the search space S of the problem $(N(s) \subset S)$. This neighborhood is defined by swap, insertion, and 2-opt operators. The probabilities of performing the swap, insertion, and 2-opt moves are fixed at 1/3, 1/3 and 1/3, respectively. The details of these three operators are as follows.

a. *Swap*
   The swap neighborhood consists of solutions that can be obtained by swapping the positions of two customers, either in the same route or in two different routes.
b. *Insert*
   The intra-insertion neighborhood operator removes a customer from its current position and then inserts it after another customer. The inter-insertion operator inserts the customer after another customer in another route.
c. *2-Opt*
   2-opt operator is used both for improving the route and the depot used in the solution. The 2-opt procedure is done by replacing two node-disjoint arcs with two other arcs.

At each iteration, the next solution s' is generated from $N(s)$ and its objective function value is evaluated. Let Δ denote the difference between $obj(s)$ and $obj(s')$, that is $\Delta = obj(s') - obj(s)$. The probability of replacing s with s', given that $\Delta > 0$, is $\exp(-\Delta/Kw)$. This is accomplished by generating a random number $r \in [0, 1]$ and replacing the solution s with s' if $r < \exp(-\Delta/Kw)$. K denoted the decreasing rate for inertia weight (w). Meanwhile, if $\Delta \leq 0$, the probability of replacing s with s' is 1. S_{best} records the best solution found so far as the algorithm progresses.

3.3 Path-relinking Operator

To transform the initial solution s into the guiding solution s' based on P_{best} and G_{best}, the PR operator repairs from left to right the broken pairs of *solution(s)*, creating a path of giant tours with non-increasing distance to *solution(s')*. The operator used is forward strategy by swapping from the solution s to s'. The procedure looks first for a customer i with different positions in s and s'. If i is found, it is permuted with another node to repair the difference. Do swap until the $\delta(s, s')$ is null. The path from s to s' is then guided by the Hamming distance δ, equal to the number of differences between the nodes at the same position in s and s', i.e., $\delta(s, s') = \sum_{i=1,|s|} s_i \neq s'_i$.

4 Computational Result

The proposed PSO-PR was tested on 12 R1 datasets of Solomon (1987). Each of these instances has 100 nodes which are randomly distributed with tight time windows. Vehicle capacity is 200. The algorithm was coded in Visual C++ and run on a personal computer with an iCore7 processor. The parameter setting is given in Table 1. Table 2 provides the traveled distance (TD) and the number of vehicles used (NV) in the solutions obtained by the proposed PSO-PR, respectively. This table also includes the results obtained by Repoussis et al. (2006) for comparison purpose.

Overall the proposed algorithm is effective in solving OVRPTW. It reduces the total travel distances in the solutions obtained by Repoussis et al. (2006) for all the test problems. The average total distance (TD) improves 30.1 % compared to the results of Repoussis et al. (2006). However, the number of vehicles used is slightly more than that reported in Repoussis et al. (2006). The average gap is 0.92 vehicle. This result is reasonable considering the trade-off between the total travel distance and the number of vehicles used.

Table 1 Parameters setting

Variable	Setting
Swarm size	10
w	0.01
$c1$	1
$c2$	1
k	0.992
Iteration number	500

Table 2 Comparison of PSO-PR versus other method on Solomon's R1 data set

Dataset	Repoussis et al. TD	NV	PSO-PR TD	NV	TD gap (%)	NV gap
r101	1479.9	19	1129.71	17	−31.0	−2
r102	1501.59	18	1008.2	16	−48.9	−2
r103	1281.52	13	911.239	14	−40.6	1
r104	1021.73	10	802.219	11	−27.4	1
r105	1285.94	14	1010.04	16	−27.3	2
r106	1294.88	12	950.115	15	−36.3	3
r107	1102.7	11	895.087	14	−23.2	3
r108	898.94	10	785.205	11	−14.5	1
r109	1150.42	12	889.806	12	−29.3	0
r110	1068.66	12	815.539	12	−31.0	0
r111	1120.45	11	913.005	13	−22.7	2
r112	966.64	10	750.85	12	−28.7	2
Average		12.67		13.58	−30.1	0.92

5 Conclusions

In this paper, we consider a variant of VRP, the OVRPTW, to resolve a practical issue of routing 'hired' vehicle fleets. We develop a discrete PSO-PR algorithm that integrates three basic components of PSO and the path-relinking concept. The local search and path-relinking used in the proposed PSO-PR algorithm significantly improve the solution in terms of total travel distance of vehicles. However, based on the comparison with an existing method for OVRPTW, the total vehicles used are increasing slightly.

References

Ioannou G, Kritikos M, Prastacos G (2001) A Greedy Look-Ahead Heuristic for the Vehicle Routing Problem with Time Windows. J Oper Res Soc 52(5):523–537

Kennedy J, Eberhart R (1995) Particle swarm optimization. In: Neural Networks, 1995. Proceedings, IEEE International Conference on Nov/Dec 1995, pp 1942–1948 vol.1944. doi:10.1109/ICNN.1995.488968

Repoussis PP, Tarantilis CD, Ioannou G (2006) The open vehicle routing problem with time windows. J Oper Res Soc 58(3):355–367

Repoussis PP, Tarantilis CD, Braysy O, Ioannou G (2010) A hybrid evolution strategy for the open vehicle routing problem. Comput Oper Res 37(3):443–455. doi:10.1016/j.cor.2008.11.003

Sariklis D, Powell S (2000) A heuristic method for the open vehicle routing problem. J Oper Res Soc 51(5):564–573

Solomon MM (1987) Algorithms for the vehicle routing and scheduling problems with time window constraints. Oper Res 35:254–265

Xiaohui H, Eberhart RC, Yuhui S (2003) Swarm intelligence for permutation optimization: a case study of n-queens problem. In: Swarm Intelligence Symposium, 2003. SIS '03. Proceedings of the 2003 IEEE, 24–26 April 2003, pp 243–246. doi:10.1109/SIS.2003.1202275

Application of Economic Order Quantity on Production Scheduling and Control System for a Small Company

Kuo En Fu and Pitchanan Apichotwasurat

Abstract Struggling to live in the 21 century, lots of small companies of which the production scheduling and control system (PSCS) are based on a rule of thumb. Without the precise mathematic model, the rule of thumb method may lead to inventory shortages and excess inventory. For reaching better controls of PSCS, this study intends to simulate a material requirement planning (MRP) production system with adopting an economic lot-sizing (EOQ) model to diminish varieties of costs for small sized companies under job-shop environments. The proposed EOQ adopting MRP approach is supposed to have better economics of PSCS than rule-of-thumb methods. A case analysis of a small company X in Thailand is conducted for verifying this proposition. Company X regularly relied on rules-of-thumb to handle PSCS. In this study, two kinds of materials and three types of products are traced top-down from production planning (PP), master production schedule (MPS), to MRP along a period of two months since July 2012. This study undertakes a series of data collection in the field. The analyzed results indicate that the presented model really reach more economics than rule-of-thumb methods for company X.

Keywords Production scheduling and control system (PSCS) · Economic order quantities (EOQ) · Material requirement planning (MRP)

1 Introduction

Presently, lots of small manufacturing firms of which the production scheduling and control system (PSCS) are still based on rules of thumb. The rule-of-thumb method usually cannot obtain better saving of cost due to the lack of a precise

---

K. E. Fu (✉) · P. Apichotwasurat
Institute of Industrial Management, Taiwan Shoufu University, Tainan, Taiwan
e-mail: gordonfu@tsu.edu.tw

P. Apichotwasurat
e-mail: m100314006@tsu.edu.tw

mathematical model. Economic lot sizing (ELS) is well-known in the area of production scheduling and inventory control because of varieties of application such as aggregate production planning and material requirement planning (MRP) (Lee et al. 1986; Taleizadeh et al. 2011; Wangner and Whitin 1958; Steven 2008; Silver and Peterson 1985). Thus, this study intends to apply ELS on PSCS system for better control of both replenishment/production scheduling and cost savings. ELS is raised to determine lot sizes and has two versions. One is fixed lot size that is EOQ. The other is dynamic or called heuristic (i.e., not fixed) lot size, such as Wagner-Whitin, Silver-Meal, and part period balancing heuristics (Min and Pheng 2006). These dynamic heuristics desire to find an optimum schedule not only lot size that satisfies given demand at minimum cost. Although these heuristics can achieve optimum costs, EOQ is still widespread (Cárdenas-Barrón 2010; Choi and Noble 2000; Taleizadeh et al. 2011) since fixed quantity is simple for operation and maintenance.

This study applies an EOQ model for satisfying customer demands with permission of backorders in a MRP productive system which includes MPS and MRP. The backorder can be suppressed by placing orders from outside suppliers who can deliver the same quality of ordered products as well. A case is studied by applying EOQ for determining the lot sizes of materials purchase in MRP and product manufacturing in MPS, respectively. In this case, three products and two materials are demonstrated. Finally, the studied result indicates that the EOQ model has better cost savings than rules of thumb that the company regularly used.

2 Literature Review

EOQ model was first derived by Harris in 1913 for determining order sizes under continuous demand at a minimum cost (Cárdenas-Barrón 2010). Later, the EOQ model was used in MRP for replacing the scheme of lot-for-lot under enterprise environment with time dependent demand. Afterwards, dynamic lot sizing heuristics, such as, Wagner-Within algorithm, Silver-Meal heuristic, Least Unit Cost, and Part Period Balancing were proposed for attaining optimum cost (Silver and Peterson 1985; Wangner and Whitin 1958).

In PSCS, both Just-in-Time (JIT) and Material Requirement Planning (MRP) are good at keeping adequate inventory levels to assure that required materials are available when needed. JIT developed in 1970s is based on elimination of system waste and applies primarily to repetitive manufacturing processes. MRP which begins around 1960s is used in a variety of industries with a job-shop environment and especially useful for dealing with system uncertainty. MRP is a push system, while JIT is considered as a pull system. In this study, MRP is suitable for application on the studied case of company X.

Regarding economic lot sizing on MRP productive systems, lots of the studies were presented. Biggs (1979) presented a study which was conducted by computer simulation to determine the effects of using various sequencing and lot-sizing rules on various performance criteria in a multistage, multiproduct, production-inventory system with MRP setting. Liberatore (1979) used MRP and EOQ model to relate with material inventory control with concerning safety stock. Lee et al. (1987) evaluated the forecast error in a productive system with master production scheduling and MRP settings. They suggested that within MRP environments the predictive capabilities of forecast-error measures are contingent on the lot-sizing rule and the product-components structure. Taleizadeh et al. (2011) explored the utility of MRP and Economic Order Quantity/Safety Stock (EOQ/SS) through case study on a chemical process operation.

3 Application of EOQ on a MRP Productive System

An EOQ model for a productive system with MRP settings is described herein. First, the relevant concepts of EOQ, including assumption, notation, and formula are presented. Subsequently, the application of EOQ on a MRP productive system is presented.

3.1 EOQ

EOQ model is utilized in this study for company X. The assumption of EOQ is indicated as follows (Steven 2008; Cargal 2003)

1. The demand rate is constant and deterministic.
2. Annual demand (D) is known.
3. No allowing (or relaxed to no concerning) shortage cost.
4. Costs include.
   - Setup cost (K)
   - Holding cost (H)
5. There is known constant price per unit (no concerning discount, no changing with time).

This model is the simplest among variants of EOQ model can be given as follows:

$$\mathrm{EOQ} = \sqrt{\frac{2 \times \mathrm{K} \times \mathrm{D}}{\mathrm{H}}} \qquad (1)$$

where
EOQ represents ordering quantity in unit of material (or product),
H   represents holding cost per unit material (or product) per unit time,
K_o represent cost per ordering setup,
D_o represent annual demand in unit of materials

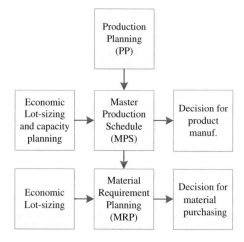

Fig. 1 The scheme of EOQ for a basic MRP productive system

3.2 The Application of EOQ on a MRP Productive System

MRP is the tool for PSCS that MRP assists to answer the basic questions of what to order, how much to order, when to order and when the delivery should be scheduled (Lee 1999). For a basic MRP productive system, three levels of operation planning can be generated. It starts with PP, then MPS and MRP in the last. PP is independent for aggregate planning of production along a medium-range period. MPS is in the following that EOQ is used in daily production scheduling for end products. The last level is MRP in that order quantity and time of placing orders for materials are planned. These three levels for the basic MRP production system are as shown in Fig. 1.

4 Case Analysis

4.1 Case Description

Company X is a small manufacturing firm which produces furniture parts such as hanger bolts, screws and rubber rings in Thailand. Capital is one million baht. There are about twenty-five persons in the company. The input materials are a variety of steels and plastics. In this company, it is the rule of thumb that is used to control the scheduling and lot sizes of both production and replenishment. Only one person in the company is responsible for ordering materials. Before ordering materials, this purchase person needs to check customer orders and stock levels of materials. Regarding manufacturing, there are eighteen persons in total. Generally, one production person takes responsibility for one work station. After a batch of production is completed, products need to be sent out for filling some chemicals to reserve.

In this study, a Hanger bolt is simulated as an example. The production of Hanger bolts, consist of three steps of machining through three respective work stations. In the first work station, materials of steel coils are fed into cutting machines for making specific sizes. In the last two work stations, fine machine thread is made on screws first and then rough machine thread is made. Passing through these three work stations, a Hanger bolt is completed. The production rate of Hanger bolt each day is about 10,000 units. As for the setup of work stations for changeover of manufacturing different products, it needs to spend 3 h at each changeover.

4.2 Estimation of EOQ for Company X

Estimation of EOQ requires knowing holding cost and setup cost first. Based on the data collection from company X, holding cost can be expressed as annual interest rate which is the sum of storage, handling, loss, taxes, and opportunity cost as shown in Table 1. The annual interest rate represents the weighting of cost for each unit product (or materials) during one year. Totally, the annual interest rate for the holding cost in product manufacturing is 12 percentages. The annual interest rate for the holding cost in material purchase is 17 percentages. The cost of each unit product during one year is 0.7 baht. Therefore, the holding cost in product manufacturing for company X is 0.084 baht in one year. For setup cost in product manufacturing, 3 h and one person is needed to setup machines. In machine setup, the mold does not change unless alternative product is to be manufactured. The setup cost is 187.5 baht at each time of machine setup as shown in Table 1.

Table 1 Cost estimation for determining EOQ of Hanger bolt

Cost items	Production scheduling	Materials ordering
Holding cost	*12 %*	*17 %*
Storage	2 %	4 %
Handling	1 %	2 %
Loss	1 %	1 %
Taxes	7 %	7 %
Opportunity cost	1 %	3 %
In total	0.084 baht per year per unit product	4.76 baht per year per kg
Setup cost	*187.5 baht*	*90 baht*
Setup time	3 h	1 h
Telephone	–	12 baht (4 times, average)
Order form	–	3 baht
EOQ	51526.6921 units of product (51,000 units of product is chosen)	368.0739 kg (370 kg is chosen)

Based on the cost estimation in both product scheduling and material ordering of Hanger bolt as shown in Table 1, EOQ_o (EOQ in material ordering) and EOQ_p (EOQ in production scheduling) can be calculated. To calculate EOQ_p for Hanger bolt, the annual demand is 5,94,720 units with the holding cost 0.084 baht per unit per year. For setup of work stations, it needs to spend 187.5 baht in average. According to Eq. (1), EOQ_p can be calculated and is 51526.6921. However, 51526.6921 must take integer value. In this study, 51,000 are chosen for representing one production lot size, as shown in Table 1. In the same way, EOQ_o can be obtained for material ordering as shown in Table 1 as well. The calculated result is 368.0739 kg for EOQ_o. Finally, 370 kg is chosen for indicating the ordering lot size instead of 368.0739 kg.

4.3 Adopting EOQ in MPS and MRP

In previous sections, MPS is based on production planning (PP). A PP for three products of planned production of company X is shown in Table 2. Table 2 shows the weekly production planning of Hanger bolt which consists of forecasted demand, planned production, and planned inventory.

A master production schedule (MPS) is a production scheduling that represented the second level for disaggregation of the overall schedules into specific products (Steven 2008). The daily MPS of company X for Hanger bolt 6 × 40, Screw 7 × 1, and Screw 6 × 1−1/4 are all created in this study for further analysis of cost. It should be noted that the MPS data have two groups, i.e., historic and EOQ adopting, for subsequent comparisons.

MRP will generate a set of net requirements that must be met if the MPS is maintained or kept on scheduled. For company X, the period of material procurement is larger than that of production scheduling. The planning period in MRP is suggested to be based on weeks. Besides, there is no necessity for this study that bill of material (BOM) is used in MRP for defining relationship between end item and subcomponents. Net requirements in MRP are directly determined through unit conversion. The week-based MRP of Hanger bolt 6 × 40, Screw 7 × 1, and Screw 6 × 1−1/4 are all created in this study for further analysis of cost as well.

4.4 The Cost Analysis of the Productive System with MRP Settings for Company X

According to the analysis described above, the total cost for both historic MPS and EOQ adopting MPS of Hanger bolt 6 × 40″ in July–August can be generated and expressed as solid and dotted lines, respectively, as in Fig. 2. It is obvious from Fig. 2 that the total cost of EOQ adopting MPS is lower than that of historic MPS

Table 2 Weekly production planning of Hanger bolt

Week		1	2	3	4	5	6	7	8	9
Forecasted demand		7,500	15,000	10,000	11,000	10,100	26,000	5,000	10,000	8,000
Planned production	Hanger bolt 6 × 40	0	35,500	0	0	57,000	0	0	0	0
	Screw 7 × 1	5,000	9,000	17,000	46,000	10,000	7,600	0	34,000	0
	Screw 6 × 1–1/4	20,000	10,000	45,000	0	0	11,000	9,000	0	5,000
Planned inventory		4,400	24,900	14,900	3,900	50,800	24,800	19,800	9,800	1,800
	11,900									

Fig. 2 Comparison of total costs between historic MPS and EOQ adopting MPS

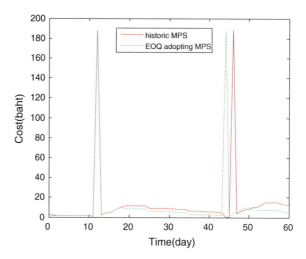

Fig. 3 Comparison of total costs between historic MRP and EOQ adopting MRP

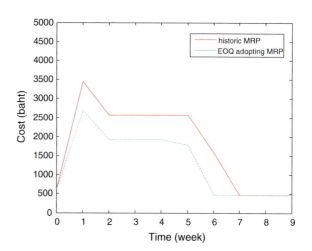

in average. Similarly, the total cost for both historic MRP and EOQ adopting MRP of Hanger bolt 6 × 40″ in July–August can be generated and expressed as solid and dotted lines, respectively, as in Fig. 3. It is obvious from Fig. 3 that the total cost of EOQ adopting MRP is lower than that of historic MRP as well.

5 Conclusions

Economic order quantity (EOQ) which can balance setup cost and holding cost can be easily used in materials ordering and production scheduling especially when the adopting company is small. Many small companies may lend EOQ to improve their controls of PSCS. This study thus suggests such simple application of EOQ to company X in Thailand to gain the benefits of cost savings. In future work, other methods of dynamic lot sizing such as Silver-Meal Heuristic can be applied to compare with the EOQ method.

References

Biggs JR (1979) Heuristic lot-sizing and sequencing rules in a multistage production-inventory system. Decis Sci 10:96–115
Cárdenas-Barrón LE (2010) An easy method to derive EOQ and EPQ inventory models with backorders. Comput Math Appl 59:948–952
Choi S, Noble JS (2000) Determination of economic order quantities (EOQ) in an integrated material flow system. Int J Prod Res 38(14):3203–3226
Cargal JM (2003) The EOQ inventory formula. Math Sci, 1.31 edn
Lee Angela (1999) A study of production management (manufacture). Process Protocol II:1–14
Lee TS, Adamee EE, JR Ebert (1986) Forecasting error evaluation in material requirement planning (MRP) production-inventory systems. Manag Sci 32(9):1186–1205
Lee TS, Adam EE, Ebert JR (1987) An evaluation of forecast error in master pro duction scheduling for material requirements planning systems. Decis Sci 18:292–307
Liberatore MJ (1979) Using MRP and EOQ/safety stock for raw material inventory control: discussion and case study. Interfaces 9(2):1–6
Min W, Pheng L-S (2006) EOQ, JIT and fixed costs in the ready-mixed concrete industry. Int J Prod Econ 102:167–180
Taleizadeh AA, Widyadana AG, Wee HM, Biabani Jahangir (2011) Multi products single machine economic production quantity model with multiple batch size. Int J Ind Eng Comput 2:213–224
Steven N (2008) Production and operation analysis. 6th Revised edn
Silver EA, Peterson R (1985) Decision system for inventory management and production planning, 2nd edn. Wiley, New York
Wangner HM, Whitin TM (1958) Dynamic Version of the Economic Lot Size Model. Manage Sci 5:89–96

CUSUM Residual Charts for Monitoring Enterovirus Infections

Huifen Chen and Yu Chen

Abstract We consider the syndromic surveillance problem for enterovirus (EV) like cases. The data used in this study are the daily counts of EV-like cases sampled from the National Health Insurance Research Database in Taiwan. To apply the CUSUM procedure for syndromic surveillance, a regression model with time-series error-term is used. Our results show that the CUSUM chart is helpful to detect abnormal increases of the visit frequency.

Keywords CUSUM chart · Enterovirus syndrome · Regression analysis · Syndromic surveillance

1 Introduction

The two major epidemic peaks for enterovirus (EV) diseases in Taiwan occur in May to June and September to October yearly according to the historical statistics from Centers for Disease Control in Taiwan (Taiwan CDC). In 1998 the EV infection caused 78 deaths and 405 severe cases in Taiwan (Ho et al. 1999). Early detection of outbreaks is important for timely public health response to reduce morbidity and mortality. By early detecting the aberration of diseases, sanitarians can study or research into the causes of diseases as soon as possible and prevent the cost of the society and medical treatments. Traditional disease-reporting surveillance mechanisms might not detect outbreaks in their early stages because laboratory tests usually take long time to confirm diagnoses.

Syndromic surveillance was developed and used to detect the aberration of diseases early (Henning 2004). The syndromic surveillance mechanism is to

---

H. Chen (✉) · Y. Chen
Department of Industrial and Systems Engineering, Chung-Yuan University, Chungli, Taoyuan, Taiwan
e-mail: huifen@cycu.edu.tw

collect the baseline data of prodromal phase symptoms and detect the aberration of diseases from the expected baseline by placing the variability of data from the expected baseline. Such surveillance methods include the SPC (statistical process control) based surveillance methods, scan methods and forecast-based surveillance methods (Tsui et al. 2008). See Sect. 2 for literature review.

In this work, we apply the CUSUM residual chart for detecting the abnormal increases of EV-like cases in Taiwan. Since the daily visits of the EV-like syndrome are time series data with seasonal effect, we use a regression model with an time-series error term to model the daily counts from ambulatory care clinic data. The residuals are then used for the CUSUM chart to detect unusual increase in daily visits. The test data are the 2003–2006 ambulatory care clinic data from the National Health Insurance Research Database (NHIRD) in Taiwan.

This paper is organized as follows. In Sect. 2, we review related literature. In Sect. 3, we summarize the data, propose a regression model whose error term follows an ARIMA model, and construct the CUSUM chart using the residuals. The conclusions are given in Sect. 4.

2 Literature Review

We review here the syndromic surveillance methods including the forecast-based, scan statistics, and SPC-based methods.

The forecast-based methods are useful to model non-stationary baseline data before monitoring methods can be applied. Two popular forecasting methods are time-series and regression models. Goldenberg et al. (2002) used the AR (Auto Regressive) model to forecast the over-the-counter medication sales of the anthrax and built the upper prediction interval to detect the outbreak. Reis and Mandl (2003) developed generalized models for expected emergence-department visit rates by fitting historical data with trimmed-mean seasonal models and then fitting the residuals with ARIMA models. Lai (2005) used three time series models (AR, a combination of growth curve fitting and ARMA error, and ARIMA) to detect the outbreak of the SARS in China.

Some works fitted the baseline data with a regression model first and then fitted the residuals with a time-series model because the baseline data may be affected by the day of the week and/or holiday factors. Miller et al. (2004) used the regression model with AR error to fit the influenzalike illness data in an ambulatory care network. The regression terms include weekend, holiday and seasonal adjustments (sine and cosine functions). Therefore, they used the standardized CUSUM chart of the residuals for detecting the outbreak. Fricker et al. (2008) applied the adaptive regression model with day-of-the-week effects using an 8-week sliding baseline and used the CUSUM chart of the adaptive regression residuals to compare with the Early Aberration Reporting System (EARS). They showed that the CUSUM chart applied to the residuals of adaptive regressions performs better than the EARS method for baseline data with day-of-the-week effects.

The scan statistics method is widely used in detecting the clustering of diseases. Scan statistics methods can be used in temporal, spatial and spatiotemporal surveillance. Heffernan et al. (2004) applied the scan statistic method to monitor respiratory, fever diarrhea and vomiting syndromes by the chief complaint data of the emergency department. They used this method in the citywide temporal and the spatial clustering surveillances. Han et al. (2010) compared CUSUM, EWMA and scan statistics for surveillance data following Poisson distributions. The results showed that CUSUM and EWMA charts outperformed the scan statistic method.

Recently the control charts have been applied in health-care and public-health surveillance (Woodall 2006). The SPC methods were first applied in the industrial statistical control (Montgomery 2005). Since the Shewhart chart is insensitive at detecting small shifts, CUSUM and exponentially weighted moving average (EWMA) charts are more commonly used in public health surveillance than the Shewhart chart. Hutwagner et al. (1997) developed a computer algorithm based the CUSUM chart to detect salmonella outbreaks by using the laboratory-based data. Morton et al. (2001) applied Shewhart, CUSUM and EWMA charts to detect and monitor the hospital-acquired infections. The result shows that Shewhart and EWMA work well for bacteremia and multiresistant organism rates surveillance and that CUSUM and Shewhart charts are suitable for monitoring surgical infection. Rogerson and Yamada (2004) applied a Poisson CUSUM chart to detect the lower respiratory tract infections for 287 census tracts simultaneously. Cowling et al. (2006) adopted the CUSUM chart with 7-week buffer interval for monitoring influenza data form Hong Kong and the United States and compared with time series and regression models. Woodall et al. (2008) show that the CUSUM chart approach is superior to the scan statistics.

3 Methods

3.1 Data Source

The data used in this study are the 2003–2006 daily counts (i.e. the number of daily visits) of EV-like cases for 160,000 people sampled from the National Health Insurance Research Database (NHIRD) by the Bureau of National Health Insurance, Taiwan. Patients' diagnoses in NHIRD were encoded using the ICD-9-CM (International Classification of Diseases, 9th Revision, Clinical Modification Reference) code. In this study, the ICD-9 codes of the EV-like syndrome are adopted from Wu et al. (2008) as listed in Appendix A.

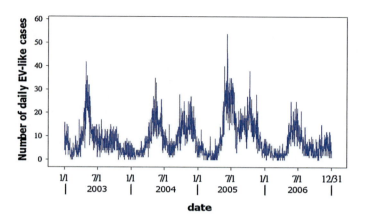

Fig. 1 The daily counts of the EV-like cases from 2003 to 2006

3.2 Data Summary

Here we summarize the daily counts of EV-like cases from 2003 to 2006 with population size 160,000. Figure 1, the run chart of the daily counts, shows that the daily counts are time-series data with seasonal variation. In general, the major epidemic peak occurs in May and June and a smaller peak occurs in September and October. Among the four years, the epidemic peaks are highest in 2005 and lowest in 2006. The day-of-the-week effect also exists. For the age effect, since more than 80 % of the EV-like cases are children younger than 6 years old, we do not consider the age effect in this study.

3.3 CUSUM Charts

Since the daily counts are time series data with seasonal variation, we use the regression model with an ARIMA error term to fit the daily counts of the EV-like cases. For normality, we first use the Box-Cox transformation to transform the daily counts data. The predictor variables are set based on the day-of-the-week, month-of-the-year, and trend effects.

The residuals calculated from the fitted regression model with an ARIMA error term can be used to construct an upper one-sided standardized CUSUM chart (Montgomery 2005) for detecting abnormal increases in daily counts of EV-like cases. Like Miller et al. (2004), we set the control limits so that the in-control average run length is 50.

To illustrate the surveillance method, we use the 2003 and 2004 daily counts data to fit a regression model and then use the 2005 data to construct CUSUM charts and forecasted values.

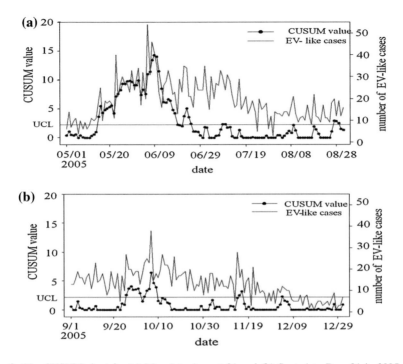

Fig. 2 The CUSUM chart for (**a**) May 1 to August 31 and (**b**) Sept. 1 to Dec. 31 in 2005

Figure 2 contain the upper one-sided standardized CUSUM charts for two periods—May 1 to August 31 in Subfigure (a) and September 1 to December 31 in Subfigure(b)—containing epidemic peaks in 2005. The number of EV-like cases and upper control limit (UCL) of the CUSUM chart are also shown. Figure 2 shows that the epidemic outbreak that occurred in May 2005 is detected quickly by the CUSUM chart. The smaller epidemic outbreak occurring at the end of September 2005 is also detected.

Using the fitted regression model, we can construct the 1-steps-ahead forecast value. Figure 3 compares the 2005 actual daily numbers of EV-like cases with its forecasted values for the period (May 1 to December 31) including high seasons. The x-axis is the date and y-axis is the actual (black line)/forecasts (gray line) number of daily EV-like cases. Figure 3 shows that the difference is higher during the peak of infection than in the low season.

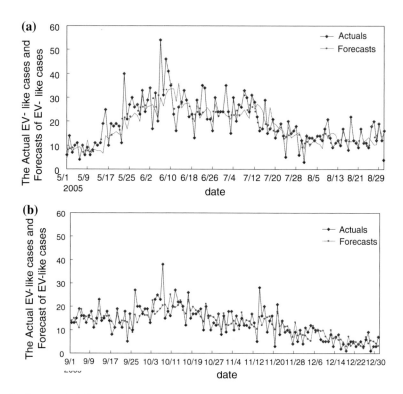

Fig. 3 The actual daily EV-like cases and its forecasts for year 2005: May 1 to Aug. 31 in Subfigure (**a**) and Sept. 1 to Dec. 31 in Subfigure (**b**)

4 Conclusions

This paper discusses the implementation of CUSUM residual charts for monitoring daily counts of EV-like cases. The population size is 160,000. Before using the CUSUM chart, we fit a regression model with an ARIMA error term to the daily counts data. The numerical results indicate that the CUSUM residual chart seems to work well in showing unusual increases in daily counts of EV-like cases.

Our fitted regression model is based on historical data of the past two years. The time window can be longer so that more data can be used for model fitting. The shortage though is that the coefficient estimates would have larger variance and hence the prediction interval would be wider. Furthermore, the behavior of daily counts may not be the same each year, using historical data occurring long ago may hurt the prediction accuracy for the future observations.

Acknowledgments This study is based in part on data from the National Health Insurance Research Database provided by the Bureau of National Health Insurance, Department of Health and managed by National Health Research Institutes. The interpretation and conclusions contained herein do not represent those of Bureau of National Health Insurance, Department of Health or National Health Research Institutes in Taiwan.

Appendix A: The ICD-9-CM code of EV-like syndrome

In this study, we adopt the EV-like syndrome definitions from Wu et al. (2008). The ICD-9 codes are listed below.

ICD9	Description
074	Specific diseases due to Coxsackie virus
079.2	Coxsackie virus: infection NOS
047.0	Coxsackie virus: meningitis
074.0	Herpangina Vesicular pharyngitis
074.1	Epidemic pleurodynia, Bornholm disease, Devil's grip; Epidemic: myalgia, myositis
074.2	Coxsackie carditis
074.20	Coxsackie carditis unspecified
074.21	Coxsackie pericarditis
074.22	Coxsackoe endocarditis
074.23	Coxsackie myocarditis, Aseptic myocarditis of newborn
074.3	Hand, foot, and mouth disease Vesicular stomatitis and exanthem
074.8	Other specified diseases due to Coxsackie virus, Acute lymphonodular pharyngitis

References

Cowling BJ, Wong IOL, Ho L-M, Riley S, Leung GM (2006) Methods for monitoring influenza surveillance data. Int J Epidemiol 35:1314–1321

Fricker RD Jr, Hegler BL, Dunfee DA (2008) Comparing syndromic surveillance detection methods: EARS' versus a CUSUM-based methodology. Stat Med 27:3407–3429

Goldenberg A, Shmueli G, Caruana RA, Fienberg SE (2002) Early statistical detection of anthrax outbreaks by tracking over-the-counter medication sales. Proc Natl Acad Sci USA 99:5237–5240

Han SW, Tsui K-L, Ariyajunya B, Kim SB (2010) A comparison of CUSUM, EWMA, and temporal scan statistics for detection of increases in Poisson rates. Qual Reliab Eng Int 26:279–289

Heffernan R, Mostashari F, Das D, Karpati A, Kulldorff M, Weiss D (2004) Syndromic surveillance in public health practice, New York City. Emerg Infect Dis 10:858–864

Henning KJ (2004) What is syndromic surveillance? Morb Morality Wkly Rep 53(Supplement):7–11

Ho M, Chen ER, Hsu KH (1999) An epidemic of enterovirus 71 infection in Taiwan. Medicine 341:929–935

Hutwagner LC, Maloney EK, Bean NH, Slutsker L, Martin SM (1997) Using laboratory-based surveillance data for prevention: an algorithm for detecting Salmonella outbreaks. Emerg Infect Dis 3:395–400

Lai D (2005) Monitoring the SARS epidemic in China: a time series analysis. J Data Sci 3:279–293

Miller B, Kassenborg H, Dunsmuir W, Griffith J, Hadidi M, Nordin JD, Danila R (2004) Syndromic surveillance for influenzalike illness in an ambulatory care network. Emerg Infect Dis 10:1806–1811

Montgomery DC (2005) Introduction to statistical quality control, 5th edn. Wiley, New York

Morton AP, Whitby M, Mclaws M, Dobson A, Mcelwain S, Looke D, Stackelroth J, Sartor A (2001) The application of statistical process control charts to the detection and monitoring of hospital-acquired infections. J Qual Clin Pract 21:112–117

Reis B, Mandl KD (2003) Time series modeling for syndromic surveillance. BMC Med Inform Decis Mak 3:2

Rogerson PA, Yamada I (2004) Approaches to syndromic surveillance when data consist of small regional counts. Morb Morality Wkly Rep 53:79–85

Tsui K-L, Chiu W, Gierlich P, Goldsman D, Liu X, Maschek T (2008) A review of healthcare, public health, and syndromic surveillance. Qual Eng 20:435–450

Woodall WH (2006) The use of control charts in health-care and public-health surveillance. J Qual Technol 38:89–104

Woodall WH, Marshall JB, Joner MD Jr, Fraker SE, Abdel-Salam A-SG (2008) On the use and evaluation of prospective scan methods for health-related surveillance. J R Statist Soc A 171(1):223–237

Wu TSJ, Shih FYF, Yen MY, Wu JSJ, Lu SW, Chang KCM, Hsiung C, Chou JH, Chu YT, Chang H, Chiu CH, Tsui FCR, Wagner MM, Su IJ, King CC (2008) Establishing a nationwide emergency department-based syndromic surveillance system for better public health responses in Taiwan. BMC Public Health 8:18

A Study on the Operation Model of the R&D Center for the Man-Made Fiber Processing Industry Headquarter

Ming-Kuen Chen, Shiue-Lung Yang and Tsu-Yi Hung

Abstract The global strategy layout for the activities of R&D is naturally the important decision making for the survival and competition of business entities in Taiwan. Based on these background and motives, this research aims to investigate and construct the reference models for the operation of R&D center within the business headquarters and we also focus our effort onto the textile sub-industry—Man-made fiber processing industry. This research is adopted with IDEF0 for processing analysis and model construction methodology. During the construction, the processes associated with the R&D center within business headquarters is constructed via literature survey and practical expert interview. By means of the factor analysis, we can extract the key items of ICOM (Input, Control, Output, Mechanism) from the activities related to the R&D processes. This research result will be meant to depict the overall profile for the operation of R&D centers from the textile industries. Hopefully, we can offer the reference basis for the planning phase of business headquarter R&D center so that the planning can thoroughly meet the demand of business strategy execution and management characteristics.

Keywords Business Headquarter · R&D Center · Factor Analysis · IDEF0

M.-K. Chen (✉) · T.-Y. Hung
Graduate Institute of Services and Technology Management, National Taipei University of Technology, Taipei, Taiwan
e-mail: mkchen@ntut.edu.tw

T.-Y. Hung
e-mail: tyhung@ntut.edu.tw

S.-L. Yang
Department of Management Information Systems, National Chengchi University, Taipei, Taiwan

1 Introduction

This research is aimed to investigate and construct the reference models for the R&D centers of business headquarters. In overview of the past research about business headquarters (Ghoshal 1997; Kono 1999; Young 1998), most research efforts are spent onto the major advanced nations located in USA, Europe and Japan for the internal management operation investigation of the local business entities. This is unavailable to offer the effective aid to plan the strategy and operation of business headquarter.

The structure of this paper is as follows. Firstly, we make the systematic arrangement for the literature review. Next, the IDEF0 is meant for the methodology of process analysis and model construction. Thirdly, the questionnaires are used for further researching with the respondents mainly from managerial levels and R&D staffs from the textile sub-industries—the Man-made Fiber Processing Industry. Finally, by using the factor analysis, we can extract the key items of ICOM (Input, Control, Output, Mechanism) from the activities related to the R&D processes.

2 Literature Review

2.1 Business Headquarter

Ghoshal (1997) proposed that the current business globalization seems to be an inevitable issue for the multinational business entities. Birkinshaw et al. (2000) suggested that the task for business headquarters is not only intended to ensure the future development and growth of business but also add the values of business. Thus, Young (1998) supposed that within the process for business globalization, how to create a powerful business headquarter for the effective management and control over the overseas branching offices has become the critical factor for successful business management. Kono (1999) supposed the business headquarter must consist of below three functions: business strategy projection, clearly creating the professional management department, developing the business core competence, and providing central service with the available affairs. Goold et al. (2001) revealed that a wholesome and powerful business head quarter must experience with three stages for execution: the role playing to act as the most minor headquarter, the role for value creation and service share. Regarding the management strategies of business management headquarters, Desouza and Evaristo (2003) had proposed three strategies available for business headquarters: the business headquarter centrally unify the commands and executions; the business headquarter can command the regional units for execution; and the regional units can further command the subordinated districts for execution.

To sum up, the business head quarter is exactly the business center for the decision making of globalization and the operation base for value creation. It is well performing its diversified functions and outstretches the global strongholds for logistic management so that it is available for business entities to create much more values.

2.2 The Globalization Management for R&D

Pearce (2003) revealed that the investment for overseas R&D is functioned to execute the company strategies and the relevant efforts shall be paid during the posterior stage of product life cycle by accompanying with exodus of production lines together. Dunning (1992) supposed when the multinational company is outstretching its R&D activities to offshore areas, it will naturally bring with the ownership advantage edging over its competitors with much more favorable trends to keep or enhance its own competence. Kuemmerle (1997) used to classify the overseas R&D strongholds as: the Home-Base-Augmenting Site and the Home-Base-Exploiting Site. Ronstadt (1978) classified the R&D units into three types: Support Laboratory (SLs), Locally Integrated Laboratories (LILs), and Internationally Inter-dependent Laboratories (IILs). Bartlett and Ghoshal (1995) classified the R&D activities according to the locations of R&D and globalization with the regional correlation as below: R&D Center, Local R&D and Local R&D and Multi-national Application and Globally Linked R&D Centers.

Within the issue of R&D globalization management, Gassmann and Zedtwitz (1999) proposed five strategies for R&D globalization: ethnocentric centralized R&D, geocentric centralized R&D, R&D centralized R&D, polycentric decentralized R&D and integrated R&D network.

3 The Globalization Management for R&D of Man-Made Fiber Processing Industry

3.1 The Design of Business Headquarter

Based on the proposals of (Goold et al. 2001; Kono 1999) for researching, we have developed the orientation for this research. The structures of R&D centers and processes are mainly based on the proposals of Cooper and Kleinschmidt (1997) to create the R&D processes and structures suitable for man-made fiber processing industries. We also adopt the proposal from Gassmann and Zedtwitz (1999) for R&D globalization management models to use as reference so that we can understand the types and models suitable for man-made fiber processing industries and also realize the interaction between R&D center and other functional centers during different stages. How to effectively integrate global units and quickly response to the regional demands for best balance is exactly the ultimate pursuit for headquarter. This research is proposed by referring to the works of (Goold et al. 2001; Kono 1999). The management centers included within the business headquarter is divided into two modules of basic function and value creation. The contents related to each module are: (1) Basic Function Module: It is mainly aimed to keep the basic functions and activities inevitable for the normal operation of business such as Financial Management Center, Human Resource Management Center. (2) Value Creation Module: This module contains of the creation and execution of organizational strategy, the relevant activities about development of core competence, etc.: (a) Proprietary Management Center; (b) R&D Management Center; (c) Marketing Management Center; (d) Production Center; (e) Procurement Center.

3.2 The Design for the R&D Center Within Man-Made Fiber Processing Industries

This research is amassed with the research results from (Bartlett and Ghoshal 1995; Cooper and Kleinschmidt 1997; Dhillon 1985; Kuemmerle 1997; Marquis and Myers 2004; Ronstadt 1978). We propose that the flow chart models for man-made fiber processing industries' R&D can be divided into 5 stages of operational processes: (1) Commodity Projection Stage; (2) Design Test & Projection Stage; (3) Mass Production Stage; (4) Marketability Stage; (5) The Resource Output during Processing Stages.

3.3 Questionnaire Sample Design

Within the first part of questionnaires, it is combined with the business headquarter functional structures mentioned by Sect. 3.2 within this research and the product development processes. We also design the questionnaire contents with the purpose to investigate the interaction for each center ruled by business headquarters for the product development processes within the man-made fiber processing industries. The questionnaire option contents consist of three levels of response: high involvement, general involvement and no involvement. The responding results of this part can be used as the basic reference for IDEF0 structure models and the design contents and results of questionnaires are shown as Table 1.

In the descriptive statistics, [★] means the highest identification degree from a single operation. Namely, the center within the corresponding operation will naturally come with the absolute predomination right. ★ means the identification degree above 70 %. ● means the identification degree between 69 and 60 %. By using this result, we can combine the conclusions from the 2nd part of questionnaires and

Table 1 The interaction for the R&D processes of man-made fiber industries

| Process item | Tasking activity | Business head quarter ||||||| Each region ||
		Financial center	Marketing center	Procurement center	R&D Design center	Production center	R&D units	Production units
Commodity projection stage	Fiders market trend analysis	★	[★]		★			
	Regional innovation creation & selection		[★]		●		●	
	Facilities/know-how feasibility analysis		●		[★]	●		●
	Competitor analysis		[★]		●		●	
	Financial/costs evaluation	[★]			★			
	Environmental protection regulation		●	●	[★]			
	Proposing the new product strategy		●		★			
	Ensure the market target—global market		[★]		●			
Design test & projection stage	Development concept and projects		[★]	●	●			
	Choices main suppliers and sub-suppliers	●	[★]	●	★	●	★●●●●●	●
	Physical analysis and design		★	●	[★]	●	●	●
	Production process programming		●		[★]	★		
	Silk sampling				★	●●	●	
	Verification and quality examination				[★]	★	●	
	Definition for QC processes				[★]	●	●	
Mass production stage	Composed of production system design for manufacturing centers				●	[★]	●	●
	Small quantities of marketing and trial sales		[★]		★	★	●	●
	Feedback and revision projects		★		[★]	●	●●	●
	Improvement for manufacturing processes				[★]	★	●●	●
	Production evaluation (time, capacity, predict prices, cost)		●		★	[★]	●	●
	Normal mass production				[★]	[★]	●●	●●
	Technique transfer				[★]	★	●●	●●

(continued)

Table 1 (continued)

Process item	Tasking activity	Business head quarter						Each region	
		Financial center	Marketing center	Procurement center	R&D Design center	Production center	R&D units	Production units
Commodity marketability stage	Commodity introduction of processing silk (global or regional markets)		★		★			
	Market and customer complaint response (global or regional markets)		★		●			
	Sales feedback analysis		★		★		●	
	The production improvement for processing silk, function correction and the functional development	●	★		[★]		●	●
Waste output processing stage	Record of waste output of all stages and analysis			●	★	●		
	Feedback to research procedure			●	[★]	●		
	Amends the textile specification and alternative project selection				★		●	

construct the operation model for the R&D center of fiber weaving industries by IDEF0 with the detailed analysis shown in Table 1.

From Table 1, we find that within the commodity projection stage of man-made fiber processing industries, the marketing center is responsible to audit the budget regulation for local and overseas strongholds form conglomerate members and the filtering for marketing strategies. The remaining operations are all under the duty of marketing centers within the projection and execution. Within the design test and projection stages, the R&D center is authorized with the leadership of technology and R&D activities. The main operation items are naturally focused on initial sample trial production and cloth applicability. Except the projection for global (regional) marketing strategies, the remaining operations are all under the control of R&D center for projection and execution. Within the mass production stage, the marketing center, after the completeness of small-quantity trial production, will send the fiber samples to customers for trial sales. After receiving the responses from the result analysis of trial sales and customer complaints, the R&D center must modify and correct the product design by focusing on the analysis results and customer complaints till the company finally receives the confirmed orders from customers. The production system design will be transferred to the regional R&D units. The regional production unit will finish the trial production of small quantities and trail sales, feedback and correction review, improvement for QC process, production timing, capability, price forecast, cost evaluation and the finalized mass production. Within the commodity marketability stage, the R&D center controls the operations like information analysis and design correction with the purposes partially for increasing the market acceptability of current products and partially for the new development reference in preparation for the next stage. Within the waste processing stage, the regional production center will lead the relevant operations for removal, recycle and R&D feedback. As in the production for large quantities, the production center will cause much major (cloth) and minor (labeling/packing materials) material residue. It is naturally required for further improvement on production patterns, mark design, and the re-design for material consumption structure so that the factory can fulfill its responsibility and give positive feedback to the social public.

4 The Analysis of the R&D Center at a Business Headquarter

4.1 Sample Analysis

There were 35 sets of valid samples collected. This research data was collected through on face-to-face interview and telephone interviews. We mainly adopted with the factor analysis to extract the ICOM key items from each R&D process with the results shown as Table 2. Within the questionnaire, we investigate the

Table 2 Sample analysis

Product development stage	Input/Output/Control item		The variance value of variables (%)
Commodity projection stage	Input item	Specifications development proposal	41.378
		Budget allocation project	26.643
	Control item	Equipment/Capacity analysis report	36.321
		The environmental protection codes from various national governments	33.694
	Output item	Projects	71.239
Design test &projection stage	Input item	Technology & production process improvement report	34.548
		Material testing report	28.962
	Control item	Capital demand & risk evaluation	73.621
	Output item	Physical analysis & technique transfer specifications	35.256
		Laboratory examination reports	29.592
Mass production stage	Input item	Alternative material report	46.266
		Regional production & sales coordination plan	30.289
	Control item	Texturing machine operation specifications	71.23
	Output item	Production layout & parameter setting for production process	45.533
		Production efficiency & product reliability analysis table	27.872
Commodity marketability stage	Input item	Marketing Tactic and Trademark patent proposal	64.419
	Control item	The environmental protection codes from various national governments	65.678
	Output item	Regional production & sales coordination plan	37.143
		Design alternation measures	29.363

55 question items by means of Principal Components Analysis so that we extract the factor dimensions for each product development stage.

Within the product projection stage, there are totally 5 variables included within the Input Items, 3 variables within Control Items and 6 variables within Output Items. Within the design test and projection stage, the Input Items totally include 3 variables. Control Items consist of 5 variables and the Output Items include 7 variables. Within the mass production stage, there 4 variables included within Input Items, 3 variables within Control Items and 8 variables within Output Items. Within the commodity marketability stage, there are 3 variables included within Input Items, 3 variables within Control Items, and 5 variables within Output Items. Within the waste output stage, there are 2 variables for each item. Thus, within the data analysis, each variable is viewed as the key item.

4.2 The Operation Modes of the R&D Center

The interview questionnaire analysis is extracted from the items such as input, output and control variable during the product developing procedures. It also accounts the interactions between the headquarter and its regional units during the new product R&D stages (cf. Table 1) by adopting the IDEF0 flow models to establish the reference models for the R&D center of the man-made fiber processing headquarter.

5 Conclusions

The results of this research depict the overall operation profile for R&D and design centers of man-made fiber processing industries. This study is adopted with IDEF0 for processing analysis and model construction kits. The industries of artificial fiber especially emphasize the R&D of basic and applicable operations. Under the product projecting stages, it is required to integrate the information from various markets and thus to create the proposals for product development. However, the operation of plausibility evaluation is vital for manufacturing facilities.

This article focuses on various industrial aspects to insightfully know about the operation contents of strategic projections and each function centers when the business entities are engaging in projection. It is available to use other analysis kits to demonstrate the operational reference models of each function center at a business headquarter to subsequently work as the future research orientation.

References

Bartlett CH, Ghoshal S (1995) Building the entrepreneurial corporation: new organizational processes, new managerial tasks. Eur Manag J 13(2):139–155

Birkinshaw J, Thilenius P, Arvidsson N (2000) Consequences of perception gaps in the headquarters—subsidiary relationship. Int Bus Rev 9(3):321–344

Cooper RG, Kleinschmidt EJ (1997) Winning businesses in product development: the critical success factors. Res Technolo Manag 39(4):18–29

Desouza K, Evaristo R (2003) Global knowledge management strategies. Eur Manag J 21(1):62–67

Dhillon BS (1985) Quality control, reliability, and engineering design. Marcel Dekker, New York

Dunning J (1992) Multinational enterprises and the globalization of innovatory capacity. Wiley Chichester

Gassmann O, Zedtwitz M (1999) New concepts and trends in international R&D organization. Res Pol 28(2–3):231–250

Ghoshal S (1997) The individualized corporation: an interview with Sumantra Ghoshal. Eur Manag J 15(6):625–632

Goold M, Pettifer D, Young D (2001) Redesigning the corporate centre. Eur Manag J 19(1):83–91

Kono T (1999) A strong head office makes a strong company. Long Plann 32(2):225–236

Kuemmerle W (1997) Building effective R&D capabilities Abroad. Harv Bus Rev 75(2):61–71

Marquis RG, Myers S (2004) NEC Folds Japan R&D Activities into Two Units. Elect Eng 6

Pearce R (2003) Industrial research institute's 5th annual R& D leaderboard. Res Technolo Manag 46(6):157–178

Ronstadt RC (1978) International R&D: the establishment and evolution of R&D abroad by seven US multinationals. J Int Bus Stud 9(1):7–24

Young JD (1998) Benchmarking corporate headquarters. Long Plann 31(6):933–936

Planning Logistics by Algorithm with VRPTWBD for Rice Distribution: A Case BULOG Agency in the Nganjuk District Indonesia

Kung-Jeng Wang, Farikhah Farkhani and I. Nyoman Pujawan

Abstract This paper addresses a vehicle routing problem with time windows encountered in BULOG (Government National Agency) specialized for distribution of rice with subsidy by government in Nganjuk East Java. It concern the delivery of rice with subsidy from central BULOG to home family that have been chosen by government as poor family in Nganjuk, delivery from central to warehouse, and from warehouse to government district also from district to village (the poor family target living). The problem can be considered as a special vehicle routing problem with time windows, with bender's decomposition as solver to minimize the total cost of distribution with still consider about time delivery and total of vehicle used. Each village is visited by more than one vehicle at one time delivery. Two mixed-integer programming models are proposed. We then proposed a Genetics Algorithm (GA) and exact method of VRPTWBD, and these approaches are tested with compare the result from four experiments with real data from BULOG, such as: Exact VRPTW, exact VRPTW-BD, Naïve GA, and Genetics Algorithm-BD.

Keywords BULOG · Vehicle routing problem · Time windows · Bender decomposition

---

K.-J. Wang (✉) · F. Farkhani
Department of Industrial Management, National Taiwan University of Science and Technology, Taipei, Taiwan, Republic of China
e-mail: kjwang@mail.ntust.edu.tw

F. Farkhani
e-mail: farikhah.farkhani@gmail.com

F. Farkhani · I. N. Pujawan
Department of Industrial Engineering, Institute Technology of Sepuluh Nopember, Surabaya, East Java, Indonesia
e-mail: pujawan@gmail.com

1 Introduction

Poverty in Indonesia is very complex; an increasing number of poor people can go up very sharply at a time of the economic crisis in Indonesia. The numbers of poor peoples in the city are generally lower than the number of poor people who live in the village. Therefore, Indonesia's government made policies to build the institutions main food subsidies; it is rice institutions of rice for rural areas known with the name of BULOG (National Agency of rice distribution).

There are many various constraints and problems faced by BULOG, including (1) There is no planning in the distribution process to balance with the existing resources (example: vehicle, driver) by demand changes every year due to increase the number of poor families. Now currently planning done by manual method which estimates individual leadership responsible for the delivery of the goods, so this is based on willingness. (2) There is delay occurs time during the process of delivery of rice to villages destination, as well as trucks are sent late back to the warehouse because there is no supervision tight. (3) There is a variety of criminal acts during trip distribution shipments of rice because there is no checking the amount of rice being sent, for example (a) A rice is not sent to the target, that poor family. But, lost by someone that not take a responsibility when delivery shipments, (b) there's some rice sold by illegal activity with high price to other people, or the targeted people should pay more that subsidy price to the shipper, (c) the distribution of rice was not evenly distributed across the target the targeted villages, still there is some poor families that include in the data of government did not get the subsidy of rice because of delivery shipment system.

Nganjuk is a district which in 2012 had a high crime rate for subsidized shipping problems to the area that has been determined, it is because the district Nganjuk very difficult to access reach because it is located in the mountainous region and there is no public transportation that can be used as a transport facility for distribute subsidized rice to stricken remote villages. And the government statistic report in 2012, Nganjuk is a top three of highest total of poor families in East Java.

In this research we try to solve the BULOG problems, with the real data from BULOG we can try to solve this case with Vehicle routing problem with time windows algorithm approach, and use upper and lower bound by bender's decomposition to more specific calculate the minimum cost not only taken by distance, the vehicle, the driver, but also the total demand that sub district region should pay on it with subsidy price. We use the Genetic Algorithm to solve the problem, and improve the Genetic Algorithm with Bender's Decomposition.

2 Solution Methodology

In this paper we use VRPTW approach (Toth and Vigo 2002) to solve BULOG case and especially for cost we use Bender's Decomposition to minimize the total cost of distribution (Van Roy 1986), and for solution method we use Genetic Algorithm compute the results.

2.1 Permutation Coding

Candidate selections of objects for the VRPTW can be represented by permutations of integers that represent the objects. A decoder scans the objects in their order in a permutation and places in the VRPTW every object that fits. The sum of the included objects' individual and joint values is the permutation's fitness. A mutation operator independently swaps values at random positions in a chromosome; the maximum number of swaps is a parameter of the operator. Crossover is alternation (Larranaga et al. 1999), which interleaves two parent permutations, preserving only the first appearance of each integer.

2.2 Heuristics Steps

Heuristic initialization seeds the GA's population with one chromosome generated by placing a random object in the VRPTW, then repeatedly appending to the permutation the object the VRPTW can accommodate that has the largest value density relative to the objects already chosen.

Similarly, *heuristic alternation* builds one offspring from two parents, beginning with the first object in the first parent. It the repeatedly examines the next unexamined objects in both parents. If only one fits, it joins the offspring. If neither fits, neither joins the offspring. If both fit, the next one to join the offspring is the object with the larger value density relative to the objects already listed in the offspring. In both operations, when the VRPTW can accommodate no more objects, the chromosome is completed by listing the unused objects in random order.

2.3 Genetics Algorithm Procedure

Genetic algorithms are inspired by Darwin's theory about evolution. Solution to a problem solved by genetic algorithms is evolved. Algorithm is started with a set of solutions (represented by chromosomes) called population. Solutions from one population are taken and used to form a new population. This is motivated by a hope, that the new population will be better than the old one. Solutions which are selected to form new solutions (offspring) are selected according to their fitness—the more suitable they are the more chances they have to reproduce.

The genetic algorithm proposed in this paper combines the following features: a permutation chromosome, exact fitness computation by splitting, improvement by local search, diversity of the population and multi-start with partially replaced population. These ingredients for designing efficient GA for vehicle-routing like problems.

In this paper, the chromosome is a permutation of all villages that give order of visits in different routes. An exact split algorithm will be presented to split the permutation into sub-strings of village to be visited by a vehicle and the vehicle return to warehouse then BULOG.

Algorithm 1: the Genetics Algorithm used

1. Generate an initial population Ω of chromosomes
Main GA exploration phase
2. Select two parents P1 and P2 roulette wheel selection;
3. Crossover (P_1, P_2);
4. Evaluate the two resulting children by Splitting;
5. Repeat 2–4 if no child is feasible. Otherwise, select randomly a feasible child C;
6. Improve C by Local Search, with probability P_m;
7. Insert C in Ω to replace a randomly selected individual among the half worst of Ω, if C is not the current worst and C has a distinct fitness value than those in Ω;
8. Repeat 2–7 for N_1 iterations or till N_2 iterations without improving the current best;
End of the main GA phase
9. Restart the main GA phase 2–8 with a partially replaced population, for N_3 phases or till N_4 phases without improving the best solution

---

The overall structure of the GA is illustrated in Algorithm 1. It starts with the generation of an initial population with insertion of good heuristic solutions to be presented. The central part of our GA is an incremental GA exploration phase in which only one chromosome is replaced at each iteration. It starts with the selection of two parents by roulette wheel selection.

Then the next step is applied to generate two child chromosomes. These child chromosomes are evaluated by the exact split algorithm. The selection and crossover operations are repeated till obtaining a feasible child chromosome, i.e. a chromosome for which a feasible split solution exists. With some probability, this feasible child chromosome is further improved by Local Search to be presented. The new child chromosome is inserted in the current population under two conditions: (1) it is better than the current worst and (2) it does not have identical fitness as an existing chromosome.

It has been proven that perverting the diversity of GA population can diminish the risk of premature convergence (Sorensen and Sevaux 2006). A simple and stricter rule is imposed in this paper to keep the diversity of the population, i.e. the fitness of any two feasible chromosomes must be different. For this reason, a child chromosome C is inserted during each main GA phase only if it has a different fitness than existing individuals. Diversity is also checked in the generation of the initial population and the partially replaced population for restart.

The remaining of this Section is devoted the detailed presentation of the fitness evaluation, generation of initial solutions, and local search.

Lai et al. (2010) introduced a hybrid Benders/Genetic algorithm which is a variation of Benders' algorithm that uses a genetic algorithm to obtain "good" sub problem solutions to the master problem. Lai and Sohn (2011) conducted a study applying the hybrid Benders/Genetic algorithm to the vehicle routing problem. Below is a detailed description of the algorithm.

Step 1. Initialization. We initialize the iteration counter k to zero, select initial trial values for the vector of binary variables Y which selects the plants to be opened.
Step 2. Primal Subsystem. We evaluate the value of v(Y) by solving a transportation linear programming problem whose feasible region is independent of Y.
Step 3. Generation of Benders'Cut. We compute a new linear support using the dual solution of the transportation sub problem and increment k by 1.
Step 4. Primal Master system by GA. A trial location plan Y is to be computed by implementing a GA whose solution delivers both a feasible investment plan and a lower bound to the minimal cost for the equivalent program.

    4a Initialization. We initialize the variable Y as a string of binary bit with the position #i corresponding to the plant #i. We generate initial population and their fitness functions are evaluated as well.

    4b Genetic Operations. We perform a standard single-point crossover approach. The mutation operation to guarantee the diversity of the population is performed as well. The current population is replaced by the new population through the incremental replacement method.

    4c Termination. We terminate the GA if no improvement within 100 iterations. The traditional method branch and bound, which were used in the master problem. It will search the solution space in parallel fashion and take advantage of the "easy" evaluation of the fitness function.

3 Computational Result

In this section we present the result of computational experiments for BULOG case with VRPTW BD and Genetics algorithm BD algorithm described in previous section. The proposed method algorithm is coded in Java NetBeans IDE 7.3 on a laptop with Intel(R) Core(TM) i5-3210M CPU @ 2.50 GHz 2.50 GHz, RAM 8.00 GB with System type, x64-based processor, under the windows 8 operating system.

Table 1 Computational result of exact VRPTW—Bender's decomposition

Number of chromosome = 100

Single case different number of iteration	Exact VRPTW—Bender's Decomposition					CPU (seconds)
	Total time delivery	Number of truck	Total cost	% found	
500	3,944 min	256	IDR 11,157,595,500.00	10.37 %	5
600	3,893 min	256	IDR 11,157,519,000.00	10.34 %	6
700	3,887 min	256	IDR 11,157,510,000.00	10.30 %	7
800	3,884 min	256	IDR 11,157,505,500.00	9.97 %	8
900	3,881 min	256	IDR 11,157,501,000.00	9.83 %	10
1,000	3,878 min	256	IDR 11,157,496,500.00	9.80 %	11
...	...	...	...	...	...
3,500	3,767 min	256	IDR 11,157,307,500.00	9.19 %	37
4,000	3,752 min	256	IDR 11,157,307,500.00	9.18 %	43
6,000	...	...	...	...	...
6,080	OOM	OOM	OOM	OOM	OOM

3.1 Parameter Selection

Parameter selection may influence the quality of the computational results. Thus, an extensive computational testing was performed to determine the appropriate value of experimental parameters. The following parameter as follows;

$$C \quad 20, 40, 60, 80, 100$$
$$I \quad 100, 200, 300, 400, 500, 600, \ldots, 6100$$

where, C is a total generate of chromosome and I as number of iteration, and the results indicate that the best parameters C is 100 and I is 4,000 total generate of chromosome.

3.2 Results of Computation

This is the results of computation by coded, and in this section OOM means "Out of Memory", the complete results as follow.

3.2.1 Exact Method

See Table 1.

3.2.2 Meta-Heuristics

See Table 2.

4 Conclusions

We have demonstrated that VRPTW and Benders' decomposition algorithm for solving the BULOG case by Genetic Algorithm method can be accelerated substantially when the master problem is solved. The Benders/GA algorithm is a variation of Benders' algorithm in which, instead of using a costly branch-and-bound method, a genetic algorithm is used to obtain "good" sub problem solutions to the master problem. The computational result by Java shows that the algorithm is effective to solve the BULOG case. The results imply that the algorithm is much more practical when only near-optimal solutions are required. Future work could extend the proposed algorithm to other location problems.

Table 2 Computational result of genetic algorithm—Bender's decomposition

Number of chromosome = 100

Single case different number of iteration	Genetic algorithm—Bender's decomposition			
	Total time delivery	Number of truck	Total cost	% found	CPU (seconds)
500	3,845 min	256	IDR 11,157,600,000.00	10.11 %	8
600	3,863 min	256	IDR 11,157,751,500.00	10.14 %	10
700	3,848 min	256	IDR 11,157,947,000.00	10.08 %	14
800	3,836 min	256	IDR 11,158,133,500.00	9.98 %	15
900	3,830 min	256	IDR 11,158,224,500.00	9.93 %	17
1,000	3,824 min	256	IDR 11,158,315,500.00	9.75 %	17
...	...	...	...	...	...
5,000	3,719 min	256	IDR 11,160,998,000.00	9.13 %	92
5,500	3,695 min	256	IDR 11,161,022,000.00	9.13 %	102
6,000	3,689 min	256	IDR 11,161,513,000.00	9.13 %	111
6,080	OOM	OOM	OOM	OOM	OOM

Acknowledgments This work is partially supported by Prof Kung Jeng-Wang as the advisor; the authors would like to thank anonymous reviewers for their detailed and constructive comments that help us to increase the quality of this work.

References

Lai M, Sohn H (2011) A hybrid algorithm for vehicle routing problems. working paper
Lai M, Sohn H, Tseng T, Chiang C (2010) A hybrid algorithm for capacitated plant location problems. Expert Syst Appl 37:8599–8605
Larranaga P, Kuijpers CMH, Murga RH, Inza I, Dizdarevic S (1999) Genetic algorithms for the traveling salesman problem: a review of representations and operators. Artificial Int Rev 13:129–170
Sörensen K, Sevaux M (2006) MAPM: Memetic algorithms with population management. Comput Oper Res 33:1214–1225
Toth P, Vigo D (2002) The vehicle routing problem. SIAM monographs on discrete mathematics and applications, Philadelphia
Van Roy TJ (1986) A cross decomposition algorithm for capacitated facility location. Oper Res 34:145–163

A Systematic and Innovative Approach to Universal Design Based on TRIZ Theories

Chun-Ming Yang, Ching-Han Kao, Thu-Hua Liu, Hsin-Chun Pei and Yan-Lin Lee

Abstract Continual expansion of the population and its diversity has increased the demand for products that take into account the elements of universal design. Universal design has been widely studied; however, a lack of all-encompassing systematic design processes makes it difficult for designers and engineers to transform universal design intent into product realization. This paper proposes a systematic approach to UD based on TRIZ theories. We begin with a UD assessment to identify design problems, followed by a PDMT analysis to determine possible directions through which to resolve the problems. The directions were mapped directly to Effects for appropriate resolutions. For complex problems, it is suggested that Functional Attribute Analysis be employed to analyze the problems, before searching for resolutions from the effects. A case study was conducted to demonstrate the effectiveness and efficiency of the proposed approach in the design and development of products with greater usability, accessibility, and creativity.

Keywords TRIZ · Knowledge effects · Functional attribute analysis · Universal design

---

C.-M. Yang (✉) · C.-H. Kao · T.-H. Liu · H.-C. Pei · Y.-L. Lee
Department of Industrial Design, Ming Chi University of Technology, 84 Gungjuan Rd, Taishan District, New Taipei, Taiwan, Republic of China
e-mail: cmyang@mail.mcut.edu.tw

C.-H. Kao
e-mail: kaoch@mail.mcut.edu.tw

T.-H. Liu
e-mail: thliu@mail.mcut.edu.tw

H.-C. Pei
e-mail: peimimimi@gmail.com

Y.-L. Lee
e-mail: leeu04xup6@hotmail.com

1 Introduction

Significant improvements in healthcare and the standard of living have increased the average lifespan of Taiwanese leading to a higher proportion of elderly individuals. According to statistics from the Ministry of the Interior (2013), Taiwan qualified as an aging society in 1993. The percentage of elderly people stood at 11.2 % in 2012 and is continually increasing. The aging index has risen by 32 % points in the past ten years and although these figures are significantly lower than the 100 % in developed countries, it is still considerably higher than the 20.6 % in developing countries (Ministry of the Interior 2013). These changes in the structure of the population have led to increased demand for products that are broadly accessible and flexible. In endeavoring to satisfy user requirements, product designers must remain aware of changes to the physical senses and psychological factors of users. It is only through the appropriate application of universal design (UD) that products that are adaptive, assistive, and accessible can be produced. Unfortunately, many current products based on UD do not fully meet the actual needs of users. Designers must be able to anticipate the needs of the end user, considering that user experience in interacting with a product determines its success or failure. Designers should also consider whether the product satisfies the principles and concepts emphasized in UD. Therefore, this study endeavors to develop an appropriate methodological model that could be referred to by designers or other professionals in the process of product design.

This research proposed an innovative, systematic approach to Universal Design based on TRIZ theories. The proposed approach started with a UD assessment to locate the design problems, followed by a PDMT analysis to generate potential problem solving directions. The directions were mapped directly to Knowledge Effects for appropriate resolutions. For complex problems, Functional Attribute Analysis is suggested to help analyze the directions, before searching for resolutions from effects. Finally, a case study was provided.

2 Literature Review

2.1 Universal Design

The concept of universal design was first proposed in 1970 by Ronald L. Mace. Although initially focused on improving the environments inhabited by physically disabled people, UD was later expanded to include broader design concepts in the development of products and services for use by as many people as possible. In recent years, interest in UD has grown and a number of researchers have expressed their viewpoints on the concept. Steinfeld and Danford (2006) indicated that UD is a normative concept used in the design of products, environments, services, and communications systems. Nieusma (2004) stated that designers should consider what types of design can be universalized to accommodate a wide audience, including disadvantaged or disabled groups. To define and clarify UD, the Center for Universal Design at North Carolina State University proposed the most representative and frequently quoted/applied principles of UD, serving as the means to assess the accessibility of current products and incorporate the UD philosophy into the early stages of design (Center for Universal Design 1997). Based on the seven UD principles, Nakagawa (2006) developed an evaluation system, called Product Performance Program or PPP, comprising 37 guidelines for research on UD products. He also proposed three supplementary principles to compensate for the inadequacies of the existing seven principles. These principles provide definitive criteria and objectives in the design of accessible products.

2.2 Triz

TRIZ is the Russian acronym for the "Theory of Inventive Problem Solving" and proposed by Genrich Altshuller. By studying a large number of patents, Altshuller and his research team formed the fundamental principles of TRIZ theory. In studying the evolution of innovation and invention, Altshuller (1999) proposed that humans frequently encounter three types of obstacles in their thought processes: psychological inertia, limited domain knowledge, and the trial and error method. As a reliable innovations theory, TRIZ is an important tool that enables one to progress beyond established paths of thought without becoming lost in a mental maze. TRIZ is a problem-solving instrument capable of systematizing and generalizing problems (Terninko et al. 1998; Busov et al. 1999; Mann 2007). It includes situational analysis, contradiction analysis, substance-field analysis, the ARIZ problem-solving system, 40 inventive principles, separation principles, the evolution patterns of the technical system, and scientific/technological effects. Although TRIZ includes a wide range of techniques, its main problem solving method focuses on Functionality, Contradiction, Ideality, Resources, and Shifting Perspective (Domb 1999; Ideation International 2006; Mann and Hey 2003).

2.3 Knowledge Effects and Functional Attribute Analysis

Altshuller claimed that the best solutions generally involve the incorporation of knowledge from other domains; therefore, he proposed an enormous effects database, including physical, chemical, and geometric characteritics, to accomplish this purpose (Altshuller 1984; Mann and Hey 2003). Stimulating innovation requires interdisciplinary knowledge, and the point of an effects database is to facilitate thinking that transcends current views by considering whether the problem at hand is similar to other problems solved in the past or in the various domains (Mann 2002; Litvin 2005). A relatively small number of knowledge effects applications, such as CREAX's effects database, have been developed; however, no comprehensive application procedure has been established. As a result, the maximization of knowledge effects has yet to be realized.

Functional attribute analysis (FAA), first proposed by Miles in 1960, is a systematic approach to describing the relationships among system components, particularly those with complex functions. Mann (2002) claimed that establishing the necessary functions and attributes of the current system should be the first step in innovative design and provide a number of different perspectives to facilitate the derivation of useful rules. FAA helps to define the relationships among system components and can overcome the influence of time-dependent issues during analysis so that users can reveal the core of the problem (Sung 2009). After identifying the functions or attributes required by the system, the TRIZ effects database is then used to generate solutions capable of achieving the objectives (Sung 2009).

3 A TRIZ-Based Product Design

This study established a systematic process of product development and innovation using TRIZ as the core. The distinguishing feature of the proposed approach is its provision of a comprehensive set of procedures to facilitate the development of products using FAA and the effects database of TRIZ tools. The detail of this TRIZ-based systematic UD process is described as follows.

Step 1. This study first administered a PPP questionnaire to rate the target product, to identify the universal principles most in need of improvement as problem points. Initially, it is preferable to have the participants actually operate the product before filling out the PPP questionnaire. The questionnaire contains 37 question items in which satisfaction is expressed using a Likert scale with the scores of 40 points, 30 points, 20 points, 10 points, and 0 points; the highest 40 points indicating the most satisfaction and the lowest 0 points indicating the least. The UD principle with the lowest score indicates the area in which the product is least universal. This principle is then set as the design target and incorporated into the process of analyzing the PDMT problem in the following step.

Step 2. We then applied the four processing steps to propel the problem-solving process forward sequentially. Purpose, direction, method, and tool (PDMT) is a preliminary problem-solving approach used to extrapolate the direction of design development through deduction (Chen 2008). Targeting the product requirements requires applying the first three phases of PDMT to deduce the purpose (P), direction (D), and method (M). Possible directions are then determined according to the purpose to serve as a preliminary design reference.

Step 3. This step involves seeking a solution. One method is to match the results of the PDMT analysis in the previous step directly to the function database to search for feasible or usable solutions one at a time. Thus, a manual search of the database is used to derive ideal solutions. Another approach is to first conduct FAA and then match the components, functions, and attributes to the effects database. With consideration of time-dependent scenarios, the problem can be divided into both static and dynamic scenarios. After initiating analyses of the components and writing process of the problem system, the key components and functions of the system are used to identify the necessary functions, the attributes to be manipulated, and the improvements to be applied to the system. Finally, these are matched to the online effects database to search for a solution. This method is significantly less time-consuming and makes more effective use of the effects tools.

Step 4. Finally, the solution derived from the effects database is applied to the design concept in order to create an innovative product with UD.

4 Results and Discussion

4.1 Product Design Analysis

The research subject in this study was a regular ballpoint pen. From the pen questionnaire, we perceived that the UD principle of "Low physical effort" is the main shortcoming in pen design. We applied "Low physical effort" for PDMT, subsequently determined four possible directions for the problem and proceeded to generate the eleven terms. The results of PDMT deduction are displayed in Table 1. We then matched the methods discovered by PDMT directly to the effects

Table 1 Deduction results from PDMT

UD principle	Purpose	Direction	Method
Low physical effort	Not be fatigue in long-term used	Weight	Material
			Friction
			Plasticity
		Ergonomics	Size and shape
			Change in shape
			Pen grasp
		In use	Not to used paper
			Not to used pen
		Alert	Vibration
			Sound
			Time to reminder

Table 2 Matching results from the function and attributes databases

Directions	Methods	TRIZ Tool: knowledge effects
Weight	Material	/
	Friction	F13 control friction
	Plasticity	F16 energy transfer
Ergonomic	Size and shape	F19 change size of object
	Change in shape	F23 modify volume
	Pen grip	/
In use	Not to used pen	F06 control displacement of object
	Not to used paper	F13 control friction
Alert	Vibration	E47 resonance
	Sound	/
	Time to reminder	F01 measure temperature

Table 3 Analysis of writing tool component

Function attribute analysis: components of writing tool	
Storage space	Desk, pen holder, pencil case, shirt pocket
Pen grasp	Normal tripod grasp, thumb wrap, cross thumb grasp, static tripod grasp, thumb tuck grasp, thumb wrap
Pen	Tip, refill, shaft, casing, ball, ink tube, plastic grip, cap
Writing surface	Horizontal, vertical

database to seek a suitable effects solution. The results of effects matching are presented in Table 2. The directing matching method quickly provides an effective solution; however, the results are also influenced by personal experience. When one wishes to think outside the box or when the problem is more complex, FAA can be applied. After analyzing the attributes and functions of a pen, we divided the process of pen usage into three stages: the grasping of the pen, pressing the end to eject the pen point, and writing with the pen on paper. The third was most directly related to the problem that we wanted to solve. Through FAA, the results are shown in Table 3. We then analyzed the dynamic process of users using pens, which require them to apply force to write. To achieve the process of writing, the component applying the force is the hand, the component subjected to the force is the shaft of the pen, and a holding action is used to grip the solid shaft. We thus matched the function of hold with the status of the shaft, solid, and derived 20 solutions from the online effects database (CREAX 2013), including adhesive, adsorption, hot-pressing, and osmotic pressure.

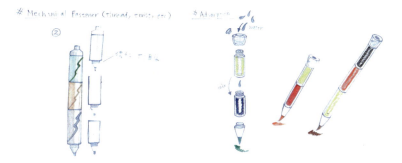

Fig. 1 Pen with replaceable casing (*left*) and palette pen (*right*)

4.2 Application of Effects Principles to Pen Design

From the solutions obtained from the effects database, we selected two principles, which are Mechanical Fastener and Adsorption, for the development of design concepts.

- Application of Mechanical Fastener in pen design—pen with replaceable casing (Fig. 1): Users can select from among a number of casings of different shapes according to their writing habits and type of grasp. Using screw patterns on the casing and shaft, users could assemble their pen at different angles to prevent issues of spinning and fatigue.
- Application of Adsorption in pen design—palette pen (Fig. 1): This study replaced the traditional ink used in pen refills with a powdered substance to alleviate the issue of ink smearing on the hands or paper during writing. Before using this pen, users inject water into the pen to absorb the color. Users can select shafts with different colors and mix a wide range of colors when combining them.

5 Conclusion

This study applies TRIZ tools to UD principles to construct an innovation process that is more systematic, efficient, and easier to use. This process deconstructs a design problem by means of identifying its goals, directions, and initial steps as well as the system's components, attributes, and functions. Thus, the "designers" of other fields may also use this process to solve their own problems. The case study demonstrates that the TRIZ tools offer some level of step-by-step procedures in the application of UD principles to product development. The Effect Database used in this study contains knowledge from many different disciplines, thus allowing the designers to "think outside of the box" and be "inspired" by these principles and knowledge. For professionals in the domain of product design and development, this model provides fresh stimulus and creativity, helping them to design more innovative, flexible, and assistive products.

Acknowledgments The authors are grateful for support of the National Science Council, Taiwan under grant NSC101-2221-E-131-002-MY2.

References

Altshuller G (1984) Creativity as an exact science. Gordon and Breach, New York
Altshuller G (1999) The innovation algorithm: TRIZ, systematic innovation and technical creativity. Technical innovation center, Worcester
Busov B, Mann D, Jitman P (1999) Case studies in TRIZ: a novel heat exchanger. TRIZ J. http://www.triz-journal.com/archives/1999/12/. Accessed 9 Jan 2012
Chen CH (2008) Introduction to TRIZ and CREAX. Workshop (Pitotech Co., Ltd.). 2 Oct 2008
CREAX (2013) Function/Attribute database. https://www.creax.com. Accessed 13 Mar 2013
Domb E (1999) Tool of classical TRIZ. TRIZ J. http://www.triz-journal.com/archives/1999/04/default.asp. Accessed 13 Jan 2012
Ideation International (2006) History of TRIZ and I-TRIZ. http://www.ideationtriz.com/history.asp. Accessed 21 Nov 2012
Litvin SS (2005) New TRIZ-based tool—function-oriented search (FOS). TRIZ J. http://www.triz-journal.com/archives/2005/08/04.pdf. Accessed 24 Nov 2012
Mann D (2002) Axiomatic design and TRIZ: compatibilities and contradictions. Paper presented at the second international conference on axiomatic design. Cambridge MA, 10–11 June 2002
Mann D (2007) Hands-on systematic innovation. Tingmao, Taipei
Mann D, Hey J (2003) TRIZ as a global and local knowledge framework. TRIZ J. http://www.triz-journal.com/. Accessed 20 Nov 2012
Ministry of the Interior (2013) Population structure. http://www.moi.gov.tw/stat/news_content.aspx?sn=7121&page=1. Accessed 9 Mar 2013
Nakagawa S (2006) Textbook for universal design. Longsea press, Taipei
Nieusma D (2004) Alternative design scholarship: working toward appropriate design. Des Issues. doi: 10.1162/0747936041423280
Steinfeld E, Danford S (2006) Universal design and the ICF. Paper presented at the promise of ICF conference on the living in our environment, living in our environment, Center for Inclusive Design and Environmental Access (IDEA Center), Vancouver, 4–5 June 2006
Sung MH (2009) Systematic innovation: an introduction to TRIZ. Tingmao, Taipei
Terninko J, Zusman A, Zoltin B (1998) Systematic innovation-introduction to TRIZ. CRC Press, USA
The Center of Universal Design (1997) http://www.ncsu.edu/project/design-projects/udi/. Accessed 3 Nov 2012

Wireless LAN Access Point Location Planning

Sung-Lien Kang, Gary Yu-Hsin Chen and Jamie Rogers

Abstract With the fast-growing demand for mobile services, where to place the access points (APs) for providing uniformly and appropriately distributed signals in a wireless local area network (WLAN) becomes an important issue in the wireless network planning. Basically, AP placement will affect the coverage and strength of signals in a WLAN. The number and locations of APs in a WLAN are often decided on the basis of the trial-and-error method. Based on this method, the network planner first selects suitable locations to place APs through observation, and then keeps changing the locations to improve the signal strength based on the received signal. Such process is complicated, laborious and time-consuming. To overcome this problem, we investigate the back-propagation neural network (BPNN) algorithm to improve over the traditional trial-and-error method. Without increasing the number of APs, our approach only needs to adjust AP locations to overcome weak signal problems and thus increase the signal coverage for the Internet connection anywhere within the area. In our experiment, we established a WLAN on a C campus. Our experiments also indicate that placing APs according to the BPNN provided better signal coverage and met the students' demands for connecting to the Internet from anywhere in the classroom.

Keywords Access point placement · Wireless LAN planning · Back-propagation neural network · Threshold

---

S.-L. Kang (✉) · G. Y.-H. Chen
Industrial and Systems Engineering, Chung Yuan Christian University, Chung Li, Taiwan, Republic of China
e-mail: slkang@mail.chihlee.edu.tw

G. Y.-H. Chen
e-mail: yuhsin@cycu.edu.tw

J. Rogers
UT-Arlington, Industrial and Manufacturing Systems Engineering, Arlington, TX 76019, USA
e-mail: jrogers@uta.edu

1 Introduction

With the advent of network and mobile technology era, the public have become more and more relying on the electronic devices, such as personal device assistant (PDA), laptop computers, tablet PC, and smart phones. The traditional fixed landline networks have no longer met most people's needs; mobility, always-on-demand, and wide band-width are the essential characteristics of latest trends in network connectivity. Therefore, the craving for the wireless local area network (LAN) has grown in leaps and bounds, resulting in the demand for high quality of signal strength and reception. The IEEE 802.11 protocol, a set of standards for implementing wireless local area network, with the flexibility and integration with the current network setting, is becoming the primary choice of network technology for schools, corporations, and public areas (IEEE 1997). Case in point, according to the network statistics in May 2011 at C School with an estimated total student population of 10,000, one out of two students accessed the wireless network each day. From the case, it is quite obvious that the wireless network demand from the students is quite strong.

Although wireless networks have numerous advantages over the traditional fixed line networks, the signal propagation in a wireless environment is significantly more complex. Among those the signal interference, signal fading, and path loss are some issues that network planners must carefully consider when it comes to the planning and configuration of wireless networks. The insulation materials in buildings, interior layout settings, and surroundings can easily lead to the exposure to those types of issues. The setup layout of access points (APs), a device that allows wireless devices to connect to a wired network such as WiFi, would impact the financial costs, number of people accessing the wireless network, and the coverage areas. Typically, the information technology specialists or network planners would place more than one access point on each floor of buildings to ensure the network coverage is adequate. However, the network planners typically decide number of APs and selection of locations based on their preferences or experiences, or so-called "trial-and-error" method. The traditional approach is to select the coverage area of a particular signal as the benchmark to determine the AP number and their locations (Cheung 1998; Chiu 1996; Panjwani 1996). Due to the structural considerations of buildings and surroundings, the network planners need to consider several layout options for placing APs in order to improve the network coverage.

This paper is divided into five sections. Section 1 provides the introduction, background as well as the motivation. Section 2 discusses the related work and literature review; this section can be further divided into two sub-sections: (Sect. 2.1) the application of wireless network applications and standards, (Sect. 2.2) the concepts relevant to the solution proposed in this study. Section 3 provides the case under study, constraints and the environment setting. Section 4 presents the result analysis from the experiment. Section 5 summarizes and concludes the entire research and proposes future work.

2 Related Work

2.1 Wireless LAN and Access Point

The Institute of Electrical and Electronic Engineers (IEEE) in 1997 defined the wireless LAN in the 802.11 technical specification (IEEE 1997). At the beginning the specification only covered 2 Mbps, and extended to both 802.11b at 11 Mbps and 802.11a at 54 Mbps later in 1999. The frequency bandwidth covered by 802.11 Wireless LAN belongs to the Industrial Scientific Medical Band (ISM) high frequency bandwidth in the frequency spectrum. 802.11 covers the frequency bandwidth (2.4–2.4835 GHz); 802.11b covers the frequency bandwidth (2.4–2.4835 GHz); 802.11a encompasses the frequency bandwidth (5.150–5.850 GHz). In Comparison of both 802.11a and 802.11b, the latter is poor in terms of signal penetration and high in costs. Consequently, in 2003 the third version of specification, 802.11 g covering bandwidth 54 Mbps (2.4–2.4835 GHz), has been brought to fruition. Even with high throughput at 54 Mbps, the WLAN standards around that time were half-duplex and with high overhead in the packet data; the WiFi network can easily reach its maximum throughput for downloading a file around 17–18 Mbps. Therefore, IEEE in 2009 came up with the latest version 802.11n for handling bandwidths from 54 to 300 Mbps, allowing the throughput of wireless LAN to be at least as high as the fixed land-line (around 100 Mbps) (Table 1).

The wireless LAN in general consists of two types of equipments: wireless transceiver and access point (AP). The wireless transceiver is a laptop or mobile device with WiFi capability. The main components of an AP include a wireless transceiver and fixed land-line network card. The primary function of an AP is the bridge or interface between the wired and wireless networks. All wireless transceivers must be connected to the access point in order to access the services from the wired networks.

The traditional approach to select the suitable location for the access point is done firstly through the observation. Next, after several runs of measuring the signal strength of the WLAN, the network engineers applied to the algorithms to

Table 1 Wireless LAN 802.11 specifications

Protocol	Release	Frequency bandwidth (GHz)	Actual throughput (average) (Mbit/s)	Actual throughput (maximum) (Mbit/s)	Indoor range (m)	Outdoor range (m)
802.11	1997	2.4–2.5	1	2		
802.11a	1999	5.15–5.35/5.47–5.725/ 5.725–5.875	25	54	30	45
802.11b	1999	2.4–2.5	6.5	11	30	100
802.11 g	2003	2.4–2.5	25	54	30	100
802.11n	2009	2.4 or 5	300	600	70	250

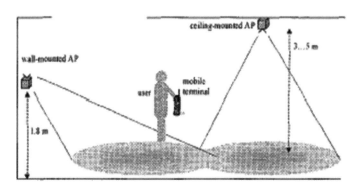

Fig. 1 Impact of deployment methods on the coverage area (Unbehaun and Zander 2001)

determine the network layout. The entire process is overly repetitive, tedious and inefficient (Kamenetsky and Unbehaun 2002). Hills (2001) proposed another method that could be applied for quick deployment and very suitable for indoor use. However, the method could not handle the problem of obstacles (such as a walls or trees). Wang (2003) analyzed the characteristics of AP and relationship between the obstacles and signals and came up with three AP deployment methods: 90-degree angle split deployment method, 60-degree angle split method and various half-diameter ratio method. Those methods provide the guidelines for quick access point deployment without going through tedious procedures and measurements, resulting in cost saving and quality improvement.

The main concern over the placement of access points is the signal separation of the access points—the network engineers strive for wide coverage area from each AP. The typical place for AP is at the highest point of a space. For the indoor deployment of AP is normally at the highest point directly overhead. With the signal projecting from top to the bottom, the coverage area naturally forms a circle as shown in the Fig. 1. In terms of signal quality and reliability, the ceiling-mounted AP will score higher than the wall-mounted AP (Unbehaun and Zander 2001). Nevertheless, due to other factors such as easy access, wiring, appearance and so on, the network engineers may choose to mount the AP on the wall instead.

2.2 Artificial Neural Network

In the past some researchers have applied the artificial neural network (ANN) and other heuristic methods to investigate the impact factors on the signal quality of wireless LAN. In this study the authors applied the neural network approach to investigate the effective placement of APs in order for the mobile devices to get connected to the networks. In the following description, the basic introduction on the concept of artificial neural network will be provided.

The artificial neural network was first proposed by two mathematicians, McCulloh and Pitts, in 1943. Modeled after human's brain, the neural network is consists of interconnected nodes or neurons with inputs and output and can be used for information processing or computation. In 1957 Perceptron was developed in artificial neural network for solving classification problems. In 1969 the ANN entered the "dark period" when Perceptron was proven to solve only the linear classification problems. Not until 1985 have the interests in the neural network picked up again when the Hopfield and back-propagation network were introduced. Through the application and modification of hidden layers and units, the ANN can be used to solve more complex problems. The neural network can be divided into two types: supervised and unsupervised learning. To evaluate the quality of the neural network, the indicator, root mean square error (RMSE), is typically used; the smaller RMSE is, the better is the quality of the neural network. Before selecting the neural networks, the RMSE of a training sample is first analyzed. If the RMSE of the training sample cannot indicate the quality of the network, the RMSE of the testing sample must also be taken into consideration. The objective is to have small yet converged RMSEs of both training and testing samples to reflect the actual scenario. Figure 2 illustrates an artificial neural network.

3 Case Study and Methodology

In this research, the researchers studied the case of *C school* as an example for wireless LAN deployment. The general classroom floor plan on campus is shown in Fig. 2. Classroom dimensions are 8.8 m long and 7.3 m wide. The classroom layout and structural materials are just like typical classrooms. General furniture including study desks and chairs are placed in the classroom. There is no obstruction above the desks and chairs. Flanking the classrooms are several large pane windows which allow high signal penetration. The front and back of the classroom consists of walls 0.6 m thick. Currently, there are six classrooms on each floor of a particular building on campus. Two wireless APs are mounted 3.3 m above the classroom hallway at the top of a column 2.7 m away from the classroom for students to access the Internet. Although the current wireless signals

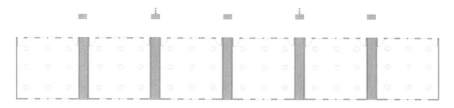

Fig. 2 The floor plan of classroom

are sufficient to cover six classrooms, existing blind spots still block students' access to the Internet anywhere in the classroom. The researchers wished to improve the signal coverage through an efficient selection of wireless AP locations and minimum required APs.

The researchers detected the wireless AP signal strength with the WiFi monitoring software, WirelessMon, and designated 9 detection points in each classroom. The researchers also set a signal strength threshold τ and designates if the signal strength is less than the threshold, the wireless communication cannot be established (Fig. 3).

$$\forall k \in A, Si(k) \geq \tau \qquad (1)$$

where

A	the experimental area,
τ	signal strength threshold,
$Si(k)$	signal strength at the k point,
k	the coordinates above the signal strength threshold.

The experiment can be divided into two scenarios. The first scenario is to simulate the actual testing objective function value through the back propagation network. The researchers divided 55 data sets into training and testing sets. Three input variables are considered: distance, door and wall obstruction coefficients. The output variable is the signal strength. By varying the hidden layers as well as the two parameters, learning rate and momentum, the researchers expected to find a pattern in which the RMSE is within the acceptable range.

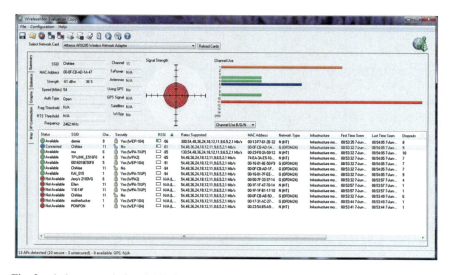

Fig. 3 wirelessmon wireless LAN detector

Table 2 Results after the experiment

AP location	Structure	Learning rate	Momentum	Sample size	Training RMSE	Testing RMSE	Over threshold	Under threshold
(9.7, −2.7) (−9.1, −2.7)	(3, 4, 2, 1)	0.1	0.9	14	0.032	0.026	41	14
(−13.8, −2.7) (14.4, −2.7)	(2, 5, 3, 1)	0.1	0.8	17	0.074	0.073	51	4

In the second scenario, the researchers moved the AP placements, hoping to the change will improve overall signal strength and help pass the threshold. Due to the factors such as the appearance, the researchers could only place APs on 5 designated wall columns; from those 5 columns, the researchers tried to determine the effective AP deployment locations.

4 Result Analysis and Discussion

Based on the aforementioned experimental assumption, the authors fed the data and actual signals into the Q-NET software which executed the back propagation network to train a model that reflected the real-world scenario. The threshold value was set at −80 dBm; if the signal strength exceeded the threshold value, the wireless communication is not established; otherwise, the wireless communication is established. The experimental results are shown in Table 2.

The table provides the before and after AP locations and their related data. The structure is set according to the whole neural network model: input node and first hidden layer node, second hidden layer node and output node. The first three are set empirically to allow more accurate prediction. In regards to the training and testing values, the researchers divide the data set into 2:1 or 7:3 ratio based on existing literatures. The ANN simulation showed that 41 data passing the threshold with AP location before the location update and 14 data failed to pass the threshold; the entire RMSE is 0.032. After the AP location update, 51 passed the threshold and only 4 failed; the entire RMSE is 0.074.

5 Conclusion

The research aims at improving the classroom AP deployment at C school to provide students a coverage area with sufficient signal strength for network connection. At the hallway and classrooms flanked with pan windows, the proposed wireless AP placement can achieve better network coverage with sufficient signal strength for network connection based on the ANN simulation.

In this study the researchers collected the data from the standard wireless AP with the transmitted (TX) power at 3 dB; for future work, APs with powers at other dB values can also be investigated. Additionally, how to find the balance point between the number of APs and the construction costs is also something to be pondered about.

Acknowledgments This work is partially supported by the grant NSC 100-2221-E-033-030. The authors are grateful to the anonymous reviewers for their comments.

References

Cheung KW (1998) A new empirical model for indoor propagation prediction. IEEE Trans Veh Technol 47(8):996–1001

Chiu CC (1996) Coverage prediction in indoor wireless communication. IEICE Trans. Commun. 79(9):1346–1350

Hills A (2001) A large-scale wireless LAN design. IEEE Commun Mag 39(11):98–107

IEEE (1997) WIRELESS LAN medium access control(MAC) and physical layer(PHY)specification. IEEE STD 802.11 IEEE Society, Piscataway, NJ, USA

Kamenetsky M, Unbehaun M (2002) Coverage planning for outdoor wireless LAN systems. In: IEEE International Zurich Seminar on Broadband Communications, Zurich, Switzerland. IEEE Society, pp 491–496

Panjwani MA (1996) Interactive computation of coverage regions for wireless communication in multifloored indoor environments. IEEE J Sel Areas Commun 14(3):420–429

Unbehaun M, Zander J (2001) Infrastructure density and frequency reuse for user-deployed wireless LAN system at 17 GHz in an office environment. IEEE Int Conf Commun 2535–2539

Wang CT (2003) IEEE 802.11 WLAN access point (AP) deployment. Master, National Central University, ChungLi, Taiwan

The Parametric Design of Adhesive Dispensing Process with Multiple Quality Characteristics

Carlo Palacios, Osman Gradiz and Chien-Yi Huang

Abstract In recent years, electronic industry trend moves toward miniaturizing its electronic components. Nevertheless, the functionality of these electronic components has grown stronger in order to keep up with the world's technology advances and demand. The electronics manufacturing companies are always searching for new and better ways to make their products more efficiently, while protecting the environment in accordance to the restriction of hazardous substance (RoHS) directive, which was announced by the European Union. However, the implementation of such directive generated some setbacks in the electronics assembly process with surface-mount technology. This paper focuses on the quality performance of the component's glue adhesion strength, considering multiple quality characteristics such as the vertical and horizontal thrusts at low and high temperatures. The Taguchi method assisted in designing and implementing the experiments. Process factors considered include the dispensing position, thermoset temperature and reflow conveyor speed. The multi-criteria analysis methods, the order preference by similarity to the ideal solution (TOPSIS) together with the principal component analysis (PCA), are employed to analyze the experimental data acquired. The optimal process parameters are thus determined. Lastly, a confirmation test is conducted to verify the results of the optimal process scenario.

Keywords Quality control · Parametric design · Surface mount technology

---

C. Palacios (✉) · O. Gradiz · C.-Y. Huang
Department of Industrial Engineering and Management, National Taipei University of Technology, 1, Sec. 3, Zhongxiao E. Rd, Taipei 10608, Taiwan, Republic of China
e-mail: Sless_28@hotmail.com

O. Gradiz
e-mail: osman1430@hotmail.es

C.-Y. Huang
e-mail: jayhuang@ntut.edu.tw

1 Introduction

1.1 Background

The modern civilized society; electronic products have become an integral part of human life. In recent years, the global electronics industry have had a rapid development and progress, the manufacturers are committed to product innovation and process improvement. China's electronics industry has gradually become in the world's main production and supply, and thus led to the middle reaches of the electronic components industry and the vigorous development of the upstream raw material.

The entire information, communication, and consumer electronics industry, Printed Circuit Board (PCB) is indeed one of the indispensable components, all electronic products are required to be used PCB, The application areas including information, communications and consumer products, the main application of products including TV, VCR, computer peripherals, fax machines, notebook computers, mobile phones and chipsets.

The electronic products in recent years have been moving towards the light, thin, short, small, meaning even more functions are placed in the same area of the circuit board, to maintain the same functionality, but the area is narrow. In response to market demands, the development of the miniaturization of electronic components lead to the developed of a surface-mount technology (SMT) to replace the traditional parts plug-in process (the Dual the Inline Package; DIP).

1.2 Rationale

The Restriction of Hazardous substance (RoHS) directive was announced in 2003 by the European Union; which gradually made lead-free solder replace the traditional lead based solder paste, On the other hand the electronics industry's technical factors were subject to limitations, the welding technology face lead-free reliability problems. For example, SAC305 (Sn96.5 %/Ag3.0 %/Cu0.5 %) lead-free environmentally friendly solder makes easy the SMT process temperature significantly elevated, which cause the circuit plate board-bending phenomenon; using a tin–bismuth alloy (Bi 58 %/Sn42 %; also known as low-temperature lead-free solder paste) can contra rest the plate bending phenomenon, but the tin–bismuth alloy with hard and brittle nature restricts its use only to less reliability and more flexible requirements consumer electronics product. Using low temperature solder paste and the dispensing baking reflow process causes reflow temperature decreases and may affect the adhesive strength of the solder paste. Therefore, to break through the technical aspects of these limitations, there is need of in-depth understanding of the product composition (raw materials, alloy and process set) and preventive measures and parametric design.

1.3 Objectives

- Use the technique for order of preference by similarity to ideal solution (TOPSIS) and Principal component analysis (PCA) statistical methods, to determine the optimal process parameters combination.
- In accordance with the best combinations of parameters from the methods applied, present a set of test samples, simultaneously conduct a series of thrust experiments and obtain from each of these a set of optimal parameters combination.
- Direct the process' optimal parameters combination to conduct a experiment confirmation data analysis, to verify whether or not the outcome of measuring the quality characteristics fall within the confidence interval, to asses if the conclusions are correct, that is to see if the adhesive dispensing process parameters combination make it difficult for the components to detach.

1.4 Research Structure

To establish research direction, we start by using certain tools of the Taguchi method into planning and execution experiment. Analyzing the data, we prepare the information in order to use the technique of order of preference similarity to Technique for Order of Preference by Similarity to Ideal Solution (TOPSIS) with principal component analysis (PCA) to determine the optimal parameter combination, accordingly prepare sample tests of the optimal process parameter combination, consequently conduct a confirmation experiment to assess its quality characteristics optimal process parameter combinations, finally conclusions and recommendations.

2 Experiment Planning and Analysis

2.1 Experiment Preparation

The Taguchi's method orthogonal array tool will set the best combination array between the four factors and the three different levels to analyze and also signal to noise ratio will normalize data to make data less sensitive to noise. The factors to take into account, the first one is the Dispensing position: Consider the glue concentration at the bottom of the components. If the glue is slightly exposed at the bottom side of the components, the components side could withstand external force better. The second factor is the crimping time: increase the pressing time of the components in the dispensing process to ensure the full contact of glue and components. Crimping long time may result in the glue overflow at the bottom of the components. Set the time in 0.00 s, 0.01 s, and 0.02 s. Thirdly the Glue thermosetting temperature: select the low-temperature of solder paste melting

Table 1 Factors and levels

Factors/Levels	Level 1	Level 2	Level 3
Dispensing position (DP)	The element is covered	1/3 Expose	1/2 Expose
Crimping time (CT)	0.00 s	0.01 s	0.02 s
Thermosetting temperature (TT)	140 °C	150 °C	160 °C
Reflow conveyor speed (RCS)	1 M/min	1.15 M/min	1.3 M/min

point of 138 °C, the reflow of thermostat temperature zone are chosen in the 3 different levels 140, 150, and 160 C, respectively. And lastly the Reflow oven conveyor speed: conveyor speed will affect the time of the PCB stay in the reflow zone and affect the glue adhesion strength after baking. Table 1 has a more visual appreciation of how the factors and levels are interconnected.

2.2 Experiment Process

Initially we must prepare and produce the experimental sample of the Printed Circuit Board (PCB) using the Surface Mount Technology (SMT) process Fig. 1, which in itself consists of four manufacturing process. Primarily we have to set the Glue point position for adhesive components using the machine's computer, secondly Component adhesive dispensing, thirdly the placement of the components on to the PCB and to finish we have the reflow of PCB in the oven.

2.3 Thrust Measurement Specification

Vertical and horizontal thrust (VTh, HTh) will be conducted for the components in room temperature and high temperature (RT, HT) using a hand held thrust-meter Fig. 2.

Fig. 1 PCB preparation

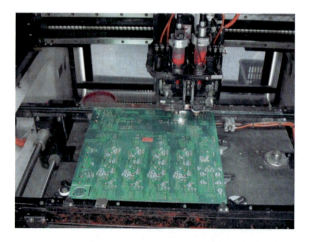

Fig. 2 Thrust measurement. *Angle* maintain 30 degrees relative to the board. *Temperature* room temperature was controlled at 20–25 degrees. Oven temperature was set at 210 °C for 90 s (temperature 170–190 °C)

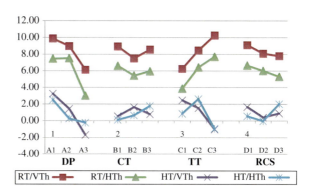

Fig. 3 Reaction chart

2.4 Thrust Experimental Data Analysis

Using Taguchi's signal to noises ration bigger the better formula given by Eq. 1

$$S/N = -10\log\left[\frac{1}{n}\sum_{i\sim 1}^{n}\left(\frac{1}{y_i^2}\right)\right] \quad (1)$$

Which was the appropriate formula for this research to gain the most robust ratio; the data acquired on the experiments is analyzed and processed according to the established steps to find the S/N ratio for each of our relevant quality characteristics where we obtain the data that will be the starting point for our parameter optimization. In addition to the reaction of each quality characteristic for the four factors reflected on Fig. 3.

3 Parameter Optimization

3.1 TOPSIS

First the signal to noise ratio data table is represented in a matrix form, and then we construct the normalized decision matrix, transforming the various attributes dimension into non-dimensional attributes, which allows comparison across the attributes. The Third step is to construct the weighted normalized decision matrix using the given weights, for room temperature thrust: 0.17 and High temperature thrust: 0.33. Next we determine the positive and negative ideal solution, after doing so we must determine the relative closeness to the ideal solution. To finalize we establish the optimal parameter combination using the orthogonal array of Taguchi method.

3.2 PCA

To begin we must Transform a number of posible correlated variables into a small number of uncorrelated variables to do this we must first standarize the S/N ratio data, secondly we need to obtaining a correlation coefficient matrix, accordingly the characteristics value $[\lambda]$, and eigenvalues $[C]$ are attained. Followed by obtaining a principal component point $[\Phi\ ip]$ and multi-quality characteristics indicator $[\Omega\ i]$ MPCI. To finish we determine the optimal parameter combination using the orthogonal array of Taguchi method.

3.3 Optimal Process Parameters

The combination of the optimal process parameters obtained in the two kinds of analysis methods (TOPSIS and PCA) are shown in Table 2, further the production of four PCB samples, and the foregoing steps for dispensing adhesive strength thrust test.

Table 2 Optimal process parameters

	Dispensing position	Crimping time (s)	Thermosetting temperature (°C)	Reflow conveyor speed (M/min)
TOPSIS	The element is covered	0.02	150	1.3
PCA	1/3 Expose	0.00	160	1

3.4 Confirmation Experiment and Analysis

According TOPSIS and PCA the combination of the optimal process parameters obtained, the production of four PCB samples to confirm the experiment. Consequently we predict the optimal conditions for the SN ratio, in order to not overestimate optimal conditions, we will only use the factors of strong effect and use the square of the weak factors as merge error, obtaining Table 3. Accordingly In order to effectively estimate the value of each observation, we must verify the prediction of the quality characteristics of the S/N ratio with a 95 % confidence interval using Eq. 2.

$$CI_1 = \sqrt{F_{\alpha:1,v_2} \times V_e \times \left(\frac{1}{n_{eff}}\right)} \qquad (2)$$

where,
$F\alpha{:}1{,}v2$ F value with significance level α,
α Significance level, confidence level $= 1-\alpha$,
V2   Pooled Error Variance degrees of freedom,
Ve   (Pooled Error Variance) Obtained through ANOVA analysis,
n_{eff} Valid observations.

The results show that the quality characteristics of the measurement results falls within the confidence interval, since the impact of the addition of the control factor model has established parameters, and there was no significant interaction. The analytical data is shown in Table 4.

Table 3 Total SN ratio

Factors	SN
Room temp/Vertical thrust	11.8167
Room temp/Hori. thrust	9.21407
High temp/Vertical thrust	4.72090
High temp/Hori. thrust	4.28661

Table 4 Confidence interval compliance

Quality characteristics (push force measured)	Predictive value (S/N ratio)	95 % confidence interval	TOPSIS measured value (S/N ratio)	PCA measured value (S/N ratio)
RT/VTh	11.82	9.25–14.38	9.30	9.46
RT/HTh	9.21	6.96–11.46	8.40	7.79
HT/VTh	4.72	2.56–6.88	2.64	2.11
HT/VTh	4.28	0.84–7.72	0.94	0.84

4 Conclusion and Recommendation

Firstly in order to solve the problem of multiple quality characteristics (the analog production process medium wave soldering process molten tin disturbance caused by the multi-directional erosion on the PCB components and may cause PCB handling stress, so the experiment consider room temperature vertical, horizontal push and high temperature vertical, horizontal push our four quality characteristics.), the research takes advantage of Technique for Order Preference by Similarity to Ideal Solution (TOPSIS) and the statistical method of Principal Component Analysis (PCA), respectively, decided the process of optimal combination of parameters. By the above method of optimal combination of parameters to produce a test sample and thrust experiments confirm experimental data analysis, and finally for optimum process parameters combination, to assess the results of the measurement of each quality characteristic falls within the confidence interval, in order to verify the obtained conclusion is correct that this point dispensing process parameter combinations is not easy to make the components fall off.

After the experimental results, we get TOPSIS with four quality characteristics of the PCA method S/N ratio both fall into the 95 % confidence interval, that the two methods can solve the issue of multiple quality characteristics and these two sets of points dispensing process parameter combinations are not easy to make the components off. Secondly, the experimental results show that multiple quality characteristics can be of TOPSIS with the PCA method to solve the problem, so we recommend that manufacturers can do further comparison (experimental), and so are two ways to choose the better parameter combinations, to reduce the shedding of electronic components probability.

Our point dispensing process as much as quality parameters design based on Taguchi experimental design and planning experiments, use the Order Preference Law and main component analysis to solve multiple quality characteristics, but on the analysis of multiple quality characteristics optimization still there are many other possible methods, for example, data envelopment analysis, genetic algorithms, fuzzy theory, etc., so the future can learn more about the advantages and disadvantages of each method, and compare and discuss the proposed two design methods.

References

Cheng YP (2008) Usage patterns tree constructed SMT solder paste printing process quality control mode. National Cheng Kung University, Industrial and Information Management Master Program, Shuo essays

Hu HJ (2011) Taguchi method is used to investigate the optimization process of the solder paste stencil. Thesis, Department of Electronic Engineering, National Kaohsiung University of Applied Sciences

Huang GY, Caixin L (2006) Green and Pb-free solder stencil printing process parameter optimization. J Technol XXI: 227–236

Li KJ (2007) Application Taguchi tin furnace welding process parameters optimization. Master's thesis, Department of Industrial Engineering and Enterprise Management, Huafan University

Su CD (2002) Quality project. Chinese Society for Quality

Su RR (2004) Taguchi method in the 0201 process parameter optimization of passive components. Thesis, Industrial Management, University Huafan

Wang Q (2011) Improve SMT dot dispensing process. Thesis, Department of Industrial Engineering and Enterprise Information, Huafan University

Wu PY (2012) Using neural networks and TOPSIS SMT point optimization of the parameters of the thrust of the dispensing process. Thesis Hua Fan University, Industrial engineering and management information system 2012; Environmental point of the dispensing process R&D and innovation parameters design. Thesis, Department of Industrial Engineering and Management, National Taipei University of Technology

Yan JY (2011) Multiple quality characteristics Taguchi method to construct a key factor selection algorithm. Master's thesis, Asia-Pacific Industrial and Business Management, National Kaohsiung University

Ye PX (2005) Solder paste printing process parameter optimization. Thesis, Department of Business Management, Shuter University of Science and Technology

The Shortage Study for the EOQ Model with Imperfect Items

Chiang-Sheng Lee, Shiaau-Er Huarng, Hsine-Jen Tsai and Bau-Ding Lee

Abstract The traditional economic order quantity (EOQ) model assumes that all the ordered items are "perfect" enough to be consumed by the customers, but some of these items could be impaired or damaged in the process of production or transportation. For such items, we might call them "imperfect items" and they need to be considered in the EOQ model. In this paper, we focus the study on the shortage problem for the EOQ model with imperfect items since the shortages are inevitably caused by the imperfect items. Consider that the imperfect probability p is a random variable with probability density function $f(p)$ and the good quantity Y comes from the Hypergeometric distribution. Consequently, the sufficient condition to ensure the occurrence of non-shortages is obtained and a mathematical discussion is provided to construct an optimal inventory model for dealing with the possibilities of the shortage problem.

Keywords EOQ model · Imperfect items · Incoming inspection · Shortage

---

C.-S. Lee (✉) · B.-D. Lee
Department of Industrial Management, National Taiwan University of Science and Technology, Taipei, Taiwan, Republic of China
e-mail: cslee@mail.ntust.edu.tw

S.-E. Huarng
Department of Statistics, Iowa State University, Ames, IA, USA

H.-J. Tsai
Department of Information Management, Fu-Jen Catholic University, New Taipei, Taiwan, Republic of China
e-mail: sai.fju@gmail.com

1 Background

The traditional EOQ model has been widely used and studied by many researchers today. It assumes that all the ordered units are "perfect" enough to be consumed completely by the customers (Silver et al. 1998) and that the shortage problem will not occur. In general, there is an incoming inspection process (Montgomery 2001) when the ordered items are delivered. A fair guess for the reason of inspection process is imperfect items. Consequently, shortage problems could happen when the percentage of defective items is large and it results in extensive studies.

Salameh and Jaber (2000) incorporated "imperfect items" into EOQ inventory model and assumed that the imperfect probability p is a random variable with probability density function $f(p)$. To avoid of shortages, they assumed that the number of good quantity $N(y,p)$ is least equal to the demand during the screening time and found the optimal order quantity under this assumption.

Based on the thought of imperfect items and EOQ inventory model built by Salameh and Jaber (2000), Rezaei (2005) proposed the assumption of "shortage backordering" for EOQ model to find the optimal order quantity under the consideration of shortages. Chang (2004) regarded both p and D, demand per unit time, as fuzzy numbers to build a Fuzzy EOQ model, then used the method of defuzzification and ranking method for fuzzy numbers to get the optimal order quantity. Chung and Huang (2006) considered the EOQ model with the idea of permissible credit period and aimed for the optimal order cycle and quantity. Yu et al. (2005) developed an optimal ordering policy for a deteriorating with imperfect quality and included partial backordering for shortage problem, they tried to build up a model for generality and solve the optimal policy under various conditions.

Although most authors developed their researches from the EOQ model built by Salameh and Jaber (2000), Papachristos and Konstantaras (2006) pointed out some unrealistic assumptions under the modified EOQ model from Salameh and Jaber (2000) and the one of the unrealistic assumptions is non-shortages. Regrettably, they did not provide the correct conditions for non-shortage and a remedial method to solve the problem.

2 Introduction to Salameh and Jaber's (2000) EOQ Model

Since our study for EOQ model with imperfect items is based on Salameh and Jaber (2000), we adopt some of the assumptions and the definitions of the notations in that paper, we first introduce Salameh and Jaber's (2000) EOQ model and it follows the definitions of the notations and assumptions:

The Shortage Study for the EOQ Model with Imperfect Items

y	the order size of a lot
c	purchasing price per unit
K	fixed cost of placing an order
p	percentage of defective items in a lot
x	screening rate, the number of units which can be screened per unit time
u	unit selling price of good quality
v	unit selling price of defective items, $v < c$
t	the total screening time of y units, $t = y/x$
D	demand per unit time
$N(y,p)$	the number of good quality of y units, $N(y,p) = y(1-p) \geq Dt$
T	cycle length
d	unit screening cost
$f(p)$	probability density function of p
h	holding cost per unit time

- A lot of size y is delivered instantaneously at the purchasing price c per unit and an fixed ordering cost K
- Each lot will receive a 100 % screening process with screening rate x and screening cost d per unit
- The defective probability p is a random variable with probability density function $f(p)$
- The defective items will be kept in the stock and sold as a single batch at the price v per unit before next shipment
- To avoid shortages, It is assumed that the number of good quantity $N(y,p)$ is at least equal to the demand during screening time t, i.e., $N(y,p) = y(1-p) \geq Dt$

Denote $TR(y)$ as the total revenue per cycle, then $TR(y)$ will be the sum of selling prices of good items and defective items. i.e., $TR(y) = uy(1-p) + vyp$. Let $TC(y)$ be the total cost per cycle, hence $TC(y)$ is the total of the fixed cost of placing an order K, purchasing price per cycle cy, unit screening cost per cycle dy, and holding cost per cycle h $(y(1-p)T/2 + py^2/x)$, and it is given as $TC(y) = [K + cy + dy + h\ (y(1-p)T/2 + py^2/x)]$. Therefore, the total profit per cycle $TP(y)$ equals the total revenue per cycle minus the total cost per cycle. i.e.,
$TP(y) = TR(y) - TC(y) = uy(1-p) + vyp - [K + cy + dy + h(y(1-p)T/2 + py^2/x)]$. The total profit per unit time $TPU(y)$ is obtained by dividing the total profit per cycle length T. Since $T = (1-p)y/D$, we have $TPU(y) = D\ (u - v + hy/x) + D\ (v - hy/x - c - d - K/y)\ (1/(1-p)) - hy(1-p)/2$.

The *objective* of Salameh and Jaber's (2000) EOQ model is to find the optimal order quantity y^* such that y^* will maximize the expected value of the total profit per unit time $E(TPU(y))$ over the domain of y. Since p is a random variable with the a known density function, the random variable y is continuous by definition a known density f is continuous by definition underunction, hence we have

$$E(TPU(y)) = D(u - v + hy/x)$$
$$+ D(vhy/x - c - d - K/y)E(1/(1-p)) - hy(1 - E(p))/2 \quad (1)$$

Besides, the random variable y is continuous by definition under Salameh and Jaber's (2000) EOQ model, then $E(TPU(y))$ is concave and y^* could be obtained by taking the first and the second derivatives of Eq. (1). After some straightforward mathematical computations, it can be demonstrated that there exists a unique value of y^* which maximize Eq. (1) and

$$y^* = \{[2K\,DE(1/(1-p))]/[h(1-E(p)) - D(1 - E(1/(1-p))/x)]\}^{1/2}. \quad (2)$$

Note that $y^* = [2\,K\,D/h]^{1/2}$ if $p = 0$ and the Salameh and Jaber's (2000) EOQ model reduces to traditional EOQ model.

3 Mathematical Model

The present study takes the definitions of notations mentioned above and gives some different thought about the assumptions of random variables and y. Consider a lot of size y delivered instantaneously, it is obviously that the items (or productions) are countable and the random variable y is a positive integer. The assumption of continuous y is unrealistic under Salameh and Jaber's (2000) EOQ model, hence $E(TPU(y))$ is not a concave function of y and the optimal order quantity y^* cannot be gotten by taking the first and the second derivatives of it.

Recall that the number of good quality of y units $N(y,p)$, $N(y,p) = y(1-) \geq p$ Dt implies that $p \leq 1 - D/x$, i.e., the demand per unit time D must be less or equal to the number of units which can be screened per unit time x. In the real life, there is no guarantee of that. Since p is a random variable, it does not ensure that the inequality $p \leq 1 - D/x$ can be always held under all circumstances. The inequality guarantees no shortages for all combinations of (x, D) will be

$$x - y \max_{p \in (a,b)}\{p\} \geq D \quad (3)$$

or

$$x - yb \geq D, \quad (4)$$

where b is the maximum failure probability that could occur in the lot of size y. For example, let $y = 500$ units/lot, $x = 110$ units/day, $D = 100$ units/day, and there are total 20 defective items found in this lot. From the example, it exists a shortage problem if there are 11(or more) defective items found in the first day's screening process even that the failure probability $p = 20/500 = 4\,\%$ satisfies the inequality $p \leq 1 - D/x$.

It follows assumptions and definitions of notations used in our study.

- G the number of good quality items
- s unit shortage cost, $s > c$
- q number of sampling times in the screening process
- r percentage rate in the last small batch (number of items in the last batch/x), $0 < r \leq 1$.
- Y_i the number of good quality at ith batch in the screening process, $i = 1, 2, \ldots, q$, $\sum_{i=1}^{q} Y_i = G$

- Let $t = \frac{y}{x} = \left[\frac{y}{x}\right] + \left(\frac{y}{x} - \left[\frac{y}{x}\right]\right) = n + r$, here $[\,]$ be denoted as the Gaussian symbol, $n = \left[\frac{y}{x}\right]$ is the biggest integer less or equal to y/x, $0 \leq r < 1$.
- Let q represents the number of small batches inspected in screening process per cycle, hence, $q = \begin{cases} n+1, & \text{if } r \neq 0 \\ n, & \text{if } r = 0 \end{cases}$.
- In the study, we assume that good items for the ith batch is Y_i, $i = 1, 2, \ldots, q$, and that the sum of all these values must be equal to the number of good items G, i.e., $\sum_{i=1}^{q} Y_i = G$. In statistics, $Y_1, Y_2, \ldots,$ and Y_{q-1} can be regarded as random variables and they are randomly selected from the finite population of size y. Note that the last batch Y_q cannot be treated as a random variable since Y_q can be directly computed from $Y_q = G - \sum_{i=1}^{q-1} Y_i$.
- From the discussions mentioned above, the good quantities $Y_1, Y_2, \ldots,$ and Y_{q-1} can be regarded as random variables which come from Hypergeometric distribution. Therefore, $pr(Y_1 = y_1 | p) = \dfrac{\binom{G}{y_1}\binom{N}{x - y_1}}{\binom{y}{x}}$,

where $y_1 = Max\{0, G + x - y\}, 1, 2, \ldots, Min(x, G)$.

$$pr(Y_2 = y_2 | p, Y_1 = y_1) = \dfrac{\binom{G - y_1}{y_2}\binom{N - (x - y_1)}{x - y_2}}{\binom{y - x}{x}},$$

where $y_2 = Max\{0, G - y_1 + x - (y - x)\}, 1, 2, \ldots, Min(x, G - y_1)$.

$pr(Y_{q-1} = y_{q-1} | p, Y_1 = y_1, Y_2 = y_2, \ldots, Y_{q-2} = y_{q-2})$

$$= \dfrac{\binom{G - y_1 - \cdots - y_{q-2}}{y_{q-1}}\binom{N - ((q-2)x - y_1 - \cdots - y_{q-2})}{x - y_{q-1}}}{\binom{y - (q-2)x}{x}},$$

where $y_{q-1} = Max\{0, G - y_1 - \ldots - y_{q-2} + x - [y - (q-2)x]\}$,
1, 2,..., $Min(x, G - y_1 - \ldots - y_{q-2})$.

We use $q = 2$ as an example to find the optimal policies for the EOQ model. In this example, we have only one random variable y_1 and there are x units at the first inspection and $y-x$ units at the second one. After the first inspection, the inventory level are $y-y_1$ if $y_1 < d$, $y-d$ if $y_1 \geq d$. Also after the second inspection, the inventory level are $y-y_1-\underline{rd}$ if $y_1 < d$, $y-d-\underline{rd}$ if $y_1 \geq d$, where $\underline{rd}$ denotes the rounded value of rd.

Let $TCPUT_1(y_1)$ be the total cost per unit time of shortages at the first inspection and $TCPUT_2(y_1)$ the total cost per unit time of non-shortages at the first inspection. Moreover, we can get that

$$TCPUT_1(Y_1) = \frac{\left\{K + cy + dy + h\left[\frac{1 \times (y+y-Y_1)}{2} + \frac{r \times (y-Y_1+y-Y_1-\overline{rD})}{2} + \frac{(y - Y_1 - \overline{rD} - N)^2/D}{2}\right] + (D - Y_1) \times s\right\}}{t + \frac{(y-Y_1-\overline{rD}-N)}{D}}$$

(5)

and

$$TCPUT_2(Y_1) = \frac{\left\{K + cy + dy + h\left[\frac{1 \times (y+y-D)}{2} + \frac{r \times (y-D+y-D-\overline{rD})}{2} + \frac{(y - D - \overline{rD} - N)^2/D}{2}\right]\right\}}{t + \frac{(y-D-\overline{rD}-N)}{D}}$$

(6)

Furthermore, we consider the situation for $(q-1)$ of random variables. After similar computational procedure, the expected total cost per unit time when p is fixed can be written as:

$$ETCPUT|_p = \sum\sum\ldots\sum f(y_1, y_2, \ldots, y_{q-1}|p) \times [TCPUT_i(y_1, y_2, \ldots, y_{q-1}|p)], i = 1, \ldots, 2^{q-1},$$

(7)

where $f(y_1, y_2, \ldots, y_{q-1}|p)$ is the joint hypergeometric probability of $(y_1, y_2, \ldots, y_{q-1})$ when p is given, and

The Shortage Study for the EOQ Model with Imperfect Items

$$f(y_1, y_2, \ldots, y_{q-1}|p) = \frac{\binom{G}{y_1}\binom{N}{x-y_1}}{\binom{y}{x}} \times \frac{\binom{G-y_1}{y_2}\binom{N-(x-y_1)}{x-y_2}}{\binom{y-x}{x}}$$

$$\times \cdots \times \frac{\binom{G-y_1-\cdots-y_{q-2}}{y_{q-1}}\binom{N-((q-2)x-y_1-\cdots-y_{q-2})}{x-y_{q-1}}}{\binom{y-(q-2)x}{x}}$$

$$= \frac{\binom{G}{y_1, y_2, \ldots, y_{q-1}}\binom{N}{x-y_1, x-y_2, \ldots, x-y_{q-1}}}{\binom{y}{x, x, \ldots, x}},$$

$$y_i = 0, 1, \ldots, x \text{ for } i = 1, 2, \ldots, q-1, \quad \sum_{i=1}^{q} y_i = G \tag{8}$$

Actually the assumption of fixed p is unrealistic in the real-life. We now discuss the cases when p is a random variable. Assumed that K, c, $D\ldots$ are known, the expected total cost per unit time can be composed by the function of y and p:

$$ETCPUT = \sum\sum \cdots \sum f(y_1, y_2, \ldots, y_{q-1}, p) \times [TCPUT_n(y_1, y_2, \ldots, y_{q-1}|p)]$$
$$= g(y, p), \tag{9}$$

where $p \sim$ Uniform (α, β), $\beta > \alpha \geq 0$. Thus we can have

$$E[ETCPUT] = \int_\alpha^\beta h(p) \cdot ETCPUT \cdot dp = \int_\alpha^\beta h(p) \cdot g(y, p) \cdot dp = ETCPUT(y) \tag{10}$$

Note that we cannot get the integrals directly and may use the method of "Riemann sum" to approach true values of these integrals. Hence,

$$ETCPUT(y) = \int_\alpha^\beta h(p) \cdot g(y, p) \cdot dp \cong \frac{\beta-\alpha}{n} \cdot \sum_{i=1}^{n} h(p_i^*) \cdot g(y, p_i^*)$$
$$= \frac{\beta-\alpha}{n} \cdot \sum_{i=1}^{n} h(p_i^*) \cdot ETCPUT|_{p_i^*} = \frac{1}{n}\sum_{i=1}^{n} ETCPUT|_{p_i^*} \tag{11}$$

$i = 1, 2, \ldots, n$, where p_i^* is the ith range average value of equal length.

The objective of this paper is to find the optimal order quantity y^* such that ETCPUT(y^*), the expected total cost per unit time at $y = y^*$, is the minimum over domain of y. Unlike the procedure in Salameh and Jaber's (2000) EOQ model, we cannot use differentiation to find y^* because y is discrete, a numerical approach method can be utilized to find the optimal order quantity y^* in our study.

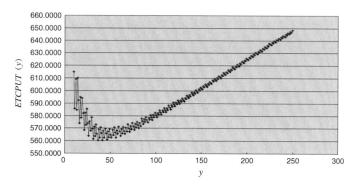

Fig. 1 The expected total cost per unit time

4 Numerical Results

The following is a numerical example that applied to our mathematical model to demonstrate the ^{method} that we discussed in Sect. 3. It follows the related parameters: $K = \$50$/cycle, $c = \$25$, $d = \$0.5$, $s = \$30$, $h = \$1$/unit, $D = 20$/unit, and $x = 50$/unit.

Let $p \sim Uniform\ (0,\ 0.04)$, the result are obtained by running Fortune codes for $y = 10, 11, \ldots, 249$ under non-shortage case, we compute the values of ETCPUT$|p(y)$ and summarize the results in Fig. 1. From Fig. 1, a convex pattern is shown and we can get the optimal order quantity $y^* = 41$ and ETCPUT$|p(y^*) = 560.3193$.

5 Conclusions

In this paper, for the EOQ model with imperfect items, we propose the assumptions of y and p to consist with the real-life situations. Based on them, a mathematical discussion is provided to construct an optimal inventory model and to ensure the occurrence of non-shortages discussed for dealing with the possibilities of the shortage problem.

References

Chang HC (2004) An application of fuzzy sets theory to the EOQ model with imperfect quality items. Comput Oper Res 31(12):2079–2092
Chung KJ, Huang YF (2006) Retailer's optimal cycle times in the EOQ model with imperfect quality and a permissible credit period. Qual Quant 40:59–77
Montgomery DC (2001) Introduction to statistical quality control. Wiley, New York

Papachristos S, Konstantaras I (2006) Economic ordering quantity models for items with imperfect quality. Int J Prod Econ 100(1):148–154

Rezaeli J (2005) Economic order quantity model with backorder for imperfect quality items. IEEE Int Eng Manage Conf II 1559191:466–470

Salameh MK, Jaber MY (2000) Economic production quantity model for items with imperfect quality. Int J Prod Econ 64(1):59–64

Silver EA, Pyke DF, Peterson R (1998) Inventory management and production planning and scheduling. Wiley, New York

Stevenson WJ (2004) Operations management. McGraw-Hill, New York

Yu CP, Wee HM, Chen JM (2005) Optimal ordering policy for a deteriorating items with imperfect and partial backordering. J Chinese Inst Indu Eng 22:509–520

Power, Relationship Commitment and Supplier Integration in Taiwan

Jen-Ying Shih and Sheng-Jie Lu

Abstract This research extends power-relationship commitment theory to investigate the impact of power and relationship commitment on supplier integration from manufacturers' perception toward their major suppliers in supply chain context in Taiwan. The power sources include expert power, referent power, legitimate power, reward power and coercive power, which can be categorized as non-mediated power and mediated power. Two types of the relationship commitment are studied, including normative relationship commitment and instrumental relationship commitment. The integration between manufacturers and suppliers (supplier integration) is measured by information integration and strategic integration. Based on a survey using data on 193 manufacturers in Taiwan, results indicate that coercive power has a positive influence on instrumental relationship commitment; however, reward power has no significant impact on any type of relationship commitment. Expert and referent power have positive impact on normative relationship commitment, while legitimate power has no significant influence on relationship commitment. Both normative and instrumental relationship commitment have positive impact on supplier integration and the former has a stronger influence than the latter. The findings can help companies enhance their supply chain integration by developing appropriate relationships with their suppliers.

Keywords Supply chain · Power · Relationship commitment · Supplier integration

---

J.-Y. Shih (✉)
Graduate Institute of Global Business and Strategy, National Taiwan Normal University, 162, He-ping East Road, Section 1, Taipei 10610, Taiwan, Republic of China
e-mail: jyshih@ntnu.edu.tw

S.-J. Lu
Department of Business Administration, Chang Gung University, 259, Wen-Hwa 1st Road, Taoyuan 333, Taiwan, Republic of China
e-mail: myronlu@pchome.com.tw

1 Introduction

The increasing global competition has enhanced awareness of the importance of supply chain management in global businesses (Flint et al. 2008). Nowadays, the focus of supply chain management is not only on cost reduction but it also extends to business agility so as to meet the requirements of challenging business environments (Lee 2004), which shifts the emphasis of supply chain integration to quick and effective response to the volatile environments. Supply chain integration involves information sharing, planning, coordinating and controlling materials, parts and finished goods at the strategic, tactical and operational levels (Stevens 1989). It has been found to be important in enhancing material flow, product development, delivery speed, dependability and flexibility (Carter and Narasimhan 1996; Boyer and Lewis 2002; Flynn and Flynn 2004). Thus, supply chain integration is also an important factor in enabling supply chain management.

Past research regarding marketing channel had investigated power influences issue through the power-relationship commitment theory (Brown et al. 1995). Power plays an important role in the supply chain, and different bases of power have contrasting influences on relationship commitment to supply chain partnership. Effects of power influences on relationship commitment to supply chain partnership and subsequent effects of this partnership upon supply chain performance expose the potential of power as a tool to promote integration of the supply chain. Thus, both the power holder and the power target must be able to recognize the presence of power, and then reconcile supply chain management for power influences. Maloni and Benton (2000) examined supply chain management from the power influences perspective by extending the power-relationship commitment theory established in marketing channel literature to a supply chain context.

Zhao et al. (2008) applied the power-relationship commitment theory to study its impact on customer integration in supply chain context from the viewpoint of a manufacturer toward its main customer by investigating the relationships among perceived customer power, manufacturers' commitment to maintain customer relationship and customer integration in China. Thus, the power-relationship commitment theory was linked with supply chain integration. In addition, because high power distance and collectivism are two important characteristics of China's national culture, their work demonstrated a different pattern of power-relationship commitment in Chinese manufacturers.

However, their study only investigated the impact of power-relationship commitment on customer integration, which lacks the investigation regarding if a manufacturer's perceived power-relationship commitment toward its main supplier has influence on supplier integration. Supplier integration is also an important component in supply chain integration. Therefore, this research extends power-relationship commitment theory to investigate the impact of power and relationship commitment on supplier integration from manufacturers' perception toward their major suppliers in supply chain context in Taiwan. The purpose of this research is to investigate the impact of the exercise of supplier power on a manufacturer's relationship commitment and to investigate how a manufacturer's relationship commitment to its supplier affects supplier integration. Thus, we propose and empirically test a model that represents the relationships among supplier power, relationship commitment and supplier integration in supply chain context.

2 Theoretical Background

Brwon et al. (1995) argued that the sellers' use of different source of power may bring different retailers' relationship commitment to the channel relationship in their marketing channel study. This argument motivated supply chain research to investigate if the power-relationship commitment structure is an enabler of supply chain integration (Zhao et al. 2008). Following this viewpoint, to investigate the impact of perceived suppliers' power usage on manufacturers' commitment to the manufacturer-supplier relationship, the conceptual definition of power in this research is the ability of a supplier to influence the decisions of a manufacturer in a supplier chain. Because French and Raven (1959) classified power into five sources, which have been applied in a lot of empirical research regarding power issue (Brown et al. 1995), this research used the five-source power structure to represent power usage.

The five sources of power include expert power, referent power, legitimate power, reward power and coercive power, which can be mapped to non-mediated power and mediated power in terms of Brown et al. (1995). The definition of the five-source power is provided in Table 1, which is adapted from Zhao et al. (2008) and Maloni and Benton (2000). Non-mediated power indicates that the manufacturer decides by itself whether and how its decision is affected by the power source but not forced by the power source. Mediated power reveals that the power source uses their power to influence the decision of power target to meet the expectation of power source. Therefore, expert power, referent power and legitimate power are considered as non-mediated power. Reward power and coercive power are regarded as mediated power.

The conceptual definition of relationship commitment is the willingness of a party (manufacturer) to maintain a relationship through the investment of financial, physical, or relationship-based resources in the relationship (Morgan and Hunt 1994). In this research, it is an attitude or willingness of a manufacturer to develop and maintain a stable, long-lasting relationship. Past research classified relationship commitment as "normative relationship commitment" and "instrumental relationship commitment". The definition of relationship commitment is shown in Table 2, which is adapted from Zhao et al. (2008) and Morgan and Hunt (1994).

Based on the argument provided by Brown et al. (1995), instrumental relationship commitment is based on evaluation of costs and benefits in developing relationship; therefore, we expected that reward and coercive power would be positively related to it; however, overly exercise of the two mediated power may damage normative relationship commitment. In contrast with the use of mediated

Table 1 Power sources

Power source	Conceptual definition	Operational definition (Scale items)
Reward power	Supplier has the ability to mediate rewards to manufacturer	A. If we do not do what as our major supplier asks, we will not receive very good treatment from it B. We feel that, by going along with our major supplier, we will be favored by it on some other occasions C. By going along with our major supplier's requests, we have avoided some of the problems other customers face D. Our major supplier often rewards us, in order to get our company to go along with its wishes
Coercive power	Supplier has the ability to mediate punishment to manufacturer and manufacturer is thus forced to satisfy the requirement of supplier.	A. Our major supplier's personnel will somehow get back at us if they discover that we did not do as they asked B. Our major supplier often hints that it will take certain actions that will reduce our profits if we do not go along with its requests C. Our major supplier might withdraw certain needed services from us if we do not go along with its requests D. If our company does not agree to its suggestions, our major supplier could make things more difficult for us
Expert power	Supplier has knowledge, expertise or skills desired by the manufacturer so that the manufacturer is willing to meet the expectation of the supplier	A. Our major supplier's business expertise makes it likely to suggest the proper thing to do B. The people in our major supplier's organization know what they are doing C. We usually get good advice from our major supplier D. Our major supplier has specially trained people who really know what has to be done

(continued)

Table 1 (continued)

Power source	Conceptual definition	Operational definition (Scale items)
Referent power	Manufacturer values identification with the supplier	A. We really admire the way our major supplier runs its business, so we try to follow its lead
		B. We generally want to operate our company very similar to the way we think our major supplier would
		C. Our company does what our major supplier wants because we have very similar feelings about the way a business should be run
		D. Because our company is proud to be affiliated with our major supplier, we often do what it asks
Legitimate power	Manufacturers believe their suppliers retain natural right to influence their decisions	A. It is our duty to do as our major supplier requests
		B. We have an obligation to do what our major supplier wants, even though it isn't a part of the contract
		C. Since it is the supplier, we accept our major supplier's recommendations
		D. Our major supplier has the right to expect us to go along with its requests

Table 2 Relationship commitment typology

Power source	Conceptual definition	Operational definition (scale items)
Normative relationship commitment	Manufacturer maintain a long-term relationship based on mutual commitment (e.g., trust) and sharing	A. We feel that our major supplier views us as an important "team member," rather than just another customer B. We are proud to tell others that we are a customer for our major supplier C. Our attachment to our major supplier is primarily based on the similarity between its values and ours D. The reason we prefer our major supplier to others is because of what it stands for, its values E. During the past year, our company's values and those of our major supplier have become more similar F. What our major supplier stands for is important to our company
Instrumental relationship commitment	Manufacturer is willing to be influenced by supplier on the expectation of receiving favorable reactions from supplier	A. Unless we are rewarded for it in some way, we see no reason to expend extra effort on behalf of our major supplier B. How hard we work for our major supplier is directly linked to how much we are rewarded by it

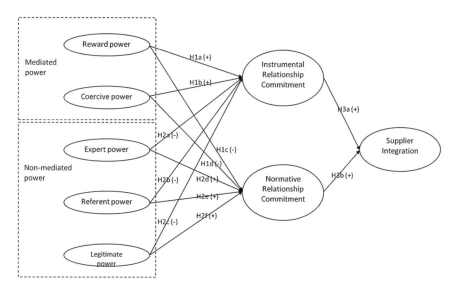

Fig. 1 Conceptual model

power, the use of non-mediated power may enhance positive attitudes toward partner relationship which, in turn, enhance normative relationship commitment. Therefore, expert, referent and legitimate power may demonstrate positive relationships with normative relationship commitment; however, the use of non-mediated power may lower instrumental relationship commitment. The reason is that the more power holder uses non-mediated power to influence channel members (power target), the more it focuses upon long-term relationship, which is a kind of intrinsic factor, therefore, extrinsic factors such as rewards and punishments become less important. Accordingly, the hypotheses of this research are demonstrated in a conceptual model shown in Fig. 1.

Zhao et al. (2008) propose that both types of relationship commitment are positively related with supply chain integration based on traction cost theory and social exchange theory. Because supply chain integration is created by cooperative, mutually beneficial partnerships with supply chain members, relationship commitment thus plays a very important role in supply chain integration. In this research, we examine influence of two types of relationship commitment on supplier integration.

3 Research Methodology

This research used purposive sampling approach to select subject companies from a sampling frame, which is composed of several directories provided by "China Productivity Center", "Metal Industries Research and Development Center",

"Small and Medium Enterprise Administration, Ministry of Economic Affairs", and "Industrial Development Bureau, Ministry of Economic Affairs". A total of 511 possible valid participants were contacted through survey mailings. The follow-up calls yielded a total of 222 respondents. Twenty-nine of the responses were deemed unusable due to incompleteness, missing data, or excessive response of "no opinion". The cleansing yielded a sample of 193 responses with a response rate of 37.77 %.

Manufacturers' responses to the survey questions (parts of them are shown in Tables 1 and 2; survey items of supplier integration construct are available upon request.) indicated the extent of their agreement with instrument items. Each question sought responses based on a Likert scale from 1 (strongly disagree) to 7 (strongly agree). A strict process for instrument development was employed to develop and validate the survey instrument, modeled on previous empirical literatures (Chen and Paulraj 2004; Garver and Mentzer 1999; Min and Mentzer 2004). The content validity for the survey instrument was refined through literature review, in-depth interview conducted with experts and researchers and iterative pilot testing prior to data collection. All of the instrument items for related constructs were developed from an extensive review of the academic and practitioner literatures to measure the research variables.

Given the multiple dependence relationships in this research, structural equation modeling (SEM) is a suitable statistical tool to evaluate the conceptual model (Fig. 1). SEM measures multiple relationships between independent and dependent variables, extending the concept of a single dependence-based regression equation to accommodate multiple dependence relationship simultaneously in an aggregate model. We used LISREL software to estimate the causal relationships among the constructs in our research model, which is able to concurrently test all of the proposed research hypotheses.

We performed a series of analyses to test construct validity and reliability after data collection. Confirmatory factor analysis (CFA) was conducted to justify the factor structure. Construct reliability was established through analysis of composite reliability and average variance extracted. The validity of the constructs showed that the instrument items correlate with what they were intended to measure and do not correlate with other constructs.

4 Results

The SEM result is provided in Fig. 2. Reward power is not significantly related to either type of relationship commitment, while coercive power has a positive impact on instrumental relationship commitment, but has no impact on normative relationship commitment, indicating that only use of coercive power may enhance manufacturer's instrumental relationship commitment.

The path coefficients in Fig. 2 reveal that supplier's expert, referent and legitimate power have no significant impact on instrumental relationship

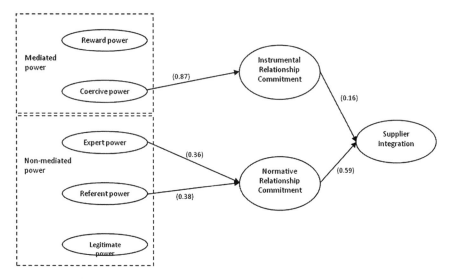

Fig. 2 Structural equation model

commitment. However, expert and referent power have positive impact on normative relationship commitment, but legitimate power has no impact on it. The influences of expert and referent power on normative relationship commitment indicate that supplier's use of these two non-mediated power may enhance manufacturer's normative relationship commitment.

Both normative and instrumental relationship commitment have positive impact on supplier integration and the former has a stronger influence than the latter. The findings can help companies enhance their supply chain integration by developing appropriate relationships with their suppliers.

References

Boyer KK, Lewis MW (2002) Competitive priorities: investigating the need for trade-offs in operations strategy. Prod Oper Manage 11(1):9–20

Brown JR, Lusch RF, Nicholson CY (1995) Power and relationship commitment: their impact on marketing channel member performance. J Retail 71(4):363–392

Carter JR, Narasimhan R (1996) Is purchasing really strategic? J Supply Chain Manage 32(1):20–28

Chen IJ, Paulraj A (2004) Understanding supply chain management: critical research and a theoretical framework. Int J Prod Res 42(1):131–163

Flint DJ, Larsson E, Gammelgaard B (2008) Exploring processes for customer value insights, supply chain learning and innovation: an international study. J Bus Logistics 29(1):257–281

Flynn BB, Flynn EJ (2004) An exploratory study of the nature of cumulative capabilities. J Oper Manage 22(5):439–458

French RP, Raven BH (1959) The bases of social power. In: Cartwright D (ed) Studies in social power. University of Michigan Press, Ann Arbor, pp 155–164

Garver M, Mentzer JT (1999) Logistics research methods: employing structural equation modeling to test for construct validity. J Bus Logistics 20(1):33–58

Lee HL (2004) The triple-a supply chain. Harvard Bus Rev 82(10):102–113

Maloni M, Benton WC (2000) Power influences in the supply chain. J Bus Logistics 21(1):49–73

Min S, Mentzer JT (2004) Developing and measuring supply chain management concepts. J Bus Logistics 25(1):63–99

Morgan RM, Hunt SD (1994) The commitment-trust theory of relationship marketing. J Mark 58(3):20–38

Stevens GC (1989) Integrating the supply chain. Int J Phys Distrib Logistics Manage 19(8):3–8

Zhao X, Huo B, Flynn BB, Yeung J (2008) The impact of power and relationship commitment on the integration between manufacturers and customers in a supply chain. J Oper Manage 26:368–388

A Search Mechanism for Geographic Information Processing System

Hsine-Jen Tsai, Chiang-Sheng Lee and Les Miller

Abstract Geographic data is becoming a critical part of mobile applications. Public and private sectors agencies create and make geographic data available to the public. Applications can make request to download maps to help the user navigate her/his surroundings or geographic data may be downloaded for use in the applications. The complexity and richness of geographic data create specific problems in heterogeneous data integration. To deal with this type of data integration, a spatial mediator embedded in a large distributed mobile environment (GeoGrid) has been proposed in earlier work. The present work looks at a search mechanism used in the spatial mediator that utilizes an algorithm to support the search of the data sources in response to application's request for maps. The algorithm dynamically evaluates uncovered region of the bounding box of the request in an attempt to search for a minimal set of data sources.

Keywords Information processing · Data integration · Geographic data · Mediator

---

H.-J. Tsai (✉)
Department of Information Management, Fu-Jen Catholic University, New Taipei, Taiwan, Republic of China
e-mail: tsai.fju@gmail.com

C.-S. Lee
Department of Industrial Management, National Taiwan University of Science and Technology, Taipei, Taiwan, Republic of China
e-mail: cslee@mail.ntust.edu.tw

L. Miller
Department of Computer Science, Iowa State University, Ames, USA
e-mail: lmiller@cs.iastate.edu

1 Introduction

Every local system designer tends to develop the database that can meet his/her organization's specific needs. It results in a great deal of diversity in a multiple heterogeneous data sources environment. In this environment, a variety of sources and applications use different data models, representations and interfaces. System designers need to develop integrated systems that allow users to access and manage information from multiple heterogeneous data sources. One reason for such need has been that environments for data access have changed from centralized data systems into multiple, distributed data sources. This need becomes even more critical in the mobile computing applications. In the mobile grid, geographic data plays an especially important role. A mobile computer in the mobile grid can either be a user or a creator of geographic data. Mobile applications can make use of downloaded maps to help the user navigate her/his surroundings or geographic data may be downloaded for use in the application. To support this view of geographic data use in the mobile grid, we see the need for an infrastructure embedded in the grid that supplies the tools necessary to locate the appropriate geographic data, has the capacity to modify the data to fit the application and the device it is being used on, and to direct the movement of the data to and from the field. One of the key capabilities of such an infrastructure is to locate data sources that can generate maps which can cover the region of the user's request.

Data quality is an important issue in the geographic data integration. It is easier to maintain a better data quality when fewer data sources are involved in the integration process. A reasonable goal of a geographic data integration system is to locate as minimal number of data sources needed as possible. This paper proposes a dynamic process that attempts to create maps from minimal number of data sources. Specifically, we present an algorithm that tends to find the minimum number of maps to cover the bounding box of a request.

In the next section we briefly look at the related work. Section 3 briefly looks at the GeoGrid model. Section 4 introduces the spatial mediator and the search algorithm. Finally, Sect. 5 provides some concluding remarks.

2 Related Work

The complexity and richness of geographic data create specific problems in heterogeneous data integration. Geographic data is very diverse and dynamic. The geospatial information may be unstructured or semi-structured, and usually there is no regular schema to describe them. As the amount of geographic data grows, the problem of interoperability between multiple geographic data sources becomes the critical issue in the developing distributed geographic systems. Many approaches have been proposed to provide solutions to this problem. The concept of a spatial mediator has gained increasing interest in projects dealing with distributed geographic data (Zaslavsky et al. 2004; Park and Ram 2004). A mediator is build to provide a uniform interface to a number of heterogeneous data sources. Given a user query, the mediator decomposes it into multiple local sub-queries and sent them to the appropriate data sources, merges the partial results and reports the final answer to the user. There is an increasing demand for geospatial information services to support interoperation of massive repositories of heterogeneous geospatial data on Internet. VirGIS is a mediation platform that utilizes an ontology and provides an integrated view of geographic data (Essid et al. 2006). The architecture of a mediator is developed to enable a user to query spatial data from a collection of distributed heterogeneous data sources. They use GML (Geography Markup Language) [OpenGIS] in their implementation to facilitate data integration.

Data quality and metadata are crucial for the development of geographic information Systems. A group of researchers have developed spatial mediation system that focus on spatial quality issues. Lassoued et al. (2007) proposes a quality-driven mediation approach that allows a community of users to share a set of autonomous, heterogeneous and distributed geospatial data sources with different quality information. A common vision of the data that is defined by means of a global schema and a metadata schema is shared by users. The QGM (Quality-driven Geospatial Mediator) supports efficient and accurate integration of geospatial data from a large number of sources (Thakkar et al. 2007). It features an ability to automatically estimate the quality of data provided by a source by using the information from another source of known quality. QGM represents the quality information in a declarative data integration framework, and exploit the quality of data to provide more accurate answers for user queries.

The work done in this paper expands on the work described in (Tsai 2011) that proposes a spatial mediator embedded in a large distributed mobile environment (GeoGrid) (Miller et al. 2007; Tsai et al. 2006). The spatial mediator takes a user request from a field application and uses the request to select the appropriate data sources, constructs subqueries for the selected data sources, defines the process of combining the results from the subqueries, and develops an integration script that controls the integration process in order to respond to the request.

3 GeoGrid Model

To represent the relationship between the components in GeoGrid, it is helpful to visualize GeoGrid as a directed graph G(N,E). N is a set of nodes with some processing power focused on supporting the GeoGrid infrastructure. The edges in the edge set E represent the communication links that tie the components of GeoGrid together. Figure 1 provides a simple illustration of the GeoGrid infrastructure (Miller and Nusser 2002).

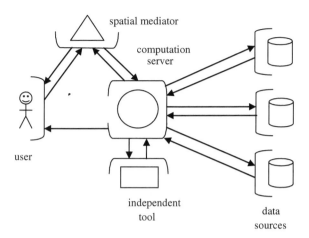

Fig. 1 A simple GeoGrid graph showing the nodes and the data flow for a mediated data request

The process of using the infrastructure starts with a user application in the field sending a request to the spatial mediator. The spatial mediator in turn uses the request to locate the data sources (via LIVs) needed to generate the requested spatial object and form the integration script that defines integration tools. The integration script is passed to the computation server where it is used to manage the process of obtaining the data and creating the spatial object(s). The resulting spatial object (e.g., a map) is then passed back to the user's application.

3.1 Data Sources

The basic structure of a data source is given in Fig. 2. The local interface view (LIV) (Yen and Miller 1995) is designed to export data from the data source into the GeoGrid environment. The number and type of LIVs is a local decision dependent on how the local information manager wants to share the available data within GeoGrid.

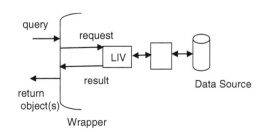

Fig. 2 Data source node layout and request/data flow for retrieval

4 Spatial Mediator

Data is evaluated by the spatial mediator as a set of spatial objects defined through local interface views (LIVs). While spatial objects can take on many forms, the focus here is on spatial objects that define maps. Each data source contributes one or more LIVs, where each LIV can be used to generate a map.

The mediation process starts with the spatial mediator determining the LIVs that are capable of responding to all or part of the incoming request. To do this, the spatial mediator makes use of the LIV registration data. A Facts database is used to identify LIVs that satisfy keywords, theme, or property requirements in the request. Location requirements can take either symbolic (e.g., a city name) or point/bounding box values.

Two lists of LIVs are created to allow the mediator to partition the LIVs into those that cover the requested map area and those that overlap part of the requested map area. The motivation for the two lists is to allow our algorithms to first examine the quality of any complete cover LIVs (if they exist) and only go to the process of generating grouping when they are required. To find the needed LIVs the mediator starts by examining the complete cover LIVs. When no covering LIVs can fulfill the request, the mediator switches to the list of partial covering LIVs. To locate LIVs, the mediator utilizes a search algorithm to locate needed LIVs. The search algorithm continues until either the collection of uncovered bounding boxes is empty or the requested map area is totally covered and the current map grouping is returned.

A grouping is recursively defined as consisting of a tool type name and a list of objects such that each object is either an LIV or a grouping. Once an acceptable map grouping has been generated, it needs to be converted to a syntax recognizable script to the computation server. The tool types must be replaced with actual tool names available in the computation server. Other information related to located LIVs such as the address of the data source site (e.g., IP address, url), and the layout of the spatial object(s) generated by the data source are also included. We call the resulting script the integration script. As mentioned in Sect. 3, the integration script is used to manage the process of obtaining the data and creating the spatial object(s). The resulting spatial object (e.g., a map) is then passed back to the user's application.

This paper highlights a search algorithm used by the mediator that tends to find the minimum number of LIVs to cover the bounding box of a request. The algorithm maintains a list of LIVs whose bounding boxes overlap part of the bounding box of the request which is referred as PartialCoverageList The search algorithm is described in the following section.

4.1 The Search Algorithm

The following algorithm tends to find the minimum number of LIVs to cover the bounding box of the incoming request. A bounding box is an area defined by two longitudes and two latitudes and specified by the set of coordinates which represents the left-bottom and right-top of the bounding box. The left-bottom point is the minimum longitude and minimum latitude. The right-top point is specified by the maximum longitude and maximum latitude of the bounding box. Given a set of LIVs that partially overlap the request bounding box and a list of LIVs whose bounding boxes overlap part of the request bounding box, the algorithm starts the search by attempting to cover from the left-top corner of the request bounding box to the right-bottom corner of the bounding box.

Before introducing the algorithm, some symbols are defined as follows.

1. Let P be the set of corner points of the region that has not be covered by the located LIVs. Initial values for P is the set of four corner points of the request bounding box. For example, $P = \{x_1 = (a_1, b_1), x_2 = (a_2, b_1), x_3 = (a_1, b_2), x_4 = (a_2, b_2)\}$, where x_1 and x_4 are the left-top and right-bottom corner points of the request bounding box respectively. That is, a_1 is the minimum longitude and b_1 is the maximum latitude of the bounding box, a_2 and b_2 are the maximum longitude and minimum latitude respectively (see Fig. 3).
2. Let $S_i = (s_{i1}, s_{i2}, s_{i3}, s_{i4})$ be the set of four longitude/latitude coordinates of the bounding box with (s_{i1}, s_{i2}) for the left-top corner point and (s_{i3}, s_{i4}) for the right-bottom corner point.

The algorithm is described as follows.

Step 1. Determine the left most point with maximum latitude ℓ_2 and the right-bottom point $r = (r_1, r_2)$ from the point set P. Set $k = 0$ at beginning.
Step 2. Find all LIVs from the PartialCoverageList that covers the point $\ell = (\ell_1, \ell_2)$ and define this set to be $S = \{S_1, S_2, \ldots, S_N\}$.
Step 3. Determine the map S_j by finding the largest inside area below the latitude ℓ_2, that is,

$$S_j = \max_{i \in S}\{(\min\{s_{i3}, r_1\} - \ell_1) * (\min\{s_{i4}, r_2\} - \ell_2)\}$$ If there are more than one LIV with the same area, then any one of them is chosen.

The reason to compute the area below the latitude ℓ_2 is to reduce the overlapping area.

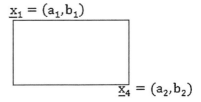

Fig. 3 A request bounding box with (a_1, b_1) and (a_2, b_2)

A Search Mechanism for Geographic Information Processing System

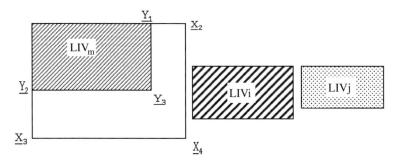

Fig. 4 Covered status of the request bounding box after algorithm finds a LIV_m

Step 4. Set k = k+1 and examine the following two cases.
(a) Delete the corner point from P if the point is covered by the above LIV, and
(b) Add the new corner points to P if they are generated by the new LIV.

If the set P becomes empty after Step 4, then the algorithm stops. Otherwise, it goes to Step 1.

The following example illustrates the search algorithm.

If the bounding box of the request is specified by (a_1, b_1) and (a_2, b_2), then the initial values are set as follows: $P = \{x_1 = (a_1, b_1), x_2 = (a_2, b_1), x_3 = (a_1, b_2), x_4 = (a_2, b_2)\}$, $\ell = (\ell_1, \ell_2) = (a_1, b_1)$, $r = (r_1, r_2) = (a_2, b_2)$ and k = 0.

Figure 4 shows the covering status of the request bounding box after algorithm finds a LIV (i.e. LIV_m) that cover the left-top corner of the request bounding box which is shaded under the diagonal line. Set P has new value which is $\{Y_1, Y_2, Y_3, X_2, X_3, X_4\}$. The algorithm then locates two LIVs that cover point Y_2. The algorithm selects LIVi over LIVj since the area of LIVi is greater than the area of LIVj. Since P is not empty, the algorithm continues until P becomes empty.

5 Conclusions

The GeoGrid structure for supporting the use of geographic data within a mobile computing environment has been designed and prototyped. GeoGrid makes use of a set of intelligent components, highlighted by a spatial mediator for providing data source selection, appropriate query construction for the chosen data sources and creation of the necessary geographic data manipulation operations to integrate the results of the queries if more than one data source is needed. In particular, a search algorithm that tends to find minimum number of maps to cover the bounding box of request from the user in GeoGrid is proposed.

References

Essid M, Colonna F, Boucelma O (2006) Querying mediated geographic data sources. Adv Database Technol 3896:1176–1181

Lassoued Y, Essid M, Boucelma O, Quafafou M (2007) Quality-driven mediation for geographic data. In: Proceedings of QDB, pp 27–38

Miller L, Nusser S (2002) Supporting geospatial data in the field. In: Proceedings of the 11th GITA conference on geographic information systems for oil and gas, pp 139–151

Miller L, Ming H, Tsai H, Wemhoff B, Nusser S (2007) Supporting geographic data in the mobil computing environment. In: Proceedings of parallel and distributed computing systems (PDCS-2007), ISCA 20th international conference Sept, Las Vegas, Nevada, pp 24–26

Park J, Ram S (2004) Information systems interoperability: What lies beneath? ACM Trans Inf Syst (TOIS) 22(4):595–632

Thakkar S, Knoblock C, Ambite J (2007) Quality-driven geospatial data integration. Paper presented at the 15th international symposium on advances in geographic information systems, ACM GIS, pp 44–49

Tsai H (2011) A spatial mediator model for integrating heterogeneous spatial data. Iowa State University, Dissertation

Tsai H, Miller L, Ming H, Wemhoffm B, Nusser S (2006) Combining spatial data from multiple data sources. In: Proceedings of the ISCA 19th international conference on computer applications in industry and engineering, pp 89–94

Yen C, Miller L (1995) An extensible view system for multidatabase integration and interoperation. Integ Comput Aided Eng 2:97–123

Zaslavsky H, Tran J, Martone M, Gupta A (2004) Integrating brain data spatially: spatial data infrastructure and atlas environment for online federation and analysis of brain images. Biological data management workshop (BIDM 2004) in conjunction with 15th international workshop on database and expert systems applications (DEXA'04), Zaragosa, Spain, Aug/Sept 2004, pp 389–393

Maximum Acceptable Weight Limit on Carrying a Food Tray

Ren-Liu Jang

Abstract This study was to simulate two ways to carry a food tray, waist-level carry and shoulder-high carry, to determine their maximum acceptable weight of load (MAWL) and to suggest a proper weight for banquet servers or workers in restaurants. Twenty college students were participated in this study. The MAWL on shoulder-high carry was 3.32 ± 0.47 kg for male and 2.78 ± 0.35 kg for female, respectively. The MAWL on waist-level carry were 2.57 ± 0.26 kg for male and 2.21 ± 0.35 kg for female. There were significantly differences between gender and carry methods. On average, the MAWL on waist-level carry was 22 % less than that on shoulder-high carry. The MAWL of female on waist-level carry was 14 % less than that for male while the MAWL of female on shoulder-high carry was 16 % less than that for male. The results suggested banquet servers or workers in restaurants should consider the proper way to deliver the food when it gets heavy.

Keywords MAWL · Food tray

1 Introduction

Banquet servers were responsible for good interactions with guests such as serving guests in a friendly and efficient manner. Meanwhile, they were expected to be attentive to the guest needs and make them feel welcome and comfortable. During the banquet, servers needed to maintain proper dining experience, deliver and remove courses, fulfill customer needs, replenish utensils and glasses. As results, servers normally walked for long periods of time, possibly extended distances in a banquet service. In food service, there were two basic hold positions to correctly

---

R.-L. Jang (✉)
Department of Industrial Engineering and Management, Ming Chi University of Technology, New Taipei, Taiwan, Republic of China
e-mail: renliuj@mail.mcut.edu.tw

carry a food tray: the waist-level carry and the shoulder-high carry. The waist-level carry put the food tray's weight in the dominant hand with a fully extended wrist. The shoulder-high carry was more difficult to perform because the food tray's weight was put on one hand and the other serving as a stabilizer. Both ways involved shoulder abduction, internal rotation, and external rotation with supination motions of the dominant hand.

Wrist posture, repetition, tendon force and wrist acceleration are the four major risk factors to cumulative trauma disorders (CTDs) and carpal tunnel syndrome (CTS) (Alexander and Pulat 1995; Armstrong 1983; Williams and Westmorland 1994). Symptoms of CTS included numbness, tingling, loss of strength or flexibility, and pain.

Carpal tunnel syndrome is a cumulative trauma disorder that develops over time when hands and wrists perform repetitive movements. For example, workers moving their hands and wrists repeatedly and/or forcefully in their tasks may lead to CTDs (Marras and Schoenmarklin 1993). The flexion/extension acceleration of wrist can establish relative high risk levels of CTDs for hand-intensive, highly repetitive jobs (Schoenmarklin et al. 1994; Silverstein et al. 1986). In the meat packing industry, the jobs that require repetitious, hand-intensive work were found to have higher incidence of CTDs (Marklin and Monroe 1998).

Snook et al. (1995) used psychophysical methods to determine maximum acceptable forces for various types and frequencies of repetitive wrist motion. In the study repetition rates of 2, 5, 10, 15 and 20 motions per minute were set to each flexion and extension task. Maximum acceptable torques was determined for the various motions, grips, and repetition rates without dramatic changes in wrist strength, tactile sensitivity, or number of symptoms.

2 Methods

Considering banquet servers carrying a food tray as a hand-intensive work, this study was to investigate the work and simulate two ways to carry a food tray, the waist-level carry and the shoulder-high carry (Fig. 1), and to determine their maximum acceptable weight of load and suggest a proper weight for workers in restaurants. Maximum acceptable weight limit is determined by the workers, as the highest acceptable workload, which can be performed comfortably based on their perceived efforts.

This psychophysical approach was assumed that people could integrate and combine all stresses into their subjective evaluation of perceived stress. This method in assessing a carrying-tray task required subjects to adjust the weight of a load to the maximum amount they can perform under a certain condition for a time period without feeling strained or becoming unusually tired, overheated, weakened or out of breath.

During the experiment, participants were instructed to follow the psychophysical method to determine the MAWL and make adjustment. After the MAWL

Fig. 1 Waist-level carry and shoulder-high carry

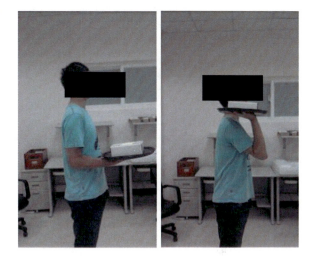

was determined, the participant was required to carry that weight and walk to a table 13.5 m away (Fig. 2). After the walk, if he disagrees on the chosen weight as the MAWL, the participant can make adjustment again until another MAWL was determined.

Twenty college students were participated in this study. Ten female students' average age ranged from 21 to 28 years, with an average of 22.5. Height ranged from 152 to 170 cm, with an average of 161.2. Weight ranged from 48 to 65 kg, with an average of 54.7. Ten male students' average age ranged from 21 to 24 years, with an average of 21.9. Height ranged from 160 to 183 cm, with an average of 174.6. Weight ranged from 55 to 88 kg, with an average of 66.2.

Fig. 2 Walking in the waist-level carry and the shoulder-high carry

3 Results

The MAWL on shoulder-high carry was 3.32 ± 0.47 kg for male and 2.78 ± 0.35 kg for female, respectively. The MAWL on waist-level carry were 2.57 ± 0.26 kg for male and 2.21 ± 0.35 kg for female. There were significantly differences between gender and carry methods using the analysis of ANOVA. On average, the MAWL on waist-level carry was 22 % less than that on shoulder-high carry. The MAWL of female on waist-level carry was 14 % of less than that for male while the MAWL of female on shoulder-high carry was 16 % of less than that for male.

4 Conclusions

This study was to simulate two ways to carry a food tray, the waist-level carry and the shoulder-high carry, and to determine their maximum acceptable weight of load. The results suggested workers in restaurants should consider the proper way to deliver the food when it gets heavy. To prevent high muscle force, the following practices were suggested:

1. Use shoulder-high carry to deliver heavier food tray.
2. Carry fewer plates in the tray at a time.
3. Make two trips or ask other severs to help with large orders.
4. Reduce travel with trays by using tray carrying carts.

References

Alexander DC, Pulat BM (1995) Industrial ergonomics: a practitioner's guide. Industrial Engineering and Management, Norcross

Armstrong TJ (1983) An ergonomics guide to carpal tunnel syndrome. American Industrial Hygiene Association, Fairfax

Marklin RW, Monroe JF (1998) Quantitative biomechanical analysis of wrist motion in bone-trimming jobs in the meat packing industry. Ergonomics 41(2):227–237 (Feb 1998)

Marras WS, Schoenmarklin RW (1993) Wrist motions in industry. Ergonomics 36(4):341–351 (Apr 1993)

Schoenmarklin RW, Marras WS, Leurgans SE (1994) Industrial wrist motions and incidence of hand/wrist cumulative trauma disorders. Ergonomics 37(9):1449–1459 (Sept 1994)

Silverstein BA, Fine LJ, Armstrong TJ (1986) Hand wrist cumulative trauma disorders in industry. Br J Ind Med 43(11):779–784 (Apr 1986)

Snook SH, Vaillancourt DR, Ciriello VM, Webster BS (1995) Psychophysical studies of repetitive wrist flexion and extension. Ergonomics 38(7):1488–1507 (Jul 1995)

Williams R, Westmorland M (1994) Occupational cumulative trauma disorders of the upper extremity. Am J Occup Ther 48(5):411–420 (May 1994)

Fatigue Life and Reliability Analysis of Electronic Packages Under Thermal Cycling and Moisture Conditions

Yao Hsu, Wen-Fang Wu and Chih-Min Hsu

Abstract Many previous researches on electronic packages focused on assessment of package lives under a single state of stress such as vibration, thermal cycling, drop impact, temperature and humidity. The present study considers both effects of thermal cycling and moisture on the fatigue life of electronic packages. The influence of moisture on thermal fatigue life of a package is investigated in particular. A Monte Carlo simulation algorithm is employed to make the result of finite element simulation close to reality. Samples of variables consisting of different package sizes and material parameters from their populations are generated and incorporated into the finite element analysis. The result of a numerical example indicates the thermal-fatigue failure mechanism of the electronic packages is not affected very much by the moisture. However, the mean time to failure of the package does decrease from 1,540 cycles to 1,200 cycles when moisture is taken into consideration.

Keywords Electronic packages · Fatigue life · Moisture diffusion · Reliability

---

Y. Hsu (✉)
Department of Business and Entrepreneurial Management, Kainan University, No. 1, Kainan Road, Luchu, Taoyuan 33857, Taiwan, Republic of China
e-mail: yhsu@mail.knu.edu.tw

W.-F. Wu
Department of Mechanical Engineering and Graduate Institute of Industrial Engineering, National Taiwan University, No. 1, Sec. 4, Roosevelt Road, Taipei 10617, Taiwan, Republic of China
e-mail: wfwu@ntu.edu.tw

C.-M. Hsu
Department of Mechanical Engineering, National Taiwan University, No. 1, Sec. 4, Roosevelt Road, Taipei 10617, Taiwan, Republic of China
e-mail: r99522525@ntu.edu.tw

1 Introduction

The reliability of electronic packages has become an important issue because it would directly determine their life expectancy. Because of the differences in the coefficient of thermal expansion amongst the materials constituting an electronic package, the package would suffer fatigue damage due to thermal stress when it is under thermal cycling loading. Many researches have been conducted on the responses and reliability of electronic packages under thermal cycling conditions (Li and Yeung 2001; Cheng et al. 2005; Kim et al. 2002; Lau and Lee 2002). When considering the moisture condition, a few studies have focused on how moisture affects the stress/strain fields and the failure mechanics in use of numerical simulation (Yi and Sze 1998; Kim et al. 2007). In the literature review, works concerning the fatigue life of electronic packages under both of thermal cycling and moisture conditions have been rarely seen. In addition, many works discussing about parameter uncertainties and how they affect the life of electronic package have been published (Evans et al. 2000; Wu and Barker 2010). Though many articles have discussed the uncertainties coming from geometry or material property, geometric and material uncertainties were usually considered separately. In summary, the purpose of this paper is to analyze the fatigue life and reliability of electronic packages when they are subjected to thermal cyclic and moisture loadings while considering the uncertainties of both geometry and material properties.

2 Finite Element Analysis

In Finite Element Analysis (FEA), the following assumptions were made. (1) The materials of the package are isotropic and the stress and strain relations for each material are identical under tension and compression. (2) There are no stresses for components in the model at the initial state of 25 °C, and residual stresses during the packaging process are disregarded. (3) The displacements of component in the direction of the model's symmetrical plane are zero. (4) All contact planes of the model are in perfect contacts. (5) The temperature and moisture are the same both inside and outside of the model at the same time instance, and situation that temperature changes with space is not taken into consideration.

The model structure considered in this paper is a wafer-level chip-scale package having a size of $9 \times 9 \times 0.508$ mm^3, connected underneath to a 10×10 lead-free ball grid array of Sn-3.9Ag-0.6Cu and a $12 \times 12 \times 1.55$ mm^3 PC board made of FR-4 epoxy glass. The diameter and the height of the solder ball are 0.3 mm and 0.44 mm, respectively, with an interval of 0.85 mm between centers of two solder balls. The model structure is shown in Fig. 1.

The material properties are shown in Table 1. All materials were assumed to be linearly elastic except for the lead-free solder ball. The following Garofalo-

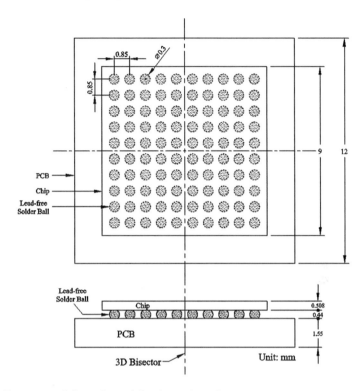

Fig. 1 Geometry and dimensions of the electronic package

Arrhenius creep equation was adopted for the eutectic Sn-3.9Ag-0.6Cu lead-free solder ball

$$\frac{d\varepsilon}{dt} = C_1[\sinh(C_2\sigma)]^{C_3}\exp\left(-\frac{C_4}{T}\right) \quad (1)$$

in which C_1 to C_4 are material constants and their values are shown in Table 2.

Because of symmetry of the packaging structure, for simplicity and time saving, only one-half of the whole package was modeled with appropriate boundary conditions setting. On the cut surface, displacements in x, y and z directions were constrained to prevent the structure from rigid-body motion. All the rest of the surfaces without setting boundary conditions were assumed to free surfaces.

Table 1 Material properties (Jong et al. 2006)

Material	Young's modulus (MPa)	Poison's ratio	CTE (ppm/K)
Chip	131,000	0.3	28
Solder ball	49,000	0.35	21.3
PCB	27,000	0.39	18

Table 2 Creep coefficients of lead-free solder (Lau and Dauksher 2005)

Coefficients	C_1 (1/s)	C_2 (1/MPa)	C_3	C_4 (K)
Value	500,000	0.01	5	5,802

Test condition G in JESD 22-A 104C of the JEDEC STANDARD was adopted for the thermal cyclic simulation on electronic packages for the reliability evaluation. It was performed at temperatures ranging from −40 to 125 °C and the reference temperature is 25 °C. Time for ramp up, dwell, ramp down and dwell loading is 900, 600, 900, 600 s, respectively, and therefore, one complete thermal cycle test is 3,000 s in total.

Regarding the moisture loading, this paper followed IPC/JEDEC Moisture Sensitivity Levels in JEDEC standard specification that requires the package to be put in the environment of 85 °C and relative humidity (RH) of 85 % for 168 h. Conditions of no moisture on the inside of the package and 85 % RH on the package's outer surfaces at initial stage were assumed. The parameters of D and C_{sat} needed in the simulation are tabulated in Table 3. It is noted that effect of thermal expansion was not considered herein during calculation when the structure is subjected to moisture loading.

3 Case Study

Finite element simulation was performed for two phases in the present study. In the first phase (Phase I), simulation was carried out under the condition that only thermal cyclic loading was applied to the model structure. In the second phase (Phase II), moisture diffusion simulation was carried out first, the simulation was subsequently performed under thermal cyclic loading as that made in Phase I.

The accumulated creep strain contour of the electronic package at the end of the simulation of Phase I case is illustrated in Fig. 2. Its maximum value is 0.0612 and it is located in the outmost solder ball as shown in Fig. 2. As for simulation of phase II case, the simulation results such as strain field distribution are nearly the same as those in Phase I case except the maximum value of the accumulated creep strain changes to 0.0786. It could be suggested that the moisture contributed little to the mechanical behaviors of electronic packages subjected to thermal cyclic loading. In addition, the fatigue lives of the package can be obtained from the

Table 3 Parameters for moisture diffusion equation (Cussler 2009)

Material	D (m^2/s)	C_{sat} (kg/m^3)
Chip	1×10^{-14}	1
Solder ball	1×10^{-14}	1
PCB	8.56×10^{-13}	38.4

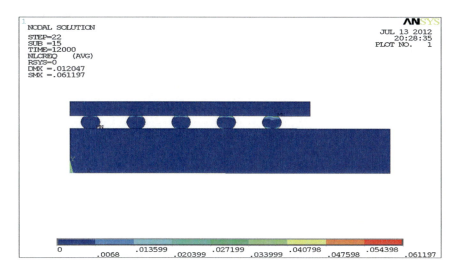

Fig. 2 Contour of accumulated creep strain of the package structure

simulations with the help of calculated creep strain and the prediction rule expressed as follows;

$$N_f = (C' \varepsilon_{acc})^{-1} \qquad (2)$$

where N_f is the number of cycles to failure, C' is the material constant and ε_{acc} is the accumulated creep strain per cycle (Syed 2004).

3.1 Uncertainty Consideration

In the present study, nine geometric parameters and material properties of the package, namely radius of the solder ball, heights of the chip and PCB, Young's moduli of the solder ball, chip and PCB, coefficients of thermal expansion of the solder ball, chip and PCB were chosen as design parameters used for considering uncertainties. All of these nine random variables were assumed to be distributed normally and each one has a mean (nominal) value and a coefficient of variation (c.o.v.) of 3 %. Their mean values and corresponding standard deviations are tabulated in Table 4. A Monte Carlo simulation was made along with the FEA to analyze the strain/stress fields of electronic packages when design parameters have randomness. A random sample of size 40 of the uncertainty parameters was generated, and then forty outcomes of the maximum accumulated creep strain were obtained from running FEA. After substituting these strains into Eq. (2), forty fatigue lives of the electronic package were obtained.

Table 4 Statistical characteristics of geometric parameter and material property

Design parameter	Mean value	Standard deviation
Solder ball radius (mm)	0.15	0.0045
Chip height (mm)	0.508	0.01524
PCB height (mm)	1.55	0.0465
Young's modulus of the solder ball (MPa)	49,000	1,470
Young's modulus of the chip (MPa)	131,000	3,390
Young's modulus of the PCB (MPa)	27,000	810
Coefficient of thermal expansion of the solder ball (ppm/K)	21.3	0.639
Coefficient of thermal expansion of the chip (ppm/K)	2.8	0.084
Coefficient of thermal expansion of the PCB (ppm/K)	18	0.54

3.2 Fatigue Life and Reliability Analysis

Predicated fatigue lives can be further analyzed in statistical way. In this research, the fatigue lives in Phase I and II cases were plotted in normal, lognormal and Weibull probability papers, respectively. Chi square tests were used to determine the best goodness-of-fit. The parameters of these three probability distributions can be estimated by least-squared regression or curve fitting techniques. In this study, the Chi square test statistics of normal, lognormal and Weibull distributions are all more than the critical value at the significant level of 0.01, which means fatigue life data cannot be appropriately fitted by any of three empirical distributions. However, the Weibull distribution fits the fatigue life data best. This study then adopted the mixed Weibull distribution for a try to fit the data and eventually has a good fitness (Rinne 2009). Once the probability density function of the fatigue life is determined, the corresponding reliability function can be obtained and is shown in Fig. 3. The mean time to failure (MTTF) of the electronic package subjected to thermal cyclic loading studied in this paper is about 1,540 cycles whereas it is

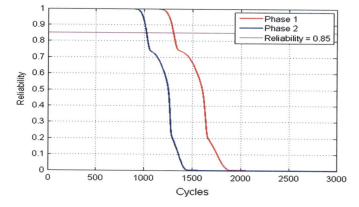

Fig. 3 Comparison of reliability curves of electronic packages

1,200 cycles when more extra moisture loading adds to the package. It can be indicated that moisture does have a negative impact on the thermal fatigue life of packages even if it doesn't seem to affect much the thermal failure mechanism and mechanical behaviors as stated in the preceding paragraph. Furthermore, when reliability of 0.85 was considered, the corresponding fatigue lives are nearly 1,300 and 1,021 cycles for Phase I and Phase cases, respectively.

4 Conclusions

This study intended to investigate how the moisture and thermal cyclic loadings affect the thermal fatigue life and reliability of electronic packages. To be closer to reality, finite element simulation was performed with a Monte Carlo simulation to consider uncertainties of geometry and materials of electronic packages. This analyzing approach proposed in this paper enables to produce the random-distributed fatigue lives of electronic packages and therefore, reliability can be analyzed quantitatively. Under assumptions made in the present paper, several conclusions can be drawn as follows:

1. No matter Phase I or Phase II cases, the numerical results show that the maximum accumulated creep strain occurs in the location of the interfaces between the outmost solder balls and materials connected to them. It is consistent with those results mentioned by other researchers.
2. There are no obvious differences found on the mechanical behaviors of electronic packages in Phase I and II cases. It could be suggested that moisture doesn't have much influence on the thermal fatigue failure mechanism of a package.
3. To the wafer-level chip-scale package considered in this study, the thermal mean time to failure of the package drops about 22 % when the moisture is taken into account. Moreover, the reliability is 0.85 at the thermal cycles of 1,300 when not considering the moisture, whereas for the same reliability, the thermal cycle shortens to 1,021 cycles when considering the moisture. It could be concluded that moisture accelerates the thermal failure of electronic packages to the extent.

Acknowledgments This work is supported by the National Science Council of Taiwan under Grant No. NSC 100-2221-E-002-052. The authors appreciate very much the financial support.

References

Cheng HC, Yu CY, Chen WH (2005) An effective thermal-mechanical modeling methodology for large-scale area array typed packages. Comput Model Eng Sci 7:1–17

Cussler EL (2009) Diffusion: mass transfer in fluid systems. Cambridge University Press, New York

Evans JW, Evans JY, Ghaffarian R, Mawer A, Lee KT, Shin CH (2000) Simulation of fatigue distributions for ball grid arrays by the Monte Carlo method. Microelectron Reliab 40:1147–1155

Jong WR, Chen SC, Tsai C, Chiu CC, Chang HT (2006) The geometrical effects of bumps on the fatigue life of flip-chip packages by Taguchi method. J Reinf Plast Compos 25:99–114

Kim DH, Elenius P, Barrett S (2002) Solder joint reliability and characteristics of deformation and crack growth of Sn-Ag-Cu versus eutectic Sn-Pb on a WLP in a thermal cycling test. IEEE Trans Electron Packag Manuf 25:84–90

Kim KS, Imanishi T, Suganuma K, Ueshima M, Kato R (2007) Properties of low temperature Sn–Ag–Bi–In solder systems. Microelectron Reliab 47:1113–1119

Lau JH, Lee SW (2002) Modeling and analysis of 96.5Sn-3.5Ag lead-free solder joints of wafer level chip scale package on buildup microvia printed circuit board. IEEE Trans Electron Packag Manuf 25:51–58

Lau J, Dauksher W (2005) Effects of ramp-time on the thermal-fatigue life of snagcu lead-free solder joints. In: Proceedings of the 55th IEEE electronic components and technology conference 2005

Li L, Yeung BH (2001) Wafer level and flip chip design through solder prediction models and validation. IEEE Trans Compon Packag Technol 24:650–654

Rinne H (2009) The Weibull distribution: a handbook. CRC Press, Boca Raton

Syed A (2004) Accumulated creep strain and energy density based thermal fatigue life prediction models for SnAgCu solder joints. Electron Componen Technol Conf 1:737–746

Wu ML, Barker D (2010) Rapid assessment of BGA life under vibration and bending, and influence of input parameter uncertainties. Microelectron Reliab 50:140–148

Yi S, Sze KY (1998) Finite element analysis of moisture distribution and hygrothermal stresses in TSOP IC packages. Finite Elem Anal Des 30:65–79

Clustering-Locating-Routing Algorithm for Vehicle Routing Problem: An Application in Medical Equipment Maintenance

Kanokwan Supakdee, Natthapong Nanthasamroeng and Rapeepan Pitakaso

Abstract This research is aimed to solve a vehicle routing problem for medical equipment maintenance of 316 health promoting hospitals in Ubon Ratchathani which conducted by maintenance department of Ubon Ratchathani Provincial Health Office by using clustering-locating-routing technique (CLR). We compared two different methods for clustering. The first method applied the sweep algorithm (SW-CLR) for clustering the health promoting hospital to 4 clusters and each cluster includes 79 hospitals. The second method used district boundary (DB-CLR) for clustering the hospital to 25 clusters. After that, load distance technique was used to determine a location of maintenance center in each cluster. Finally, saving algorithm was applied to solve the vehicle routing problem in each cluster. Both SW-CLR and DB-CLR can reduce transportation cost effectively compared with traditional route. The SW-CLR reduced overall annually maintenance cost 52.57 % and DB-CLR reduced cost 37.18 %.

Keywords Clustering-locating-routing · Vehicle routing problem · Medical equipment maintenance · Sweep algorithm · Saving algorithm

---

K. Supakdee (✉) · R. Pitakaso
Department of Industrial Engineering, Faculty of Engineering, Ubon Ratchathani University, Ubon Ratchathani, Thailand
e-mail: ksupakdee@hotmail.com

R. Pitakaso
e-mail: enrapapi@mail2.ubu.ac.th

N. Nanthasamroeng
Graduate School of Engineering Technology, Faculty of Industrial Technology, Ubon Ratchathani Rajabhat University, Ubon Ratchathani, Thailand
e-mail: nats@ubru.ac.th

1 Introduction

Nowadays, transportation cost was increased due to scarcity of fossil fuel. Therefore, logistics played an important role in both private and public sectors. If the firm could manage logistics activities effectively, they would have the competitive advantage among their rivalry.

Ubon Ratchathani Provincial Health Office (URPHO) had an objective to develop a standard for public health administration in the province. One of their activities was to take care and repair of medical equipments in 316 health promotion hospitals. Maintenance staffs had to travel with random routing from URPHO office to the hospitals. From information gathered in 2012, total distance of 150 routes was 22,945.20 km and cost 481,574.95 Thai-baht (THB).

This research is aimed to solve a vehicle routing problem for medical equipment maintenance of 316 health promoting hospitals in Ubon Ratchathani which conducted by maintenance department of Ubon Ratchathani Provincial Health Office by using clustering-locating-routing technique (CLR). We compared two different methods for clustering. The first method applied the sweep algorithm (SW-CLR) for clustering. The second method used district boundary (DB-CLR) for clustering the hospital.

2 Literature Review

2.1 Vehicle Routing Problem

In classical Vehicle Routing Problem (VRP), the customers are known in advance. Moreover, the driving time between the customers and the service times at each customer are used to be known. The classical VRP can be defined as follow: Let $G = (V, A)$ be a graph where $V = \{1 \ldots n\}$ is a set of vertices representing *cities with the depot* located at vertex 1, and A is the set of arcs. With every arc (i, j) $i \neq j$ is associated a non-negative distance matrix $C = (c_{ij})$. In some contexts, c_{ij} can be interpreted as a *travel cost* or as a *travel time*. When C is symmetrical, it is often convenient to replace A by a set E of undirected edges. In addition, assume there are m available vehicles based at the depot, where $m_L < m < m_U$. When $m_L = m_U$, m is said to be fixed. When $m_L = 1$ and $m_U = n - 1$, m is said to be free. When m is not fixed, it often makes sense to associate a fixed cost f on the use of a vehicle. The VRP consists of designing a set of least-cost vehicle routes in such a way that:

1. each city in V{1} is visited exactly once by exactly one vehicle;
2. all vehicle routes start and end at the depot;
3. some side constraints are satisfied.

The formulation of the VRP can be presented as follow.

Let x_{ij} be an integer variable which may take value $\{0, 1\}$, $\forall\{i, j\} \in E\backslash\{\{0, j\}: j \in V\}$ and value $\{0, 1, 2\}$, $\forall\{0, j\} \in E$, $j \in V$. Note that $x_{0j} = 2$ when a route including the single customer j is selected in the solution.

The VRP can be formulated as the following integer program:

$$\text{Minimise} \sum_{i \neq j} d_{ij} X_{ij} \tag{1}$$

Subject to:

$$\sum_j x_{ij} = 1, \quad \forall i \in V, \tag{2}$$

$$\sum_i x_{ij} = 1, \quad \forall j \in V, \tag{3}$$

$$\sum_i x_{ij} \geq |S| - v(S), \{S : S \subseteq V \setminus \{1\}, |S| \geq 2\}, \tag{4}$$

$$x_{ij} \in \{0, 1\}, \forall \{i, j\} \in E;\ i \neq j \tag{5}$$

In this formulation, (1), (2), (3) and (5) define a modified assignment problem (i.e., assignments on the main diagonal are prohibited). Constraints (4) are sub-tour elimination constraints: $v(S)$ is an appropriate lower bound on the number of vehicles required to visit all vertices of S in the optimal solution.

2.2 Sweep Algorithm

The sweep algorithm is a method for clustering customers into groups so that customers in the same group are geographically close together and can be served by the same vehicle. The sweep algorithm uses the following steps.

1. Locate the depot as the center of the two-dimensional plane.
2. Compute the polar coordinates of each customer with respect to the depot.
3. Start sweeping all customers by increasing polar angle.
4. Assign each customer encompassed by the sweep to the current cluster.
5. Stop the sweep when adding the next customer would violate the maximum vehicle capacity.
6. Create a new cluster by resuming the sweep where the last one left off.
7. Repeat Steps 4–6, until all customers have been included in a cluster.

2.3 Saving Algorithm

In the method of Clarke and Wright (1964), saving of combining two customers i and j into one route is calculated as:

$$S_{ij} = d_{oj} + d_{jo} - d_{ij} \tag{6}$$

where d_{ij} denotes travel cost from customer i to j. Customer "0" stands for the depot.

3 Mathematical Modeling

A mathematical model was formulated from the vehicle routing problem model. The objective function was focus on the capacitated vehicle routing problem (CVRP) which was calculate from health promoting hospitals, time for maintenance, labor cost, fuel consumption, maintenance costs and distance. The objective of CVRP is to minimize the traveling cost. The capacitated vehicle routing problem can be modeled as a mixed integer programming as follows:

$$\text{Minimize} \quad \sum_{i=0}^{N}\sum_{j=0}^{N}\sum_{K=1}^{K} C_{ij} X_{ij}^{k} \tag{7}$$

$$\text{Subject to} \quad \sum_{i=0}^{N}\sum_{j=0}^{N} X_{ij}^{K} d_i \leq Q^k \quad 1 \leq k \leq K, \tag{8}$$

$$\sum_{i=0}^{N}\sum_{j=0}^{N} X_{ij}^{K}(c_{ij} + S_i) \leq T^k \quad 1 \leq k \leq K, \tag{9}$$

$$\sum_{j=1}^{N} X_{ijk} = \sum_{j=1}^{N} X_{ijk} \leq 1 \quad for \, i = 0 \tag{10}$$

$$\text{and} \quad k \in \{1, \ldots, k\},$$

$$\sum_{i=0}^{N}\sum_{j=0}^{N} X_{ij}^{K} \leq K \quad for \, i = 0, \tag{11}$$

where C_{ij} is the cost incurred on customer i to customer j, K the number of vehicles, N the number of customers, the S_i the service time at customer i, Q_k the loading capacity of vehicle k, T_k the maximal traveling (route) distance of vehicle k, d_i the demand at customer i, $X_{ij}^{K} \in 0$ and 1 ($i \neq j;\ i,j \in 0, 1, \ldots, N$). Equation (6) is the objective function of the problem. Equation (7) is the constraint of loading capacity, where $X_{ij}^K = 1$ if vehicle k travels from customer i to customer j directly, and 0 otherwise. Equation (8) is the constraint of maximum traveling distance. Equation (9) makes sure every route starts and ends at the delivery depot. Equation (10) specifies that there are maximum K routes going out of the delivery depot.

4 Test Problem

The test instance used in this research was formulated form a real case study of 316 health promoting hospitals into 25 districts in Ubon Ratchathani.

5 Clustering-Locating-Routing Technique

The test instance used in this research was formulated form a real case study of 316 health promoting hospitals in Ubon Ratchathani.

5.1 Sweep Algorithm Clustering-Locating-Routing Technique

Figure 1.

5.1.1 Data

The data used in this research, including the time for maintenance, labor cost, fuel consumption, maintenance costs and distance.

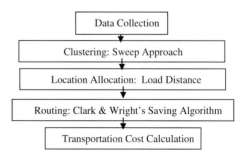

Fig. 1 SW-CLR procedure

5.1.2 Sweep Algorithm

The clustering the health promoting hospital to 4 clusters and each cluster includes 79 hospitals was shown in Fig. 2.

5.1.3 Choosing the Right Location to Find a Way with the Load-Distance Technique

Cluster 1 New location is Ban Khonsanhealth promoting hospital
Cluster 2 New location is Donjikhealth promoting hospital
Cluster 3 New location is Maiphattanahealth promoting hospital
Cluster 4 New location is Ban Kham health promoting hospital

5.1.4 Saving Value

Saving value was calculated by using Eq. (6). Some of calculation results were shown in Table 1.

5.1.5 Routing of Saving Algorithm

Route to health promoting hospital, all health cluster (Table 2), a total distance of 7,514.7 km is the fourth cluster (Table 3)

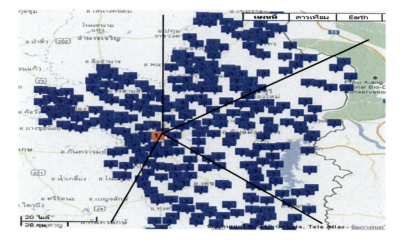

Fig. 2 Ubon Ratchathani Provincial Health Office by using clustering

Table 1 Example of saving value

	d_{0i}	d_{j0}	d_{ij}	Saving value
S1,2	18.1	47.6	42.6	23.1
S1,3	18.1	38.4	20.6	35.9
S1,4	18.1	26.6	8.9	35.8

Table 2 Example of route to health promoting hospital

Route	Sequence	Total distance (km)
1	0-24-32-9-41-0	77.8
2	0-6-10-47-79-0	75.4
3	0-36-42-11-50-0	84.1

Table 3 Total cost of fourth cluster

Cluster	Number of tours	Distance (km)	Total cost (THB)
1	18	1,620.5	51,252.55
2	20	2,237.8	59,548.47
3	21	1,976.2	60,376.44
4	20	1,680.2	56,302.12
Total	79	7,514.7	227,479.58

5.1.6 Transportation Cost

It can be calculated as Eq. (12) below

$$\text{Transportation cost} = [(C_f + C_m)d] + [(C_l r)] + [(C_h(r-1)] \qquad (12)$$

5.2 District Boundary Clustering-Locating-Routing Technique

District boundary clustering-locating-routing technique (DB-CLR) procedure was similar to SW-CLR procedure except the clustering method. In DB-CLR, district boundary was use to cluster health promoting hospital to 25 clusters. Each cluster contained different number of hospital as shown in Table 4 (Fig. 3).

5.3 Comparison Result

The comparison between SW-CLR and DB-CLR was shown in Table 5. From the result, SW-CLR contributed the lower total cost than DB-CLR algorithm 15.58 %

Table 4 Clustering result for with district boundary

No.	Province	Number of hospital	No.	Province	Number of hospital
1	Mueang	17	14	Na Tan	7
2	Don Mot Daeng	4	15	Khemarat	11
3	Lao SueaKok	7	16	DetUdom	26
4	Tan Sum	8	17	Buntharik	17
5	Phibun Mangsahan	20	18	Na Chaluai	10
6	Sirindhorn	8	19	Nam Yuen	13
7	Khong Chiam	11	20	Muang Sam Sip	24
8	Warin Chamrap	19	21	KhueangNai	31
9	Sawang Wirawong	7	22	Nam Khun	7
10	Trakan Phuet Phon	29	23	Thung Si Udom	7
11	Si Mueang Mai	17	24	Na Yia	6
12	Pho Sai	11	25	Samrong	13
13	Kut Khaopun	9			

Fig. 3 DB-CLR procedure

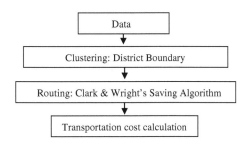

Table 5 Comparison between SW-CLR and DB-CLR

	Traditional	SW-CLR	DB-CLR
Total cost (THB)	481,574.95	227,479.58	298,497.85
Decrease (%)	–	52.76	37.18

6 Conclusions

We compared two different methods for clustering. The first method applied the sweep algorithm (SW-CLR) for clustering the health promoting hospital to 4 clusters and each cluster includes 79 hospitals. The second method used district boundary (DB-CLR) for clustering the hospital to 25 clusters. After that, load distance technique was used to determine a location of maintenance center in each cluster. Finally, saving algorithm was applied to solve the vehicle routing problem in each cluster. Both SW-CLR and DB-CLR can reduce transportation cost effectively compared with traditional route. The SW-CLR reduced overall annually maintenance cost 249,774.84 bath 52.57 % and DB-CLR reduced cost 176,652.44 bath 37.18 %

Reference

Clarke G, Wright JW (1964) Scheduling of vehicles from a central depot to a number of delivery points. Oper Res 12:568–581

Whole-Body Vibration Exposure in Urban Motorcycle Riders

Hsieh-Ching Chen and Yi-Tsong Pan

Abstract Twenty-two male and twenty-three female motorcycle riders performed ninety test runs on six 20-km paved urban routes. Root mean square of acceleration, 8-hour estimated vibration dose value ($VDV_{(8)}$), and 8-hour estimated daily static compression dose (S_{ed}) were determined in accordance with ISO 2631-1 (1997) and ISO 2631-5 (2004) standards. The analytical results indicated that over 90 % of the motorcycle riders revealed $VDV_{(8)}$ exceeding the upper boundary of health guidance caution zone (17 m/s$^{1.75}$) recommended by ISO 2631-1 or S_{ed} exceeding the value associated with a high probability of adverse health effects (0.8 MPa) according to ISO 2631-5. Over 50 % of the motorcycle riders exceeded exposure limits for VDV and S_e within 3 h. Significantly greater exposure levels were observed in male participants than in female participants for $VDV_{(8)}$ ($p < 0.05$) and S_{ed} ($p < 0.005$). The health impacts of WBV exposure in motorcycle riders should be carefully addressed with reference to ISO standards.

Keywords ISO 2631 standards · Motorbike · Health risk

1 Introduction

Commonly reported health effects caused by whole-body vibration (WBV) exposure include discomfort, musculoskeletal problems, muscular fatigue, reduced stability, and altered vestibular function (Seidel 1993; Wasserman et al. 1997;

H.-C. Chen (✉)
Department of Industrial Engineering and Management, Taipei University of Technology, No. 1, Sec. 3, Zhongxiao E. Road, Taipei 10608, Taiwan, Republic of China
e-mail: imhcchen@ntut.edu.tw

Y.-T. Pan
Institute of Occupational Safety and Health, Council of Labor Affairs, Executive Yuan, No. 99, Lane 407, Hengke Road, Xizhi, New Taipei 22143, Taiwan, Republic of China
e-mail: yitsong@mail.iosh.gov.tw

Bongers et al. 1988; Griffin 1998). Several studies have indicated that long-term WBV exposure is associated with early spinal degeneration (Frymoyer et al. 1984), low back pain, and herniated lumbar disc (Bovenzi and Zadini 1992; Boshuizen et al. 1992).

Motorcycles are a common transportation mode in Asia. An estimated 33 million motorcycles are used in mainland China, 18 million in Vietnam, 13.8 million in Taiwan, 13 million in Japan, 5.8 million in Malaysia, 2.7 million in Korea, and 1.8 million in the Philippines (IRF 2006). Most motorcycles in Taiwan can be categorized as scooters (i.e., no clutch, seated riding position) or motorbikes (equipped with clutch, straddled riding position). These motorcycles generally have 125 cc engines or smaller, are ridden on shoulders or in reserved lanes, and are convenient for accessing driving lanes. Although motorcycles are typically used only for short-distance transport, they are the main transportation mode for workers such as postal workers, delivery workers, and urban couriers. Consequently, these occupations are likely associated with high WBV exposure.

Health problems associated with WBV exposure in motorcycle riders are often overlooked despite the potentially large size of the population. Chen et al. (2009) reported high WBV exposure in twelve male motorcycle riders traveled on a 20.6 km rural–urban paved road according to ISO 2631-1 (1997) and ISO 2631-5 (2004) standards. However, the generalizability of the experimental results in that study is limited since they analyzed only male subjects riding on one specific route.

This study measured the WBV exposure of motorcycle riders while riding on urban routes with standard paved surfaces. The vibration exposure in motorcycle riders was compared with the upper boundary of health guidance caution zone (HGCZ) recommended by ISO 2631-1 (1997) $\left(\overline{VDV}\right)$ and with the limit value associated with a high probability of adverse health effects according to ISO 2631-5 (2004) $\left(\overline{S_{ed}}\right)$.

2 Method

2.1 Participants and Vehicles

This study analyzed 45 university students (23 male and 22 female) who volunteered to participate in motorcycle riding tests. All subjects rode their own motorcycles, and all had at least 2 years of motorcycle riding experience.

Table 1 presents the detailed characteristics of the participants and their vehicles. The male participants were significantly taller and heavier than the female participants ($p < 0.001$, t test) were. Males were also significantly older (3.2 years) in age, and had more years of riding experience as well as more years of vehicle ownership (3.6 and 1.9 years, respectively) ($p < 0.001$, t-test) than the female participants did. The number of male participants who rode 125 cc motorcycles (90.9 %) was significantly higher than the number of female participants who rode 125 cc motorcycles (30.4 %) ($p < 0.001$, Mann-Whitney).

Table 1 Characteristics of the participants and motorcycle (N = 45)

Participant						Motorcycle			
Gender	Age (year)	Height (cm)	Weight (kg)	Experience (year)		Engine size(cc)	Wheel size(in)	Years (year)	Manufacturer (n)
Male (n = 22)	24.5 (2.6)	172.9 (6.8)	72.5 (14.9)	6.7 (2.8)		125 * 20 100 * 2	10	5.7 (2.1)	Yamaha (5) Sanyang (11) Kymco (6)
Female (n = 23)	21.3 (2.5)	160.2 (3.8)	52.8 (8.0)	3.1 (2.4)		125 * 7 100 * 16	10	3.8 (3.0)	Yamaha (4) Sanyang (11) Kymco (8)
Gender diff. p-value	<0.001 (t-test)	<0.001 (t-test)	<0.001 (t-test)	<0.001 (t-test)		<0.001 (Mann-Whitney test)	–	0.001 (t-test)	–

2.2 Equipment and Field-Testing Procedure

A triaxial ICP seat pad accelerometer (model 356B40, Larson Davis Inc., USA) was employed to measure vibrations transmitted to the seated human body as a whole through the supporting surface of the buttock. Seat pad outputs were connected to a 3-channel amplifier (model 480B21, PCB Piezotronics Inc., USA). The outputs of the amplified signals were recorded on a portable data logger which acquires three analog signals each at a rate of 5 k samples/s. The logger can continually store collected data on a 2 GB compact flash (CF) memory card up to 2 h.

A GPS device was used to provide geographical information as participants rode on an assigned route. Vocal messages from the GPS were transmitted to participants via earphones. The view of a motorcycle rider was recorded by a portable media recorder and a minicamera. The recorded video was synchronized with the logged acceleration data using a remote transmitter, which sent radio frequency signals at the beginning and end of the riding task.

The test runs were performed on six 20-km paved urban routes. All routes started from a cafe located in downtown Taichung City and followed various main roads of Taichung City before returning to the cafe. The standard testing procedure for the riding test was explained, and detailed instructions were given to all participants before the test. Each motorcycle rider performed two different riding tasks on randomly assigned routes. Each motorcycle rider wore a helmet with a mini-camera taped on its frontal side and carried a portable media recorder in a case with a shoulder strap (Fig. 1). Each rider also wore a backpack containing the signal amplifier, the data logger, and a rechargeable battery set. The backpack weighed approximately 1.5 kg. The GPS device was affixed with a suction cup and tape to the control panel of test motorcycle.

No speed limits were imposed on participants. However, the subjects were instructed to comply with urban speed limits. Each participant was asked to remain seated throughout the riding task. To avoid harmful shocks, the riders were instructed to either avoid or to slowly drive over any potholes, manhole covers, humps, or uneven road surfaces. Each task required 50–60 min. The duration of each task was measured from the time the participant exited the cafe parking lot to the time that the participant returned.

2.3 Data Analysis

'Viewlog' software programmed with LabVIEW 7.0 (National Instruments, USA) was applied to download the logged data from the CF card, and to combine the data with the taped video. 'Viewlog' consists of calibration, vibration analysis, script interpretation and batch processing modules to facilitate analysis and processing of data in bulk. The vibration analysis module, developed under a research contract of Taiwan IOSH (2007), evaluates WBV exposure for both ISO 2631-1

Fig. 1 Experimental apparatus

(1997) and ISO 2631-5 (2004). The module was specifically designed to perform batch computing and export the results to a user-defined MS Excel template. The Excel report employed an embedded macro program to calculate the estimated 8-h *RMS*, *VDV* and S_e. Detailed data processing and artifact removal procedures were documented in previous study (Chen et al. 2009).

The statistical analysis was performed with SPSS 10 for Windows. Gender differences in participant and motorcycle characteristics were assessed by independent *t*-test for all scale variables and by Mann-Whitney test for the nominal variable (engine size). According to both Pearson and Spearman rho correlations, participant age, body height and weight, riding experience, and vehicle years were all significantly correlated (r range, 0.401–0.952, p < 0.01). Therefore, only riding experience was applied as an independent variable in ANOVA analyzes. The differences among all measurements obtained in the motorcycle ride tests were compared by univariate ANOVA using gender, route, and engine size as fixed factors and riding experience and measurement period as covariates. A *p* value less than 0.05 was considered statistically significant.

3 Results

Male participants completed the test much faster than female participants did (male = 50.2 ± 6.6 min, female = 56.9 ± 11.1 min) ($p < 0.05$, Table 2). However, the measurement period was not significantly affected by route and engine size.

Table 2 Mean ± SD [range] of RMS, VDV$_{(8)}$, S$_{ed}$, and riding period

Gender	Engine size(cc)	N	(1997)		(2004)	
			RMS§ (m/s^2)	VDV$_{(8)}$* (m/s$^{1.75}$)	S$_{ed}$** (MPa)	Riding period* (s)
Male	125	40	0.81 ± 0.13 [0.56 – 1.11]	23.49 ± 4.20 [16.16 – 34.57]	1.17 ± 0.41 [0.68 – 2.49]	3062 ± 418 [2,280 – 4,050]
	100	4	0.88 ± 0.03 [0.85 – 0.91]	24.35 ± 1.46 [23.57 – 26.38]	1.51 ± 0.21 [1.23 – 1.75]	2,903 ± 250 [2,640 – 3,240]
Female	125	14	0.72 ± 0.08 [0.59 – 0.90]	19.57 ± 2.78 [16.50 – 26.79]	0.88 ± 0.20 [0.65 – 1.26]	3,403 ± 551 [2,280 – 4,260]
	100	32	0.80 ± 0.11 [0.56 – 1.02]	22.05 ± 3.04 [16.84 – 28.37]	1.08 ± 0.36 [0.64 – 2.61]	3,419 ± 721 [2,580 – 5,370]

*$p < 0.05$, **$p < 0.005$; significant gender difference (ANOVA)
$p < 0.05$, significant engine size effect (ANOVA)
§ $p < 0.01$, significant riding period effect (ANOVA)

The analytical results indicated that acceleration in the z-axis generated the most severe total *RMS* and *VDV* levels. Therefore, *RMS* and *VDV* vibration parameters in this study were determined by frequency-weighted acceleration in the z dominant axis. Significant gender differences were observed in *VDV*$_{(8)}$ ($p < 0.05$) and S_{ed} ($p < 0.005$, Table 2). Nevertheless, only *RMS* revealed a significant association with measurement period ($p < 0.01$) and engine size ($p < 0.05$). Route and riding experience revealed no associations with vibration parameters. In all 20-km trials, vibration exposures exceeded the lower boundary of HGCZ recommended by ISO 2631-1 (1997) or the value associated with a low probability of adverse health effects according to ISO 2631-5 (2004); 32 and 2 % trials exceeded $\overline{S_{ed}}$ and $\overline{VDV}$ (upper boundary), respectively (Fig. 2). In approximately 97 % of the trials, the 8-h predicted vibration exposures exceeded either $\overline{VDV}$ or $\overline{S_{ed}}$, and in 83 % trials, they exceeded both (Fig. 2). Statistical results revealed both linear and nonlinear (power) regression lines through the conjunction of the two caution zones and 8-h predicted regression lines revealed a tendency toward increased *VDV*.

For most motorcycle riders, daily motorcycle use is typically for short periods. Therefore, allowable durations (T_a) for *RMS*, *VDV*, and S_e to reach corresponding values of $\overline{RMS}$, $\overline{VDV}$, and $\overline{S_{ed}}$ were computed for each riding task. Figure 3 shows the cumulative probability of T_a to reach $\overline{RMS}$, $\overline{VDV}$ and $\overline{S_{ed}}$. Analytical results of S_e and *VDV* showed that 50 % exposures were limited to about 2 and 3 h, respectively. Nevertheless, T_a to reach $\overline{RMS}$ was significantly longer than that to reach $\overline{VDV}$ and $\overline{S_{ed}}$. For 1-h daily exposure, 35 and 9 % motorcycle riders were restricted according to ISO 2631-5 and ISO 2631-1 standards, respectively (Fig. 3). Besides the difference between the two ISO standards, T_a also revealed significant gender differences. Fifty percent of male participants but only 25 % of female participants revealed T_a less than 1.2 h according to S_e. Similarly, 50 % of male participants had T_a less than 2.3 h according to *VDV*, but less than 35 % female participants had T_a less than 2.3 h.

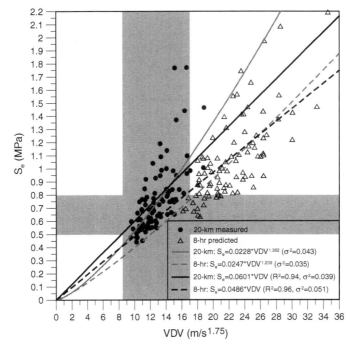

Fig. 2 VDV-S_e scatter plot of all tasks, with the caution zones (*grey bands*) of ISO 2631-1 and ISO 2631-5 (*solid symbol* 20-km measured, *hollow symbol* 8-h predicted)

Fig. 3 Cumulative probability distribution of allowable duration (T_a) to reach limit values of $\overline{VDV}$ and $\overline{S_{ed}}$ according to ISO 2631-1 (1997) and ISO 2631-5 (2004) standards

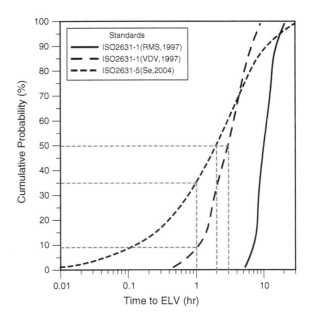

4 Discussion

This study suggests that, after prolonged traveling on typical paved roads, vibration exposure may be greater in motorcycle riders than in 4-wheel vehicle drivers. Previous studies report WBV exposure of 0.17–0.55 m/s^2 *RMS* among urban taxi drivers in Taiwan (Chen et al. 2003) and 0.34–0.56 m/s^2 *RMS* among highway transport truck drivers (Cann et al. 2004). In this research, significantly increased WBV exposure (ranged 0.56–1.11 m/s^2 *RMS*) was observed in urban motorcycle riders. To compare WBV exposure of motorcycle riders in this study with that of industrial vehicle drivers of previous studies, experimental measurements were extrapolated to predict 8-h exposure of *VDV* and S_e. Rehn et al. (2005) indicated that $VDV_{(8)}$ in operators of forwarder vehicles were in the 2.92–18.9 m/s$^{1.75}$ range, and in all tested operators, T_a reached $\overline{VDV}$ for longer than 8 h. Eger et al. (2008) estimated the 8-h WBV exposure in seven operators of load-haul-dump mining vehicle after 7 h operation and 1 h resting and found that $VDV_{(8)}$ was 12.38–24.67 m/s$^{1.75}$, and S_{ed} was 0.21–0.59 MPa. The $VDV_{(8)}$ (range 16.16–34.57 m/s$^{1.75}$) and S_{ed} (range 0.64–2.49 MPa) obtained from motorcycle rides exceeded above mentioned vehicles and many other passenger and industrial vehicles reported by Paddan and Griffin (2002).

Chen et al. (2009) reported that, after 8 h of exposure, over 90 % of all motorcycle rides would produce $VDV_{(8)}$ (mean 23.1 m/s$^{1.75}$) and S_{ed} (mean 1.15 MPa) values over the corresponding $\overline{VDV}$ and $\overline{S_{ed}}$, which indicates a high probability of adverse health effects. The current study revealed a similar conclusion for motorcycle riding in urban areas. Notably, the use of fundamental *RMS* evaluation (ISO 2631-1 1997) probably underestimates the health risks of multiple shocks, especially after prolonged exposure. Therefore, certain Asian countries that top use WBV guidelines solely based on frequency-weighted accelerations and *RMS* index may overlook the potential health risks caused by WBV exposure in motorcycle riders.

The conclusion drawn in this study may not be applicable to all motorcycle riding conditions. The experimental setup was limited to the use of young subjects (<30 years) and main roads of Taichung city. The possible effects of age on speed and personal riding characteristics as well as the effects of road condition on vehicle vibration may limit the ability to generalize the experimental results to other riding cases. Additionally, the testing conditions were limited to single riders and the most common motorcycle engine sizes (100 and 125 cc) in Taiwan. Heavier loads can also affect riding speed, acceleration, and vibration characteristic of the system. Therefore, the WBV results for heavy-duty motorbikes or motorcycles ridden with a passenger or additional weight may differ from those reported here.

The experimental results of this study indicate the need for caution in occupations in which motorcycles are the major transportation mode (e.g., postal workers, police officers, delivery workers, and some urban couriers). Further research in the dose–response relationship of these workers should also be explored further to confirm the findings of this study.

Acknowledgments The authors wish to thank the Institute of Occupational Safety and Health for financially supporting this research study (IOSH97-H317).

References

Bongers PM, Boshuizen HC, Hulshof TJ, Koemeester AP (1988) Back disorders in crane operators exposed to whole-body vibration. Int Arch Occup Env Health 60:129–137

Boshuizen HC, Bongers PM, Hulshof CT (1992) Self-reported back pain in fork-lift truck and freight-container tractor drivers exposed to whole-body vibration. Spine 17:59–65

Bovenzi M, Zadini A (1992) Self-reported low back symptoms in urban bus drivers exposed to whole-body vibration. Spine 17:1048–1059

Cann AP, Salmoni AW, Eger TR (2004) Predictors of whole-body vibration exposure experienced by highway transport truck operators. Ergonomics 47:1432–1453

Chen JC, Chang WR, Shih TS, Chen CJ, Chang WP et al (2003) Predictors of whole-body vibration levels among urban taxi drivers. Ergonomics 46:1075–1090

Chen HC, Chen WC, Liu YP, Chen CY, Pan YT (2009) Whole-body vibration exposure experienced by motorcycle riders—An evaluation according to ISO 2631-1 and ISO 2631-5 standards. Int J Ind Ergon 39:708–718

Eger T, Stevenson J, Boileau P-É, Salmoni A, Vib RG (2008) Predictions of health risks associated with the operation of load-haul-dump mining vehicles: Part 1—Analysis of whole-body vibration exposure using ISO 2631-1 and ISO 2631-5 standards. Int J Ind Ergon 38:726–738

Frymoyer JW, Newberg A, Pope MH, Wilder DG, Clements J, MacPherson B (1984) Spine radiographs in patients with low-back pain. An epidemiology study in men. J Bone Joint Surgery Am Vol 66:1048–1055

Griffin MJ (1998) General hazards: vibration. Encyclopedia of occupational health and safety (International Labour Organization Geneva), 50.2–50.15

International Organization for Standardization (1997) ISO2631-1 Mechanical vibration and shock–evaluation of human exposure to whole-body vibration. Part 1: general requirements. ISO, Geneva

International Organization for Standardization (2004) ISO2631-5 Mechanical vibration and shock–evaluation of human exposure to whole-body vibration. Part 5: method for evaluation of vibration containing multiple shocks. ISO, Geneva

IRF (2006) World road statistics 2006—data 1999 to 2004 International Road Federation (IRF), Geneva, Switzerland

Paddan GS, Griffin MJ (2002) Effect of seating on exposures to shole-body vibration in vehicles. J Sound Vibr 253:215–241

Rehn B, Lundström R, Nilsson L, Liljelind I, Jörvholm B (2005) Variation in exposure to whole-body vibration for operators of forwarder vehicles—aspects on measurement strategies and prevention. Int J Ind Ergon 35:831–842

Seidel H (1993) Selected health risks caused by long-term whole-body vibration. Am J Ind Med 23:589–604

Taiwan IOSH (2007) Development of software and field testing tools for assessing whole-body vibration with shocks. Taiwan Institute of Occupational Safety and Health (IOSH), Council of Labor Affairs, Taiwan. Contract Report No. IOSH 96-H318 (in Chinese)

Wasserman DE, Wilder DG, Pope MH, Magnusson M, Aleksiev AR, Wasserman JF (1997) Whole-body vibration exposure and occupational work-hardening. J Occup Env Med 39:403–407

Analysis of Sales Strategy with Lead-Time Sensitive Demand

Chi-Yang Tsai, Wei-Fan Chu and Cheng-Yu Tu

Abstract Nowadays, more and more customers order products through non-conventional direct sales channel. Instead of purchasing from retail stores, customers place orders directly with the manufacturer and wait for a period of lead-time for the ordered items to be delivered. With limited production capacity, it is possible that the manufacturer is unable to deliver all the orders in regular delivery time if too many orders are placed in a period of time. The delivery lead-time may become longer than expected and that can have an impact on future demand. This study assumes customer demand is sensitive to the length of lead-time. That is, demand decreases as the actual delivery lead-time in the previous period becomes longer. Mathematical models are constructed for the considered multiple-period problem. Numerical experiments and sensitivity analysis are conducted to examine how lead-time sensitive demand affects system behaviors. It is observed that the system behaviors heavily depend on the size of initial demand. It is also found that when initial demand is greater than the fixed capacity, the cumulated profit of the manufacturer may increase in the beginning. However, it will decline in the long run.

Keywords Lead-time sensitive demand · Direct sales channel · Pricing

---

C.-Y. Tsai (✉) · W.-F. Chu · C.-Y. Tu
Department of Industrial Engineering and Management, Yuan Ze University,
135 Yuan-Tung Rd, Chung-Li, Tao Yuan, Taiwan
e-mail: iecytsai@saturn.yzu.edu.tw

W.-F. Chu
e-mail: s995414@mail.yzu.edu.tw

C.-Y. Tu
e-mail: s1005422@mail.yzu.edu.tw

1 Introduction

The market of direct sales has been growing significantly in the past years. Instead of purchasing from retail stores, more and more customers purchase products directly from manufacturers. Usually, the sales price of a product purchased from the direct sales channel is lower due to the reduced costs by bypassing the retailer. It is a motivation for customers to select direct sales channel. Therefore, it is common that pricing strategies are applied in direct sales channel system to increase demand. In addition, one major difference between the direct sales channel and the traditional retail sales channel is that it takes a period of lead time to deliver the ordered product to the customer. With finite production capacity, the manufacturer can be overwhelmed by excessive demand and fails to deliver within the promised lead time. As the delivery lead time lengthens, customers are discouraged and future demand may drop.

This study attempts to develop a mathematical model for the direct sales channel system with lead-time-sensitive and price-sensitive demand. The model is used to analyze the behaviors of the system and explore the use of pricing strategy to maximize profits.

In a direct sales channel, manufacturers sell products directly to customers through the internet, their owned stores or mail order (Kotler and Armstrong 1994). Lead time is the time it takes from when customers confirm their order through the internet until customers receive the ordered products (Hua et al. 2010). Reducing delivery lead time may reduce the resistance of customers to direct purchase (Balasubramanian 1998). Liao and Shyu (1991) proposed an inventory model where lead time is considered as a decision variable. Ben-Daya and Raouf (1994) extended their model and considered both lead time and order quantity as decision variables with shortages prohibited. Ouyang et al. (1996) further extended their model to allow shortages. Hua et al. (2010) constructed a dual-channel (direct and retail) system and identified the optimal price and lead time that maximize profit.

The rest of the paper is organized as follows. Section 2 describes the considered direct sales channel system and the constructed mathematical model. A base instance is presented. Section 3 provides the conducted numerical experiment. The last section concludes the paper.

2 Problem Description and Model

This study considers a direct channel system in a multi-period setting. The system contains a manufacturer who produces a single product. It takes orders directly from customers and delivers order to customers. The production capacity per period of the manufacturer is assumed to be finite and fixed. If demand of a period cannot be satisfied in the same period, it is backordered. The available total amount of the product the manufacturer can supply in a period is the production

Analysis of Sales Strategy with Lead-Time Sensitive Demand

capacity. If the available quantity is enough to cover the backordered quantity and the demand of that period, the length of time to deliver is one period. For demand that is backordered, the delivery lead time becomes longer.

It is assumed that demand is sensitive to the actual length of the delivery lead time. To be more precise, the amount of demand in a period is set to be a function of the actual length of the lead time for demand in the previous period. If the actual lead time is longer than the promised length of one period, demand in the next period declines.

2.1 Model

The following notation will be used.

a	Primary demand
α	Demand sensitivity to lead time
b	Demand sensitivity to price
k	Backorder cost per unit per period
Q	Production capacity per period, in units
u	Production cost per unit
D_i	Demand in period i
L_i	Lead time in period i
O_i	Backorders in period i
q_i	Production quantity in period i
P	Sales price per unit

Demand from customers is assumed to be sensitive to the sales price and the lead time. It is expressed as

$$D_i = D_0 - \alpha D_0 \left(1 - \frac{1}{L_{i-1}}\right), \tag{1}$$

where $D_0 = a - bP$ is the base demand. The demand in period i decreases as the lead time in the previous period increases.

The production quantity in period i can be written as $q_i = \min\{Q, D_i + O_{i-1}\}$. Backorders in period i is $O_i = D_i + O_{i-1} - q_i$, where $O_0 = 0$. The actual length of the lead time in period i is determined as

$$L_i = 1 + \frac{O_i}{Q}. \tag{2}$$

The basic lead time is one period. The second term on the right hand side of the equation represents the expected length of time to fill the current backorders. The actual lead time is equal to the basic lead time when there are no backorders. It becomes longer with larger backorder quantity. It is assumed that $L_0 = 1$.

Costs that occur in each period include production cost and backorder cost. Revenue comes from taking orders from customers. Profit per period then can be written as $\pi_i = PD_i - (uq_i + kO_i)$. If the system operates n periods, total profit is

$$\pi_n^T = \sum_{i=1}^{n} \pi_i - uO_n, \qquad (3)$$

where the last term is the production cost of the backorders at the end of period n.

When the base demand, D_0, is not greater than the production capacity per period, Q, the delivery lead time in period 1 is one period. Thus, the demand in the next period will be equal to the base demand. It then can easily be seen that demand per period will remain at the base demand. Therefore, the following analysis will focus on the case when the base demand is greater than the production capacity per period ($D_0 \geq Q$).

2.2 Base Instance

A base instance is presented to demonstrate the behavior of the system. The parameter setting of the base instance is shown in Table 1. Figure 1 compares the demand per period at the two sales prices, 500 and 480.

When the sales price is 500, the base demand is equal to the production capacity. There are no backorders and lead time is always at one period. Thus,

Table 1 Parameter setting of base instance

Parameters	Value
a	105,000
b	200
α	0.3
Q	5,000
k	10
u	380

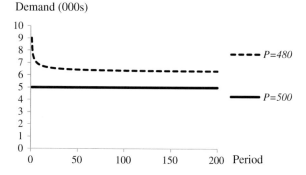

Fig. 1 Comparison of demand per period

demand period remains at the same level. With a sales price of 480, the base demand is 9,000, which is higher than the production capacity. Since backorders occur, lea time becomes longer. As a result, demand per period drops but remains above the production capacity.

Figure 2 illustrates the comparison of the total profits at the two sales prices. As can be seen, the total profit increases linearly when the sales price is 500. It is because demand per period is steady at 5,000 at this sales price and there are no backorders. When the sales price is 480, the total profit goes up in the first place as well, even at a higher increment rate. The increment slows down gradually, though, until the total profit reaches a highest level. It then starts dropping afterwards and keeps dropping at an accelerated speed. Demand is boosted by a lower sales price. However, since demand per period becomes greater than the production capacity per period, backorders occur and accumulate. The total profit is rising at a faster speed in the beginning due to the revenue generated by the boosted demand. The delivery lead time is then prolonged due to the growing backorders. The demand per period decreases and so does the revenue per period. Furthermore, the total profit is eaten away gradually by the growing backorder cost and eventually may become negative.

The trend indicates that it is possible to create greater profit by lower the sales price and attract more demand. However, there is a risk with this strategy when the production capacity is limited. The delivery lead time can be prolonged due to the overwhelming demand and the backorders become too much of a burden for the system to bear.

3 Numerical Experiment

Numerical experiment is conducted to further explore the behaviors of the system constructed in the previous section. First, sales price setting strategy with the consideration of profit maximization is discussed. Next, sensitivity analysis on the various parameters is carried out.

Fig. 2 Comparison of total demand
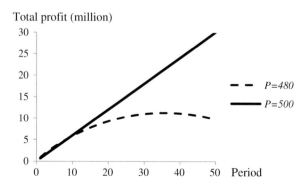

3.1 Pricing Strategy Analysis

Analysis is performed on how the sales price affects the system. The parameter setting of the base instance is used and the sales price is set at 500, 480, 460, 440 and 420, respectively. Table 2 shows the comparison.

When the sales price is set at 480, the highest total profit is 11,268.498. It is reached after 36 periods. The total profit starts falling afterwards and reaches zero when the system operates for 74 periods. The fall continues after that and the total profit becomes negative. As shown, the highest total profit increases as the sales becomes closer to the base price, which is 500. The best number of periods and the break-even number of periods are greater as well.

Table 3 compares the total profit in each period under various sales prices in more detail. Values in bold face are the greatest total profit in the same period. During period 1 to period 4, the total profit is the greatest with the sales price of 460. It is the greatest with the sales price of 480 from period 5 to period 9 and with the sales price of 500 after period 10. It implies that the best sales price setting depends on how long the system is intended to run. If the system is intended to run for only a short duration of time, a lower sales price should be set in order to quickly attract more demand. If it is planned to run the system for longer duration, a higher sales price is preferred.

3.2 Sensitivity Analysis

Sensitivity analysis is conducted to further explore the behaviors of the system. The value of one parameter is changed while all the other parameter values are fixed as in the basic instance except the sales price which is set at 480. The sales

Table 2 Comparison under various sales prices

Sales price	Highest total profit	Best number of periods	Break-even number of periods
480	11,268,498	36	74
460	5,599,691	15	30
440	3,192,569	9	17
420	1,464,201	5	10

Table 3 Total profits in each period under various sales prices

Sales price	Period											
	1	2	3	4	5	6	7	8	9	10	11	12
500	60	120	180	240	300	360	420	480	540	600	660	720
480	86	157	222	283	341	396	448	497	544	589	632	673
460	96	167	230	285	335	379	418	453	482	507	526	541
440	90	150	199	239	271	294	310	318	319	312	298	277
420	68	105	130	144	146	138	120	91	52	3	−55	−124

Table 4 Parameter setting in sensitivity analysis

Parameter	Value				
α	0	0.3	0.5	0.7	0.9
b	10	50	100	150	20
k	0	10	20	30	
u	300	340	380	320	

Table 5 Results of sensitivity analysis

Parameter	Highest total profit	Best number of periods	Break-even number of periods
$\alpha \uparrow$	$\uparrow$	$\uparrow$	$\uparrow$
$b \uparrow$	$\downarrow$	$\downarrow$	$\downarrow$
$k \uparrow$	$\downarrow$	$\downarrow$	$\downarrow$
$u \uparrow$	$\downarrow$	$\downarrow$	$\downarrow$

price is set at this level such that the base demand is greater than the production capacity. Table 4 lists the parameter setting in the sensitivity analysis. The results are summarized in Table 5.

When demand sensitivity to lead time increases, the highest total profit, the best number of periods and the break-even number of periods all increase as well. It is because the base demand is higher than the production capacity. Backorders accumulate and thus lead time becomes longer. With greater demand sensitivity to lead time, demand per period drops more quickly to the level where the manufacturer can better manage with its limited production capacity. On the other hand, demand per period is reduced in a slower pace if customers are not very sensitive to lead time. The fast-accumulating backorders quickly erodes the profit. The total profit starts declining earlier and quickly reaches zero. Demand sensitivity to price has the opposite effect on the system. As it increases, the base demand grows and the highest total profit, best number of periods and break-even number of periods all drop. The backorder cost per unit per period and the production cost per unit have inference on the system in the same fashion. All the three indexes go down as they rise.

4 Conclusions

This study considers a multi-period direct sales channel system. A single manufacturer produces a single product with finite production capacity. Demand for a single product is price sensitive and lead time sensitive. A mathematical model is constructed for the system and numerical experiment is conducted for the purpose of analyzing its behaviors. It is found that if the base demand does not surpass the production capacity, demand per period is stable as the delivery lead time maintains at the promised level and thus profit per period is constant. When the base demand is greater than the production capacity, total profit may grow more quickly in the beginning. However, the increment slows down gradually and eventually the total profit starts declining. It is suggested that if the system is intended to operate for a long duration of time, the sales price should be set such that the base demand matches the production capacity. On the other hand, if the planned operating duration is short, e.g. the sold product has a short life cycle, a lower sales price can be set to quickly attract more demand and boost profit in a short term.

The setting of the relationship between the demand per period and the lead time length is primitive and simple. A setting that better grasps the nature of the demand sensitivity to lead time can be developed and applied to the proposed model in future research work.

References

Balasubramanian S (1998) Mail versus mall: a strategic analysis of competition between direct marketers and conventional retailers. Mark Sci 17:181–195

Ben-Daya M, Raouf A (1994) Inventory models involving lead time as decision variable. J Oper Res Soc 45:579–582

Hua GW, Wang SY, Cheng TCE (2010) Price and lead time decisions in dual-channel supply chains. Eur J Oper Res 295:113–126

Kotler P, Armstrong G (1994) Principles of marketing. Prentice Hall, New Jersey

Liao CJ, Shyu CH (1991) An analytical determination of lead time with normal demand. Int J Oper Prod Manag 11:72–78

Ouyang LY, Yeh NC, Wu KS (1996) Mixture inventory model with backorders and lost sales for variable lead time. J Oper Res Soc 47:829–832

Order and Pricing Decisions with Return and Buyback Policies

Chi-Yang Tsai, Pei-Yu Pai and Qiao-Kai Huang

Abstract This paper considers application of return and buyback policies in a supply chain system. The system contains a manufacturer and a retailer. The manufacturer produces and sells a single product to the retailer and promise to buy back all the remaining units from the retailer at the end of the selling season. The retailer orders from the manufacturer before the season and applies a return policy to its customers with full refund. Customer demand is assumed to be sensitive to the retail price. Before the beginning of the selling season, the manufacturer offers the wholesale price and the buyback to the retailer. The retailer then determines the order quantity and the retailer price. Two types of control are investigated. Under decentralized sequentially decision making, both the manufacturer and the retailer make their decisions to maximize their own profit. Under centralized control, all the decisions are jointly made to maximize the profit of the whole system. Mathematical models under the two types of control are constructed and optimal order and pricing decisions are derived. The optimal decisions and the resulting profits are compared. We show that centralized control always generates higher overall profit.

Keywords Return policy · Buyback policy · Supply chain · Price-sensitive demand

---

C.-Y. Tsai (✉) · P.-Y. Pai · Q.-K. Huang
Department of Industrial Engineering and Management, Yuan Ze University,
135 Yuan-Tung Road, Chung-Li, Tao Yuan, Taiwan
e-mail: iecytsai@saturn.yzu.edu.tw

P.-Y. Pai
e-mail: s1005407@mail.yzu.edu.tw

Q.-K. Huang
e-mail: s1015411@mail.yzu.edu.tw

1 Introduction

People's everyday life is linked to supply chains. A supply chain is a system and network of organizations, people, technology, activities, information and resources that participate in the production, product or service delivery, and product sale from the supplier to the retailer and from the retailer to the customer. In a supply chain system, there is management or control of materials, information, and finances as they move in a process from suppliers to customers.

In a supply chain system, customers may feel unsatisfied with the product they purchase from retailers for a variety of reasons. For example, the product is not suitable for them, not meeting their expectations or tastes, such as its color, shape, function, size in the case of T-shirts. Because of these conditions, customer's willingness to purchase the product may decline. In order to increase customer's willingness to purchase the product, the retailer may adopt customer product return policy. Under this policy, customers can return their purchased product to the retailer and get full or partial refund. To deal with the returned product, the manufacturer may adopt the buyback policy where the retailer can "sell back" the product returned by customers and unsold products (unsold inventories) to the manufacturer. The manufacturer buys back those products with a pre-agreed buyback price. Product buyback often occurs in the special seasons such as Christmas and Thanksgiving.

Cooper and Ellram (1993) stated that a supply chain consists of all parties involved in fulfilling a customer request, directly or indirectly. Hines et al. (1998) pointed out that supply chain strategies require a system view of the linkages in the chain that work together efficiently to create customer satisfaction at the end point of delivery to the consumer. Toktay et al. (2000) stated that customer return rates ranged from 5 to 9 % of sales for most retailers. Vlachos and Dekker (2003) considered a newsvendor-style problem with returned products to be resold only in the single period. Chen and Bell (2009) showed that given a wholesale price, an increasing customer return will reduce retailer's profit dramatically. Davis et al. (1995) stated that part of the motivation for retailers to offer a liberal returns policy is the fact that the retailer can share the cost of customer product returns with the supplier/manufacturer through a buyback policy. Kandel (1996) showed that the manufacturers and the retailers use a buyback policy to deal with demand information asymmetry. Tsai and Peng (2013) considered a single-period supply chain system where the manufacturer applies a buyback policy and the retailer accepts product returns from customers. The optimal pricing and order policy under decentralized was derived.

The rest of the paper is organized as follows. Section 2 describes the considered supply chain system and the mathematical model. The optimal pricing and order policies under decentralized and centralized controls are derived in Sect. 3. A numerical example is displayed in Sect. 4. The last section concludes the paper.

2 Problem Description and Model

This research follows the work of Tsai and Peng (2013) and discusses about the decision making in single-period supply chain system. The system contains a manufacturer and a retailer. The manufacturer adopts a buyback policy and the retailer applies a product return policy. At the beginning of the sales period, the manufacturer sets the wholesale price of its product and the buyback price. The retailer then places an order with the manufacturer and set the retail price. The manufacturer produces the quantity ordered by the retailer and delivers. During the period, customers purchase the product from the retailer. After purchasing the product, if customers feel unsatisfied with the product, they can return it to the retailer. The retailer will accept the returned product unconditionally and give full refund to the customer for the product returned. After the sales period ends, if there are units remaining at the retailer, including unsold units and units returned from customer, the retailer will repackage those units and sell them back to the manufacturer. The manufacturer buys back those units at the promised buyback price. After buying back those inventories, the manufacturer will sell all of them at the salvage price. The described supply chain system and its operations are shown in Fig. 1.

The following notation will be used.

P_r Retail price per unit
P_w Wholesale price per unit
P_b Buyback price per unit set
Q_r Quantity ordered by the retailer to the manufacturer
C_u Production cost per unit for the manufacturer
C_h Handling cost per unit by the retailer for handling sold back units to the manufacturer
P_s Salvage price per unit for the manufacturer
α Customer product return rate, $0 \leq \alpha \leq 1$
D_r Customer demand, products demand by the customer
a Primary demand
b Price sensitivity

It is assumed that the salvage price per unit is less than the production cost per unit, or the manufacturer would benefit simply by producing and selling at the salvage price. There is no limit on how many products the retailer can order. If the order quantity is greater than or equal to the customer demand, the selling volume is the customer demand; if the customer demand is greater than or equal to the order quantity, the selling volume is the order quantity.

It is assumed customer demand is retail-price sensitive. The customer demand is expressed as

$$D_r = a - bP_r. \tag{1}$$

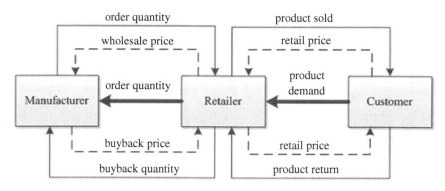

Fig. 1 Supply chain model framework

The costs incurred at the manufacturer include production cost from producing the quantity ordered by the retailer and cost of buying back all the remaining units from the retailer at the end of the period. The total cost of the manufacturer can be expressed as

$$TC_M = \begin{cases} Q_r C_u + P_b(Q_r - (1-\alpha)D_r) & \text{if } Q_r \geq D_r, \\ Q_r(C_u - \alpha P_b) & \text{otherwise.} \end{cases} \quad (2)$$

The manufacturer's revenue comes from the payment from the retailer for the ordered quantity and the salvage of the buyback units. It can be written as

$$TR_M = \begin{cases} Q_r P_w + P_s(Q_r - (1-\alpha)D_r) & \text{if } Q_r \geq D_r, \\ Q_r(P_w - \alpha P_s) & \text{otherwise.} \end{cases} \quad (3)$$

Equation (4) calculates the total cost of the retailer which includes payment to the manufacturer for the ordered quantity, cost of handling returned product and full refund to customers. The retailer's revenue contains the sales to customers and the buyback from the manufacturer. It can be expressed as Eq. 5.

$$TC_R = \begin{cases} Q_r P_w + \alpha D_r(P_r + C_h) & \text{if } Q_r \geq D_r, \\ Q_r(P_w + \alpha(P_r + C_h)) & \text{otherwise.} \end{cases} \quad (4)$$

$$TR_R = \begin{cases} Q_r P_r + P_b(Q_r - (1-\alpha)D_r) & \text{if } Q_r \geq D_r, \\ Q_r(P_r - \alpha P_b) & \text{otherwise.} \end{cases} \quad (5)$$

The respective profits of the manufacturer and the retailer are the difference between their own total revenue and total cost.

3 Optimal Control Policies

In this section, the optimal pricing and order policies are developed. Two types of control are considered. Under decentralized control (Tsai and Peng 2013), the manufacturer firstly sets the wholesale price and the buyback price. The retailer then determines the retail price and the order quantity. The two parties make their decisions to maximize their own profits. This paper further derives the optimal pricing and order policy for the system under centralized. That is, the decisions are made to maximize the profit of the system.

3.1 Optimal Decisions Under Decentralized Control

Under decentralized control, the retailer determines the retail price and the order quantity based on the wholesale price and the buyback price offered by the manufacturer, as well as customer demand that is sensitive to the retail price, and the product return rate. It is assumed that the manufacturer has full information of the entire supply chain system and it knows how the retailer makes decisions. It is a sequential decision making process where the manufacturer needs to makes its decisions first and the retailer makes its own decisions based on the manufacturer's offer. Therefore, the manufacturer needs to anticipate the retailer's move and incorporated it into its own decision making.

Given the wholesale price, buyback price and retailer price, the retailer would order a quantity that matches the demand if the following condition (marked as condition 1) holds. Otherwise, it would not order from the manufacturer as no profit would be gained.

$$P_r - P_w + \alpha(P_b - P_r - C_h) > 0 \tag{6}$$

In the case where this condition holds, the optimal retail price for the retailer can be obtained, as Eq. 7.

$$P_r = \frac{a}{2b} + \frac{P_w - \alpha(P_b - C_h)}{2(1-\alpha)} \tag{7}$$

Knowing how the retailer would make its decisions, it can be shown that the optimal wholesale price and buyback price for the manufacturer need to satisfy the following equation. Apparently, the manufacturer has to set the wholesale price and the buyback price such that condition 1 holds in order for the retailer to place an order.

$$P_w - \alpha P_b = \frac{1}{2}\left[\frac{a(1-\alpha)}{b} + C_u - \alpha(P_s + C_h)\right] \tag{8}$$

3.2 Optimal Decision Under Centralized Control

Under centralized control, all the decisions are jointly determined with the single objective of maximizing total profit of the whole supply chain system. It can be derived that the optimal order quantity is the demand if the following condition (marked as condition 2) holds, and zero otherwise.

$$(1 - \alpha)P_r + \alpha(P_s - C_h) - C_u > 0 \tag{9}$$

When condition 2 holds, the optimal retail price can be written as Eq. 10. As can be seen, it does not involve the wholesale price and the buyback price. These two prices are the transfer prices in the supply chain system who transfer profit between the manufacturer and the retailer within the system. They determine how the total profit of the system distributes between the manufacturer and the retailer but do not affect the total profit of the system.

$$P_r = \frac{a}{2b} + \frac{C_u - \alpha(P_s - C_h)}{2(1 - \alpha)} \tag{10}$$

It can be shown mathematically that under decentralized control, the optimal profit of the manufacturer is greater than that of the retailer. Furthermore, the optimal retail price under centralized control is always lower than that under decentralized control. It implies that the optimal order quantity under centralized control is greater than that under decentralized control. In addition, the optimal total profit of the system under centralized control is greater than that under decentralized control.

4 Numerical Example

A numerical example is created for the purpose of illustration. Table 1 shows the parameter setting of this example. Table 2 lists the optimal decisions, the resulting profit of the two parties and the system under centralized control and decentralized control. The example is designed such that condition 2 holds. Under decentralized control, the wholesale price and the buyback price are set at 14.375 and 12, respectively. The two prices are set such that Eq. 8 and condition 1 hold. Under

Table 1 Parameter setting of numerical example

Parameters	Value
a	500
b	20
α	0.05
P_s	3
C_u	4
C_h	1

Table 2 Optimal decisions and profits

Value	Decentralized control	Centralized control	Centralized control
Wholesale price	14.375	14.375	13.000
Buyback price	12.000	12.000	12.000
Optimal retail price	19.78	14.55	14.55
Optimal order quantity	104.47	208.95	208.95
Retailer's profit	518.45	0.00	287.30
Manufacturer's profit	1036.90	2073.80	1786.50
System's profit	1555.35	2073.80	2073.80

centralized control, two sets of the wholesale price and the buyback price, (14.375, 12) and (13, 12), are applied.

It can be seen clearly that under centralized control the optimal retail price, the optimal order quantity and the optimal profit of the system are the same regardless what wholesale price and buyback price are used. The two prices do not affect the optimal order quantity, the optimal retail price and the optimal system's profit. They only determine how the profit of the system is distributed between the manufacturer and the retailer.

When comparing the results under the two types of control, it can be found that the optimal order quantity and the optimal retail price under decentralized control are both lower than those under centralized control. Under decentralized control, the retailer tends to set a higher retail price in order to protect its own profit. It leads to a lower level of customer demand and thus a lower order quantity. In addition, centralized control generates higher system's profit compared to that under decentralized control.

5 Conclusions

This study considers a single-period supply chain system with a manufacturer and a retailer. The retailer accepts product return from customers with full refund. The manufacturer promises to buy back all the remaining units from the retailer at the end of the period. Customer demand is assumed to be retail-price sensitive. At the beginning of the period, the wholesale price, buyback price, retail price and quantity ordered by the retailer are to be determined. Assuming under centralized control, the optimal pricing and order policy are developed with the objective of maximizing the profit of the system. The condition on whether an order should be placed with the manufacturer is also derived. The results are compared to the optimal policy under decentralized control (Tsai and Peng 2013). It is shown that under centralized control, the wholesale price and the buyback price serve as transfer prices that do not affect the profit of the system and thus the optimal retail price and order quantity. The two prices only control how the system's profit is allocated to the manufacturer and the retailer. In addition, it can be mathematically proven that centralized control always generates higher profit of the system when compared to decentralized control. A numerical example is presented to further illustrate the result.

As it is shown the optimal overall profit of the system is never achieved with decentralized control. The issue of supply chain coordination rises for systems under decentralized control. The design of contracts that coordinate the system and are accepted by both the manufacturer and the retailer would be an interesting extension of this research.

References

Chen J, Bell PC (2009) The impact of customer returns on pricing and order decisions. Eur J Oper Res 195:280–295

Cooper MC, Ellram LM (1993) Characteristics of supply chain management and the implications for purchasing and logistics strategy. Int J Logist Manag 4:13–24

Davis S, Gerstner E, Hagerty M (1995) Money back guarantees in retailing matching products to consumer tastes. J Retailing 71:7–22

Hines P, Rich N, Bicheno J, Brunt D, Taylor D, Butterworth C, Sullivan J (1998) Value stream management. Int J Logist Manag 9:25–42

Kandel E (1996) The right to return. J Law Econ 39:329–356

Toktay LB, Wein LM, Zenios SA (2000) Inventory management of remanufacturable products. Manag Sci 46:1412–1426

Tsai CY, Peng PH (2013) Sequential decision making in a supply chain with customer return and buyback policies. Paper presented at the 3rd international forum and conference on logistics and supply chain management, Bali, Indonesia, pp 27–29

Vlachos D, Dekker R (2003) Return handling options and order quantities for single period products. Eur J Oper Res 151:38–52

Investigation of Safety Compliance and Safety Participation as Well as Cultural Influences Using Selenginsk Pulp and Cardboard Mill in Russia as an Example

Ekaterina Nomokonova, Shu-Chiang Lin and Guanhuah Chen

Abstract The aim of this study was to assess the effectiveness of the job demands–resources (JD–R) model in explaining the relationship of job demands and resources with safety outcomes (i.e., workplace injuries and near-misses). We collected self-reported data from 203 pulp and paper production workers from Pulp and Cardboard Mill which is located in Russia during the period 2000–2010. The results of a structural equation analysis indicated that job demands (psychological and physical demands) and job resources (decision latitude, supervisor support and coworker support) could affect safety performance and safety compliance, and thus influence the occurrence of injuries and near-misses and whether the cultural influences play a significant role in both safety compliance and safety participation.

Keywords Safety participation · Safety compliance · Job demands · Job resources · Cultural influences

---

E. Nomokonova (✉) · S.-C. Lin
Department of Industrial Management, National Taiwan University of Science and Technology, 43 Keelung Road, Sect. 4, Taipei 106, Taiwan, Republic of China
e-mail: nkaterina.s@gmail.com

S.-C. Lin
e-mail: slin@mail.ntust.edu.tw

G. Chen
Foundation of Taiwan Industry Service, Industrial Safety Division, No. 41, Alley 39, Ln. 198, Siwei Road, Taipei, Taiwan, Republic of China
e-mail: bill@ftis.org.tw

1 Introduction

Russia has predominantly a huge Pulp and Paper Industry. The Russian paper industry is the second largest in the world. The Pulp and paper industry graph in Russia has continued to show an upward trend since 1999.

According to RAO (Russian Association of Pulp and Paper Mills and Institutions) Bumprom, Russia's Pulp and Paper Industry has increased its paper production by 5.4 % (about 3.9 million tons) and the production of cardboard, which also includes production of container board by 7 % (about 2.9 million tons) since 2004.

As there are a lot of mills in this industry, on this research we will analyze one factory only and then compare it with Pulp and Cardboard Mill in Taiwan, which is one of the top three pulp factories among the world for potential cultural difference, in order to see if there are any cultural influence for safety performance and safety compliance.

2 Background Information

Selenginsk pulp and cardboard mill was created in 1973. In 1992, in accordance with privatization program of state and municipal enterprises of Republic of Buryatia. JSC "Selenginsk PCM" is a diverse enterprise, but its main activity makes it related to the pulp and paper industry. Since 1990 the Company operates the world's only system of a closed water cycle which enables to exclude discharge of industrial wastewater plant.

The company organized integrated conversion of pulp into the final product, that is, a cardboard container. Selenginsk Mill is a city-forming enterprise that provides more than 2.1 thousand of Selenginsk' population with jobs and, therefore, is a major source of replenishment of the local budget via fees and taxes. The territory of the mill is more than 400 ha and its annual containerboard production averages approximately 100,000 tone, sacks and paper bags production around 6 million and corrugated production around 60 million square meters.

Pulp and paper manufacturing can also be very hazardous due to massive weights and falling, rolling, and/or sliding pulpwood loads. Workers may be struck or crushed by loads or suffer lacerations from the misuse of equipment, particularly when machines are used improperly or without proper safeguards. Other causes of multiple deaths included electrocution, hydrogen sulphide and other toxic gas inhalation, massive thermal/chemical burns and one case of heat exhaustion. The number of serious accidents associated with paper machines has been reported to decrease with the installation of newer equipment in some countries. In the converting sector, repetitive and monotonous work, and the use of mechanized equipment with higher speeds and forces, has become more common.

Although no sector-specific data are available, it is expected that this sector will experience greater rates of over-exertion injuries associated with repetitive work.

Today, many industries and indeed regulatory agencies still focus completely on common safety performance measures such as lost time injury frequency rate and number of lost days in an effort to measure safety performance. Unfortunately, such indicators just measure failure to control and give no indication of risk management effort, which may take time to come to fruition. A lot of researches have been done in order to investigate safety performance or safety-related components for the last quarter century.

In more recent years, there has been a trend toward conceptualizing safety performance as multi-dimensional in the occupational safety research literature. For example, Griffin and Neal (2000) have conceptualized two types of employee safety performance: safety compliance and safety participation. Safety compliance corresponds to task performance and includes such behaviors as adhering to safety regulations, wearing protective equipment, and reporting safety-related incidents. Safety participation is parallel to contextual performance and focuses on voluntary behaviors that make the workplace safe beyond prescribed safety precautions, including taking the initiative to conduct safety audits and helping co-workers who are working under risky conditions. There are three safety–specific determinants such as job demands, job control and social support.

3 Hypotheses

The objective of this study is to investigate how these three characteristics simultaneously influence the work performance on the example of Selenginsk Pulp and Cardboard Mill in Russia, and whether the cultural influences play a significant role in both safety compliance and safety participation. In the present study, the job demands–resources (JD–R) framework (Bakker et al. 2003a) is used to explore how job demands and resources affect safety outcomes, specifically near-misses and injuries, thus we have made three hypotheses, as stated below, for further investigations into our objective.

Hypothesis 1 Job resources are positively correlated with safety compliance.

Although the Job demands resources (JD–R) model does not propose relationship between job demands and engagement, there have been inconsistent and unexpected results. In the area of occupational safety, however, challenge demands may actually constrain employees' progress toward workplace safety as stress can compel individuals to focus narrowly on only a few specific aspects of the work environment or objectives (Hofmann and Stetzer 1996; Weick 1990). For instance, under increased performance pressure or time constraints, employees might utilize more "short cut" work methods, perceiving there is not always enough time to follow safety procedures and resulting in a higher frequency of unsafe behaviors (Hofmann and Stetzer 1996). A recent meta-analysis found that hindrance demands such as risks and hazards, physical demands, and complexity were negatively associated with safety compliance (Nahrgang et al. 2011). Taken together, we expect that both challenge demands and hindrance demands will be negatively related to safety compliance.

Hypothesis 2 Job demands are negatively correlated with safety compliance.

In general, the JD–R model proposes that a state of exhaustion leads to increased absenteeism as a result of illness and decreased in-role performance (Bakker et al. 2003b, 2005). Burnt-out employees are more likely to commit mistakes and injure themselves, because of a depletion of their mental and physical energy (Nahrgang et al. 2011). Thus, we expect that emotional exhaustion will be positively associated with near-misses and injuries in the workplace.

Hypothesis 3 Safety compliance is positively correlated with safety outcomes (safety performance).

Existing studies support the dual psychological processes proposed by the JD–R model; related data suggest that job demands and resources can predict important organizational outcomes by such dual pathways. In the literature on occupational safety, a direct correlation between job demands/resources and safety outcomes has been found; situational constraints, time pressures and work overload have all shown positive relationships with injuries and near-misses (Goldenhar et al. 2003; Jiang et al. 2010; Nahrgang et al. 2011; Snyder et al. 2008). Conversely, supportive environments and job autonomy have shown negative associations with injuries and near-misses (Barling et al. 2003; Goldenhar et al. 2003; Jiang et al. 2010; Nahrgang et al. 2011).

4 Methods

A survey was administered to pulp and paper workers from Selenginsk Pulp and Cardboard Mill in Russia. The core job responsibility of these workers is to ensure effective execution and completion of the pulp and paper production plan by working as a crew. Specifically, their job duties are to inspect and resolve equipment failure, to operate and maintain all kinds of equipment (e.g., starting or shutting paper machine line), to adjust or replace mechanical components according to workflow, to handle emergency, and so on. During work hours, these workers might be subject to various risks, such as harsh physical environments, being injured by machines, being struck by objects, falling from a high place, fire, and chemical corrosion.

The present study examined 203 reports of occupational accidents and fatalities in the industry from 2000 to 2010, as recorded in the occupational accidents database of the Selenginsk Pulp and Cardboard Mill. The major factors affecting pulp and cardboard accidents were: most people who got injuries are men (74 %), at the age from 26 to 30 (18 %), accident type (fall down, trapping by shifts and influence by moving objects), unsafe acts mostly failure to use safeguard measures and warnings. This percentage of male respondents may seem usual for pulp and paper workers. One explanation is that male workers had a lower response rate than female workers.

5 Proposed Measurements

5.1 Job Demands

In the present study, job demands were defined as physical demands (4 items) and psychological demands (7 items), using the Job Content Questionnaire (JCQ); Karasek et al. (1998). Two factors supported the use of this measure. First, physical exposure (e.g., working with heavy equipment) is the most frequent stressor experienced by pulp and paper production workers. Second, the psychological demands subscale includes mental and cognitive workloads and time constraints, which have demonstrated positive relationships with injuries and near-misses (Goldenhar et al. 2003; Jiang et al. 2010; Nahrgang et al. 2011; Snyder et al. 2008).

5.2 Job Resources

According to the literature, supportive environments and job autonomy have both shown negative associations with injuries and near-misses (Barling et al. 2003; Goldenhar et al. 2003; Jiang et al. 2010; Nahrgang et al. 2011). Thus, in the current study, job resources were measured by decision latitude (9 items), supervisory support (5 items) and coworker support (6 items), according to the JCQ. As a variable often used to describe control on the job (i.e., autonomy; Westman 1992), decision latitude was divided into the two theoretically distinct, although often highly correlated, sub dimensions of skill discretion and decision authority (Karasek et al. 1998).

All items from the JCQ were rated by the respondents using a 4-point scale: strongly disagree (1), disagree (2), agree (3), and strongly agree (4).

5.3 Safety Compliance

A 3-item scale by Neal and Griffin was adopted to assess safety compliance in terms of core safety activities that should be carried out by employees to maintain workplace safety. All items were rated on a 5-point scale, with responses ranging from 1 (strongly disagree) to 5 (strongly agree).

5.4 Safety Outcomes

Self-reported injuries and near-misses were used to assess safety outcomes. Specifically, injuries were indicated by a summation of the employees' responses regarding whether any major body parts had been injured during the past year, including the head, neck, eyes, shoulders, arms, wrists, hands, upper back, lower back, legs, ankles, feet and other (Goldenhar et al. 2003; Jiang et al. 2010). Participants were also asked to recall the total number of near-misses (i.e., incidents that could have resulted in an injury, but did not) that they had experienced during the past year (Goldenhar et al. 2003; Jiang et al. 2010).

5.5 Analyses

Maximum likelihood structural equation modeling (SEM) will be used to test our proposed model with Lisrel 8. Based on Anderson and Gerbing's suggestions, a two-step approach to SEM was adopted. First, the factor structure of the variables in this study was tested to confirm that the model specifying the posited structure of the underlying constructs would fit the observed data. Second, the proposed structural model was compared with an alternative model to assess which one better accounted for the covariances observed between the model's exogenous and endogenous constructs.

References

Bakker AB, Demerouti E, de Boer E, Shaufeli WB (2003) Job demands and job resources as predictors of absence duration and frequency. J Vocat Behav 62:341–356

Bakker AB, Demerouti E, Euwema MC (2005) Job resources buffer the impact of job demands on burnout. J Occup Health Psychol 10:170–180

Barling J, Kelloway EK, Iverson RD (2003) High-quality work, job satisfaction, and occupational injuries. J Appl Psychol 88:276283

Goldenhar LM, Williams LJ, Swanson NG (2003) Modelling relationships between job stressors and injury and near-miss outcomes for construction labourers. Work Stress 17:218–240

Griffin MA, Neal A (2000) Perceptions of safety at work: a framework for linking safety climate to safety performance, knowledge, and motivation. J Occup Health Psychol 5:347–358

Hofmann DA, Stetzer A (1996) A cross-level investigation of factors influencing unsafe behaviors and accidents. Pers Psychol 49:307–339

Jiang L, Yu G, Li Y, Li F (2010) Perceived colleagues' safety knowledge/behavior and safety performance: safety climate as a moderator in a multilevel study. Accid Anal Prev 42:1468–1476

Karasek R, Brisson C, Kawakami N, Houtman I, Bongers P, Amick B (1998) The job content questionnaire (JCQ): an instrument for internationally comparative assessments of psychosocial job characteristics. J Occup Health Psychol 3:322–355

Nahrgang JD, Morgeson FP, Hofmann DA (2011) Safety at work: a meta-analytic investigation of the link between job demands, job resources, burnout, engagement, and safety outcomes. J Appl Psychol 96:71–94

Snyder LA, Krauss AD, Chen PY, Finlinson S, Huang Y (2008) Occupational safety: application of the job demand–control–support model. Accid Anal Prev 40:1713–1723

Weick KE (1990) The vulnerable system: an analysis of the Tenerife air disaster. J Manage 16:571–593

Westman M (1992) Moderating effect of decision latitude on stress–strain relationships does organizational level matter? J Organiz Behav 13:713–722

Identifying Process Status Changes via Integration of Independent Component Analysis and Support Vector Machine

Chuen-Sheng Cheng and Kuo-Ko Huang

Abstract Observations from the in-control process consist of in-control signals and random noise. This paper assumes that the in-control signals switch to different signal types when the process status changes. In these cases, process data monitoring can be formulated as a pattern recognition task. Time series data pattern recognition is critical for statistical process control. Most studies have used raw time series data or extracted features from process measurement data as input vectors for time series data pattern recognition. This study improves identification by focusing on the essential patterns that drive a process. However, these essential patterns are not usually measurable or are corrupted by measurement noise if they are measurable. This paper proposes a novel approach using independent component analysis (ICA) and support vector machine (SVM) for time series data pattern recognition. The proposed method applies ICA to the measurement data to generate independent components (ICs). The ICs include important information contained in the original observations. The ICs then serve as the input vectors for the SVM model to identify the time-series data pattern. Extensive simulation studies indicate that the proposed identifiers perform better than using raw data as inputs.

Keywords Time series data pattern · Independent component analysis · Support vector machine

---

C.-S. Cheng (✉) · K.-K. Huang
Department of Industrial Engineering and Management, Yuan Ze University,
135 Yuan-Tung Road, Chung-Li 32003, Taiwan
e-mail: ieccheng@saturn.yzu.edu.tw

K.-K. Huang
e-mail: m9223016@gmail.com

1 Introduction

Statistical process control (SPC) concepts and methods have been successfully implemented in the manufacturing and service industries for decades. A process is considered out of control when a point exceeds the control limits or a series of points exhibit an unnatural pattern. Unnatural pattern analysis is important for SPC because unnatural patterns can reveal potential process quality problems (Western Electric Company 1956). Once unnatural patterns are recognized, narrowing the diagnostic search process scope to a smaller set of possible causes for investigation reduces the troubleshooting time required. Hence, accurate and timely detection and recognition of time series data patterns are important when implementing SPC.

A real manufacturing process often operates in an in-control state over a relatively long period. This study assumes that process status during an in-control state is gamma distributed. The gamma distribution is a typical case for the asymmetric distributions. Borror et al. (1999) used the gamma distribution to study the effect of a skewed distribution on monitoring performance. In addition, the impurities and particle counts of the semiconductor industry are considered as gamma distribution (Levinson 1997). However, certain conditions occasionally occur, resulting in an out-of-control state where much of the process output does not conform to requirements. Capizzi and Masarotto (2012) considered eight nonrandom patterns, including constant, intermittent, geometric, linear, mixture-20, mixture-80, cosine and chirp. Figure 1 shows these out-of-control patterns. These out-of-control patterns shows either intermittent or oscillatory behaviors.

Occasionally, measurement noise corrupts time series data patterns. The time series data pattern identifier is usually unable to accurately recognize the out-of-control time series data patterns caused by this noise. Thus, a powerful time series data pattern identifier is necessary to accurately classify all types of time series data patterns. Several approaches have been proposed for identifying time series data patterns. These include statistical approaches (Cheng and Hubele 1996; Yang and Yang 2005), rule-based system (Cheng 1989; Wang et al. 2009), artificial neural network techniques (Cheng and Cheng 2008; Jiang et al. 2009), and support vector machine (SVM) (Das and Banerjee 2010). For a more detailed discussion of how to identify time series data patterns, please refer to Hachicha and Ghorbel (2012). Most studies have used raw process data or extracted features from process measurement data as input vectors for time series data pattern recognition.

Process-measured data can be considered as linear mixtures of unknown time series data patterns where the coefficient mixing matrix is also unknown. This study improves identification by focusing on the essential patterns that drive a process. However, these essential patterns are not usually measurable or are corrupted by measurement noise if they are measurable. Independent component analysis (ICA) is a novel algorithm that is able to separate independent components (ICs) from complex signals (Hyvärinen 1999; Hyvärinen and Oja 2000). This study develops a novel approach that combines ICA and an SVM for time series data pattern identification. The proposed method applies ICA to the measurement data to generate ICs. The ICs include important information contained in the original observations. The ICs then serve as the input vectors for the SVM model to identify the time series data patterns.

2 Methodology

2.1 Independent Component Analysis

ICA is a very useful technique and suitable for data processing in manufacturing process. ICA can be used to demix the observed time series data blindly and extract the statistically independent components expressed as linear combinations of observed variables. ICA is usually used for revealing latent factors underlying sets of random variables, signals, or measurements. ICA was originally proposed to solve the problem of blind source separation, which involves recovering independent source signals after they have been linearly mixed by an unknown mixing mechanism.

The general model for ICA is that the sources are generated through a linear basis combination, where additive noise can be present. Suppose we have N statistically independent signals, $s_i(t)$, $i = 1, 2, \ldots, N$. We assume that the sources themselves cannot be directly observed and that each signal, $s_i(t)$, is a realization of some fixed probability distribution at each time point t. Also, suppose we observe these signals using $s_i(t)$ sensors, then we obtain a set of N observation signals $x_i(t)$, $i = 1, 2, \ldots, N$ that are mixtures of the sources. A fundamental aspect of the mixing process is that the sensors must be spatially separated so that each sensor records a different mixture of the sources. With this spatial separation assumption in mind, we can model the mixing process with matrix multiplication as follows:

$$x(t) = \mathbf{A}s(t). \qquad (1)$$

where $\mathbf{A}$ is an unknown matrix called the mixing matrix and $x(t)$, $s(t)$ are the two vectors representing the observed signals and source signals respectively. Incidentally, the justification for the description of this signal processing technique as blind is that we have no information on the mixing matrix, or even on the sources themselves. The objective is to recover the original signals, $s_i(t)$, from only the observed vector $x_i(t)$. We obtain estimates for the sources by first obtaining the "unmixing matrix" $\mathbf{W}$, where, $\mathbf{W} = \mathbf{A}^{-1}$. This enables an estimate, $\hat{s}(t)$, of the independent sources to be obtained:

$$\hat{s}(t) = \mathbf{W}x(t). \qquad (2)$$

2.2 Support Vector Machine

The SVM is a powerful new machine learning algorithm, which is rooted in statistical learning theory. By constructing a decision surface hyper-plane which yields the maximal margin between position and negative examples, SVM approximately implements the structure risk minimization principle. A classification task usually involves training and testing data that consist of certain data instances. Each training set instance contains one target value (class label) and several attributes (features). The goal of a classifier is to produce a model that predicts the target value of testing set data instances that are only given attributes.

In practical, constructing a separating hyper-plane in the feature space leads to a non-linear decision boundary in the input space. Time-consuming calculation of dot products in a high-dimensional space can be avoided by introducing a kernel function that satisfies $K(\mathbf{x}_i, \mathbf{x}_j) = \phi(\mathbf{x}_i) \cdot \phi(\mathbf{x}_j)$. The kernel function allows all necessary computations to be performed directly in the input space. Many types of kernel functions can be used in SVM, but two types are usually used: the radial basis function (RBF), $K(\mathbf{x}_i, \mathbf{x}_j) = \exp\{-\gamma \|\mathbf{x}_i - \mathbf{x}_j\|^2\}$, $\gamma > 0$; and the polynomial kernel of degree d, $K(\mathbf{x}_i, \mathbf{x}_j) = (\gamma \mathbf{x}_i \cdot \mathbf{x}_j + r)^d$, $\gamma > 0$. This research uses the RBF kernel function for the SVM classifier because it can analyze data that are non-linear and have multiple dimensions and require fewer parameters to be determined.

3 The Proposed ICA and SVM Scheme

This section presents a novel ICA-SVM scheme to identify the eight out-of-control time series data patterns (shown in Fig. 1) and a gamma distribution, in-control, time series data pattern. The nine time series data patterns are the corresponding multiple target values (class labels) in the SVM. The proposed scheme has two stages: off-line training and on-line testing. Each training set instance contains one target value (class label) and several attributes (features). However, each testing set instance has attributes and no target values because this is generated by the proposed identifier. The training and testing data have the same number of data instances, other than the testing data target values, which are predicted by the identifier.

In the training stage, time-series data are collected during an in-control state. Each data matrix row is normalized individually using the mean (μ_{in}) and standard deviation (σ_{in}) of each row. A normalized data matrix is referred to as $\mathbf{X}_{in}$. Applying the FastICA algorithm to the normalized data produces demixing Matrix $\mathbf{W}$. From (2), the ICs ($\hat{\mathbf{S}}_{in}$) at in-control conditions can be calculated using $\hat{\mathbf{S}}_{in} = \mathbf{W}\mathbf{X}_{in}$. A similar process is used to obtain ICs at out-of-control conditions, but they are normalized using in-control μ_{in} and σ_{in}. The out-of-control data matrix

Identifying Process Status Changes via Integration of Independent Component

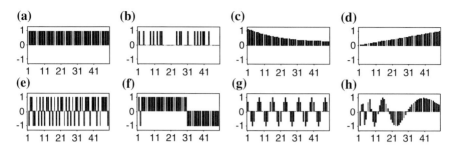

Fig. 1 Eight types of out-of-control patterns. **a** Constant, **b** Intermittent, **c** Geometric, **d** Linear, **e** Mixture-20, **f** Mixture-80, **g** Cosine, **h** Chirp

is referred to as $\mathbf{X}_{out}$. ICs ($\hat{\mathbf{S}}_{out}$) at out-of-control operation conditions are calculated using $\hat{\mathbf{S}}_{out} = \mathbf{W}\mathbf{X}_{out}$.

In order to select the target $\hat{\mathbf{S}}$ which represents the time series pattern, the non-gaussianity of each $\hat{\mathbf{S}}$ is evaluated using the Shapiro–Wilk test of normality. The null hypothesis is that $\hat{\mathbf{S}}$ has a normal distribution with mean and variance estimated from $\hat{\mathbf{S}}$, against the alternative that $\hat{\mathbf{S}}$ is not normally distributed with the estimated mean and variance. The test returns the $p-value$, computed using inverse interpolation into the table of critical values. Small values of $p-value$ cast doubt on the validity of the null hypothesis. The test rejects the null hypothesis at the 5 % significance level.

This study uses an SVM to develop time-series data pattern identifiers. The LIBSVM program (Chang and Lin 2010) in a MATLAB development environment was used to build the SVM model. Selecting an appropriate kernel is the most important SVM design decision because this defines the feature space and mapping function ϕ. Different SVM parameter kernel functions were tested. The RBF kernel was selected as the kernel function because it tends to perform better. The RBF kernel function must only tune kernel parameter γ and penalty factor C. However, there are no general rules for selecting C and γ. SVM parameters are usually selected by experimenting. To determine the two parameters, a grid search with 5-fold cross-validation experiment was used to select parameter settings that yielded the best results. Exponentially increasing C and γ sequences were tested, that is, $C \in \{2^{-5}, 2^{-4}, \ldots, 2^{5}\}$ and $\gamma \in \{2^{-5}, 2^{-4}, \ldots, 2^{5}\}$, respectively. C and γ parameter sets with the highest correct classification rates during cross-validation were selected as the optimal parameter sets. The final SVM model was built using these settings and used to test time-series data pattern identification.

During the testing stage, normalized input data were obtained before the identifier was applied. The normalized data set is referred to as $\mathbf{X}_{new}$. The new ICs ($\hat{\mathbf{S}}_{new}$) were obtained using $\hat{\mathbf{S}}_{new} = \mathbf{W}\mathbf{X}_{new}$. $\hat{\mathbf{S}}_{new}$ was fed into the trained SVM model to identify the IC pattern.

Table 1 Parameters for out-of-control patterns

Pattern type	Equations
Constant	$s(t) = \delta$
Inrermittent	$s(t) = \delta \times u(t)$, where $u(t) \sim \text{Binomial}(1, 0.2)$
Geometric	$s(t) = \delta \times (0.2 + 0.95^t)$
Linear	$s(t) = \delta \times (t/50)$
Mixture-y	$s(t) = \begin{cases} \delta, & t = 0 \\ \delta \times (-1)^{1-u(t)} x(t-1), & t \neq 0 \end{cases}$ where $u(t) \sim \text{Binomial}(1, \frac{y}{100})$
Cosine	$s(t) = \delta \times \cos(2\pi t/8)$
Chirp	$s(t) = \delta \times \cos\left(\frac{2\pi t}{2 + t/5}\right)$

4 Experimental Results

This section uses a simulation to illustrate the efficiency of the proposed scheme. Ideally, example patterns for in- and out-of-control process statuses should be developed from a real process. However, sufficient training instances of out-of-control statuses may not be easily available. Studies have generated training examples based on a predefined mathematical model and Monte Carlo simulations. This study uses an observation window with 52 data points, implying that each sample pattern consists of 52 time-series data (Table 1).

A two-dimensional vector ($\mathbf{x}(t) = [x_1(t), x_2(t)]^T$) represents the two process measurement data sets obtained simultaneously from two parallel machines. The two variables in the two-dimensional vector ($\mathbf{s}(t) = [s_1(t), s_2(t)]^T$) denotes as the source variables, $s_2(t)$ is gamma distribution and $s_1(t)$ is the process pattern. There are eight out-of-control time series data patterns and gamma distribution (in-control) time series data pattern considered as process patterns in this paper. Observations from gamma distribution are non-negative. The gamma distribution can be denoted as Gam (α, β) with the probability density function given by

$$f(x) = \frac{1}{\alpha^\beta \Gamma(\beta)} x^{\beta-1} \exp(-x/\alpha), \quad x > 0. \quad (3)$$

And the mean and variance given, respectively, by $\mu = \alpha/\beta$ and $\sigma^2 = \alpha/\beta^2$. In this paper, we considered two heavily skewed cases, Gam(0.5,1) and Gam(1,1).

It is assumed that the measured data, $\mathbf{x}(t)$, are linear combinations of unknown source patterns $\mathbf{s}(t)$. The discrete ICA time model is described as

$$\mathbf{x}(t) = \begin{bmatrix} a_{11} & a_{12} \\ a_{21} & a_{22} \end{bmatrix} \mathbf{s}(t) + v. \quad (4)$$

where $a_{11}, a_{12}, a_{21}, a_{22}$ are unknown mixing constants. Without a generality loss, it is assumed that $\sum_{j=1}^{2} a_{1j} = 1$ and $\sum_{j=1}^{2} a_{2j} = 1$. $a_{11}, a_{12}, a_{21}, a_{22}$ were randomly generated between 0 and 1; therefore, $a_{11} = 0.1736$, $a_{12} = 0.8264$, $a_{21} = 0.5482$, and $a_{22} = 0.4518$. v following normal distribution with standard deviation $\sigma_v = 0.5$.

The first in-control distribution considered in this paper is Gam(0.5,1). FastICA is first used to estimate demixing matrix **W** in the training stage. Then, in the testing stage, FastICA is applied to measurement data from the two parallel machines to estimate the two ICs. Each input instance was composed of 52 data points per observation window, and 36,000 instances were used as SVM input data to build the raw data-based time series data pattern recognition model. To demonstrate the performance of the proposed scheme, FastICA was used to preprocess the same 36,000 instances to obtain ICs that were used to build the IC-based time-series data pattern recognition model.

After applying the grid search method to the raw data-based identifier to tune the SVM parameters, the best parameter sets were $C = 0.5$ and $\gamma = 2^{-7}$. The best set of parameters for the IC-based identifier was $C = 1$ and $\gamma = 2^{-7}$. The average correct classification rates of the raw data-based identifier and IC-based identifier were 86.21 and 83.17 %, respectively. The results prove that the proposed method removes noise while preserving the main time-series data pattern information.

When the parameter α increases, the gamma distribution appears more nearly normal. To further compare the performance of the proposed ICA-SVM scheme to methods under different α values, Gam(1,1) was considered. This study also used the RBF kernel function and grid search strategy to select the best parameter sets. The best parameter sets for the raw data-based identifier and IC-based identifier were $C = 0.5$, $\gamma = 2^{-6}$ and $C = 0.5$, $\gamma = 2^{-6}$, respectively. Without the proposed ICA-SVM, the average correct classification rates of the raw data-based identifier and IC-based identifier were 86.13 and 92.48 %, respectively. The simulation experiments confirm empirically that the IC-based identifier performs better than the raw data-based identifier under Gam(1,1).

5 Conclusion

Effectively and accurately identifying time-series data patterns from an out-of-control process status is a difficult and challenging task. Most previous studies have used raw process measurement data as the input vector for time series data pattern recognition. The process signal is often mixed with unknown noise. A traditional pattern identifier has difficulty excluding noise that affects accurate pattern classification. This paper proposes a novel approach that uses ICA and SVM for time series data pattern recognition. The proposed method applies ICA to the obtained measurement data to generate ICs. The SVM model is then applied to each IC for time series data pattern recognition. This study uses nine time series data pattern groups to evaluate the performance of the proposed scheme.

The simulated experiment results show that the proposed ICA-SVM identifier produces a higher average correct classification rate than the traditional identifier when using the same raw data as inputs. The results show that the proposed scheme effectively and accurately recognizes the nine types of time-series data patterns in either in-control or out-of-control process statuses.

References

Borror CM, Montgomery DC, Runger GC (1999) Robustness of the EWMA control chart to non-normality. J Qual Technol 31:309–316

Capizzi G, Masarotto G (2012) Adaptive generalized likelihood ratio control charts for detecting unknown patterned mean shifts. J Qual Technol 44:281–303

Chang CC, Lin CJ (2010) LIBSVM 3.11: a library for support vector machines. Technical report, Department of Computer Science and Information Engineering, National Taiwan University. http://www.csie.ntu.edu.tw/~cjlin/libsvm/

Cheng CS (1989) Group technology and expert systems concepts applied to statistical process control in small-batch manufacturing. PhD Dissertation, Arizona State University, Tempe, AZ, USA

Cheng HP, Cheng CS (2008) Denoising and feature extraction for control chart pattern recognition in autocorrelated processes. Int J Signal Imaging Syst Eng 1:115–126

Cheng CS, Hubele NF (1996) A pattern recognition algorithm for an x-bar control chart. IIE Trans 29:215–224

Das P, Banerjee I (2010) An hybrid detection system of control chart patterns using cascaded SVM and neural network-based detector. Neural Comput Appl 20:287–296

Hachicha W, Ghorbel A (2012) A survey of control-chart pattern-recognition literature (1991–2010) based on a new conceptual classification scheme. Comput Ind Eng 63:204–222

Hyvärinen A (1999) Fast and robust fixed-point algorithms for independent component analysis. IEEE Trans Neural Networks 10:626–634

Hyvärinen A, Oja E (2000) Independent component analysis: algorithms and applications. Neural Networks 13:411–430

Jiang P, Liu D, Zeng Z (2009) Recognizing control chart patterns with neural network and numerical fitting. J Intell Manuf 20:625–635

Levinson W (1997) Watch out for nonnormal distributions of impurities. Chem Eng Prog 93:70–76

Wang CH, Dong TP, Kuo W (2009) A hybrid approach for identification of concurrent control chart patterns. J Intell Manuf 20:409–419

Western Electric Company (1956) Statistical quality control handbook. Western Electric Company, Indianapolis

Yang JH, Yang MS (2005) A control chart pattern recognition system using a statistical correlation coefficient method. Comput Ind Eng 48:205–221

A Naïve Bayes Based Machine Learning Approach and Application Tools Comparison Based on Telephone Conversations

Shu-Chiang Lin, Murman Dwi Prasetio, Satria Fadil Persada and Reny Nadlifatin

Abstract This paper investigates the application of hybrid Bayesian based semi-automated task analysis machine learning tool, Text Miner. The tool is still in its developing stage. Telephone's dialog conversation between call center agent and customer was used as training and testing dataset to feed in parsing based Text Miner. Preliminary results extracted from Text Miner based on the naïve Bayes approach was further compared to one open sourced machine learning program, tokenizing based Rapid Miner. Fifteen prediction words combinations were compared and the study finds that both tools are capable of processing large dataset with Text Miner performs better than Rapid Miner in predicting the relationship between prediction words and main subtask categories.

Keywords Machine learning · Naive bayes · Text miner · Rapid miner · Parsing · Tokenizing · Task analysis

S.-C. Lin (✉) · S. F. Persada · R. Nadlifatin
Department of Industrial Management, National Taiwan University of Science and Technology, 43 Keelung Road, Section 4, Taipei 106, Taiwan, Republic of China
e-mail: slin@mail.ntust.edu.tw

S. F. Persada
e-mail: d10101807@mail.ntust.edu.tw

R. Nadlifatin
e-mail: m10001834@mail.ntust.edu.tw

M. D. Prasetio
Independent Electrical Consultant, Hidrodinamika 4 T-74, Surabaya 60115, Indonesia
e-mail: prasetio_arel.corp@yahoo.com

Y.-K. Lin et al. (eds.), *Proceedings of the Institute of Industrial Engineers Asian Conference 2013*, DOI: 10.1007/978-981-4451-98-7_121,
© Springer Science+Business Media Singapore 2013

1 Introduction

Lin and Lehto's study (2009) proposed a Bayesian inference based semi-automated task analysis tool, Text Miner, to better help task analysts predict categories of tasks/subtasks performed by knowledge agents from telephone conversations where agents help customers troubleshoot their problems. The study finds that the task analysis tool is able to learn, identify, and predict subtask categories from telephone conversations. With the profusion of results obtained in previous work, more analyses needed to be done in order to identify the relationship between prediction words and main subtask category. For that reason, this study further examines and compares the performance results from Text Miner with another machine learning tool, an open sourced Rapid Miner. The prediction accuracy to main subtask categories among fifteen combinations of prediction words were carried out by both tools based on naïve Bayes method.

2 Literature Review

2.1 Task Analysis and Task Modeling

Methods of collecting, classifying and interpreting data on human performance lie at the very fundamental of ergonomics. Task analysis and task modeling especially play critical roles for interactive system design. Task analysis helps the designers to acquire an user oriented perspective of the activity, thus avoiding function oriented biases during the development process (Kirwan and Ainsworth 1992). Task modeling helps for in depths understanding, communicating, testing and predicting fundamental aspects of the human–machine system.

2.2 Naive Bayes

Naive Bayes is one of the most efficient probabilistic algorithms for machine learning (Zhang 2004) without trading much of the effectiveness. The general naïve Bayes rule is illustrated in Eq. (1), where P(B|A) is the probability of the evidence P(A) given that the hypothesis B is true, P(B) is the prior probability of the hypothesis being true prior to obtaining the evidence A.

$$P(B|A) = \frac{P(A|B) * P(B)}{P(A)} \quad (1)$$

The implementation of naïve Bayes in text classification is known as Naïve Bayes Classifier. Many researches use Naive Bayes Classifier as a core classification in their classification method. Mosteller and Wallace (1964, 1984) proposed

Hierarchical Bayes Model for contagious distribution in modern language models. Lin et al. (1999) used Naive Bayes to identify plausible causes for ship collisions in and out five major US ports. Berchiala et al. (2013) used the Naïve Bayes Classifier with selection feature in hospital to identify the severity level of children injury in order to detect the most critical characteristic of the injury that can result to the hospital treatment. Perhaps the most recent success example was Silver (2012)'s Bayesian approach to predict US presidential elections.

3 Model Used and Tool Comparison

3.1 Current Model

Lin and Lehto's (2009) proposed model is shown in Fig. 1. The model consists of 4 phases: data collection, data manipulation, machine learning tool environment, and tools evaluation. In the machine learning phase, the machine learning was used to adapt the situation from knowledge base and classify the data into several subtasks categories. An application of Text Miner was used as a semi-automated task analysis tool to help identify and predict the subtask categories based on naïve Bayes, fuzzy Bayes, and hybrid Bayes.

3.2 Machine Learning Tools Comparison

To represent the use of machine learning, a software called Rapid Miner was chosen as a comparison tool to the proposed Text Miner. Rapid Miner is an open source environment for machine learning and data mining (Mierswa et al. 2006). Similar to Text Miner, Rapid Miner provides most of the necessity functionalities for a project utilizing machine learning tool. The functionalities include data acquisition, data transformation, data selection, attribute selection, attribute transformation, learning or modeling and validation. In our study, we utilized Rapid Miner to create a prediction with the probability of the frequencies combination and subtasks categories based on naïve Bayes. The results were compared to Text Miner's naive Bayes based session. The comparison schema is shown in Fig. 2.

4 Results and Discussions

To predict the relationship between prediction words and main subtasks categories, 71 subtasks categories for a dataset of 5,184 narrative dialog datasets with more than 175.000 words were fed into both Text Miner and Rapid Miner. Two third of

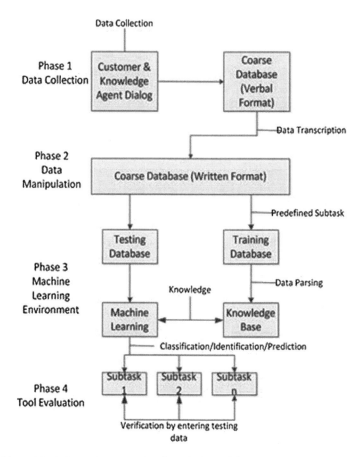

Fig. 1 Lin and Letho's semi-automated task analysis model

dataset was used as training set with the rest one third of dataset was used as testing set. Both Text Miner and Rapid Miner have to go through several phases prior to extract the combination words, as briefly described below.

For Text Miner, the first phase begins with setup initialization where the input texts were inserted in this phase, following with create database phase where the parsed dataset was classified between usable and unusable data to get the morph list. The next phase is learning phase, where Text Miner creates fifteen combination words frequency list. The last phase is cross index, where Text Miner provides the index for the possible word of each category when the combination words are presented.

For Rapid Miner, the first phase is started by tokenizing the characters into words. After the characters are set, the Rapid Miner tool filters the stop word which has similar function with separating usable and unusable words. The third phase is filtering the token which Rapid Miner will categorize the token by the

A Naïve Bayes Based Machine Learning Approach and Application Tools

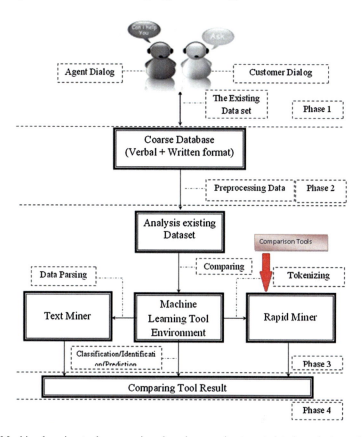

Fig. 2 Machine learning tools comparison based on semi-automated task analysis model

length of the sentences. The last phase is generating the n-gram to identify the sequence of combination words.

Table 1 summarizes the testing results after the combination words processed in both Text Miner and Rapid Miner. As illustrated in Table 1, the results reveal that the average probability of correction prediction in train set for Rapid Miner is 21.23 and 50.31 % for Text Miner, following with the average probability in testing set for Rapid Miner is 17.30 and 36.17 % for Text Miner. Overall, Rapid Miner has average of 19.91 % while for Text Miner, the probability is 45.58 %.

The differences between Text Miner and Rapid Miner in Table 2 were statistically significant, with the t-value equals to 8.18 and the p value equals to 0.007 with the confidence level of 95 %. This result leads to a suggestion that Text Miner statistically performs better than Rapid Miner.

Table 1 Comparison of two machine learning

Task	Value	
	Rapid miner	Text miner
Process the dataset	Tokenizing process	Parsing process
Eliminate stop words	Yes	Yes
Compute frequency	Yes	Yes
Format output	Converted excel	Converted access
Efficient for large database	Yes	Yes
Overall performance	Average prob-all: 19.91 %	Average prob-all: 45.58 %
	Average prob-train: 21.23 %	Average prob-train: 50.31 %
	Average prob-test: 17.30 %	Average prob-test: 36.17 %

Table 2 t-Test analysis result

Tools	N	Mean	StDev	SE mean	T-value	p-value
Text miner	3	44.02	7.20	4.16		
Rapid miner	3	19.48	2.00	1.15	8.18	0.007
Differences	3	22.54	5.20	3.0		

5 Conclusions

Both tools are capable of processing large dataset with parsing based Text Miner performs better than tokenizing based Rapid Miner in predicting the relationship between prediction words and main subtask categories when fifteen prediction words combinations were compared. The finding with Text Miner's robust predicted performance provides an opportunity to explore more of Text Miner's tool performance with different criteria, and the possibility to re-evaluate coarse subtask categories rather than main subtask categories to improve tool's efficiency. A comparison with other machine learning tools can also be considered in future research.

References

Berchialla P, Foltran F et al (2013) Naïve Bayes classifiers with feature selection to predict hospitalization and complications due to objects swallowing and ingestion among European children. Saf Sci 51(1):1–5

Kirwan B, Ainsworth LK (1992) A guide to task analysis. Taylor & Francis, London

Lin S, Lehto MR (2009) A Bayesian based machine learning application to task analysis. In: Wang J (ed) Encyclopedia of data warehousing and mining, classification B, 1-7, 2nd edn

Lin S, Patrikalakis NM, Kite-Powell HL (1999) Physical risk analysis of ship grounding. Navigational technology for the 21st century, 55th annual meeting, Cambridge, MA

Mierswa I, Wurst M, Klinkenberg R et al (2006) YALE: rapid prototyping for complex data mining tasks. In: Proceedings of the 12th ACM SIGKDD international conference on knowledge discovery and data mining (KDD-2006), pp 935–940

Mosteller F, Wallace DL (1964) Inference and disputed authorship: the federalist. Addison-Wesley, Reading

Mosteller F, Wallace DL (1984) Applied Bayesian and classical inference: the case of "The Federalist" papers. Springer, Berlin

Silver N (2012) The signal and the noise: why so many predictions fail-but some don't. Penguin Group US

Zhang H (2004) The optimality of naive Bayes. FLAIRS conference. doi: citeulike-article-id:370404

Evaluating the Development of the Renewable Energy Industry

Hung-Yu Huang, Chung-Shou Liao and Amy J. C. Trappey

Abstract While the world is increasingly concerned with the utilization of fossil energy and carbon emission, renewable energies have been promoted by many governments as alternatives worldwide. Wind power is currently the most mature renewable energy technology. Although Taiwan is located in a region of abundant wind power, this energy industry still faces big challenges in effectively growing the market. This study employs a variation of the hidden Markov model (HMM) to analyze the development of wind power industry in Taiwan, because the renewable energy development is determined by multiple time-series-related factors. The methodology may assist industry in making correct investment decision and provide recommendations to the government for setting suitable green energy policy.

Keywords Renewable energy · Wind power · Hidden Markov model

1 Introduction

Many nations have recognized the need to reduce dependence on traditional energy in order to avoid the associated environmental damage last several years. Governments have emphasized the application of renewable energies. In particular,

---

H.-Y. Huang (✉) · C.-S. Liao · A. J. C. Trappey
Department of Industrial Engineering and Engineering Management,
National Tsing Hua University, Hsinchu 30013, Taiwan
e-mail: smileking081700@gmail.com

C.-S. Liao
e-mail: csliao@ie.nthu.edu.tw

A. J. C. Trappey
e-mail: trappey@ie.nthu.edu.tw

wind power is currently the most mature renewable energy technology. In the past decades, some research focused on the assessment of the development of wind power industry. The research of Buen (2006), compares the role of policy instruments in stimulating technological change in Denmark and Norwegian wind industry. Camadan (2011) investigated the status and future of the wind energy industry in Turkey. Zhao et al. (2013) have adopted a strength, weakness, opportunity and threat analysis (SWOT) approach to examine both the internal and external factors that affect the competitiveness of the wind power industry in China. However, with increasing number of influential factors, these analytic approaches have become overly complicated and time consuming.

In order to encourage the public as well as industrial companies to invest in and use renewable energy, a series of measures was adopted by government in Taiwan. However, the huge budget required for the government to carry out this policy adversely affects the promotion of renewable energy. Because industrial investors and policy makers are generally concerned with profit-making opportunities and promotion of renewable energy, our aim is to develop an efficient forecasting approach for developments in the renewable energy industry. To this end, in this study, we employed HMM to assess the development of the wind power industry, due to its strong statistical foundation, computational efficiency, and ability to handle new data robustly and to predict similar patterns efficiently (Hassan and Nath 2005). HMM has been successfully used to forecast oil prices (De Souza e Silva et al. 2010), the Dow Jones Average stock index (Wang 2004) and the leading indicator composite index of real estate (Wu 2009). Because the renewable energy development is determined by multiple time-series-related factors, we use a variation of the HMM, called the hidden Markov model with multiple feature streams (HMM/MFS), to simultaneously consider distinct factors influencing the development trends of renewable energy. This model can be used to assist industrial investors when making cost analysis decisions and provide specific recommendations to the government about sustainable energy policy.

2 Modeling

2.1 Hidden Markov Model with Multiple Feature Streams

HMM (Rabiner 1989) is a doubly embedded stochastic process: the first process is a Markov chain, and the second is an observation process, whose distribution at any given time is fully determined by the current state of the Markov chain. This study considered the development of wind power industry as being determined by a number of factors; therefore, we used a variation of HMM, called the Hidden Markov Model with Multiple Feature Streams (HMM/MFS) (Somervuo 1996).

Figure 1 shows an example of the HMM with two states and multiple (three) feature streams. It simultaneously considers three independent factors $\{V^1, V^2, V^3\}$

Fig. 1 An illustration of a hidden Markov model with two states and three feature streams

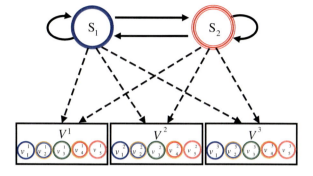

for the development of industry. In HMM, factor vectors represent observation vectors. The set of the factor V^i's factor vectors is $\{v_1^i, v_2^i, v_3^i, v_4^i, v_5^i\}$, where $1 \leq i \leq 3$. We considered how the observation streams could be fed into HMM: either by concatenating them into a single factor vector or keeping the three factor vectors separate. If vectors with different scales are concatenated into a single factor vector (e.g., CPI and crude oil price), the result would be dominated by the factors with large variance; however, the factors with small variance may play a more important role. This is called scaling problems, which may miss information. If different factor vectors are kept separate, scaling problems can be avoided (Somervuo 1996).

2.2 Dataset

Our forecasting methodology can be used to analyze time-series data. Thus, the data set of interest is the history of the development of wind power. In this study, the installed monthly capacity of wind power (BEMOEA)[1] as the hidden state in the development of this industry. To select related factors as the observations of our model, which potentially fit the states of a development trend, we studied the relevant factors about the wind power industry. From the interviews with the Green Energy and Environment Research Laboratories of ITRI experts, as well as related background and literature, we selected five related factors as the observations of our model; include the steel price index (Finance),[2] the Consumer Price Index (CPI) (BEMOEA), the crude oil price (Mundi index),[3] the export price of

---

[1] Bureau of Energy of Ministry of Economic Affairs (BEMOEA). Taiwan, ROC. http://www.moeaboe.gov.tw/ (in Chinese).

[2] Finance. http://big5.ifeng.com/gate/big5/app.finance.ifeng.com/data/indu/jgzs.php?symbol=5/ (in Chinese).

[3] Mundi index. http://www.indexmundi.com/ (in English).

coal from Australia (Mundi index) and the electricity price (Taiwan Power Company).[4]

For data transformation, first, HMM has M distinct observation symbols, which represent the physical outputs of the system being modeled. To predict the installed capacity for wind power, it is necessary to assign one symbol for each monthly installed capacity. Therefore, the classes of hidden states and observations in training data were labeled through Wang's approach (Wang 2004) and transformed them into variables with the discrete data as the input of HMM.

We divided the number of hidden states into three states based on the total length of the change rate per period of installed capacity. The three states are similar to the monitoring indicators announced by the Council for Economic Planning and Development (CEPD).[5] The mapping of the change rate into symbols for the HMM, described hereafter. The three states are the three signals of the monitoring indicator. If the change rate per period of installed capacity is less than X, the signal is Blue, between X–Y Green, and more than Y Red, where $Y > X$. The signals of the monitoring indicators are: Blue shows that the development of the wind power industry is slumping; Green represents stability; and Red indicates a boom. Similarly, the data of observations can be transformed into discrete data in a similar manner.

3 Experiments

In this experiment, there were two stages to construct the model: model training and model predicting. Model training (2010/12-2012/10), was to determine the parameters of the HMM model. The parameters were used for the structure inference. Next, model predicting (2012/11-2013/01), was to predict the future installed capacity of wind power

To further validate our method, we employed HMM by the Viterbi algorithm to predict the state of the wind power industry in Taiwan. The next monthly state (November 2012) depended only on the industry state in October 2012. The real state transit was from (Green) to (Green). The predicted result of the HMM with a single factor is shown in Table 1. Consider the observation of the crude oil price. If the state in October 2012 is (Green), there is a 56 % chance that the next state is (Green); this probability value was the highest. And consider the observation of the electricity price. If the state in October 2012 is (Green), there is a 75 % chance that the state of next month is (Red). Result of predicted analyses show that when we consider these observations of crude oil price, coal price and steel price index, these models correctly predicted the industry trend than CPI and electricity price. It shows that these three factors have a higher degree of influence for wind power industry of Taiwan.

---

[4] Taiwan Power Company. Taiwan, ROC. http://www.taipower.com.tw/ (in Chinese).

[5] Council for Economic Planning and Development (CEPD). The monitoring indicators. Taiwan, ROC. http://www.cepd.gov.tw/m1.aspx?sNo=0000160/ (in Chinese).

Table 1 HMM of wind power industry in Taiwan from 2012/11 to 2013/01

	2012/11 State	2012/12 State	2013/01 State
Single factor			
Crude oil price	G (56 %)	G (69 %)	G (73 %)
Coal price	G (93 %)	G (90 %)	G (81 %)
Steel price index	G (73 %)	G (43 %)	G (41 %)
CPI	G (61 %)	G (64 %)	G (51 %)
Electricity price	R (75 %)	G (55 %)	G (83 %)
Multiple factors	G (100 %)	G (100 %)	G (100 %)
Real state	G	G	G

R Red(Boom), *G* Green(Stability), *B* Blue(Downturn)

On the other hand, the predicted probability of multiple factors was much closer to the real state than single factor. For this case, the probability of correct prediction in HMM/MFS is significantly larger, compared with that in HMM with a single factor.

4 Conclusions

Developing renewable energy has become the focus of national policy due to energy shortages. For government or companies who want to support the renewable energy industry, there are budget and policy concerns for the former and investment capital and income concerns for the latter. Therefore, in this study we have used HMM and HMM/MFS to analyze the developmental movements of the renewable energy industry. According to our experimental result, we concluded that HMM/MFS is an effective forecasting model for predicting the trends of industrial development. Because the Taiwanese renewable energy industry is an emerging industry, data collection has limitations. However, when this problem is overcome, it would be of great interest if the forecasting model could provide more accurate prediction. Finally, it would be worthwhile to apply the proposed methodology to other renewable energy industries for future investigation.

References

Buen J (2006) Danish and Norwegian wind industry: the relationship between policy instruments, innovation and diffusion. Energy Policy 34(18):3887–3897
Camadan E (2011) An assessment on the current status and future of wind energy in Turkish electricity industry. Renew Sustain Energy Rev 15(9):4994–5002
De Souza e Silva EG, Legey LFL, De Souza e Silva EA (2010) Forecasting oil price trends using wavelet and hidden Markov models. Energy Econ 32(6):1507–1519
Hassan MR, Nath B (2005) Stock Market forecasting using hidden markov model: a new approach. In: Proceeding of the 5th international conference on intelligent systems design and applications, pp 8–10

Rabiner LR (1989) A tutorial on hidden markov models and selected applications in speech recognition. Proc IEEE 77(2):257–285

Somervuo P (1996) Speech Recognition Using Context Vectors And Multiple Feature Streams. Dissertation, Helsinki University of Technology

Wang SC (2004) Investment Decision Support with Dynamic Bayesian Networks. Dissertation, National Sun Yat-sen University

Wu YL (2009) Studies on the Predicting Power of Leading Indicator of Taiwan's Real Estate Cycle—Hidden Markov Model Analysis. Dissertation, National Pingtung Institute of Commerce

Zhao ZY, Hong Y, Zuob J, Tian YX, Zillantec G (2013) A critical review of factors affecting the wind power generation industry in China. Renew Sustain Energy Rev 19:499–508

On-Line Quality Inspection System for Automotive Component Manufacturing Process

Chun-Tai Yen, Hung-An Kao, Shih-Ming Wang and Wen-Bin Wang

Abstract The automotive industry in a nation not only forms an economic base, but also plays an important role for safety and convenience in the society. In average, a vehicle is composed of more than thousands components, and the quality of each component is definitely critical. Automotive component manufacturers, as suppliers to automotive manufacturers, are forced to emphasize the quality inspection for their products. Hence, this research proposes an online quality inspection system for vehicle component manufacturing industries. The system is composed of three subsystems, which are real-time machine condition monitoring, supply chain management, and production information management. The transparency of the production process can further be analysed to infer the quality of product on production line. The proposed system is introduced to a Taiwanese vehicle component manufacturing industry for validating its capability to reduce the effort spent on inspection work, the rate of waste, and to improve overall equipment efficiency (OEE).

C.-T. Yen (✉)
Institute of Industrial Engineering, National Taiwan University, Taipei, Taiwan, Republic of China
e-mail: ctyen516@gmail.com

H.-A. Kao
NSF I/UCRC for Intelligent Maintenance Systems, University of Cincinnati, Cincinnati, OH, USA
e-mail: kaohn@mail.uc.edu

S.-M. Wang
Department of Mechanical Engineering,
Chung Yuan Christian University, Chung-Li, Taiwan, Republic of China
e-mail: shihming@cycu.edu.tw

W.-B. Wang
Innovative DigiTech-Enabled Applications and Services Institute, Institute for Information Industry, Taipei, Taiwan, Republic of China
e-mail: wwbjoe@iii.org.tw

Keywords Production optimizations · Online quality inspection · CNC machine monitoring · Manufacturing information management

1 Introduction

The automotive industry in a nation not only forms an economic base, but also plays an important role for safety and convenience in the society. In average, a vehicle is composed of more than thousands components, and the quality of each component is definitely critical. Automotive component manufacturers, as suppliers to automotive manufacturers, are forced to emphasize the quality inspection for their products. For instance, a transmission shaft of a vehicle is required to have total inspection and the inspection result should be within $\mu \pm 6\sigma$, where μ is specification value and σ is defined standard deviation. Besides, since the sales performance of automotive industries vary a lot, the needs for automotive component also alternate through seasons. The common requirement of Just-in-time from customers also makes automotive component manufacturers to focus on how to response to customers' orders quickly and precisely.

To aid automotive component manufacturers in solving the above-mentioned needs from customers, this research proposes an online quality inspection system for automotive component manufacturing process. The system is composed of three subsystems, which are real-time machine condition monitoring subsystem, supply chain management subsystem, and production information management subsystem. The transparency of the production process can further be analysed to infer the quality of online product. The proposed system is introduced to a Taiwanese vehicle component manufacturing industry, and validated its capability to reduce the effort spent on inspection work, the rate of waste, and to improve overall equipment efficiency (OEE).

2 Information Communication Technology Based Solution Framework

For automotive component manufacturing process, a system including three subsystems, which are *real-time machine condition monitoring subsystem*, *supply chain management subsystem*, and *production information management subsystem*, is proposed based on information communication technology. The architecture of this system is shown in Fig. 1. First of all, the controller signal and sensory data can be extracted automatically from online process by *real-time machine condition monitoring subsystem*, which can reveal the condition of machine operation, the NC program of the machine, and the corresponding control parameters in real time.

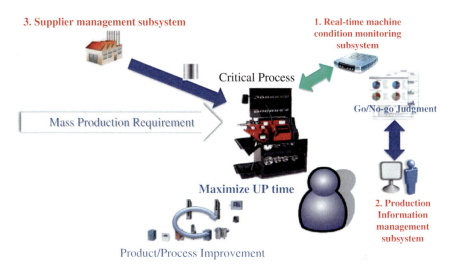

Fig. 1 Architecture of online quality inspection system

Secondly, by analysing these signals, a quality inference engine is utilized to assess the quality and stability of the current process, in order to achieve online quality inspection. The quality information, which is computed by online intelligent inspector, will be mapped with original production recipe (control parameters and production history). The quality information of a product will be stored, sorted, and streamlined according to its work order identification, batch number, and ERP tag by the *production information management subsystem*. Finally, the manufacturing operators can make query and retrieve related information for optimization of the manufacturing process. The *supply chain management subsystem* can support to communicate with suppliers and seamlessly feedback scraps and their impacts for process improvement.

3 Real-Time Machine Condition Monitoring

To acquire the operation condition from machines, architecture for extracting controller signal from CNC systems is designed. A *controller signal extractor* and *real-time monitoring module* are integrated in an industrial PC. Each machine is equipped with a controller and RJ-45 Internet cable is used to connect between controller and the signal extractor through the Ethernet Internet protocol.

The *controller signal extractor* is implemented by C# and .Net Framework and utilize FOCAS library API provided by FANUC controllers. It includes a parameter extractor (connect with controllers and extract parameters), rule configuration module (write rules and trigger points for parameter extraction), command receiver (receive the commands from real-time monitoring module and configure accordingly), and an information sender (transfer extracted signals to real-time monitoring module). The *real-time monitoring module* is developed by Java and web programming library. It includes some basic components like server interface, user authorization, Model-View-Controller application, and central management unit.

Since the sampling frequency of the *controller signal extractor* is high, the communication between the *controller signal extractor* and the *real-time monitoring module* is frequent. To efficiently process the data set and at the same time provide a real-time, stable data extraction performance, an asynchronous transmission mechanism is designed and adopted. Once the *controller signal extractor* extract the signal data, the data set will be sent to a message queue instead of monitoring module directly. The message queue is a high performance buffer space and can be easily customized to receive and store the data. Once the data set is reserved in the message queue, the *real-time monitoring module* will use its Listener to register to the message queue. When there are messages waiting in the message queue, the Listener will be triggered to extract the messages from the queue and parse them. The parsed information will then be sent to *real-time monitoring module* to analyze and become visualized information through user interface.

4 Online Quality Inspection System

In this research, we propose to develop rules for diagnose the quality of production by using real-time signals generated from manufacturing processes. The error estimation and cutting monitoring can be integrated into online process in order to infer the quality information. This method can save the cost spent on inspection, and also avoid the uncertainty of manual inspection. Besides, the integration of error estimation results and production history can let manufacturers examine the relationship among design, operation and control parameters and further improve the production quality.

The signals are extracted from manufacturing process based on *real-time machine condition monitoring module*. The feature values are then analysed and served as input for model training. There are three development steps: (1) Signal processing and feature extraction; (2) Quality inspection rule and threshold establishment; (3) Validation. The experiments are firstly conducted in order to collect the vibration signal, control signal, and corresponding air gage quality inspection data. The data set is then used to analyse the correlation of vibration and cutting accuracy. Based on a history data set, rules and thresholds are defined for online precision diagnosis. After the model is learned from training data set, the system is implemented by the integration of each subsystem, and validated using real production line.

In this research, the vibration signal, which is collected from an accelerometer (X and Z directions) during cutting process, is mainly used to infer the quality information since the sensitivity of vibration signals is suitable for the detection of abnormal cutting process. The amplifier will adjust the acquired signals. Other than vibration signals, NC code information is also collected for understanding of the current operation of the machine. The air gage measurement results are then utilized to correlate the vibration and size distribution of work pieces. After combining these three types of input data, the model for quality inspection can then be learned.

To implement the real-time machine condition monitoring, a ServBox is proposed to be a service-oriented device to enable the functions. ServBox is a service channel that provides synchronized communication between the CNC machine tool and the remote service platform, so called ServAgent, inside a factory. With both ServBox and ServAgent, machinery industry could implement innovative services and machine network via remote cloud computing service to achieve optimal productivity and enhance the partnership with customers. Through ServBox, we can change the traditional model of service communication, and achieve an active and fast two-way communication with service platform.

Nowadays, ServBox can extract more than 4,000 parameters from Fanuc i-series controllers, including 6 main categories: historical information, CNC File data, NC Program, Spindle servo, library handle.

To achieve online quality inspection goals, a platform for increase productivity and collaborative manufacturing with suppliers is implemented. Here, the proposed platform is so called ServAgent. When there are more and more machine tools in a factory, ServAgent can grasp the machine tools' real-time condition, monitoring operation and detecting production quality and performance. In addition, it can connect to machine tools only when needed and avoid data-leakage problem. A win–win situation is created for automotive component manufacturers and their suppliers by integrating the solution and production line based on the mechanism of production management.

ServAgent leverages the machining information extracted from machine tools to unveil the machining data that in the past is concealed in the mysterious manufacturing process. It collects productivity information from production lines and forms a machine service network to help customers manage their production line in real-time and more intelligent, and finally the productivity and utilization can be improved. While the domain-specific knowledge of machining process is accumulating, it can also support customer to find out the weakness in the current manufacturing process, which means the quality of their manufacturing is also increased. Besides, the knowledge collected can used to be an expert or reference and provided to customers. Before the abnormal condition occurs, take the actions to avoid the possible damages and stabilize the production quality level. Finally, customers can manage the overall productivity and efficiency of their equipment in the factory. It offers customers the flexibility and knowledge toward their equipment and factory.

5 Experiments

To validate the feasibility of the proposed system, an experiment is designed and conducted in automotive manufacturing facilities. To find out which signals and features are reliable for quality inspection, a turning machine is used as test-bed and the relationship between signals and turning tool is evaluated. A CMOS camera also records the turning tool wear condition during each experiment in order to quantify tool wear condition.

The experimental process is as follows:

1. Use clay to cover the spindle to avoid the lubricant interfere the accelerometer. The sensor is then connected to an amplifier and DAQ system. Through ServBox, the sensory signal and control signals are then be extracted.
2. Test if the signals from the accelerometers are table and correct.
3. Install a work piece and set the controller.
4. Start the NC program to activate the cutting process.
5. Observe the signals extracted. If the signals are normal and stable then the experiments can be started.
6. Cutting 20 times and 10 times for 2 repetitions each. So, there will be totally 4 set of experimental data.

The results are shown in Fig. 2. Figure 2e is the cutting vibration of cycle 1–10, while Fig. 2f is cycle 11–20. It can be observed that the latter vibration condition is 0.015 g larger than previous one. Compared with Fig. 2c, the wear condition after 20 cutting cycles is 1.5–2 times of the 10-cutting-cycle condition, and after 30 cutting cycles the wear condition is even worse. At the same time, the wear condition at the middle area on the blade expands as cutting cycles pass. The reason is that after the cutting tool wears, the leaning angle of the blade will change, which forces the friction between debris and blade increases. From the results of the experiments, it can be concluded that the cutting vibration can be used to monitor if the turning tool reaches the threshold and can cause unwanted quality. After the training data set is collected, an automatic rule-learning

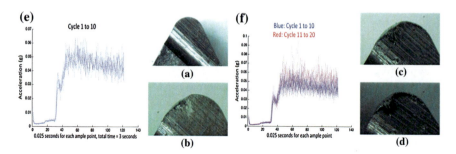

Fig. 2 Cutting vibration signals of cycle times. **a** New tool, **b** After 10 cycles, **c** After 20 cycles, **d** After 30 cycles, **e** Cycle 1 to 10, **f** Cycle 11 to 20

On-Line Quality Inspection System 1037

Fig. 3 Result of 'Go' case from online quality inspection system

algorithm develops rules and thresholds for quality inspection. The model is then introduced in the system as a core analyzer.

The experiments period is three months and totally 7,286 work pieces are validated. The air gauge measurement results are used as ground truth to evaluate the performance of proposed system. The results shown in Fig. 3 are from the same experiment with qualified work piece. The online quality inspection system can successfully diagnose it is accepted during manufacturing process. That is, the quality information can be detected before metrology stage. On the other hand, Fig. 4 show results from one of the failed products. The system can detect the work piece might have unwanted quality during the process. The process can be adjusted immediately instead of waiting until knowing bad inspection results. The overall experimental result shows no difference between the judgment of the proposed online quality inspection system and the results from the air gage metrology, which proves the feasibility of the system.

Fig. 4 Result of 'No-Go' case from online quality inspection system

6 Conclusions

This research proposes an online quality inspection system for automotive component manufacturing process. The system is composed of three subsystems, which are real-time machine condition monitoring subsystem, supply chain management subsystem, and production information management subsystem. The vibration signals are used to evaluate the process and tool condition, and the models are used to diagnose the quality and work piece in real time. After the training data set is collected, an automatic rule-learning algorithm develops rules and thresholds for quality inspection. The overall system is introduced to an automotive component manufacturing process to validate its feasibility. The experimental results from 7,286 work pieces prove its capability of distinguish good and bad quality of work piece during the process. The proposed method can benefits automotive component manufacturers to understand the quality information earlier and improve current process immediately. Eventually, it can optimize the manufacturing process and also the collaboration among manufacturers, suppliers and customers.

Acknowledgments This study is conducted under the "Precision Machine Equipment Characteristic Service Development—Trial Service Operation for Machine Tool Agile Service Center Project" of the Institute for Information Industry which is subsidized by the Ministry of Economy Affairs of the Republic of China.

References

Carrilero MS, Bienvenido R, Sa'nchez JM, A'lvarez M, Gonza'lez A, Marcos M (2000) A SEM and EDS insight into the BUL and BUE differences in the turning processes of AA2024 Al–Cu alloy. Int J Mach Tools Manuf 42:215–220
Dohner JL, Lauffer JP, Hinnerichs TD, Shankar N, Regelbrugge M, Kwan CM, Xu R, Winterbauer B, Bridger K (2004) Mitigation of chatter instabilities in milling by active structural control. J Sound Vib 269(1–2):197–211
Doi M, Masuko M et al (1985) A study on parametric vibration in chuck work. Bull JSME 28(245):2772–2781
Hahn RS (1954) On the theory of regenerative chatter in precision grinding operations. Trans ASME, B 76(1):593–597
Hook CJ, Tobias SA (1963) Finite amplitude instability—a new type of chatter. In: Proceedings of 4th MTDR, pp 97–109
Inamura T, Senda T, Sata T (1977) Computer control chattering in turning operation. Ann CIRP 25(1):181–186
Kao A et al (2011) iFactory cloud service platform based on IMS tools and servo-lution. World Congress on Engineering Asset Management, Cincinnati
Marui E et al (1983) Chatter vibration of lathe tools. Trans ASME, B 105:107–133
Shinobu K, Etsuo M, Masatoshi H, Yamada T (1986) Characteristice of chatter vibration of lathe tools. Mem Fac Eng, Nagoya, 38(2):208–215
Tlusty J, Spacek L (1954) Self-excited vibrations in machine tools. Nakladateistvi CSAV Prague, Czech

Using Six Sigma to Improve Design Quality: A Case of Mechanical Development of the Notebook PC in Wistron

Kun-Shan Lee and Kung-Jeng Wang

Abstract Wistron has transformed from private brand to OEM/ODM after reorganizing in 2003, witnessing phenomena that profits of personal computer and its peripherals decline day by day, and domestic manufacturers had gradually moved to China mainland because unit cost is not competitive, continued to survive with the help of low processing cost and then expanded the production base. Therefore, companies of this case considered to fully carry out Six Sigma program to improve the plan in order to solve their troubles in 2005. The company of the case in 2007/Q2 grew quickly in NB business, but its defective products caused a huge loss to the company. Consequently, a project team was established to consider carrying out preventive measures and design improvement during product R&D. As a whole, the individual case will discuss subjects as follows: (1) The core processes of Six Sigma and the opportunity of using tools shall be understood in order to get the maximum benefit and enhance the capability of solving problems. (2) Performance comparisons before and after Six Sigma design are imported. (3) The process mode and standard design norms of new product R&D are constructed through research of actual cases.

Keywords Six Sigma · Research and Design · DMAIC · Notebook PC

K.-S. Lee (✉)
Department of Industrial Management, National Taiwan University of Science and Technology, No. 43, Section 4, Keelung Road, Da'an District,
Taipei City 106, Republic of China
e-mail: andy_lee@wistron.com

K.-J. Wang
School of Management Department Assistant Head, National Taiwan University of Science and Technology, No. 43, Section 4, Keelung Road, Da'an District, Taipei City 106, Republic of China
e-mail: kjwang@mail.ntust.edu.tw

1 Individual Case

Wistron transformed from private brand to OEM/ODM after reorganizing in 2003. It is mainly to assemble products according to product specifications and complete detail design provided by OEM customers, or ODM customers strive for getting orders according to products designed by themselves.

In 2005, Wistron imported Six Sigma project to improve plan and began to consider how to use Six Sigma technology to improve manufacturing quality. The past OEM focuses on improvement and control of product process, whereas, ODM focuses on improvement of design process.

In the 2nd quarter of 2007, a customer complained to Jeff (leader of league), secretary of the NB business division via telephone about quality control. The extra charges for defective products become increasingly high, so the company suffered from a big loss (the more the goods are shipped, the higher the loss will be).

1.1 Company Develops Highly but Falls into Reworking Trouble

In 2007/Q4, the NB business grew quickly. Its turnover increased by 6 times compared to the year before, and shipping quantity of Notebook PC was up to 7 times. The tendency was getting better, but customers required returning goods or reworking the defective goods locally because of the poor quality. As a result, company suffered from a big loss because of a great amount of reworking cost.

"Why cannot we use preventive measures to improve design quality during R&D? How do they optimize design to meet customer demands? What should we do?" Andy, manager of R&D department, gazed at the scrap/reworking cost report and was lost in through after receiving the indication of Jeff (Fig. 1).

2 Case Company and Industrial Profile

Wistron is the OEM/ODM of notebook (NB) ranked top 3 in Taiwan. In 2008, it was aim to ship more than 2,000 notebooks. Wistron is also one of the professional designers and OEMs of the biggest information and communication product in the world, which currently has 3 R&D support centers, 6 manufacturing bases, 2 global service centers, and more than 60,000 professionals and solid global service network in the world. Products cover notebooks, desktop computers, servers, network household appliances, cable and wireless data communications, digital consumer electronics, etc.

Using Six Sigma to Improve Design Quality

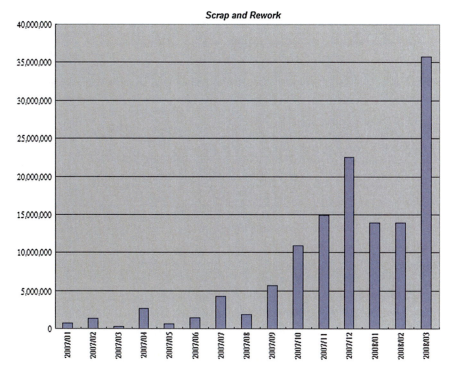

Fig. 1 2008 CoPQ (cost of poor quality) scrap/reworking cost

3 Establishment of Project Team and Improvement Items

The work started with definition of scope. Jeff faced up with quality problem, and the company suffered from high cost of reworking every month. He hence changed the work focus to Andy and thought: "under the leadership of Andy, Six Sigma is imported at the beginning of research and development, the preventive measures and design quality improvement are imported, and possible defective causes are analyzed and found to effectively control abnormal change and minimize the reject ratio and meanwhile reduce manufacturing cost. This way, will the core problem about the more shipment the higher the loss be solved?"

From the business scorecard of Jeff (Table 1), we can see that his goal of KPI is to make the cost of poor quality to be 6.0 % less than the business cost in 2007. In other words, they should find a way to save cost or bring profit for company and enhance customer satisfactions.

From the figure, we can know that the scrap and reworking costs caused by structure problem (Sub Y) accounts for 62 % (cost for design change in structure accounts for 90 %, and mold repairing cost accounts for 10 %). How did Andy find out biggest problem in the shortest time and raise an effective improvement plan?

Table 1 Business scorecard

Strategy direction		Performance assessment
Strategy (2–3 years)	Preliminary in 2008	Business result (2007 KPI)
• Become a world leader in research and development and manufacture of mobile industry, and focus on innovation design and improvement of manufacturing quality	• BB planning of six sigma	• Win 500 k/month, M10 RFQ
• Continuously provide NB production line with innovation products	• Design quality improvement: reduce scrap and reworking costs	• **CoPQ < 6.0 %**
• Expand handheld production line and become the leader	• Assembly design improvement	• All items will be shipped timely according to the plan
• Become the leader in NB industry share price/EPS		

It was the toughest time, but the team finally reached a compromise and made improvement from quality after discussing intensively. The scrap and rework costs (CoPQ < 6.0 %) should be reduced to meet the performance goal of the league leader (other department managers also felt relaxing.). Finally the "costs for structure design change and reworking" were selected as the subject and the object for importing Six Sigma for quality improvement. After confirming the project subject, a series of activities were carried out. The reworking cost was caused by poor structure design (Small y), so the next target to improve can be formulated (Fig. 2).

Member of the project team had got ready and formulated project schedule Focusing on key problems, they used Six Sigma DMAIC and tools to analyze and solve the problem, and expected that quality improvement could be proved from the financial report after 6 months.

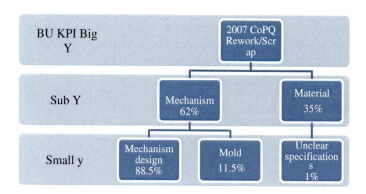

Fig. 2 Tree-diagram drill-down

The financial statement analysis (Fig. 3) clearly shows that the reworking cost for design change has been 98.5 % improved, scrap of structures material has been 78 % improved, and the CoPQ scrap/reworking cost after improving structure quality of the total project plan has been 83 % improved. The indirect benefits after carrying out the improvement program include: A standard design process is established so that colleagues have a good interactive and communicating channel, and as a result the overall efficiency is enhanced. The production per unit is increased, so the factory production cost MOH is reduced to provide the rate of equipment utilization. The reject ratio of production is reduced and the factory process is stable, so the production stock and material stock are reduced. Design documents are standardized to reduce mistakes so as to provide all staff with references. The design quality is enhanced, the assessment standard is established and product confidence level is increased.

The appearance of the completed new product and the key point for improving structure design quality of the new product. The DMAIC technique shall be used to find out key problems of the structure design and key factors affecting Y1, Y2 and Y3. The design scorecard and new process preventive measures shall be imported at the beginning of research and development of new products. The causes of defective products shall be actively analyzed and found out so Wistron can effectively control variation, minimize reject ratio, and meanwhile reduce manufacturing cost, the company can also prevent reoccurrence of defects and mistakes in order to lower down reworking cost caused by poor structure design at the same time (Figs. 4, 5).

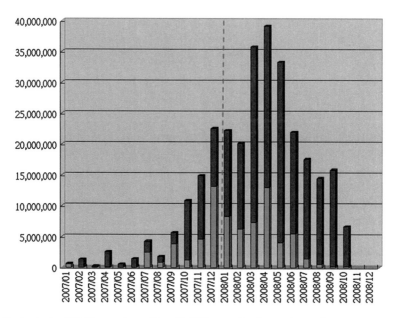

Fig. 3 Analysis of 2008 scrap/reworking (CoPQ) cost after improving project

Fig. 4 Appearance of completed new product

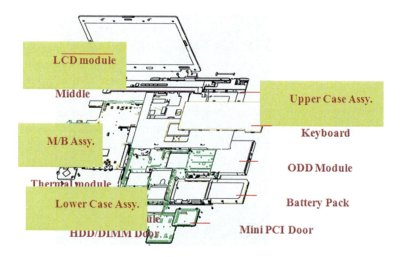

Fig. 5 Key point for improving structure design quality of new product

4 Road to Future Quality Improvement

The final goal of Six Sigma activity lies in business customers and creating all-around customers. When promoting the aforesaid steps, the real demands of customers must be understood, and the key processes and quality items should be confirmed based on customer viewpoint.

When members were indulging in celebrating, Jeff and Andy stepped into the office. Everyone guessed what will Jeff talk to Andy?

References

Breyfogle FW III (2003) Implementing six sigma: smarter solutions using statistical method, 2nd edn. Wiley Europe, New York

Harry M, Schroeder R (2000) Six sigma: the breakthrough management strategy revolutionizing the world's top corporations. Currency Doubleday, New York

Kuei C, Madu CN (2003) Customer-centric six sigma quality and reliability management. Int J Qual Reliab Manage 20(8):954–964

Lin Y (2003) Six sigma will determine your orders and promotion. Bus Week 795:110–112

Mazur GH (1993) QFD for service industries-from voice of customer to task delopyment. The 5th symposium on quality function deployment, Michigan

Wistron, website: http://www.wistron.com.tw/, 2012

Zheng R, Guo C (2001) Six sigma constructs enterprise competitive advantage. J Manage 326:76–79

Estimating Product Development Project Duration for the Concurrent Execution of Multiple Activities

Gyesik Oh and Yoo S. Hong

Abstract Many companies adopt concurrent engineering for their product development projects to reduce time to market. It is often the case that the multiple activities are overlapped in a concurrent engineering environment, while most of the product development research covers only the two-activity problems. The concept of degree of evolution has been proposed in literature to represent how close the unfinalized design of the downstream activity is to its final one, which is actually a measure of the real progress, reflecting the rework requirement due to overlapping. When more than three activities are concurrently executed, it is midstream activity whose degree of evolution is important, since it affects the rework duration for downstream and consequently the overall project duration. It is difficult to estimate the degree of evolution for midstream since involves has two uncertainties, one derived from incomplete information from upstream while the other from changing design information of itself. This paper models the degree of evolution for midstream activity taking into account the two uncertainties. On top of the model, this paper develops a methodology to calculate the project duration, which depends on the project managers' decision on information transfer frequency and overlapping ratio, when three activities are concurrently executed. This paper is expected to help firms forecast the effect of management decision about concurrent engineering dealing with overlapping among multiple activities.

Keywords Product development process · Concurrent engineering · Degree of evolution · Product development management · Overlapping · Multiple activities

G. Oh (✉) · Y. S. Hong
Department of Industrial Engineering, Seoul National University, Seoul, South Korea
e-mail: gushigi4@snu.ac.kr

Y. S. Hong
e-mail: yhong@snu.ac.kr

1 Introduction

The companies adopt concurrent engineering in their product-development projects in order to reduce time-to-market. Concurrent engineering is the one of management methods, which allows overlapping between two dependent activities which were executed sequentially. Project duration under concurrent engineering is shorter than that under sequential development. In sequential development, downstream begins its work based on the finalized information from upstream. We call them a former activity and a latter activity respectively. A latter activity depends on the information from a former activity, therefore, and begins later than a former activity. Different from sequential development, a latter activity begins its work based on the imperfect design information of a former activity since a former activity is in progress. Although a latter activity needs to rework in order to adjust on the changed information from a former activity, it finishes its work earlier than under sequential development. In this reason, firms in many industries such as automobile, airplane, software and computer apply concurrent engineering to product development project (Clark and Fujimoto 1989; Cusumano and Selby 1997; Eisenhardt and Tabrizi 1995).

In the perspective of project management, it is important to decide the overlapping ratio and information transfer frequency between activities in concurrent engineering. As overlapping ratio increases, a latter activity begins its work earlier and project is finished earlier. However, rework duration increases since preliminary information transferred to a latter activity is more uncertain and likely to be changed. In order to alleviate the effect of uncertain information from a former activity, a latter activity receives updated information from a former activity and revises its work. A latter activity is not able to increase the number of information transfer infinitely since setup time for preparing rework is required for each rework. Setup time prolongs activity duration of a latter activity. Therefore, a project manager needs to optimize overlapping ratio and information transfer frequency in order to minimize project duration.

In modeling the concurrent execution of multiple activities, the modeling of midstream activity is difficult because it involves two uncertainties, one derived from incomplete information from upstream and other from changing design information of itself. The concept of degree of evolution has been proposed to represent the certainty of design in product development process. The degree of evolution refers to how close the unfinalized design of a former activity is to its final one. It is actually a measure of the real progress reflecting the rework requirement due to overlapping, which has influence on project duration. In two-activity problem, upstream has the uncertainty caused by changing information in progress. However, in multiple-activity problem, midstream, a former activity of downstream as well as a latter activity of upstream, has two uncertainties. Since upstream is in progress, transferred information from upstream entails uncertainty. Also, design information of midstream involves uncertainty. We model the degree of evolution for midstream in consideration of two uncertainties.

2 Model Illustration

We consider a simple project which consists of three activities and aims to minimize development duration. Three activities have sequential dependency relationships. Midstream depends on upstream and downstream does on midstream respectively. Based on the basic three-activity model, we will develop a general model for multiple activities in future research.

The project duration equals to time period from the beginning of upstream to the end of downstream. It is the summation of work duration before overlapping of upstream and midstream respectively, (a, b), and whole work duration of downstream, (c), as in Fig. 1. Since a latter activity relies on information of a former activity, it is not able to begin its work before a former activity achieves a certain amount of progress. Therefore, duration (a) and (b) are not decreased below the certain duration required by a former activity to reach a progress level at which a latter activity can start. We refer the moment as possible start time to each latter activity. In terms of downstream, work duration (c) is increased at the amount of rework duration, which depends on the information quality from a former activity. The real progress of an activity is important to develop a methodology since it affects both rework duration and possible start time.

We adopt the concept, the degree of evolution, to represent the real progress of a design activity. Krishnan et al. (1998) defines the degree of evolution as how close the unfinalized design is to its final one. The degree of evolution increases and reaches to one at the end of design activity. Whereas Krishnan et al. (1998) defines the degree of evolution for one design parameter, we extend the concept to a set of design parameters. We suggest that the degree of evolution is measured as the portion of fixed design parameters, which are not changed afterwards. A design activity in product development process is a decision making task to determine design parameters (Petkas and Pultar 2005; Eppinger and Browning 2012). As more design parameters are determined by designers and design becomes certain, the degree of evolution increases from zero to one. Therefore, we can use the portion of determined design parameters as the proxy for the degree of evolution for a design task.

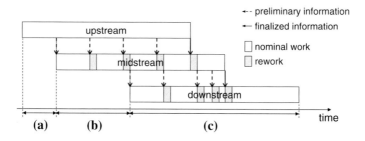

Fig. 1 Illustration of three-activity concurrent model

The degree of evolution for a former activity has an impact on rework duration of a latter activity. Since a former activity progresses, not all parameters of a former activity are fixed when it transfers design information to a latter activity. As Steward (1981) noticed, a latter activity assumes unfixed design parameters of a former activity to execute its work. When a former activity sends updated information, a latter activity needs to rework since design parameters of a former activity is different from its assumption. The more uncertain parameters based on which design is executed are, the longer rework duration is. Therefore, design execution based on information with the lower degree of evolution requires more rework for a latter activity.

Also, the degree of evolution of a former activity determines the possible start time of a latter activity. The critical degree of evolution for a former activity refers to the minimum degree of evolution based on which a latter activity is able to begin its work. Since a latter activity depends on the design information of a former activity, it requires a certain set of parameters of a former activity to begin. It is a pairwise concept between specific two dependent activities.

3 Degree of Evolution in Multi-Activity Concurrent Execution

In the perspective of upstream, the degree of evolution is the same as the portion of determined parameters. Since it is independent on other activities, its determined parameters are not changed after designers make a decision about it. Krishnan illustrates that there are various types of degree of evolution function such as linear, concave, convex and s-shape. In this research, we assume that the function of the degree of evolution is linear in order to simplify the analysis. General function types are able to be considered in future research.

Due to the uncertainty of information from upstream, the degree of evolution for midstream is lower than that performed based on the perfect information. Since midstream executes its work on the assumption of some design parameters of upstream, final parameters of upstream might be different from that assumed by midstream. If the difference is significant that what midstream has done is not compatible with updated information, midstream is required to rework the part which depends on upstream. As Fig. 2, the portion of determined parameters in midstream is decreased when it receives information from upstream and realizes the need of rework at t_2. The fixed parameters are fewer than determined parameters of midstream because of the uncertainty from upstream. The degree of evolution refers to the portion of fixed parameters which is not changed over time. Therefore, the degree of evolution for midstream is not $p_m(t_2)$ but $e_m(t_2)$ at time t_2.

The degree of evolution for midstream is important for downstream when it considers possible start time and rework duration. At t_1, downstream might begin its work since the portion of determined parameters in midstream reaches the

Estimating Product Development Project Duration

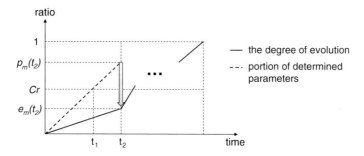

Fig. 2 Comparison between the degree of evolution and portion of determined parameters

critical degree of evolution, Cr. When midstream receives updated information from upstream at t_2, it needs to rework for determined design parameters. The portion of determined parameters decreases below the critical level. Consequently, what downstream has done on the basis of information from midstream at t_1 becomes meaningless since some of basis information from midstream changes. Therefore, the criteria of possible start time should not be the portion of determined parameters but the degree of evolution. In the same manner, rework duration for downstream needs to be calculated based on the degree of evolution for midstream.

We model the degree of evolution for midstream, considering both uncertainties. In Fig. 3, t_i represents the time when the ith information is transferred from upstream to downstream. Under this situation, the degree of evolution for midstream is calculated as Eq. (1).

$$e_m(t_i) = \{1 - (1 - e_u(t_{i-1})) \times I(u,m)\} \times f_m\left(\sum_{k=1}^{i} d(N_k)\right) \qquad (1)$$

$e_u(t_{i-1})$ refers to the degree of evolution of upstream at t_i. Therefore, $1 - e_u(t_{i-1})$ represents the uncertainty of upstream design of the latest information transfer. $e_m(t_i)$ represents the degree of evolution for midstream at t_i before midstream receives information from upstream. The right term represents the certainty of

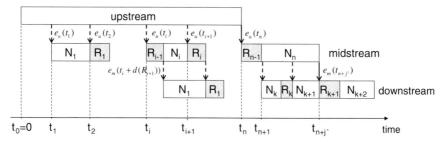

Fig. 3 Illustration of information transfer between *upstream* and *midstream*

midstream design itself. It is the degree of evolution function of sum of nominal work before $t_i, f_m(\sum_{k=1}^{i} d(N_k))$. N_k denotes the kth nominal design work. Since not all parameters of midstream depend on upstream, we refer to the portion at which midstream is dependent on upstream as the dependency of midstream on upstream, $I(u,m)$. Since midstream is affected at the portion of $I(u,m)$, uncertainty derived from upstream is $(1 - e_u(t_{i-1})) \times I(u,m)$. After rework, the design work of midstream has done based on the latest information from upstream at t_i with the degree of evolution as $e_u(t_i)$. Therefore, the degree of evolution after rework is calculated as Eq. (2).

$$e_m(t_i + d(R_{i-1})) = \{1 - (1 - e_u(t_i)) \times I(u,m)\} \times f_m(\sum_{k=1}^{i} d(N_k)) \qquad (2)$$

4 Rework Duration

Rework duration has influence on the information transfer time and project duration. As in Fig. 3, midstream transfers its updated information after its rework. In the respect of project management, synchronization of information transfer between activities facilitate activities executes their work based on the latest information. However, midstream transfers information after rework duration since we consider rework duration in product development process. Therefore, midstream has sent information after rework duration since it receives information from upstream.

Rework consists of setup time and execution time. Setup time refers to the fixed duration for preparing rework before rework execution. Since upstream sends complex design information, midstream requires time to understand design information. Then, midstream checks the difference between its assumption and fixed design parameters from upstream in order to decide the direction of rework.

Execution time refers to duration required to revise what a latter activity has done in order to adjust on the updated information from a former activity. When information is transferred to downstream at t_{i+1}, the duration of work which is done on the latest information transferred at t_i is $t_{i+1} - t_i$. Among them, the portion of dependent parameters is the dependency of midstream on upstream. When designer conducts the same design again, work rate is higher than that of nominal work because of learning effect (Ahmadi et al. 2001). This research considers the learning effect and reflects it as r. Therefore, rework duration of midstream to adjust on updated information from upstream at t_{i+1} is the sum of setup time and execution time as Eq. (3).

$$d_m(R_i) = s + \frac{(t_{i+1} - t_i) \times I(u,m) \times (e_u(t_{i+1}) - e_u(t_i))}{r} \qquad (3)$$

5 Three-Activity Project Modeling

We model the three-activity project in order to illustrate our methodology. The manager makes a decision about the overlapping ratio between upstream and midstream and the number of transfer from upstream and midstream. We denote them $\lambda(u,m)$ and $n(u,m)$ respectively. Under this decision, the ith time at which upstream transfers information to midstream, t_i, is $d_u\{1 - \lambda(u,m) \times (1 - i/n(u,m))\}$. The degree of evolution and rework duration is calculated through Eqs. (1) and (3) respectively. Checking the degree of evolution graph, the project manager perceives the possible start time of downstream.

Under the assumption of the linearity of degree of evolution function, it is possible to find the optimal start time of downstream and the optimal number of information transfer between midstream and downstream after the end of upstream. When upstream and midstream are overlapped, the time point of information transfer between midstream and downstream is fixed as the end of rework of midstream. When upstream is finished, the project modeling changes into two-activity project modeling between midstream and downstream. Since overlapping ratio is already determined before t_i, the optimal number of information transfer is derived as Eq. (4). d_m refers to nominal work duration

$$j(m,d)* = \sqrt{\frac{srd_m}{d_m(N_n)I(m,d)}} \qquad (4)$$

Due to the limitation of space, proof is omitted. Since $j(m,d)*$ might not be integer, optimal number, $j(m,d)'$, is selected by comparing adjacent two integers.

We found that the optimal start time of downstream is the earliest information transfer time from midstream after midstream reaches the critical degree of evolution. Project duration is shorter when start time is t_i than t_{i+1}. Therefore, it is optimal decision that selects the start time of downstream as the earliest possible start time of downstream.

As a result, overall project duration under the decision on information transfer frequency and overlapping ratio between upstream and downstream is represented as Eq. (5).

$$pd = (1 - \lambda(u,m))d_u + \sum_{i=1}^{cr}(d_m(N_i) + d_m(R_i)) + d_d + \sum_{k=1}^{l}d_d(R_k) \qquad (5)$$

cr represents the minimum number of rework in order to reach the critical degree of evolution between midstream and downstream. Also l refers to the number of rework of downstream.

6 Conclusion

As the time to market becomes an important factor for product development, companies execute product development activities concurrently. However, due to uncertainty of design work, it is difficult to estimate project duration when multiple activities are concurrently executed. In order to figure out uncertainties, we adopt the concept of the degree of evolution and model it on midstream activity. On the top of the model, we derive the equation which estimates the project duration based on management decision variables, information transfer frequency and overlapping ratio between activities, under three-activity project environments. In future, we will develop methodology which is able to estimate project duration under multi-activity project environments.

References

Ahmadi R, Roemer TA, Wang RH (2001) Structuring product development processes. Eur J Oper Res 130(3):539–558

Clark K, Fujimoto T (1989) Reducing the time to market: the case of the world auto industry. dmi Rev 1(1):49–57

Cusumano MA, Selby RW (1997) How microsoft builds software. Commun ACM 40(6):53–67

Eisenhardt KM, Tabrizi BN (1995) Accelerating adaptive processes: product innovation in the global computer industry. Adm Sci Q 40(1):84–110

Eppinger SD, Browning TR (2012) Design structure matrix methods and applications. The MIT Press, Cambridge

Krishnan V, Eppinger SD, Whitney DE (1998) A model-based framework to overlap product development activities. Manage Sci 43(4):437–451

Petkas ST, Pultar M (2005) Modelling detailed information flows in building design with the parameter-based design structure matrix. Des Stud 27(1):99–122

Steward DV (1981) The design structure system: a method for managing the design of complex systems. Eng Manage, IEEE Trans 28(3):71–74

Modeling of Community-Based Mangrove Cultivation Policy in Sidoarjo Mudflow Area by Implementing Green Economy Concept

Diesta Iva Maftuhah, Budisantoso Wirjodirdjo and Erwin Widodo

Abstract One of the environmental damages in Indonesia is Sidoarjo mudflow disaster causing impact significantly in various sectors. These problems of course could lead to instability for the local social dynamics and the environment as well as the global economy. Cultivating mangrove vegetation is one of the answers to overcome these problems, especially to neutralize the hazardous waste contained in the mud and definitely to rebuild the green zone in the observed area. It is necessary to conduct a research on mangrove cultivation policy in Sidoarjo mudflow area in order to support green economy concept giving benefits to the economy, environment and society. Considering several numbers of variable which have complex and causal relationships in mangrove cultivation policy in line with green economy concept, and also the developing pattern of line with the changing time make the problem to be solved appropriately with system dynamics approach which is able to analyze and assess the mangrove cultivation policy in accordance with the principles of green economy concept. Therefore, by modeling the policy of mangrove cultivation based on the concept of green economy is expected to be able to reduce carbon emissions and to create a new ecosystem that could be used by communities to give added value and selling for Sidoarjo mudflow area.

Keywords Green economy · Policy · Sidoarjo mudflow · Community-based mangrove · System dynamics

---

D. I. Maftuhah (✉) · B. Wirjodirdjo · E. Widodo
Department of Industrial Engineering, Sepuluh Nopember Institute of Technology (ITS), Surabaya, Indonesia
e-mail: diesta11@mhs.ie.its.ac.id

B. Wirjodirdjo
e-mail: wirjodirdjo@gmail.com

E. Widodo
e-mail: erwin@ie.its.ac.id

1 Introduction

Green economy concept was introduced by the United Nations Environment Programme (UNEP 2011) as the concept of economic development that gives more attention to the use of natural resources and environment and the results in improving human well-being in the social aspect. Green economy is based on knowledge of ecological economics that aims to address the interdependence of human economies and natural ecosystems and the adverse effects of human economic activities on climate change and environmental degradation (Bassi 2011). Green economy is considered capable of creating jobs in harmony with nature, sustainable economic growth, and preventing environmental pollution, global warming, resource depletion, and environmental degradation. Degradation of environmental resources that occur on these days should increase the awareness of all parties to be responsible for improving the balance of the earth which is actually just a deposit for future generations.

In general, the concept of green economy is considered positive and should be developed and implemented. However, most developing countries also stressed that the implementation of green economy vary widely, as it covers a diverse elements, and need to be adapted to the characteristics and needs of each country. The implementation of the green economy approach is for integrated modeling of environmental systems aspects, social, and economic case studies of tourism in Dominica (Patterson et al. 2004). In addition, the concept of green economy simulation has also been analyzed in the ecological farming system with system dynamics approach as the previous paper (Li et al. 2012). The concept of green economy focus on the interaction of three things, namely the concept of environmental, economic, social and based on the utilization of energy (Bassi and Shilling 2010). Meanwhile, the world has recognized the importance of sustainable development issues by integrating the concept of sustainable development as an objective guide for policy-making and development, as described by modeling the green economy policies along with the measuring instrument through system dynamics, the ecological foot print (Wei et al. 2012). However, in implementing the policy, it is still difficult to implement the commitment towards the implementation of sustainable development. Based on observations so far, there has been no research on the implementation of the green economy, which comprehensively covers all aspects of the object of observation in finding solutions to the problem.

The research on mangrove field is still in a narrow scope, such as the form of expression of the mangrove viewed in terms of ethno-biology, management, and development (Walters et al. 2008), the environmental influences on the development of mangrove (Krauss et al. 2008), as well as specific approaches used to see the development of mangrove (Guebas and Koedam 2008) and managerial implication by using community-based mangrove management (Datta et al. 2012). However, all research is still limited to specific reviews of the mangrove, no study used a holistic approach to observe and model the dynamics of mangrove development and management, such as system dynamics. Similarly, the object of this research is Sidoarjo mudflow area, for this research are limited by coastal land use in Porong River related to pollution theoretically (Yuniar 2010) and mangrove vegetation types that have the best growth at planting media in Sidoarjo mudflow (Purwaningsih 2008). Based on the previous research, there is no research conducting mangrove utilization review policies related to the green economy concept with its modeling of system dynamics. This was of course there are still opportunities to do research on mangrove policy of Sidoarjo mudflow area by taking the concept of the green economy.

Based on three gaps, such as the need of Sidoarjo mudflow area to get real solutions, utilization of mangrove that has not been well integrated, as well as the research gap of green economy utilization which is still partially, so that a more in-depth study on this issue is needed. A system dynamics modeling and policy analysis that examines the use of mangrove cultivation in the affected area of Sidoarjo mudflow in accordance with the principles given the green economy of its conditions that has not been used optimally could be proposed. Therefore, the cultivation of mangrove vegetation through green economy approach is expected to neutralize the hazardous waste levels in the mud, and is able to play a role in reducing carbon emissions in the region, as well as to create a new ecosystem that could be utilized by the local community to increase the added value and the selling for Sidoarjo Mudflow area.

2 Research Methodology

Research methodology in designing simulation models of community-based mangrove through green economy concept using system dynamics methodology is divided into four main stages. The first stage is the variable identification and conceptualization of the model using causal loops diagram that shows a causal relationship. The second stage is designing system dynamics model which is needed to carry out the simulations by formulating the simulation model and applying the scenarios. The third stage is the creation of simulation models and the policy scenario using Stella © (iSee Systems) simulation software. Stella is one of the software used to build simulation models visually using a computer and has advantages, such as many users (users) and is often used in business and academic (Voinov 1999 in Costanza and Gottlieb 1998). And the fourth stage is an analysis and interpretation of the model which is included the impact of policy scenarios application. The methodology of this research is able to be seen in Fig. 1.

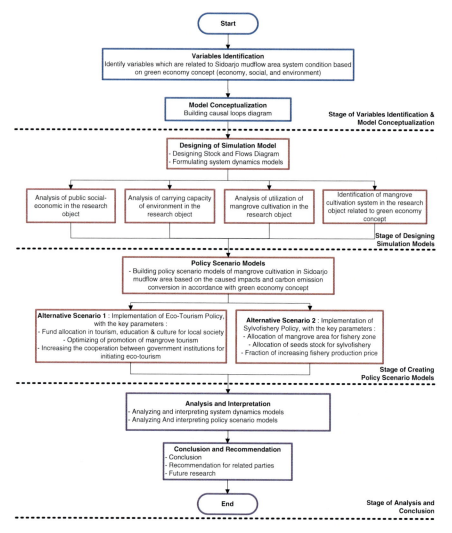

Fig. 1 Research methodology

3 Simulation Model Design

3.1 Variables Identification and Model Conceptualization

In modeling a system with system dynamics approach, is necessary to understand the elements related to identification and contributed in the development of the system, especially the stakeholders of the system. In this case, the observed object is mangrove cultivation system in Sidoarjo mudflow area. Identification of variables related to mangrove cultivation system is an important step to be done in as it

Modeling of Community-Based Mangrove Cultivation Policy

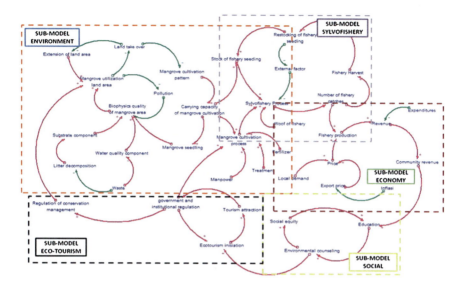

Fig. 2 Causal loops diagram

is an union of some aspects and elements based on green economy that are committed to cooperate in improving the green zone in the research object, especially the environmental carrying capacity. In general, some aspects that will be conducted in this research include environmental, economy, and social in accordance with green economy concept. Each of these elements has an important roles and function of the survival and progress of Sidoarjo mudflow area system.

Model presented by Forrester (1968) is the basic of experimental investigations that are relatively inexpensive and time-efficient than if it is conducted experiments on real systems. Conceptualization of the model begins with first by identifying the variables that interact in the system related mangrove cultivation system in Sidoarjo mudflow area based on green economy concept. The second is by creating a conceptual model of causal loops diagram and stock and flow diagram. The causal loops diagram of this research is able to be seen in Fig. 2. Causal loops diagram is made to show the main variables that will be described in the model. With the existence of causal loops, it is able to understand the relationship, and how far the influence of variables on system behavior is and how the impacts of each variable to the others are.

3.2 System Dynamics Simulation Model

System dynamics simulation model represented by stock and flow diagrams is designed based on causal loops diagram in Fig. 3. The purpose of making stock and flow diagram is to describe the interaction between variables in accordance with the logic structures used in software modeling. Modeling of the interaction of variable on the stock and flow diagrams produced three sectors that are interrelated, such as economy, environment, and social aspects. All sectors in stock and flows diagram are constructed to represent the interaction of green economy concept in mangrove cultivation system in Sidoarjo mudflow area in order to regain green zone area and increase the environmental carrying capacity. The stock and flows structure, formulation and data were designed. The three-sub models in Fig. 3 which are designed are environment, economy, and social model according to Wijaya et al. (2011) and Bassi and Shilling (2010). The design of the stock and flow diagrams are also considering the purpose of research, to know and see the development of mangrove cultivation system in the research object in developing policies that increase its environmental carrying capacity.

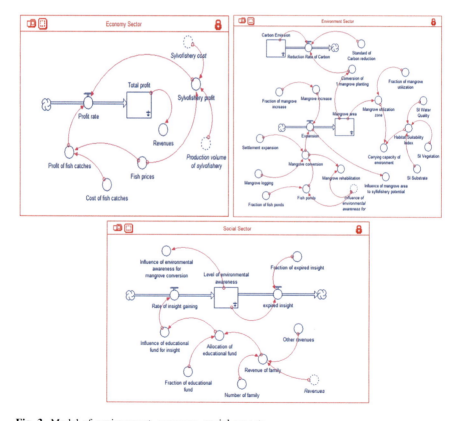

Fig. 3 Model of environment, economy, social aspects

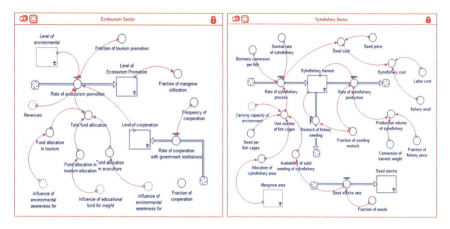

Fig. 4 Scenario model of sylvofishery and eco-tourism program

4 Policy Scenario Models

Based on created simulation models, it has been designed two alternative scenarios for supporting community-based mangrove and improving the green zone area in Sidoarjo mudflow area, such as sylvofishery and eco-tourism program (Glaser and Diele 2004 and Abidin 1999 in Datta et al. 2012) that could be seen in Fig. 4. The improvement scenarios are based on the conditions that allow it to be controlled by mangrove cultivation system based on green economy concept. Those improvement scenarios are explained below.

4.1 Implementation of Sylvofishery (Mangrove-Based Fishery)

In silvofishery development on the affected area of Sidoarjo mudflow, 20 % of area is planned to dam/watery land for fishery and 80 % is for mangrove (Harnanto 2011). The main key parameters for this policy are allocation of mangrove area for fishery zone, allocation of seeds stock for sylvofishery, and fraction of increasing fishery production price.

4.2 Implementation of Eco-Tourism Program

By programming eco-tourism in the research object, knowledge about mangrove and their resources for local community and other group at large will be increased. This program can provide educational value to always conserve the mangrove biodiversity in a sustainable manner. The main key parameters for this policy are fund allocation in tourism, education, and culture for local society, as well as optimizing of promotion of mangrove tourism, and increasing the cooperation between government institutions for initiating eco-tourism program.

5 Analysis and Conclusion

Based on the conceptual and scenario models, it could be concluded that variables of mangrove cultivation system play an important role on the regaining green zone area and increase the environmental carrying capacity in Sidoarjo mudflow area. In addition, this model could be benefit for especially environment, economy, and society. It is known that by implementing sylvofishery and eco-tourism program will produce faster recovery time in regaining the green zone and inversely with the reduction of carbon emission in the area. Moreover, these policy scenarios give managerial implications of green economy concept for local community to get some benefits from mangrove cultivation and to improve their revenues economically and greening their area environmentally. Nevertheless, it is needed to conduct future research in the same topic or related with the topic, especially for developing the simulation model in social and eco-tourism aspects and identifying variables in wider scope of research.

Acknowledgments This work is partially supported by Sidoarjo Mudflow Mitigation Institution (BPLS). We also gratefully acknowledge the helpful comments and suggestions of the reviewers, which have improved the presentation, especially for Mr. Soegiarto as Head of Infrastructure in BPLS and his friends for their supports of domain knowledge and data collection.

References

Bassi AM (2011) Introduction to the threshold 21 (T21) model for green development. Millenium Institute
Bassi AM, Shilling JD (2010) Informing the US energy policy debate with threshold 21. J Technol Forecast Soc Change 77:396–410
Costanza R, Gottlieb S (1998) Modeling ecological and economic systems with STELLA: part II. J Ecol Model 112:81–84
Datta D, Chattopadhyay RN, Guha P (2012) Community based mangrove management: a review on status and sustainability. J Environ Manage 107:84–95
Forrester JW (1968) Principle of system. Wright-Allen Press, Inc, Massachusetts
Guebas FD, Koedam N (2008) Long-term retrospection on mangrove development using trans-disciplinary approaches: a review. J Aquat Bot 89:80–92
Harnanto A (2011) The Roles of Porong river in flowing Sidoarjo Mud to the sea. BPLS-Badan Penanggulangan Lumpur Sidoarjo, Surabaya
Krauss KW, Lovelock CE, McKee KL et al (2008) Environmental drivers in mangrove establishment and early development: a review. J Aquat Botany 89:105–127

Li FJ, Dong SC, Li F (2012) A system dynamics model for analyzing the eco-agriculture system with policy recommendations. J Ecol Model 227:34–45

Patterson T, Gulden T, Cousins K, Kraev E (2004) Integrating environmental, social, and economic systems—a dynamic model of tourism in Dominica. J Ecol Model 175:121–136

Purwaningsih E (2008) The impacts of Sidoarjo Mudflow for the growth of mangrove types, such as *Avicennia marina*, *Rhizophora apiculata* and *Rhizophora mucronata*. Thesis of master program of aquaculture, University of Brawijaya, Malang

UNEP (2011) Green economy-why a green economy matters for the least developed countries. St-Martin-Bellevue, France

Walters BB, Rönnbäck P, Kovacs JM et al (2008) Ethno-biology, socio-economic and management of mangrove forests: a review. J Aquat Botany 89:220–236

Wei S, Yang H, Song J et al (2012) System dynamics simulation model for assessing socio-economic impacts of different levels of environmental flow allocation in the Weihe River Basin, China. Eur J Oper Res 221:248–262

Wijaya NI et al (2011) General use zone governance by optimizing of *Scylla serrata* Utilization in Kutai National Park on East Kalimantan Province. Dissertation in Institut Pertanian Bogor

Yuniar WD et al (2010). Analysis of coastal land use related to pollution in Porong river. J Urban Reg Plann 2(2)

Three Approaches to Find Optimal Production Run Time of an Imperfect Production System

Jin Ai, Ririn Diar Astanti, Agustinus Gatot Bintoro
and Thomas Indarto Wibowo

Abstract This paper considers an Economic Production Quantity (EPQ) model where a product is to be manufactured in batches on an imperfect production system over infinite planning horizon. During a production run of the product, the production system is dictated by two unreliable key production subsystems (KPS) that may shift from an in-control to an out-of-control state due to three independent sources of shocks. A mathematical model describing this situation has been developed by Lin and Gong (2011) in order to determine production run time that minimizes the expected total cost per unit time including setup, inventory carrying, and defective costs. Since the optimal solution with exact closed form of the model cannot be obtained easily, this paper considered three approaches of finding a near-optimal solution. The first approach is using Maclaurin series to approximate any exponential function in the objective function and then ignoring cubic terms found in the equation. The second approach is similar with first approach but considering all terms found. The third approach is using Golden Section search directly on the objective function. These three approaches are then compared in term computational efficiency and solution quality of through some numerical experiments.

Keywords EPQ model · Imperfect production system · Optimization technique · Approximation and numerical method

---

J. Ai (✉) · R. D. Astanti · A. G. Bintoro · T. I. Wibowo
Department of Industrial Engineering, Universitas Atma Jaya Yogyakarta,
Jl. Babarsari 43, Yogyakarta 55281, Indonesia
e-mail: jinai@mail.uajy.ac.id

R. D. Astanti
e-mail: ririn@mail.uajy.ac.id

A. G. Bintoro
e-mail: a.bintoro@mail.uajy.ac.id

T. I. Wibowo
e-mail: t8_t10@yahoo.co.id

1 Introduction

The problem considered in this paper had been formulated by Lin and Gong (2011) as follow. A product is to be manufactured in batches on an imperfect production system over an infinite planning horizon. The demand rate is d, and the production rate is p. The imperfectness of the system is shown on two imperfect key production subsystems (KPS) that may shift from an in-control to an out-of-control state due to three independent sources of shocks: source 1's shock causes first KPS to shift, source 2's shock causes second KPS to shift, and source 3's shock causes both KPS to shift. Each shock occurs at random time U_1, U_2, and U_{12} that follows exponential distribution with mean $1/\lambda_1$, $1/\lambda_2$, and $1/\lambda_{12}$, respectively. When at least one KPS on out-of-control state, consequently, the production system will produced some defective items with fixed but different rates: α percentage when first KPS out-of-control, β percentage when the second KPS out-of-control, and δ percentage when the both KPS out-of-control. The cost incurred by producing defective items when the first KPS is shifted, the second KPS is shifted, and both KPS are shifted are π_1, π_2, and π_{12}, respectively. The optimization problem is to determining optimal production run time τ that minimizes the expected total cost per unit time including setup, inventory carrying, and defective costs. It is noted that the unit setup cost is A and inventory carrying per unit per unit time is h.

As derived in Lin and Gong (2011), the objective function of this optimization model is given by following equations.

$$Z(\tau) = \frac{Ad}{p\tau} + \frac{h(p-d)\tau}{2} + \frac{d(\pi_1 E[N_1(\tau)] + \pi_2 E[N_2(\tau)] + \pi_{12} E[N_{12}(\tau)])}{p\tau} \quad (1)$$

$$E[N_1(\tau)] = p\alpha \left(\frac{1 - \exp[-(\lambda_2 + \lambda_{12})\tau]}{\lambda_2 + \lambda_{12}} - \frac{1 - \exp[-(\lambda_1 + \lambda_2 + \lambda_{12})\tau]}{\lambda_1 + \lambda_2 + \lambda_{12}} \right) \quad (2)$$

$$E[N_2(\tau)] = p\beta \left(\frac{1 - \exp[-(\lambda_1 + \lambda_{12})\tau]}{\lambda_1 + \lambda_{12}} - \frac{1 - \exp[-(\lambda_1 + \lambda_2 + \lambda_{12})\tau]}{\lambda_1 + \lambda_2 + \lambda_{12}} \right) \quad (3)$$

$$E[N_{12}(\tau)] = p\delta \left(\frac{\exp[-(\lambda_1 + \lambda_{12})\tau] + (\lambda_1 + \lambda_{12})\tau - 1}{\lambda_1 + \lambda_{12}} \right.$$
$$+ \frac{\exp[-(\lambda_2 + \lambda_{12})\tau] + (\lambda_2 + \lambda_{12})\tau - 1}{\lambda_2 + \lambda_{12}}$$
$$\left. - \frac{\exp[-(\lambda_1 + \lambda_2 + \lambda_{12})\tau] + (\lambda_1 + \lambda_2 + \lambda_{12})\tau - 1}{\lambda_1 + \lambda_2 + \lambda_{12}} \right) \quad (4)$$

Although the problem is a single variable optimization, the optimal solution with exact closed form of the model cannot be obtained easily. Therefore, this paper considered three approaches of finding a near-optimal solution. These approaches are then compared through some numerical experiments.

2 Approaches of Finding a Near-Optimal Solution

2.1 First Approach

In this first approach, following Maclaurin series is applied to approximate any exponential function in the objective function.

$$\exp(-\lambda\tau) \approx 1 - \lambda\tau + \frac{1}{2!}(\lambda\tau)^2 - \frac{1}{3!}(\lambda\tau)^3 \tag{5}$$

Therefore after some algebra, the objective function can be approximated as following equations.

$$Z(\tau) \approx Z_1(\tau) = \frac{Ad}{p\tau} + \frac{H\tau}{2} - \frac{B\tau^2}{6} \tag{6}$$

$$H = h(p-d) + d(\pi_1\alpha\lambda_1 + \pi_2\beta\lambda_2 + \pi_{12}\delta\lambda_{12}) \tag{7}$$

$$B = d[\pi_1\alpha\lambda_1(\lambda_1 + 2\lambda_2 + 2\lambda_{12}) + \pi_2\beta\lambda_2(2\lambda_1 + \lambda_2 + 2\lambda_{12}) - \pi_{12}\delta(2\lambda_1\lambda_2 - \lambda_{12}^2)] \tag{8}$$

From calculus optimization, it is known that the necessary condition for obtaining the minimum value of Z_1 is set the first derivative equal to zero. Applying this condition for Eq. (6), it is found that

$$\frac{dZ_1(\tau)}{d\tau} = -\frac{Ad}{p\tau^2} + \frac{H}{2} - \frac{2B\tau}{6} = 0 \tag{9}$$

Equation (9) can be rewritten as

$$2B\tau^3 - 3H\tau^2 + 6Ad/p = 0. \tag{10}$$

If the cubic term in Eq. (10) is ignored, then a near-optimal solution of the first approach can be obtained as follow

$$\tau_1^* = \sqrt{\frac{2Ad}{p[h(p-d) + d(\pi_1\alpha\lambda_1 + \pi_2\beta\lambda_2 + \pi_{12}\delta\lambda_{12})]}} \tag{11}$$

2.2 Second Approach

The second approach is developed based on Eq. (9). Another near-optimal solution can be found as the root of this equation. Bisection algorithm can be applied here to find the root of this equation (τ_2^*) with lower searching bound of $\tau_L = 0$ and upper searching bound

$$\tau_U = \sqrt{\frac{2Ad}{ph(p-d)}} \quad (12)$$

It is noted that the lower bound is selected equal to zero due to the fact that the optimal production run time have to be greater than zero. While the upper bound is selected as Eq. (12) due to the fact that the optimal production run time in the presence of imperfectness, i.e. with non negative values of α, β, and δ, is always smaller than the optimal production run time of perfect production system, i.e. with zero values of α, β, and δ. Substituting $\alpha = \beta = \delta = 0$ to Eq. (11) provides the same value as optimal production run time of classical and perfect EPQ (Silver et al. 1998), as shown in the right hand side of Eq. (12). The detail of bisection algorithm can be found in any numerical method textbook, i.e. Chapra and Canale (2002).

2.3 Third Approach

The third approach is using pure numerical method to find the minimum value of Z based on Eq. (1). The Golden Section method is applied here using the same bound as the second approach. Therefore, the searching of the optimal production run time of this approach (τ_3*) is conducted at interval $\tau_L \leq \tau_3^* \leq \tau_U$, where $\tau_L = 0$ and τ_U is determined using Eq. (12). Further details on the Golden Section method can be found in any optimization textbook, i.e. Onwubiko (2000).

3 Numerical Experiments

Numerical experiments are conducted in order to test the proposed approaches for finding the optimal production run time. Nine problems (P1, P2, ..., P9) are defined for the experiments and the parameters of each problem are presented in Table 1. The result of all approaches are presented in Table 2, which comprise of the optimal production run time calculated from each approach (τ_1*, τ_2*, τ_3*) and their corresponding expected total cost [$Z(\tau_1$*), $Z(\tau_2$*), $Z(\tau_3$*)]. Some metrics defined below are also presented in Table 2 in order to compare the proposed approaches.

In order to compare the proposed approaches, since the optimal expected total cost cannot be exactly calculated, the best expected total cost is defined as following equation

$$Z^* = \min\{Z(\tau_1^*), Z(\tau_2^*), Z(\tau_3^*)\} \quad (13)$$

Three Approaches to Find Optimal Production Run Time

Table 1 Problem parameters of the numerical experiments

Parameters	P1	P2	P3	P4	P5	P6	P7	P8	P9
p	300	300	300	300	300	300	300	300	300
d	200	200	200	200	200	200	200	200	200
A	100	100	100	100	100	100	100	100	100
h	0.08	0.08	0.08	0.08	0.08	0.08	0.08	0.08	0.08
π_1	10	10	10	10	10	10	10	10	10
π_2	10	10	10	10	10	10	10	10	10
π_{12}	12	12	12	12	12	12	12	12	12
α	0.1	0.1	0.1	0.3	0.3	0.3	0.5	0.5	0.5
β	0.1	0.1	0.1	0.3	0.3	0.3	0.5	0.5	0.5
δ	0.16	0.16	0.16	0.48	0.48	0.48	0.8	0.8	0.8
λ_1	0.05	0.15	0.25	0.05	0.15	0.25	0.05	0.15	0.25
λ_2	0.1	0.3	0.5	0.1	0.3	0.5	0.1	0.3	0.5
λ_{12}	0.02	0.06	0.1	0.02	0.06	0.1	0.02	0.06	0.1

If τ^* is the best value of production run time in which $Z^* = Z(\tau^*)$, the deviation of the solution of each approach from the best solution is calculated from following equation

$$\Delta \tau_i = \frac{|\tau_i^* - \tau^*|}{\tau^*} \times 100\% \qquad (14)$$

Furthermore, the deviation of the expected total cost of each approach from the best one can be determined using following equation

$$\Delta Z_i = \frac{Z(\tau_i^*) - Z^*}{Z^*} \times 100\% \qquad (15)$$

Regarding to solution quality, it is shown in Table 2 that the third approach is consistently providing the best expected total cost across nine test problems among

Table 2 Results of the numerical experiments

Result	P1	P2	P3	P4	P5	P6	P7	P8	P9
τ_1^*	1.7085	1.0496	0.8239	1.0496	0.6198	0.4823	0.8239	0.4823	0.3746
τ_2^*	1.8067	1.2035	1.0199	1.0897	0.6679	0.5350	0.8489	0.5104	0.4046
τ_3^*	1.7986	1.1750	0.9642	1.0877	0.6633	0.5280	0.8479	0.5084	0.4016
$\Delta\tau_1$	5.01 %	10.67 %	14.55 %	3.51 %	6.56 %	8.65 %	2.83 %	5.14 %	6.72 %
$\Delta\tau_2$	0.46 %	2.43 %	5.78 %	0.18 %	0.70 %	1.33 %	0.11 %	0.40 %	0.72 %
$\Delta\tau_3$	0.00 %	0.00 %	0.00 %	0.00 %	0.00 %	0.00 %	0.00 %	0.00 %	0.00 %
$Z(\tau_1^*)$	76.18	120.96	151.69	124.89	208.55	265.54	159.60	269.73	344.77
$Z(\tau_2^*)$	76.09	120.34	150.38	124.82	208.12	264.61	159.54	269.39	344.02
$Z(\tau_3^*)$	76.09	120.31	150.20	124.82	208.12	264.59	159.54	269.38	344.02
ΔZ_1	0.12 %	0.54 %	0.99 %	0.06 %	0.21 %	0.36 %	0.04 %	0.13 %	0.22 %
ΔZ_2	0.00 %	0.02 %	0.12 %	0.00 %	0.00 %	0.01 %	0.00 %	0.00 %	0.00 %
ΔZ_3	0.00 %	0.00 %	0.00 %	0.00 %	0.00 %	0.00 %	0.00 %	0.00 %	0.00 %

three approaches. It is also shown in the Table 2 that $Z(\tau_1^*) > Z(\tau_2^*) > Z(\tau_3^*)$, while the deviations of the expected total cost of the first and second approaches are less than 1.0 and 0.2 %, respectively. Furthermore, it is found that production run time found by the three approaches are $\tau_1^* < \tau^* = \tau_3^* < \tau_2^*$. The deviations of the first approach solution from the best solution are less than 14.6 %, while the deviations of the second approach solution from the best solution are less than 5.8 %.

These results show that the first approach is able to find reasonable quality solution of the problems although its computational effort is very simple compare to other approaches. It is also implied from these result that the Maclaurin approximation used in the second approach is effective to support the second approach finding very close to best solution of the problems, although the computational effort of the second approach is higher than the computational effort of the first approach. Since the third approach is using the highest computational effort among the proposed approaches, it can provide the best solution of the problems.

4 Concluding Remarks

This paper proposed three approaches for solving Lin and Gong (2011) model on Economic Production Quantity in an imperfect production system. The first approach is incorporating Maclaurin series and ignoring cubic terms in the first derivative of total cost function. The second approach is similar with the first approach but incorporating all terms in the total cost function, then using Bisection algorithm for finding the root of the first derivative function. The third approach is using Golden Section method to directly optimize the total cost function. Numerical experiments show that the third approach is able to find the best expected total cost among the proposed approaches but using the highest computational effort. It is also shown that the first approach is able to find reasonable quality solution of the problems despite the simplicity of its computational effort.

Acknowledgments This work is partially supported by Directorate of Higher Education, Ministry of Education and Culture, Republic Indonesia under *International Research Collaboration and Scientific Publication* Research Grant and Universitas Atma Jaya Yogyakarta, Indonesia. The authors also gratefully acknowledge the helpful comments and suggestions of the reviewers, which have improved the presentation.

References

Chapra SC, Canale RP (2002) Numerical method for engineers, 4th edn. McGraw-Hill, New York
Lin GC, Gong DC (2011) On an economic lot sizing model subject to two imperfect key production subsystems. In: Proceedings of IIE Asian conference 2011, Shanghai, China
Onwubiko C (2000) Introduction to engineering design optimization. Prentice Hall, New Jersey
Silver EA, Pyke DF, Peterson R (1998) Inventory management and production planning and scheduling, 3rd edn. Wiley, New York

Rice Fulfillment Analysis in System Dynamics Framework (Study Case: East Java, Indonesia)

Nieko Haryo Pradhito, Shuo-Yan Chou, Anindhita Dewabharata and Budisantoso Wirdjodirdjo

Abstract Food fulfillment is one of the things that affect the stability of a country. The rapid population growth but not matched by the ability of food production will be a threat in the future, there is no balance between supply and demand. In 2011, there was rice shortage in East Java, which also resulted in the rice shortage at the national level, this phenomenon is caused anomaly weather, pests, land mutation, weak network Supply Chain Management, distribution, transportation, etc. that cause dependence on rice imports higher. This research purposes are to identify the holistic process of rice fulfillment in the context of supply chain system and analyzing possible risk raised as an important variables and provide a projection capabilities in the future by using a simulation scenario. The complexity of interaction between variables and the behavior of the system considered the selection of System Dynamics methods to solve problems. The advantages using System Dynamics as tools analysis is combine qualitative and quantitative method, also model can provide reliable forecast and generate scenarios to test alternative assumptions and decisions. Finally, the research contribution is formulated policy improvements in rice fulfillment, also provide more robust sensitivities and scenarios, so this research predict the impact of major changes in strategy accurately in uncertainty condition.

Keywords Rice fulfillment · Supply chain · System dynamics · Policy

---

N. H. Pradhito (✉) · S.-Y. Chou · A. Dewabharata
Department of Industrial Management, National Taiwan University of Science and Technology, Taipei, Taiwan, Republic of China
e-mail: nieko.haryo.pradhito@gmail.com

B. Wirdjodirdjo
Deptartment of Industrial Engineering, Sepuluh Nopember Institute of Technology, Surabaya, Indonesia
e-mail: budisantoso.wirjodirdjo@gmail.com

Table 1 Land area, production and productivity Indonesian

Sector	Harvest area (ha)	Productivity (Tons/Hectare)	Production (Tons)
Indonesia	13,443,443.00	5.14	69,045,523.25
East Java	1,975,719.00	6.17	12,198,089.11

Source Indonesian Statistics Central Bureau (2012)

1 Introduction

Fulfillment food needs is one of the strategic issues that are closely related to the stability condition of a country, including Indonesia. Rice is a staple food of the population of Indonesia since 1950, and now has reached the current consumption until 95 %, but on the other hand, the growth rate of rice production reached only 3 %.

Problems of rice, is a complex issue and holistic, so that the modeling needs to be simplified but still represent locations and representative of the system. To validate this model, it was chosen as a representative of East Java province on issues of production and supply of rice. Table 1 will be provides latest condition in 2012 about harvest area, productivity and production between total in Indonesia and East Java contribution.

The complexity of interaction between variables and the behavior of the system considered the selection of System Dynamics methods to solve problems. The advantages using System Dynamics as tools analysis is combine qualitative and quantitative method, also model can provide reliable forecast and generate scenarios to test alternative assumptions and decisions (Sterman 2000).

This research purposes are to identify the holistic process of rice fulfillment in the context of supply chain system and analyzing possible risk raised as an important variables and provide a projection capabilities in the future by using a simulation scenario, benefits of the research are to gain predictive effectiveness of the plan to achieve fulfillment rice from the key points in the policy that also involves farmers, trading in the market, purchasing power and dependence on rice.

2 Previous Related Work

Food fulfillment in which the pressures of a rising world population, climate change, water shortages, the availability of quality land for crop production and the rising cost of energy are all colliding. Agriculture in the 21st century will have to deal with major alterations in the physical landscape: arable land per capita is declining, supply of water, energy costs, pest and disease problems, climate changes. The key issues in supply chain management (SCM) in agriculture area are the formation of the supply chain and its efficient coordination with objectives

Table 2 Summary of previous research area using system dynamics

Author	Years	Case
Smith	1997	Economic
Cooke	2002	Disaster
Shuoping	2005	Logistic
Ho	2006	Earthquake
Deegan	2007	Flood
Zhai	2009	Water
Yang	2009	Inventory
Yang	2009	Financial
Cui	2011	Market
Erma	2011	Capacity
Sidola	2011	Risk
Patel	2010	Safety Stock

of customer satisfaction and sustaining competency. Uses system dynamic modeling approach so previous dynamic modeling works are reviewed.

In System Dynamics context, feedback structure of a system is described using causal loops. These are either *balancing* (capturing negative feedback) or *reinforcing* (capturing positive feedback). A balancing loop exhibits goal seeking behavior: after a disturbance, the system seeks to return to an equilibrium situation (conforming to the economic notion of a stable equilibrium (Smith 2000). Table 2 will be provides several researches on other areas using System Dynamics.

3 System Dynamics Modeling and Validation

3.1 Causal Loops Diagram

Causal loop diagram is a tool to represent the feedback structure of system, it consists of variables connected by arrows denoting the causal influences among them. This diagram shows the cause and effect of the system structure. Each arrow represents a cause and effect relationship between two variables. The + and − signs represent the direction of causality. A + sign indicates can increase the result to destination variable. While the − sign indicates can decrease the result to the destination variable (Fig. 1).

However, the growth of demand will make the utilization higher. Causal loop diagrams emphasize the feedback structure of the system, it can never be comprehensive. We have to convert the causal loop diagram into flow diagram that emphasizes the physical structure of the model. It has a tendency to be more detailed than causal loop diagram, to force us to think more specifically about the system structure.

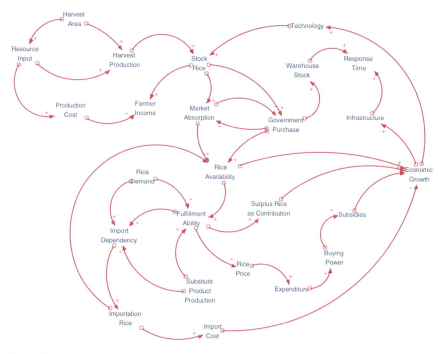

Fig. 1 Causal loops diagram

3.2 Rice Fulfillment Based on Existing Condition

In explanation of the analysis of the existing condition, Fig. 2 will be provide the mechanism of the rice supply chain will be more easily described in terms of the flow from upstream to downstream, where the earliest upstream processes called pre-harvest until in post harvest as the end of the process.

3.3 Parameter Estimation

Parameter estimation is required to develop mathematical models by utilizing data or observation from the real system. The estimation of parameters can be obtained in some ways such as statistics data, published reports, and statistical methods. Historical data are needed to see trends and forecasting analysis of the future, the behavior of the system can also be used to analyze the data validation by quantitative calculation. All the data collect from Indonesian Statistics Central Bureau and Indonesian Population Agency in East Java. Table 3 will be provides the historical data for the 10 years, more historical data, more accurately trend analysis.

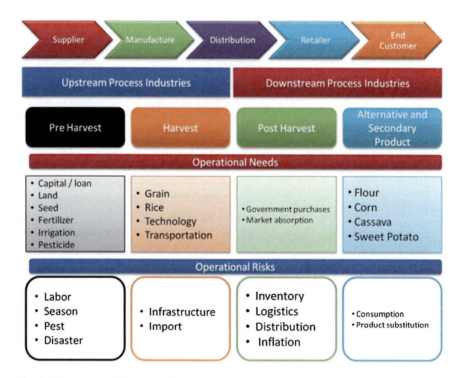

Fig. 2 Rice supply chain mechanisms

Table 3 Supply and demand sector from historical data

Years	Supply sector			Demand sector	
	Harvest area (Hectares)	Productivity (Tons/Ha)	Production (Tons)	Populations	Growth rate (%)
2003	1,695,514	5.258	8,915,013	35,130,220.16	1.16
2004	1,697,024	4.97	8,434,209	35,480,961.18	1.16
2005	1,693,651	5.318	9,006,836	35,835,204.00	3.01
2006	1,750,903	5.338	9,346,320	36,913,843.64	1.02
2007	1,736,048	5.416	9,402,436	37,290,364.85	3.01
2008	1,774,884	5.902	10,475,365	38,412,804.83	1.46
2009	1,904,830	5.911	11,259,450	38,973,631.78	2.77
2010	1,963,983	5.929	11,644,455	40,053,201.38	2.12
2011	1,926,796	5.489	10,576,183	40,902,329.25	2.88
2012	1,975,719	6.174	12,198,089	41,237, 728.35	0.82

Source Indonesian Statistics Central Bureau, Indonesian Population Agency (2012)

Table 4 Mathematical formula for supply sector model formulation

No.	Variables	Model buildings	Formulation
1.	Harvest paddy area	Stock	Paddy_Harvest_Area(t) = Paddy_Harvest_Area (t−dt) + (Paddy_Area_Rate) * dt INIT Paddy_Harvest_Area = 1,695,514
2.	Paddy area rate	Flow	Paddy_Harvest_Area*Paddy_Area_Growth
3.	Paddy area growth	Converter	Paddy_Area_Growth = GRAPH(TIME) (1.00, 0.001), (2.13, −0.002), (3.25, 0.034), (4.38, −0.008), (5.50, 0.022), (6.63, 0.073), (7.75, 0.031), (8.88, −0.019), (10.0, 0.025)

3.4 Model Formulation

Barlas (2000) give mathematical formulation example for supply and demand case, formulation of the model shows how the model is based on mathematical formulas and other quantitative approaches. Table 4 will be provide example how build formulation in mathematical model for supply sector.

The simulation based on the actual historical data for 10 years will be provide in Fig. 3 to show the behavior of the system.

3.5 Model Validation

Validation is a process of evaluating model simulation to determine whether it is an acceptable representation of the real system, historical data during the time horizon of simulation of the base model is required. For the validation, Eq. 1 is used for statistical data fitness:

$$Error\_rate = \frac{|\bar{S} - \bar{A}|}{\bar{A}}$$

where:

$$\bar{S} = \frac{1}{N}\sum_{i=1}^{N} S_i$$
$$\bar{A} = \frac{1}{N}\sum_{i=1}^{N} A_i \quad (1)$$

From the simulation, the error rate paddy production as supply sector known that the result is 1.4 % and result for population as demand sector is 1.1 %, the result less than 5 % mean that the simulation valid for represent real problem.

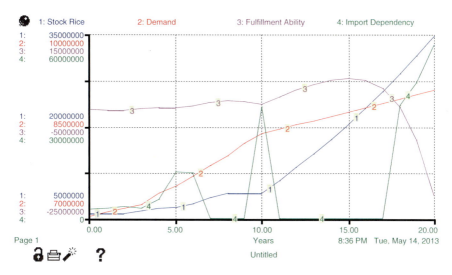

Fig. 3 Simulation result for existing condition

4 Scenario Planning

4.1 Structure Scenario

For the improvement scenario, this research provide strategy to utilize secondary source of carbohydrate food, especially for corn, cassava and sweet potato to adding the rice stock in nutrient context, Fig. 4 will be provide the simulation result for the scenario and how to improve the actual condition.

4.2 Evaluative Comparison

And from the scenario improvement from utilize secondary source of carbohydrate food, Fig. 5 will be show the improvement number by this scenario.

5 Discussion Analysis and Conclusion

This research start from identify holistic system of rice fulfillment to catch the big picture process, and then developing a System Dynamics model of rice availability in Supply Chain Management framework, the simulation projecting the rice availability in the future using a System Dynamics model. Improvement scenario

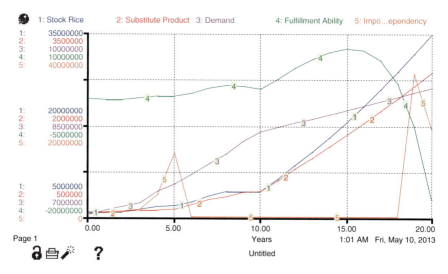

Fig. 4 Simulation result for existing condition

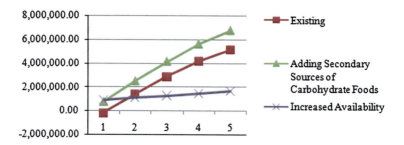

Fig. 5 Improvement number using scenario

as a better policy for achieving availability of rice by utilize secondary source of carbohydrate food as a substitute product to tackle importation rice dependency and as the recommendation, this scenario can support the fulfillment ability for 5 next years.

Acknowledgments This work is partially supported by Professor Shuo-Yan Chou, Professor Budisantoso Wirdjodirdjo, Anindhita Dewabharata and Erma Suryani. The authors also gratefully acknowledge the helpful comments and suggestions of the reviewers, which have improved the presentation.

References

Barlas Y (2000) 5.12 System dynamics: systemic feedback modeling for policy analysis

Indonesian_Statistics_Central_Bureau (2012) http://jatim.bps.go.id/e-pub/2012/prodpadipalawija2011/, http://jatim.bps.go.id/e-pub/2012/jtda2012/

Sterman JD (2000) Business dynamics. McGraw-Hill, New York

Activity Modeling Using Semantic-Based Reasoning to Provide Meaningful Context in Human Activity Recognizing

AnisRahmawati Amnal, Anindhita Dewabharata, Shou-Yan Chou and Mahendrawathi Erawan

Abstract As rapid increasing of World-Wide Web technology, ontology and pervasive/ubiquitous concept become one of the most interesting parts recently. According to ubiquitous issue, activity modeling which has capability to provide proper context information of the user becomes important part of context awareness. In order to develop required services in intelligence environment; to get deep knowledge provision framework; and precise activities relationship; context of knowledge should be determined based on socio-environment situation. Based on that background, this paper is aimed to develop activity model through activity recognition by implementing semantic based reasoning to recognize human activity.

Keywords Context-awareness · Activity modeling · Semantic-based reasoning

A. Amnal (✉) · A. Dewabharata · S.-Y. Chou
National Taiwan University of Science and Technology, Keelung Road Sec. 4,
Daan District, Taipei, Taiwan
e-mail: m10101812@mail.ntust.edu.tw

A. Dewabharata
e-mail: d10101801@mail.ntust.edu.tw

S.-Y. Chou
e-mail: sychou@mail.ntust.edu.tw

M. Erawan
Information System Departments, SepuluhNopember Institute of Technology,
Surabaya, Indonesia
e-mail: mahendra_w@its.edu.ac.id

1 Introduction

Human Activity Recognition (HAR) is researches mostly observe human actions to understand types of human activities perform within time interval. Dominantly, it observes a series of physical actions construct one physical activity (PA) and activities of daily livings (ADL) due to its benefit in healthcare application. Generally, people perform activities depend on their lifestyle, so that context-aware information developed should contains meaningful data to analyze human activities in their environment (Wongpatikaseree et al. 2012).

More than ability to recognize recent information about user activity, context-aware computing is also answering challenges in understanding dynamic changes from users. It is not only about how to adapt and sense environmental changes, but more about how to sense the context in current environments and react to such changing context automatically. Context-awareness system enables actors to interact through context-aware applications running on portable devices and other computing hardware platforms cohesively, allow software to adapt based on user location, gather actors and object nearby, and change those objects based on time (Dey and Abowd 1999). The system also facilitates the provision of information and assistance for the applications to make appropriate decisions in the right manner, at the right time, and at the right place (3Rs) (Xue and Keng Pung 2012).

In order to describe context as a whole, provide appropriate information and assist the application to make proper decisions, context modeling that involves all important entities and relation become crucial issues. The modeling of the context in this research will be based on the technology concept borrowed from semantic web, which is: the ontology. Rapid advances of world-wide web technology makes ontology become common concept that is used to represent formal knowledge and shared vocabulary as well as their properties and relationship in multiple specific domains with several advantages engaged (Noy and McGuinness 2001). Ontology concept is widely use because its ability to share common understanding of information among people or software agents, reuse domain knowledge, make domain assumptions explicit, etc. According to those advantages, context aware computing is developed from ontology and ubiquitous concept to perform computation from heterogeneous devices integrated to physical environment to make interaction between human–computer run smoothly.

Additionally, to check consistencies in the model, ontology reasoning also required to deduce knowledge from the model developed (July and Ay 2007; Wang et al. 2004). The ontology reasoning includes RDFS reasoning and OWL (Ontology Web Language) reasoning. The RDFS reasoning supports all the RDFS entailments described by the RDF Core Working Group. The OWL reasoning supports OWL/lite (OWL Web Ontology Language Overview) which includes constructs such as relations between classes (e.g. disjointness), cardinality (e.g. 'exactly one'), equality, characteristics of properties (e.g. symmetry), and enumerated classes.

User defined reasoning instead of OWL reasoning can help to provide flexible reasoning mechanism. Through the creation of user-defined reasoning rules within a set of first order logic, a wide range of higher-level conceptual context such as "what the user is doing" can be deduced from relevant low-level context. The user defined reasoning should be performed after the ontology reasoning because the facts which are receive from the ontology reasoning in the input for the user-defined reasoning.

According to the background, the objective of this study is to provide proper context information for recognizing human activity behavior viewed from the daily living and physical activities that is performed both in outdoor and indoor environment using semantic-based approach. In the end of the study, the performances of ontology and user defined reasoning show that rules deduce are consistence enough to recognize current activity of the people. Furthermore, study of this paper with some additional values in activity daily living treatment that involves more comprehensive aspects will give benefit in developing healthcare services application.

2 Literature Review

Context awareness is one of feature key in pervasive computing that enabling the application sense the changing of environments and adapt operation and behaviors adjusting available resources (Xue and Keng Pung 2012). Context awareness comes up with some notions about context data, context source, and context attribute, whereas all of them should be unity to build context awareness services.

Due to evolving nature of context aware computing, formalizing all context information by using context modeling is an important process. Context model should be able to capture set of upper-level entities, and providing flexible extensibility to add specific concepts in different application domains (Wang et al. 2004). Context model should be developed using common vocabulary that enable each domain to share common concept, but in the same time it also should be able to provide a flexible interface to define specific knowledge of application (Scott and Benlamri 2010). Previous work related to context modeling and context reasoning are have been conducted by (Wang et al. 2004) that evaluate feasibility of context modeling and context reasoning using Context Ontology (CONON).

In order to harvest deep understanding about context data, context sources, and context attributes, activity theory could be useful concept to capture important information engaged to the system. Activity theory is merely characterized as a conceptual framework to understand everyday practice in the real world depicted into set of concept and categories for communicating about nature of human activity (Kaptelinin and Nardi 1996).

Activity theory is broadly defined as a philosophical and cross-disciplinary framework for studying different forms of human practices as development processes, both individual and social levels interlinked at the same time by following

Fig. 1 The structure of human activity

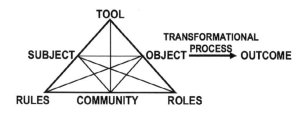

three key principles (Kuutti 1996): (a) Activities as a basic units of analysis; (b) History and development; (c) Artifacts and mediation.

Essential of activity theory can be summarized by a couple of simple diagrams supported by triangle diagram (Fig. 1).

Instead of just presenting information to human, ontology is more beneficial in supporting knowledge sharing, context reasoning, and interoperability in ubiquitous and pervasive computing system (Chen et al. 2004). Furthermore, its ability to share common understanding of the information and domain knowledge reuse simplify developer to extend domain of interest in the future, whereas domain ontology that is developed will provide specific knowledge for application and metadata or schema to subordinate instance-base that is updated when application executed (Hong et al. 2007; Noy and McGuinness 2001). Explicit knowledge of ontology can be used to reason context information by applying rules using ontology and user defined reasoning, where integration of semantic-based reasoning with ontology model allow the system to deal with rapidly changing context environment and having relatively low computational complexity (Bikakis et al. 2008).

3 Ontology-Based Approach for Activity Modeling

3.1 Activity Modeling

Generally the most common ontology-based approach for activity modeling consists of a specific semantic data from observation of indoor, outdoor, and smart home surround, such as time, user location, or object, but it still needs improvement in recognition ability for some ambiguous cases. In the effort to improve recognition ability, the model in Fig. 2 developed to presents the semantic of each class and relationship between classes through ontology into 5 domain concepts, which are: *Location* class to define environment type captured that is contains environment, indoor corridor, and smart home room; *Activity* class to describe the type of activity monitored and observed, consists of Activity Daily Living, Physical Activity, Activity Behavior, and Physical Activity Effect; *People* class to document user profile; *Sensor Manager* class to infer activity and sensor that become high level information and low level raw data; and *Wireless Network*

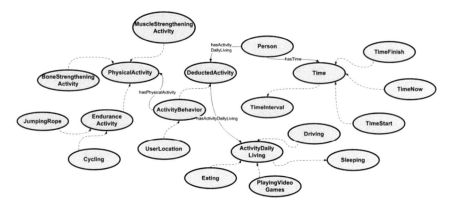

Fig. 2 Part of activity modeling using ontology-based approach

Devices class that document all wireless sensors that are used to gain context in system.

Context model is structured around a set of abstract entities as well as a set of abstract sub-classes, whereas each entity associated in its attributes and relationship that is represented in OWL as *DatatypeProperty* and *ObjectProperty* (Wang et al. 2004). According to Fig. 2, object properties that represent relationship between entities in class level can be seen more deeply in Table 1. Some sharing relationships that are occur enables system to provide extension of the new concepts in a specific domain in the future.

3.2 Ontology Reasoning

Ontology reasoning is mechanism that aimed to check consistency of context and deducing high level, implicit context from low-level, explicit context that can be processed with logical reasoning (Wang et al. 2004). To explain the role of context reasoning in activity modeling, activity daily living scenario is present to deduct user activity at home.

For the example, we'll present condition when the light in gym room is on, treadmill machine is also on, people location is at home and the position is running, then it can inferred that people is doing treadmill in gym at home. In order to get the result, ontology reasoning and user-defined resoning is conducted. According to the model, *UserLocation* class, *PhysicalActivity* Class, *Location* Class, and *WirelessNetworkDevices* Class are required to deduct high level activity related to user activity, while relationship between classes are linked using *"hasDailyLivingActivity"*, *"hasPhysicalActivity"*, and *"isCapturedOf"* properties. *UserLocation* class also used to capture people location, with property *"isLocatedIn"* link *UserLocation*class to *LocationNow* class value.

Table 1 Part of object properties list used in activity modeling

Class level

Domain	Object property	Range
DeducedActivity	HasDailyLivingActivity	ActivityDailyLiving
Person	HasDailyLivingActivity	ActivityDailyLiving
DeducedActivity	HasTime	Time
Person	HasTime	Time
DeducedActivity	IsCapturedOf	LowLevelNature
ActivityBehavior	HasPhysicalActivity	PhysicalActivity
Person	HasPhysicalActivity	PhysicalActivity
UserLocation	IsLocatedIn	LocationNow
Position	IsLocatedIn	LocationNow
NearbyServices	IsLocatedIn	LocationNow
EnvironmentServices	IsLocatedIn	Environment
IndoorCorridorServices	IsLocatedIn	IndoorCorridor
RoomServices	IsLocatedIn	Room
Position	IsLocatedIn	LocationNow
Person	IsLocatedIn	LocationNow
SensorManager	IsLocatedIn	LocationNow
Position	HasLocationBasedServices	NearbyServices
Person	HasLocationBasedServices	NearbyServices
LowLevelNature	HasLocationBasedServices	NearbyServices MobileSensors
EnvironmentServices	HasRawDataSignal	MobileSensors
IndoorCorridorServices	HasRawDataSignal	VideoCamera or Microphone
RoomServices	HasRawDataSignal	SmartBuildingSensors or VideoCamera or Microphone

Figure 3 present that *UserLocation* class also inherit property *"isCapturedOf"* from *Activity* class that means people activity is also captured by *LowLevelNature*class that get the raw data from several sensors depend on people position detected. *LowLevelNature* is a class that sense raw data signal from *WirelessNetworkDevices* class, which is linked by *"hasRawDataSignal_1"* property. This class also provides some service from *LocationBasedServices"* through*"hasLocationBasedServices"* property according to people position that is also sensed using property *"isLocatedIn"*.

Knowledge representation can be gained from OWL individual that act as Jess facts, where the rules that is determined become Jess rules (Fig. 4). Performing inference using rules from OWL knowledge base then will show the result about user activity recognition (Fig. 5).

Class and relationship rules determined then become knowledge of the context and also fact that has able to deduct using some rules we are determined. In order to accomplish the result, Jess system is used as one of rule engines that is work well with Java that is consists of a rule base, a fact base, and an execution engine to match fact base by reason OWL individual.

Activity Modeling Using Semantic-Based Reasoning

Fig. 3 *Userlocation* class using OWL reasoning rules

Fig. 4 User defined rules activity recognition

```
Jess> (defrule getActivity (object (is-a http://www.owl-ontologies.com/Ontology1368804236.owl#Location) (http://www.owl-ontologies.
com/Ontology1368804236.owl#LocationNow ?ln)) (object (is-a http://www.owl-ontologies.com/Ontology1368804236.owl#Position) (http:
//www.owl-ontologies.com/Ontology1368804236.owl#PositionP ?p)) (object (is-a http://www.owl-ontologies.com/Ontology1368804236.
owl#Room) (http://www.owl-ontologies.com/Ontology1368804236.owl#RoomName ?rn) (http://www.owl-ontologies.
com/Ontology1368804236.owl#RoomStatus ?rs)) (object (is-a http://www.owl-ontologies.com/Ontology1368804236.owl#ObjectThing)
(http://www.owl-ontologies.com/Ontology1368804236.owl#ObjectName ?on) (http://www.owl-ontologies.com/Ontology1368804236.
owl#ObjectStatus ?os)) (object (is-a http://www.owl-ontologies.com/Ontology1368804236.owl#Person) (http://www.owl-ontologies.
com/Ontology1368804236.owl#Name ?n) (http://www.owl-ontologies.com/Ontology1368804236.owl#UserLocationActivity ?la&: (= ?la ?
ln)) (http://www.owl-ontologies.com/Ontology1368804236.owl#UserPositionActivity ?up&: (= ?up ?p)) (http://www.owl-ontologies.
com/Ontology1368804236.owl#UserRoomActivity ?ra&: (= ?ra ?rn)) (http://www.owl-ontologies.com/Ontology1368804236.
owl#UserRoomActivityStatus ?ras&: (= ?ras ?rs)) (http://www.owl-ontologies.com/Ontology1368804236.owl#UserEquipmentName ?uen&:
(= ?uen ?on)) (http://www.owl-ontologies.com/Ontology1368804236.owl#UserEquipmentStatus ?ues&: (= ?ues ?os))) => (printout t ?n
"doing " ?p "in" ?rn "at" ?ln "using" ?on crlf))
TRUE
Jess> (run)
Anis Amna"doing"Running"in"PrivateGym"at"Home"using"Treadmill Machine
```

Fig. 5 Jess rule reasoning result

4 Conclusions

Our study in this paper showed that activity modeling that is developed using semantic-based approach can provide meaningful context to recognize human activity behavior in pervasive computing. Furthermore, after it's checked by using OWL and user defined reasoning as the feature of Protégé, context model is able to recognize human activity.

Acknowledgments This work is partially supported by Prof. Shuo-Yan Chou and Mr. Anindhita Dewabharata. The authors also gratefully acknowledge the helpful comments and suggestions of the reviewers, which have improved the presentation.

References

Bikakis A, Patkos T, Antoniou G, Plexousakis D (2008) A survey of semantics-based approaches for context reasoning in ambient intelligence. Commun Comput Inf Sci 11:14–23
Dey AK, Abowd GD (1999) Towards a better understanding of context and context-awareness. Comput Syst 40:304–307
Hong CS, Kim H-S, Cho J, Cho HK, Lee H-C (2007) Context modeling and reasoning approach in context-aware middleware for urc system. Int J Math Phys Eng Sci 1:208–221
Chen H, Perich F, Finin T, Joshi, A (2004) SOUPA: standard ontology for ubiquitous and pervasive applications. In: The First annual international conference on mobile and ubiquitous systems networking and services 2004 MOBIQUITOUS 2004
July O, Ay F (2007) Context Modeling and Reasoning using Ontologies. Network, pp 1–9
Kaptelinin V, Nardi BA (1996) Activity theory and human-computer interaction. Activity theory and human-computer interaction. MIT Press

Kuutti K (1996) Activity theory as a potential framework for human-computer interaction research. Context Conscious Act Theory Human–Comput Inter

Noy N, McGuinness D (2001) Ontology development 101: a guide to creating your first ontology. Development 32:1–25

Scott K, Benlamri R (2010) Context-aware services for smart learning spaces. IEEE Trans Learn Technol 3:214–227

Wang XH, Zhang DQ, Gu T, Pung HK (2004) Ontology based context modeling and reasoning using OWL. In: IEEE annual conference on pervasive computing and communications workshops 2004

Wongpatikaseree K, Ikeda M, Buranarach M, Supnithi T, Lim AO, Tan Y (2012) Activity Recognition Using context-aware infrastructure ontology in smart home domain. In: 2012 Seventh international conference on knowledge, information and creativity support systems. doi:10.1109/KICSS.2012.26

Xue W, Keng Pung H (2012) Context-Aware middleware for supporting mobile applications and services. In: Kumar A, Xie B (eds) Handbook of mobile systems applications and services. CRC Press, Boca Raton

An Integrated Systems Approach to Long-Term Energy Security Planning

Ying Wang and Kim Leng Poh

Abstract While heavy attempts have been made to evaluate energy systems, few can gain wide acceptance or be applied to various jurisdictions considering their lack of comprehensiveness and inability to handle uncertainties. This paper first proposes a MCDM approach using Fuzzy Analytic Hierarchy Process (FAHP) to assess security status in energy system. An Energy Security Index I_{ES} making comprehensive yet clear reference to the current scope of energy security is introduced as the indicator. Next, the paper proposes an integrated framework to develop long-term security improvement plan in energy system. The framework is constructed with a holistic planning cycle to evaluate energy security policies' effectiveness, project estimated I_{ES} with optimized energy portfolio, and verify the results generated. The framework is established with integrated analytical process incorporating FAHP, which better accommodates the complexities and uncertainties throughout planning. Meanwhile, the complete planning cycle enhances the tool validity. This framework would be useful in helping policy makers obtain a helicopter view of the security level of their energy system, and identify the general improvement direction. An application of the proposed framework to Singapore context shows "Reduce and Replace" strategy should be implemented and 77 % improvement in I_{ES} is expected by 2030 comparing to business-as-usual scenario.

Keywords Energy security policies · Integrated assessment · Analytical planning · Fuzzy logic · Portfolio optimization · Multi-criteria decision making

Y. Wang (✉) · K. L. Poh
Department of Industrial and Systems Engineering, National University of Singapore,
1 Engineering Drive 2, Singapore 117576, Singapore
e-mail: cynthia.wang.ying@gmail.com

K. L. Poh
e-mail: pohkimleng@nus.edu.sg

1 Introduction

As instability continues to intensify in the Middle East after the Arab Spring, security of energy sources has become one of the main concerns internationally. However, regardless of the universally agreed significance of energy security and heavy attempts in energy security measurements, these measurements can hardly gain wide acceptance because of the large amount of related issues encompassed by energy security (Martchamadol and Kumar 2012). Single-aspect indicators can hardly provide all-round assessments while existing aggregated indicators usually have restricted application to specific countries or energy sources to address specific criteria. Additionally, while there exist separate studies to assess energy security policies, no synthetical measurements are conducted to evaluate the status and the policies as a whole. These studies miss the section to project and demonstrate the effectiveness of proposed policies, which undermines the persuasiveness.

This paper aims to fill the gap by proposing a general framework with a holistic evaluation and planning cycle which is able to assess and project the long-term energy security level of a jurisdiction based on the optimized fuel mix portfolio, evaluate varies energy strategies and testify their effectiveness. An illustration of the framework is provided by applying to the Singapore context.

2 A Fuzzy MCDM Model for Energy Security Evaluation

This section proposes an Energy Security Index (I_{ES}) to help policy makers assess and understand the nation's current energy security level. The index is formulated by evaluating fuel security levels using ratings approach to AHP incorporating with fuzzy logic and then taking a weighted average based on the nation's energy portfolio. This method is believed can comprehensively evaluate the security level and well accommodate the uncertainties in human subjective judgments.

2.1 Frame of AHP Model

Figure 1 shows the complete AHP hierarchy containing criteria and corresponding rating intensities for evaluating fuel security level and the score generated is defined as Fuel Security Index (Is). The proposed model separates the assessment criteria into two sub-sets, namely, physical availability and price component (International Energy Agency 2007). While physical availability is derived based on the "Five Ss" concept (Kleber 2009), the second sub-set targeting economic concern is classified into: Setup Cost, Fuel Cost, and Operation and Maintenance (O&M) Cost.

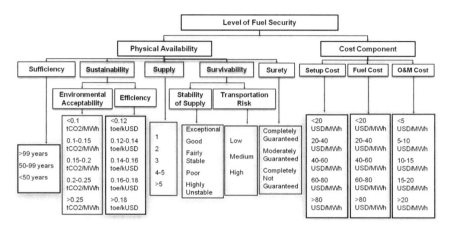

Fig. 1 Proposed AHP model for fuel security evaluation

2.2 Fuzzy AHP

Energy related decision making is a process filled with uncertainties due to the complex nature of energy security problem, including the multi-dimensional definition of the term, the contributing factors on national and international level, the fast changing energy situation around the globe, etc. Meanwhile, subjective human judgments weigh heavily in the decision making. These factors show the necessity for the model to handle ambiguity and uncertainty in subjective human assignment of quantitative weights. In this study, the incorporation of traditional AHP with fuzzy logic is proposed to be the solution. It ensures the model's capability to handle vague input so that holistic solution can be expected.

2.3 Formulation of Energy Security Index

After obtaining the Fuel Security Indexes (I_S) and electricity generation fuel mix for a specific nation, Energy security Index (I_{ES}) can be derived by using weighted average (Lim 2010) where higher value representing a higher level of energy security. The formulation is as follows:

$$I_{ES} = \sum_{k=1}^{N} p_k I_s^k, (I_{ES} \in [0, 1]). \tag{1}$$

k Index of fuel sources, $k = 1, 2, \ldots, N$
p_k The share of fuel alternative k in the total energy consumption $(\sum_{k=1}^{N} p_k = 1)$
I_s^k Fuel Security Index for fuel alternative k $(I_{ES} \in [0, 1])$

2.4 Evaluation of Energy Security Level in Singapore

Right now Singapore generates electricity with three major types of sources: natural gas, petroleum products, and biofuel (Energy Market Authority 2012). Based on Singapore context, weights for the criteria and sub-criteria are determined by performing pairwise comparisons with the triangular fuzzy number between each criterion with respect to the goal and between each sub-criterion with respect to the upstream criteria. Assuming a 0.5 confidence level, I_S is derived and Energy Security Index (I_{ES}) based on the current fuel mix in Singapore is formulated. The results are listed in Table 1.

According to the proposed measurement index, with a range of 0–1, the score of current energy security level is 0.355. The current energy security level highly relies on the security level of natural gas, since it counts for almost 79 % of the total consumption. The relative insecurity of natural gas implies there is great potential to improve Singapore's current energy security level.

3 Fuzzy Analytical Planning Approach to Energy Security Policy Decision Making and Energy Portfolio Planning

3.1 Forward–Backward Planning

The conventional AHP model deals with static situation only. However, evaluation of energy security level is a stochastic problem considering the continuous change in fuel mix and fuel properties of a nation. The gap in-between drives the requirement of an advanced method to better handle the complexity and uncertainty. This study uses forward and backward planning developed by Saaty and Kearns (Saaty 1980) to tackle the problem.

3.2 Outline of Proposed Framework

In this section, a framework with holistic evaluation and planning cycle is proposed to ensure the validity of the tool. The tool uses forward–backward planning

Table 1 Summary of model results

Alternative	Fuel security index (I_s^k)	The share of fuel alternative k (p_k) (%)
Petroleum products	0.2185	18.70
Natural gas	0.3720	78.70
Biofuel plant	0.8109	2.60
Energy security index (I_{ES}) 0.355		

incorporating fuzzy AHP in order to better accommodate complexities and uncertainties. The AHP method can help decompose a problem into sub-problems, which results in the methodology summarized in Fig. 2. In the next section, a step-by-step illustration of the proposed methodology is provided with an application in Singapore's context.

4 Application to Energy Security Improvement Strategy Formulation in Singapore

4.1 First Forward Process

The first forward process projects the most likely level of energy security. This scenario is formulated with current electricity generation fuel mix and energy strategy unchanged, say, the business-as-usual (BAU) scenario. It uses the Energy Security Index (I_{ES}) proposed in Sect. 3 as the indicator. The value is calculated to be 0.355 which indicates great improvement potential of BAU scenario. We therefore proceed to the next phase and detect possible solutions.

4.2 First Backward Process

The backward planning discovers the most effective energy security improvement strategy in the Singapore context. The proposed solutions for energy security improvement, according to Larry Hughes's research, can be categorized into three strategies (2009):

Reduce: This study solely considers reducing energy demand via energy efficiency since reduction in service level or sacrifice citizens' living standard is not desired in Singapore.

Replace: Replacing the insecure fuel sources with more secure ones can be realized by diversifying suppliers for current fuel sources, or reconstructing the infrastructures and plant to switch to a secure alternative.

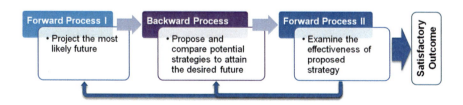

Fig. 2 Summary of proposed framework outline

Restrict: Maintenance of existing infrastructure and fuel sources, but limiting the additional demand to be generated with secure sources only.

Since energy efficiency projects are profitable, considering the limited budget available, two alternative strategies are proposed to Singapore government to select from: *Reduce and Replace*, or *Reduce and Restrict*.

On the other hand, although there exists great potential to improve energy security in Singapore as identified in the first forward analysis, Singapore has limited access to alternative fuel sources due to its small and restricted landscape. As such, the possible alternative fuel sources can be taken into account limit to: Liquefied Natural Gas (LNG), Coal Plant with Carbon Capture & Storage (CCS), and Solar Power. Implementing the fuzzy AHP model used in Sect. 3, the corresponding Fuel Security Index (I_S) is derived in Table 2. The model shows biofuel and solar power are the most secure source, while LNG and coal plant with CCS can perform better comparing to petroleum products and natural gas.

In addition, a trade-off analysis with respect to physical availability and price component is conducted to further compare among the fuel alternatives. Two efficiency frontiers are plotted for fossil fuel and renewable source respectively. According to Fig. 3, petroleum products is generally dominated and thus considered an inferior source. This implies the source is expected to be replaced once "Reduce & Replace" strategy is carried out. Otherwise, if "Reduce & Restrict" strategy is put into application, petroleum products is impossible to be the solution for new demand. Therefore, the two potential energy strategies for Singapore are defined. The first backward analysis hierarchy is structured in Fig. 4 and the outcome is listed in Table 3.

With a slightly higher score, Reduce and Replace should be preferred in the Singapore context. This may due to its larger potential to improve from the current situation. We therefore proceed to the second forward planning to examine the finding.

4.3 Second Forward Process

The second forward process aims to verify if the strategy selected previously is the most effective. The verification is conducted by comparing the updated I_{ES} when aforementioned energy strategies are implemented correspondingly.

4.3.1 Reduce and Restrict

If "Reduce and Restrict" strategy is carried out, the new energy demand beyond the current consumption level is assumed to be covered by the secure sources

Table 2 Summary of model results

Alternative	LNG	Coal plant with CCS	Solar power
Energy security index (I_{ES})	0.5149	0.4874	0.8882

An Integrated Systems Approach to Long-Term Energy Security Planning

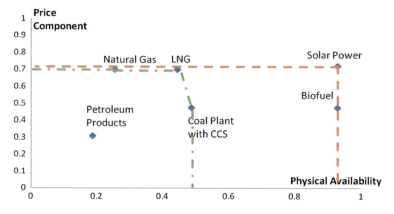

Fig. 3 Trade-off analysis of possible fuel alternatives

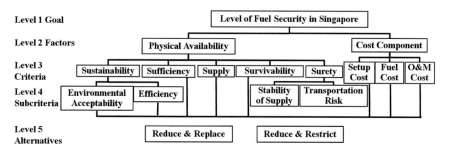

Fig. 4 Hierarchical structure of energy strategy assessment

Table 3 Summary of model results

Alternative	Reduce and replace	Reduce and restrict
Score	**0.780**	0.705

biofuel and solar power. The future energy consumption trend is captured with an uni-variant regression model of GDP versus energy consumption. Making a 20-year projection, the energy consumption level in 2030 is estimated at roughly 1.6 times the consumption level in 2010. This indicates newly constructed biofuel plant and solar power plant should cover additional 60 % of the current consumption. The optimal portfolio of the fuel mix can be determined using the linear programming.

Ideally, the optimal I_{ES} can be achieved when all new demand is covered by solar power, the fuel alternative with the highest Fuel Security Index. The model shows by 2030 Singapore's I_{ES} will experience 55 % improvement and reach the level up to 0.5498.

4.3.2 Reduce and Replace

In this scenario, a similar model is created to measure I_{ES}. To ensure a fair comparison, equivalent cost should be assumed for the two scenarios. In addition, considering the limited access to renewable energy due the geographical restrictions of Singapore, the share of renewable energy is assumed to be capped at 40 % so that to make the model more realistic. The optimal portfolio of the fuel mix can be determined using the linear programming.

Ideally, the optimal I_{ES} can be achieved when 40 % of total demand is covered by solar power, the fuel alternative with the highest I_S, while the rest 60 % of the demand is mainly supplied by LNG, the most secure fossil fuel alternative. The implementation of "Reduce & Replace" policy can help improve Singapore's I_{ES} by 77 % and reach the level up to 0.6281 by 2030.

As a result, the convergent outcome of the first backward process and second forward process validates the effectiveness and advantage of "Reduce & Replace" strategy. Therefore the forward–backward planning is completed and the best strategy is determined.

5 Conclusion

In this paper, an energy security index I_{ES} is developed with comprehensive and clear reference to the current scope of energy security. In the next phase, an integrated framework with a holistic planning cycle is proposed to evaluate the effectiveness of policies, project the estimated I_{ES}, and verify the results with optimized long-term energy portfolio. An application to the Singapore context shows that "Reduce & Replace" strategy should be implemented and 77 % improvement of I_{ES} is expected by 2030. The incorporation of fuzzy logic with AHP in the model takes into accounts the uncertainties and ambiguousness for policy makers' subjective judgment and ensures the tool's capability to handle them. Meanwhile, policy makers can better handle the complexity of the problem by adding more iteration to forward–backward planning if necessary. The proposed framework can be useful in helping policy makers to obtain a helicopter view of their nation's energy security level and identify long-term energy security planning.

Acknowledgments This work is partially supported by Department of Industrial and Systems Engineering, National University of Singapore. The authors also gratefully acknowledge the helpful comments and suggestions of the reviewers, which have improved the presentation.

References

Energy Market Authority (2012) Singapore energy statistics 2012. Singapore
Hughes L (2009) The four 'R's of energy security. Energy Policy 37:2459–2461
International Energy Agency (2007) Energy security and climate policy. Paris, France
Kleber D (2009) Valuing energy security. J Energy Secur. http://www.ensec.org/index.php?option=com_content&view=article&id=196:the-us-department-of-defense-valuing-energy-security&catid=96:content&Itemid=345. Accessed 12 Nov 2012
Lim WZ (2010) A multi-criteria decision analysis and portfolio optimization approach to national planning for long-term energy security. B. engineering dissertation, National University of Singapore
Martchamadol J, Kumar S (2012) An aggregated energy security performance indicator. Appl Energy 103:653–670
Saaty TL (1980) The analytic hierachy process. United States, New York

An EPQ with Shortage Backorders Model on Imperfect Production System Subject to Two Key Production Systems

Baju Bawono, The Jin Ai, Ririn Diar Astanti and Thomas Indarto Wibowo

Abstract This paper is an extension of the work of Lin and Gong (2011) on Economic Production Quantity (EPQ) model on an imperfect production system over infinite planning horizon, where the production system is dictated by two unreliable key production subsystems (KPS). While any shortage on the inventory of product was not allowed in the model of Lin and Gong (2011), planned shortage backorders is considered in the model proposed in this paper. The mathematical model is developed in order to determine production run time (τ) and production time when backorder is replenished (T_1) that minimizes the expected total cost per unit time including setup, inventory carrying, shortage, and defective costs. Approaches to solve the model are also being proposed in this paper, altogether with some numerical examples.

Keywords Economic production quantity model · Shortage backorders · Imperfect production system · Optimization technique · Approximation method

---

B. Bawono (✉) · T. J. Ai · R. D. Astanti · T. I. Wibowo
Department of Industrial Engineering, Universitas Atma Jaya Yogyakarta,
Jl. Babarsari 43, Yogyakarta 55281, Indonesia
e-mail: baju@mail.uajy.ac.id

T. J. Ai
e-mail: jinai@mail.uajy.ac.id

R. D. Astanti
e-mail: ririn@mail.uajy.ac.id

T. I. Wibowo
e-mail: t8_t10@yahoo.co.id

1 Introduction

Productivity is generally defined as the ratio between output and input. The input can be man, material, machine, money, and method. In manufacturing industry, where the output is tangible product, the productivity can be measured by how many or how much the output resulted. Productivity might be affected by one of the input mentioned above, such as machine. Machine is one element of the production subsystem. The machine is reliable if it can perform as good as the standard. However, in reality there is a condition where, the machine does not perform well or it is called imperfect condition. This condition might happen due to, for example, machine breakdown. As illustration, boiler breakdown in a Crude Palm Oil (CPO) industry will increase the concentration of ALB in the oil so that it will decrease the quality of CPO (Sulistyo et al. 2012). Therefore, the output of the CPO is also decreased. In other word, the productivity of the industry is decreased.

The economic production quantity (EPQ) model developed by many researchers in the past, such as Silver et al. (1998) under the assumption the production subsystem is perfect (no breakdown). However, this ideal condition is rarely happened in the real situation. If this model is applied in the real situation where the production subsystem is imperfect, then, the target production is never be reached. The EPQ model considering imperfect production subsystem have been proposed by some researchers, such as Rosenblatt and Lee (1986). They assumed that in the production system there may exist an imperfect condition where the in-control state changed to out-control state where the random time to change is assumed following exponential distribution. As the result, the system might produce defective product. Following this work, some models dealt with various additional system setting had been proposed, such as (Lee and Rosenblatt 1987, 1988, 1989; Lee and Park 1991; Lin et al. 1991; Ben-Daya and Hariga 2000; Hsieh and Lee 2005). In those previous models, the imperfectness of production system is assumed to be dictated by single key production subsystem (KPS) only. Lin and Gong (2011) recently extended the study by proposing an EPQ model where the production system is imperfect and dictated by two imperfect KPS's over an infinite planning horizon. Ai et al. (2012) continued this work by considering finite planning horizon.

While those above mentioned researches discussed on EPQ model without shortage, the research conducted by Chung and Hou (2003) extended the work of Rosenblatt and Lee (1986) by incorporating the shortage into their model. Shortage itself can be defined as the situation when the customer order arrive but the finished good are not yet available. When all customer during the stockout period are willing to wait until the finished goods are replenished then it is called as completely backorder case. Shortage is common in practical situation, when the producer is realized that its customers loyal to their product.

This paper is extending the work of Chung and Hou (2003) and Lin and Gong (2011) by combining both works into a new EPQ model that consider 2 (two) imperfect KPS and shortage. The organization of this paper is as follow: Sect. 2

describes the mathematical model development, Sect. 3 discusses the solution methodology of the proposed method, Sect. 4 presents the numerical example, followed by some concluding remarks in Sect. 5.

2 Mathematical Model

This paper considers a production lot sizing problem where a product is to be manufactured in batches on an imperfect production system over an infinite planning horizon, in which shortage of product is allowed at the end of each production cycle and all shortage is backordered. The demand rate is d, and the production rate is p. As defined in Lin and Gong (2011) and Ai et al. (2012), the imperfectness of the system is shown on two imperfect key production subsystems (KPS) that may shift from an in-control to an out-of-control state due to three independent sources of shocks: source 1's shock causes first KPS to shift, source 2's shock causes second KPS to shift, and source 3's shock causes both KPS to shift. Each shocks occur at random time U_1, U_2, and U_{12} that follows exponential distribution with mean $1/\lambda_1$, $1/\lambda_2$, and $1/\lambda_{12}$, respectively. When at least one KPS on out-of-control state, consequently, the production system will produced some defective items with fixed but different rates: α percentage when first KPS out-of-control, β percentage when the second KPS out-of-control, and δ percentage when the both KPS out-of-control. The cost incurred by producing defective items when the first KPS is shifted, the second KPS is shifted, and both KPS are shifted are π_1, π_2, and π_{12}, respectively.

The production cycle of this situation can be described as Fig. 1, in which the inventory level is increased during the production uptime (τ) at rate ($p - d$) and decreased at rate $-d$ during the production downtime. It is shown in Fig. 1, the production cycle length T can be divided into four sections, each of them with length T_1, T_2, T_3, and T_4, respectively. The backorders are replenished in Sect. 1, in which the inventory level is increased from $-B_{max}$ to 0. The inventories are accumulated during Sect. 2, in which inventory level at the end of this section is I_{max}. After that, the inventory level is decreased to 0 during Sect. 3 and the shortage happened in Sect. 4.

The optimization problem is to determining optimal production run time τ and production time when backorder is replenished T_1, that minimizes the expected total cost per unit time including setup, inventory carrying, shortage and defective costs.

In single production cycle, although shortages are exist, the number of product being produced ($p.\tau$) is equal to the demand of product ($d.T$) in that cycle. Therefore $T = p\tau/d$. If the setup cost is denoted as A, based on (1), the setup cost per unit time (C_1) can be defined as

$$C_1 = \frac{A}{T} = \frac{Ad}{p\tau} \qquad (1)$$

The average inventory per production cycle as function of τ and T_1 can be expressed as

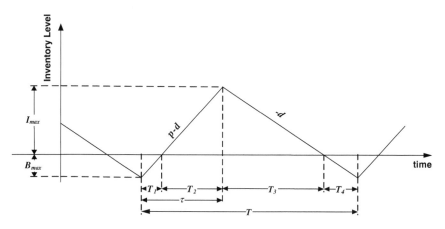

Fig. 1 Production cycle

$$\bar{I}(T_1, \tau) = \frac{(p-d)\tau}{2} - (p-d)T_1 + \frac{(p-d)T_1^2}{2\tau} \quad (2)$$

Therefore, if unit inventory holding cost per unit time is defined as h, the inventory carrying cost per unit time (C_2) can be defined as

$$C_2 = \frac{h(p-d)\tau}{2} - h(p-d)T_1 + \frac{h(p-d)T_1^2}{2\tau} \quad (3)$$

The average shortage per production cycle as function of τ and T_1 can be expressed as

$$\bar{B}(T_1, \tau) = \frac{(p-d)T_1^2}{2\tau} \quad (4)$$

Therefore, if unit shortage cost per unit time is defined as s, the shortage cost per unit time (C_3) can be defined as

$$C_3 = \frac{s(p-d)T_1^2}{2\tau} \quad (5)$$

The results from Lin and Gong (2011) are presented below to obtain the defective cost per unit time (C_4), in which the unit defective cost when the first KPS is shifted, the second KPS is shifted, and both KPS are shifted is defined as π_1, π_2, and π_{12}, respectively.

$$C_4 = \frac{d(\pi_1 E[N_1(\tau)] + \pi_2 E[N_2(\tau)] + \pi_{12} E[N_{12}(\tau)])}{p\tau} \quad (6)$$

$$E[N_1(\tau)] = p\alpha \left(\frac{1 - \exp[-(\lambda_2 + \lambda_{12})\tau]}{\lambda_2 + \lambda_{12}} - \frac{1 - \exp[-(\lambda_1 + \lambda_2 + \lambda_{12})\tau]}{\lambda_1 + \lambda_2 + \lambda_{12}} \right) \quad (7)$$

$$E[N_2(\tau)] = p\beta\left(\frac{1 - \exp[-(\lambda_1 + \lambda_{12})\tau]}{\lambda_1 + \lambda_{12}} - \frac{1 - \exp[-(\lambda_1 + \lambda_2 + \lambda_{12})\tau]}{\lambda_1 + \lambda_2 + \lambda_{12}}\right) \quad (8)$$

$$E[N_{12}(\tau)] = p\delta\left(\frac{\exp[-(\lambda_1 + \lambda_{12})\tau] + (\lambda_1 + \lambda_{12})\tau - 1}{\lambda_1 + \lambda_{12}}\right.$$
$$+ \frac{\exp[-(\lambda_2 + \lambda_{12})\tau] + (\lambda_2 + \lambda_{12})\tau - 1}{\lambda_2 + \lambda_{12}}$$
$$\left. - \frac{\exp[-(\lambda_1 + \lambda_2 + \lambda_{12})\tau] + (\lambda_1 + \lambda_2 + \lambda_{12})\tau - 1}{\lambda_1 + \lambda_2 + \lambda_{12}}\right) \quad (9)$$

Therefore, the expected total cost per unit time can be stated as

$$Z[\tau, T_1] = C_1 + C_2 + C_3 + C_4$$

$$Z[\tau, T_1] = \frac{Ad}{p\tau} + \frac{h(p-d)\tau}{2} - h(p-d)T_1 + \frac{h(p-d)T_1^2}{2\tau} + \frac{s(p-d)T_1^2}{2\tau}$$
$$+ \frac{d(\pi_1 E[N_1(\tau)] + \pi_2 E[N_2(\tau)] + \pi_{12} E[N_{12}(\tau)])}{p\tau} \quad (10)$$

3 Solution Methodology

Following Lin and Gong (2011), all exponential terms in the total cost expression can be approximate by MacLaurin series:

$$\exp(-\lambda\tau) \approx 1 - \lambda\tau + \frac{1}{2!}(\lambda\tau)^2 - \frac{1}{3!}(\lambda\tau)^3 \quad (11)$$

Therefore the total cost expression can be rewritten as

$$Z[\tau, T_1] \approx \frac{Ad}{p\tau} + h(p-d)\left[\frac{\tau}{2} - T_1\right] + \frac{(h+s)(p-d)T_1^2}{2\tau} + \frac{H\tau}{2} - \frac{K\tau^2}{6} \quad (12)$$

$$H = d(\pi_1\alpha\lambda_1 + \pi_2\beta\lambda_2 + \pi_{12}\delta\lambda_{12}) \quad (13)$$

$$K = d[\pi_1\alpha\lambda_1(\lambda_1 + 2\lambda_2 + 2\lambda_{12}) + \pi_2\beta\lambda_2(2\lambda_1 + \lambda_2 + 2\lambda_{12}) + \pi_{12}\delta(\lambda_{12}^2 - 2\lambda_1\lambda_2)] \quad (14)$$

It is well known from calculus optimization that the necessary condition for minimizing $Z[\tau, T_1]$ are the first partial derivatives equal to zero. Applying this condition for Eq. (12), it is found that

$$\frac{\partial}{\partial\tau}Z[\tau, T_1] = \frac{H}{2} + \frac{h(p-d)}{2} - \frac{K\tau}{3} - \frac{Ad}{p\tau^2} - \frac{(h+s)(p-d)T_1^2}{2\tau^2} = 0 \quad (15)$$

$$\frac{\partial}{\partial T_1} Z[\tau, T_1] = -h(p-d) + \frac{(h+s)(p-d)T_1}{\tau} = 0 \qquad (16)$$

Solving Eq. (16) for T_1, it is found that

$$T_1^* = \frac{h}{(h+s)} \tau^* \qquad (17)$$

Substituting Eq. (17) to Eq. (15), it is obtained that

$$\frac{H}{2} + \frac{hs(p-d)}{2(h+s)} - \frac{K\tau}{3} - \frac{Ad}{p\tau^2} = 0 \qquad (18)$$

If the term $K\tau/3$ is neglected or approximated as zero, it found after some algebra that

$$\tau^* = \sqrt{\frac{2Ad}{p\left[H + hs\left(\frac{p-d}{h+s}\right)\right]}} = \sqrt{\frac{2Ad}{p\left[d(\pi_1\alpha\lambda_1 + \pi_2\beta\lambda_2 + \pi_{12}\delta\lambda_{12}) + hs\left(\frac{p-d}{h+s}\right)\right]}} \qquad (19)$$

The sufficient condition of this result can be easily proven, since the Hessian matrix is positive definite.

4 Numerical Examples

To illustrate the proposed model and solution methodology, the numerical example is conducted on 8 (eight) sample problems as it is shown in Table 1.

Table 1 Parameters and solutions of sample problems

Parameters	Prob 1	Prob 2	Prob 3	Prob 4	Prob 5	Prob 6	Prob 7	Prob 8
d	200	200	200	200	200	200	200	200
p	300	300	300	300	300	300	300	300
α	0.1	0.1	0.3	0.3	0.1	0.1	0.3	0.3
β	0.1	0.1	0.3	0.3	0.1	0.1	0.3	0.3
δ	0.16	0.16	0.48	0.48	0.16	0.16	0.48	0.48
λ_1	0.05	0.05	0.05	0.05	0.15	0.15	0.15	0.15
λ_2	0.1	0.1	0.1	0.1	0.3	0.3	0.3	0.3
λ_{12}	0.02	0.02	0.02	0.02	0.06	0.06	0.06	0.06
π_1	10	10	10	10	10	10	10	10
π_2	10	10	10	10	10	10	10	10
π_{12}	12	12	12	12	12	12	12	12
A	100	100	100	100	100	100	100	100
h	0.08	0.08	0.08	0.08	0.08	0.08	0.08	0.08
s	0.16	0.24	0.16	0.24	0.16	0.24	0.16	0.24
Solutions								
τ^*	1.761	1.747	1.061	1.058	1.061	1.058	0.622	0.622
T_1^*	0.587	0.437	0.354	0.265	0.354	0.265	0.207	0.155
$Z[\tau^*, T_1^*]$	75.73	76.32	125.6	126	125.6	126	214.3	214.5

5 Concluding Remarks

This paper extend the work of Chung and Hou (2003) and Lin and Gong (2011) by incorporating 2(two) imperfect KPS considering shortage on EPQ model. Based on the numerical example it can be shown that the proposed model and its solution methodology works on 8 (eight) sample problems. Further work will be conducted to find another solution methodology approaches and to do the sensitivity analysis on the proposed model. In addition, formulating an EPQ model with 2(two) imperfect KPS considering shortage can be further investigated for finite planning horizon.

Acknowledgments This work is partially supported by Directorate General of Higher Education, Ministry of Education and Culture, Republic of Indonesia through *Hibah Bersaing* Research Grant and Universitas Atma Jaya Yogyakarta, Indonesia. The authors also gratefully acknowledge the helpful comments and suggestions of the reviewers, which have improved the presentation.

References

Ai TJ, Wigati SS, Gong DC (2012) An economic production quantity model on an imperfect production system over finite planning horizon. In: Proceedings of IIE Asian conference 2012, Singapore

Ben-Daya M, Hariga M (2000) Economic lot scheduling problem with imperfect production processes. J Oper Res Soc 51:875–881

Chung KJ, Hou KL (2003) An optimal production run time with imperfect production processes and allowable shortages. Comput Oper Res 30:483–490

Hsieh CC, Lee ZZ (2005) Joint determination of production run length and number of standbys in a deteriorating production process. Eur J Oper Res 162:359–371

Lee HL, Rosenblatt MJ (1987) Simultaneous determination of production cycle and inspection schedules in a production system. Manage Sci 33:1125–1136

Lee HL, Rosenblatt MJ (1989) A production and maintenance planning model with restoration cost dependent on detection delay. IIE Trans 21:368–375

Lee HL, Rosenblatt MJ (1988) Economic design and control of monitoring mechanisms in automated production systems. IIE Trans 20:201–209

Lee JS, Park KS (1991) Joint determination of production cycle and inspection intervals in a deteriorating production system. J Oper Res Soc 42:775–783

Lin GC, Gong DC (2011) On an Economic lot sizing model subject to two imperfect key production subsystems. In: Proceedings of IIE Asian conference 2011, Shanghai, China

Lin TM, Tseng ST, Liou MJ (1991) Optimal inspection schedule in the imperfect production system under general shift distribution. J Chin Inst Ind Eng 8:73–81

Rosenblatt MJ, Lee HL (1986) Economic production cycles with imperfect production processes. IIE Trans 18:48–55

Silver EA, Pyke DF, Peterson R (1998) Inventory management and production planning and scheduling, 3rd edn. Wiley, New York

Sulistyo ARL, Astanti RD, Dewa DMRT (2012) Rancangan Preventive Maintenance dengan Pendekatan TPM di PT Perkebunan Nusantara VII. Unpublished thesis at Universitas Atma Jaya Yogyakarta

Reducing Medication Dispensing Process Time in a Multi-Hospital Health System

Jun-Ing Ker, Yichuan Wang and Cappi W. Ker

Abstract The process of prescribing, ordering, transcribing, and dispensing medications is a complex process and should efficiently service the high volume of daily physician orders in hospitals. As demand for prescriptions continues to grow, the primary issue confronting the pharmacists is overloading in dispensing the medication and tackling the mistakes caused by misinterpretation of the prescriptions. In this paper, we compared two prescribing technologies, namely no carbon required (NCR) and digital scanning technologies to quantify the advantages of the medication ordering, transcribing, and dispensing process in a multi-hospital health system. NCR technology uses a four parts physician order form with no carbon required copies, and digital scanning technology uses a single part physician order form. Results indicated a reduction of 54.5 % in queue time, 32.4 % in order entry time, 76.9 % in outgoing delay time, and 67.7 % in outgoing transit time in digital scanning technology. Also, we present the cost analysis to justify the acquisition of the Medication Order Management System (MOMS) to implement digital scanning technology.

Keywords Prescription order handling · Medication dispensing process · Medication order management · Medical errors reduction

---

J.-I. Ker (✉)
Industrial Engineering, Louisiana Tech University, Ruston, LA 71272, USA
e-mail: Ker@latech.edu

Y. Wang
Department of Aviation and Supply Chain Management, Auburn University, Auburn, AL 36830, USA
e-mail: yzw0037@auburn.edu

C. W. Ker
Engineering and Technology Management, Louisiana Tech University, Ruston, LA 71272, USA
e-mail: cwk005@latech.edu

1 Introduction

The process of prescribing, ordering, transcribing, and dispensing medications is a complex process and should efficiently service the high volume of daily physician orders in hospitals. According to the National Community Pharmacists Association (NCPA)'s report (NCPA 2012), the number of prescriptions being dispended in a pharmacy per day in the United States rose approximately 12 % (from 178 to 201) from 2006 to 2011. Additionally, the annual sales amount of prescription drugs being prescribed is 277.1 billion in 2012, and the projected figure will reach to 483.2 billion in 2021 (Keehan et al. 2012). As demand for prescriptions continues to grow, the primary issue confronting pharmacists is overloading in dispensing medications and tackling the mistakes caused by misinterpretation of the prescriptions (Jenkins and Eckel 2012). These mistakes can lead to medication related errors, which may endanger patient safety.

Another problem for pharmacists is the long delays in receiving medications, leading to doses being administered long after the standard administration times determined by the hospital pharmacy. The common causes of delays include prescribing error, poor drug distribution practices, drug and its device related problems, and illegible or unclear handwritten prescriptions (Cohen 2007). Interpreting the faint or illegible prescription is a major cause of delay in the transcribing process. Drugs with similar looking names can be incorrectly dispensed due to unclear handwriting. Currently, the adoption of emerging technology in hospitals reveals prominent effects on improving the existing operational process of medical service. This is particularly important for pharmacies. The use of information technologies (IT), such as computerized physician order entry system, has shown promise at not only reducing prescribing related errors (Ammenwerth et al. 2008; Jimenez Munoz et al. 2011; Samliyan et al. 2008; Shawahna et al. 2011) but also decreasing dispensing delays (Chuang et al. 2012; Jenkins and Eckel 2012).

Based on previous studies, we found that the issue of reducing the medical errors through IT has been documented well. The study for improving the dispensing delays, however, still remains underspecified. To address this gap, the aim of this study was to assess the timelines of various processes (i.e., queue time, order entry time, outgoing delay time and outgoing transit time) involved in administering the medication by comparing the two prescribing technologies: no carbon required (NCR) and digital scanning technologies. Also, we present the cost analysis to justify the implementing MOMS that uses digital scanning technology.

2 Literature Review

2.1 From Drug Prescribing to Dispensing in Hospitals

The prescribing medication is the physician's most frequently used, efficacious, and most dangerous tool used, outside of surgical interventions. A prescription is a written, verbal or electronic order from a practitioner or designated agent to a pharmacist for particular medication. The prescribing process is an important component of workflow in every physician practice and hospital unit. Pharmacies in hospitals have been using different prescribing methods. The traditional approach to medication management such as handwritten prescription is inefficient and error-prone (Berwick and Winickoff 1996). Illegible or unclear prescriptions result in more than 150 million calls from pharmacists to physicians and nurses, asking for clarification, a time-consuming process that costs the healthcare system billions of dollars each year in wasted time (Institute for Safe Medication Practices 2000).

Having received the prescription by the pharmacists at the computer workstation, the pharmacist enters the prescription into the pharmacy information system, checks for any known contraindications, and dispenses the medication. Before drugs are dispensed, labels are printed and the drugs are packed with the labels in single unit packages. This process which is widely used in the hospitals in United States has become known as a unit dose drug distribution system (Benrimoj et al. 1995). This distribution system is based on a pharmacy coordination method of dispensing and controlling medication in healthcare settings (American Society of Hospital Pharmacists 1989). The advantages of this system include reduction in medication errors (Allan and Barker 1990), reduction in drug inventory costs and minimized drug wastages. However, as demand for mediations and prescriptions continue to be in explosive growth in recent years, this system cannot be satisfied for dealing with high volumes of prescription, thereby leading to dispensing delays.

2.2 Significance of IT Usage in Hospitals

In the past two decades, researchers have started to identify needed interventions and justify related unnecessary spending to enhance safety and quality of the prescribing process by employing various technologies (e.g., automation method, the use of digital label) (Kohn et al. 1999; Flynn et al. 2003). Since the study of Kohn et al. (1999) spurred interest in employing technologies to simplify the prescribing process, the adoption of the drug prescribing process has shifted from paper-based prescribing process to electronic prescribing process (Åstrand et al. 2009).

Current studies show that many IT tools or systems for prescribing and dispensing practices are available in the market and have the abilities to enhance the accuracy of drug dispensing practices and improve the efficiency of prescribing practices. For example, pharmacists at one US hospital used a computerized prescription order entry system to review all prescriptions, which alerted the prescriber and pharmacist to dosage errors and reduced misinterpretation of prescriptions (Jayawardena et al. 2007). Åstrand et al. (2009) surveyed 31,225 prescriptions by comparing the proportions of ePrescription and non-electronic prescriptions with the prescriber at the time of dispensing and found that the ePrescription method would enhance safety and quality for the patient, especially in improving efficiency and cost-effectiveness within the healthcare system.

3 Research Methodology

3.1 Hospital Selection

This study was conducted to examine the timelines of various processes involved in administering the medication process at two public hospitals in Louisiana. In 2012 Hospital A had approximately 450 licensed inpatient beds and 430,000 outpatient visits, and Hospital B had approximately 250 licensed inpatient beds and 141,000 outpatient visits. The pharmacies at both hospitals were chosen to be involved in this study, as both had different prescribing technologies. The pharmacy at Hospital A used NCR copies which were used for prescribing, and Hospital B used digital scanning technology.

3.2 Data Collection

Data was collected over a two week period from both hospital pharmacies to determine the time elapsed from when the orders were sent to the pharmacy from the nursing station to the time the medication was sent back to the nursing station from the pharmacy.

An investigator can achieve high levels of precision by increasing the sample size. Should there be constraints, because of budgetary limitations or a shortage of time, some proper sample size has to be determined. Equation (1) was used to calculate the required sample size "n" for estimating the mean. For this data, we set d to 2 s; σ was estimated to be 9.4 by the sample standard deviation; and the value of critical deviate with a 0.001 α error was 2.326. After calculating, the sample size needed to achieve this accuracy level was 107. Hence, a sample size of 110 was selected.

$$n = \frac{t_{2/\alpha}\sigma^2}{d^2} \qquad (1)$$

where
- d Desired precision (or maximum error)
- $t_{\alpha/2}$ Critical deviate for specified reliability $1-\alpha$
- σ Population standard deviation

3.3 Drug Distribution Systems Studies

The two methods of drug distribution were NCR technology and digital scanning technology in the selected pharmacy respectively. The detailed information for both methods is described in the next two sections.

3.3.1 No Carbon Required Technology

At Hospital A, the physician used NCR technology. The NCR technology is where medication orders are written on a "no carbon required" physician order form. The NCR copies have four parts altogether. The first part is the chart copy, on which the physician writes and puts on a patient's chart. The remaining three copies are the pharmacy copies on which the order is imprinted. Those copies are sent to the pharmacy through a "tubing system" by the unit secretary or nurse for the pharmacists to fill in the prescriptions.

3.3.2 Digital Scanning Technology

At Hospital B, the physician used digital scanning technology. Computerized digital scanning of original physician orders eliminates the distortion that is experienced with NCR copies. It allows the pharmacist at order entry to enlarge the image, turn the image, and change the contrast of the image, all which are used to improve readability. The important benefit of the digital scanning technology is that, it reduces turn-around time in receiving the order and it documents electronically the time is was scanned, and the time it was processed. Digital scanning technology provides an instant prioritizing of orders, placing STAT (a medical abbreviation for urgent or rush) orders at the top of the computer queue in red in order to alert the pharmacist.

3.4 Parameter Measured

To compare the two technologies, the following parameters were studied: (1) Queue time refers to the time spent when the prescription waits in the queue until it is attended or viewed by the pharmacist at the order entry station. In case of NCR technology the prescription after time stamped is placed in the queue. For digital

Table 1 The ANOVA results of each observed time

Methods		Queue time	Order entry time	Outgoing delay time	Outgoing transit time
NCR copies	Mean	654.70	224.76	38.23	129.43
(N = 110)	Variance	416,300.20	36,621.70	3877.66	6228.03
Digital scanning	Mean	297.81	151.95	8.83	41.77
(N = 110)	Variance	70,475.93	113,018	265.36	87.65
Percentage of time reduction		54.50 %	32.40 %	76.90 %	67.72 %
F value		28.78	3.90	22.95	133.82
P value		0.000***	0.049*	0.000***	0.000***

Note All tests are two-tailed. $*p < 0.05$, $**p < 0.01$, $***p < 0.001$

scanning technology, the prescription starts in the queue as soon as the prescription is scanned; (2) Order entry time refers to the time spent when the pharmacist at the order entry station starts and ends entering the prescription into the pharmacy information system. In NCR technology, this time starts when the pharmacist picks the prescription and ends after entering the prescription. For digital scanning technology, the timing starts as soon as the prescription image is viewed on the screen and ends when order entry is completed; (3) Outgoing delay time refers to the time spent when the tube enters the system but before it is sent from the pharmacy to the nursing station; (4) Oncoming transit time measures the time a tube spent in traveling from the pharmacy to the nursing station.

4 Research Results

The statistical analysis results (See Table 1) show that the digital scanning technology outperformed the NCR technology in all measures. Results indicated a reduction of 54.5 % in queue time, 32.4 % in order entry time, 76.9 % in outgoing delay time, and 67.7 % in outgoing transit time in digital scanning technology. Additionally, the results of ANOVA found significant differences in times between the NCR technology and the digital scanning technology for the prescription.

The comparison of cost analysis between NCR technology and digital scanning technology was conducted at Hospital A for a five-year period. We selected four high cost items that includes pharmacist time spent, prescription papers, the introducing cost of MOMS that utilizes the digital scanning technology, and paper shredding costs. The pharmacist time spent, denoted as T, includes time a pharmacist spent in (1) handling the tube and (2) filling, labeling, and stamping the prescription orders. Equation (2) shows the calculation of T (in hours).

$$T = [AX + (B + C)Y]/3,600 \qquad (2)$$

Table 2 The comparison of cost analysis between two technologies (five-year period)

	NCR copies	MOMS (digital scanning technologies)
Pharmacist cost	US$589,027.93 (US$54.09* 5.967 h * 365 days * 5 years)	US$0.00 (no tube to handle and all orders are scanned)
Prescription paper cost	Inpatient: US$1,800,000.00 (US$1.00/sheet*30,000 sheets/month * 60 months)	Inpatient: US$396,000.00 (US$0.22/sheet*30,000 sheets/month * 60 months)
	Outpatient: US$541,200.00 (US$0.22/sheet*41,000 sheets/month * 60 months)	Outpatient: US$541,200.00 (US$0.22/sheet*41,000 sheets/month * 60 months)
MOMS cost	US$0.00 (No new server and stations)	US$54,480.00 (for 2 servers) US$104,040.00 (for 7 stations)
Shredding cost	US$6,334.20 (US$0.102/lb*1035 lb/month*60 months)	US$0 (No prescription sheets to shred)
Total cost	US$2,936,562.13	US$1,095,720.00
Monthly cost	US$48,942.70	US$18,262.00

where
X Average number of incoming tubes per day
Y Average number of order forms per day
A Time to handle tube (in seconds)
B Time to punch each order (in seconds)
C Time to fill each order (in seconds)

During the study period, it was found that X = 500, Y = 970, A = 10 s, B = 2 s, and C = 15 s. These standard time data came from analyzing the tube time spent plus filling time spent using occurrence sampling and the stopwatch methods. These results were determined after assuming 15 % allowances. Table 2 shows the cost analysis results. Note that the total inpatient-carbon prescription order handling time was 5 h and 58 min (5.967 h) for a pharmacist that cost $54.09 per hour at both hospitals.

5 Conclusion

This study shows the use of digital scanning technology in a multi-hospital health system reduces medication dispensing time during the drug distribution process. The cost analysis also justifies the implementation of MOMS. The time and cost savings can be viewed as an opportunity to reduce delays in delivering medications, reduction of missing doses and improving service to the patients.

References

Allan EL, Barker KN (1990) Fundamentals of medication error research. Am J Hosp Pharm 47:555–571

American Society of Hospital Pharmacists (1989) Statement on unit dose drug distribution. Am J Hosp Pharm 46:2346

Ammenwerth E, Schnell-Inderst P, Machan C, Siebert U (2008) The effect of electronic prescribing on medication errors and adverse drug events: a systematic review. J Am Med Inform Assoc 15:585–600

Åstrand B, Montelius E, Petersson G, Ekedahl A (2009) Assessment of ePrescription quality: an observational study at three mail-order pharmacies. BMC Med Inform Decis 9:8

Benrimoj SI, Thornton PD, Langford JH (1995) A review of drug distribution systems: part 1- current practice. Aust J Hosp Pharm 25:119–126

Berwick DM, Winickoff DE (1996) The truth about doctors' handwriting: a prospective study. British Med J 313:1657–1658

Chuang MH, Wang YF, Chen M, Cham TM (2012) Effectiveness of implementation of a new drug storage label and error-reducing process on the accuracy of drug dispensing. J Med Syst 36:1469–1474

Cohen MR (2007) Medication errors, 2nd edn. American Pharmacists Association, Washington

Flynn EA, Barker KN, Carnahan BJ (2003) National observational study of prescription dispensing accuracy and safety in 50 pharmacies. J Am Pharm Assoc 43:191–200

Institute for Safe Medication Practices (2000) A call to action: eliminate handwritten prescriptions within 3 years. http://www.ismp.org/newsletters/actecare/articles/whitepaper.asp. Accessed 12 May 2013

Jayawardena S, Eisdorfer J, Indulkar S, Pal SA et al (2007) Prescription errors and the impact of computerized prescription order entry system in a community-based hospital. Am J Ther 14:336–340

Jenkins J, Eckel SF (2012) Analyzing methods for improved management of workflow in an outpatient pharmacy setting. Am J Health-Syst Pharm 69:966–971

Jimenez Munoz AB, Muino Miguez A, Rodriguez Perez MP, Duran Garcia ME, Sanjurjo Saez M (2011) Comparison of medication error rates and clinical effects in three medication prescription-dispensation systems. Int J Health Care Qual Assur 24:238–248

Keehan SP, Cuckler GA, Sisko AM, Madison AJ et al (2012) National health expenditure projections: modest annual growth until coverage expands and economic growth accelerates. Health Aff 31:1600–1612

Kohn LT, Corrigan JM, Donaldson MS (eds) (1999) To err is human: building a safer health system. National Academy Press, Washington

National Community Pharmacists Association (NCPA) (2012) 2012 NCPA digest in-brief. NCPA, Virginia

Samliyan TA, Duval S, Du J, Kane RL (2008) Just what the doctor ordered. Review of the evidence of the impact of computerized physician order entry system on medication errors. Health Serv Res 43:32–53

Shawahna R, Rahman Nu, Ahmad M, Debray M, Yliperttula M, Decleves X (2011) Electronic prescribing reduces prescribing error in public hospitals. J Clin Nurs 20:3233–3245

A Pareto-Based Differential Evolution Algorithm for Multi-Objective Job Shop Scheduling Problems

Warisa Wisittipanich and Voratas Kachitvichyanukul

Abstract This paper presents a multi-objective differential evolution algorithm (MODE) and its application for solving multi-objective job shop scheduling problems. Five mutation strategies with different search behaviors proposed in the MODE are used to search for the Pareto front. The performances of the MODE are evaluated on a set of benchmark problems and the numerical experiments show that the MODE is a highly competitive approach which is capable of providing a set of diverse and high-quality non-dominated solutions compared to those obtained from existing algorithms.

Keywords Multi-objective · Differential evolution · Job shop scheduling problems · Evolutionary algorithm

1 Introduction

Multi-objective optimization (MO) toward Pareto-based methods has increasingly captured interests from both practitioners and researchers due to its reflection to most real-world problems containing multiple conflicting objectives. Compared to an aggregated weighted approach, Pareto-based approaches offer an advantage for decision makers to simultaneously find the trade-offs or non-dominated solutions on the Pareto front in a single run without prejudice. Evolutionary Algorithms

W. Wisittipanich (✉)
Industrial Engineering Dept, Chiang Mai University, 239 Huay Kaew Road,
Muang District, Chiang Mai 50200, Thailand
e-mail: warisa@eng.cmu.ac.th

V. Kachitvichyanukul
Industrial and Manufacturing Engineering Dept, Asian Institute of Technology,
P.O. Box 4, Klong Luang, Pathumthani 12120, Thailand
e-mail: voratas@ait.ac.th

(EAs) is the most commonly selected solution techniques to search for the Pareto front due to their high efficiency to find good solution quality within reasonable time.

Differential Evolution (DE) algorithm, proposed by Storn and Price in (1995), is one of the latest EAs which has been widely applied and shown its strengths in many application areas. Nevertheless, only few research works have attempted to apply DE to find solutions for MO problems. Abbass et al. (2001) presented a Pareto-frontier DE approach (PDE) where only non-dominated solutions are allowed to participate in generating new solutions. However, if only few non-dominated solutions are found in the beginning, the chances of discovering new better solutions may be limited, and the solutions may get trapped at local optima. Madavan (2002) introduced the combination of the newly generated population and the existing parent population which results in double population size. The non-dominated solutions are then selected from this combination to allow the global check among both parents and offspring solutions; however, the drawback is the requirement of additional computing time in the sorting procedure of the combined population.

Although Multi-objective job shop scheduling problem (MOJSP) has been subjected to many researchers, the numbers of research works on MOJSP are still limited. Lei and Wu (2006) developed a crowding measure-based multi-objective evolutionary algorithm (CMOEA) for JSP. The CMOEA uses crowding measure to adjust the external population and assign different fitness for individuals to simultaneously minimize makespan and the total tardiness of jobs. Ripon et al. (2006) presented a jumping genes genetic algorithm (JGGA) for MOJSP. The jumping operations in JGGA exploit scheduling solutions around the chromosomes whereas the general genetic operators globally explore solution from the population. The results showed that the JGGA is capable of searching a set of well diverse solutions near the Pareto-optimal front while the consistency and convergence of non-dominated solutions are well maintained. Chiang and Fu (2006) proposed a genetic algorithm with cyclic fitness assignment (CFGA) which effectively combines three existing fitness assignment mechanisms: MOGA, SPEA2 and NSGA-II to obtain better performance and avoid rapid loss of population diversity. Lei and Xiong (2007) presented an evolution algorithm for multi-objective stochastic JSP in which archive maintenance and fitness assignment were performed based on crowding distance. Lei (2008) designed a Pareto archive particle swarm optimization (PAPSO) by combining the selection of global best position with the crowding measure-based archive maintenance to minimize makespan and total tardiness of jobs. The PAPSO is capable of producing a number of high-quality Pareto optimal scheduling plans.

This paper is inspired by the efficiency of the MODE algorithm proposed in Wisittipanich and Kachitvichyanukul (2012) for solving many continuous optimization problems. In this paper, the MODE is extended to solve MOJSP. The performances of the MODE are evaluated on a set of benchmark JSP problems and compared with existing solution methods.

2 MODE Algorithm

2.1 MODE Framework

Similar to the classic DE algorithm, MODE starts with randomly generated initial population of size N of D-dimensional vectors and evaluate objective value of each solution vector. However, in MO problems, there is no single solution but rather a set of non-dominated solutions. Therefore, the selection based on only one single solution in the classic DE algorithm may not be applicable in MO environment. The MODE framework is illustrated in Fig. 1.

The MODE adopts the concept of Elitist structure in NSGA-II (Deb et al. 2002) to store a set of non-dominated solutions in an external archive, called Elite group. In MODE, as the search progress, instead of applying the sorting procedure to every single move of a vector, the sorting is only performed on the set of newly generated trial vectors after all vectors completed one move to identify the group

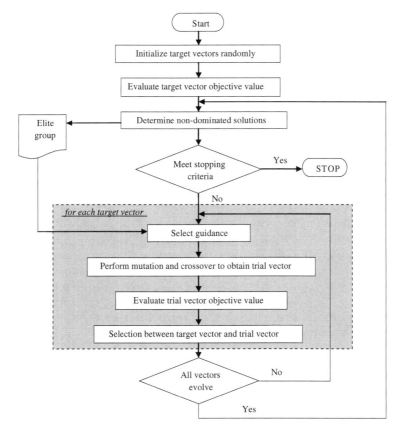

Fig. 1 MODE framework

of new non-dominated solutions. The reason is to reduce computational time. This sorting procedure applies to the group of new solutions and current solutions in the external archive and store only non-dominated solutions into an archive for the Elite group. Then, Elite group screens its solutions to eliminate inferior solutions. As a result, the Elite group in the archive contains only the best non-dominated solutions found so far in the searching process of the MODE population.

2.2 Mutation Strategies

One of the critical decisions in MO problems is the selection of candidates among the Elite group as guidance toward the Pareto frontier. In the MODE, five mutation strategies with distinct search behavior are proposed as the movement guidance of vectors in order to obtain the high quality Pareto front.

- MODE-ms1: Search around solutions located in the less crowded areas
- MODE-ms2: Pull the current front toward the true front
- MODE-ms3: Fill the gaps of non-dominated front
- MODE-ms4: Search toward the border of the non-dominated front
- MODE-ms5: Explore solution space with multiple groups of vectors

Mutation strategies MODE-Ms1 and MODE-Ms3 aim to improve the distribution of solutions on the front. Mutation strategy MODE-Ms2 intends to pull to the current front toward the true Pareto front, and mutation strategy MODE-Ms4 try to explore more solutions around the border to increase the spread of the non-dominated front. As mentioned, each mutation strategy exhibits different search behavior with its own strengths and weaknesses. Thus, mutation strategy MODE-Ms5 aims to extract the strengths of various DE mutation strategies and to compensate for the weaknesses of each individual strategy in order to enhance the overall performance by combining four groups of vectors with multiple mutation strategies in one population. For more details on each strategy, see Wisittipanich and Kachitvichyanukul (2012).

3 Multi-objective Job Shop Scheduling Problem

The classical JSP schedules a set of n jobs on m different machines subjected to the constraints that each job has a set of sequential operations and each operation must be processed on a specified machine with deterministic processing time known in advance. In this paper, the objective is to simultaneously minimize two objectives; makespan and total tardiness.

1. Makespan: makespan = max $[C_j]$ where C_j is the completion time of job j
2. Total tardiness of jobs: Total tardiness = $\sum_{j=1}^{n} \max[0, L_j]$ where L_j is the lateness of job j

In this study, solution mapping procedures presented in the paper work of Wisittipanich and Kachitvichyanukul (2010) is implemented. Random key representation encoding scheme (Bean 1994) and permutation of job-repetition (Bierwirth 1995) are adopted to generate the sequence of all operations. Then, the operation-based approach (Cheng et al. 1996) is used as decoding method to generate an active schedule. For more details on the encoding and decoding schemes, see Wisittipanich and Kachitvichyanukul (2010).

4 Computational Experiments

The performances of the MODE algorithm are evaluated using 9 benchmark problems taken from JSP OR-Library. To determine the total tardiness of a schedule, this study adopts the due date setting according to Lei and Wu (2006). The release date of all jobs is set to be 0.

In this study, the population size and number of iteration are set to 500 and 200 respectively. The scale factor value F is set to random value to retain population diversity. Crossover rate (C_r) is linearly increased from 0.1 to 0.5 to maintain the characteristic of generated trial vectors at the beginning of the search and yield more deviations for the generated trial vectors as the search progresses. The maximum members in Elite archive are set as 100. If the number of new non-dominated solutions found by the population exceeds the limit of this archive, the solutions with lower crowding distance will be removed.

4.1 Experimental Results

The performances of the MODE algorithm are compared to SPEA (Zitzler and Thiele 1999) and CMOEA (Lei and Wu 2006). Table 1 shows some of the best solutions obtained from the experiment. The notation (x, y) represents (makespan, total tardiness). It is noted that, for each instance, the non-dominated solutions are determined from 10 independent runs.

As shown in Table 1, SPEA, CMOEA, MODE-ms1, MODE-ms2, MODE-ms3, MODE-ms4, and MODE-ms5 perform well in generating a set of non-dominated solutions. The makespan values are just about 3–20 % bigger than the optimal makespan, and some makespan values are very close to the optimal results; while the total tardiness of these solutions are reasonable. In most cases, the minimum values of makespan and total tardiness of jobs found by MODE-ms1, MODE-ms-2, MODE-ms3, and MODE-ms5 are lower than those obtained from SPEA and

Table 1 Comparison results on benchmark problems

Instance	SPEA	CMOEA	MODE-ms1	MODE-ms2	MODE-ms3	MODE-ms4	MODE-ms5
ABZ5	(1306,439)	(1277,422)	(1249,367)	(1255,403.5)	(1238,408)	(1265,393)	(1251,529)
	(1316,486)	(1296,360)	(1254,362.5)	(1259,318)	(1264,434)	(1290,387)	(1256,467)
			(1266,273)	(1261,270)	(1270,398)	(1295,385)	(1258,297)
			(1275,234)	(1272,90)	(1274,393)	(1296,377)	(1261,295)
			(1281,232)	(1298,25)	(1275,355)	(1319,276)	(1269,229)
			(1282,164)	(1320,0)	(1276,348)	(1340,184)	(1280,228)
ABZ7	(790,487)	(786,350)	(718,583)	(732,712)	(739,991)	(749,1133)	(724,777)
	(783,396)	(789,465)	(722,499)	(734,630)	(740,845)	(761,1060)	(725,663)
	(794,354)	(792,353)	(723,461)	(736,609)	(748,751)	(766,1021)	(728,571)
			(724,400)	(738,597)	(753,719)	(774,890)	(730,541)
			(735,395)	(740,572)	(760,664)	(788,700)	(732,337)
ABZ8	(805,707.5)	(817,293)	(739,610.5)	(754,932)	(769,1011.5)	(790,1114.5)	(750,929.5)
	(808,624.5)	(819,552.5)	(743,495.5)	(760,905.5)	(771,981.5)	(798,929.5)	(751,600.5)
	(810,582.5)	(824,251)	(752,464.5)	(761,824.5)	(777,861.5)	(811,877.5)	(753,578.5)
			(778,452.5)	(762,759.5)	(778,833.5)	(821,859.5)	(758,521)
			(799,435)	(768,664.5)	(788,794.5)	(827,829.5)	(759,516.5)
ORB1	(1160,718)	(1160,858)	(1108,481.5)	(1132,648.5)	(1112,668)	(1187,832)	(1113,755.5)
	(1193,413)	(1161,393.5)	(1126,452.5)	(1148,384.5)	(1118,529)	(1197,791)	(1119,657)
	(1191,469)	(1188,381.5)	(1132,373.5)	(1151,204)	(1150,509)	(1198,743.5)	(1128,605.5)
			(1141,320)	(1183,190)	(1198,395)	(1203,722.5)	(1137,592)
			(1146,261.5)	(1252,170.5)	(1205,310.5)	(1243,674)	(1178,325.5)
ORB3	(1164,890)	(1167,367)	(1062,710)	(1102,844)	(1102,1176)	(1156,1181)	(1070,772)
	(1165,793)	(1173,364)	(1070,489)	(1108,716)	(1106,1104)	(1172,1081)	(1076,737)
	(1158,799)	(1191,596)	(1078,480)	(1121,693)	(1114,970)	(1197,864)	(1099,733)
			(1104,479)	(1122,578)	(1118,864)	(1203,709)	(1112,666)
			(1119,444)	(1125,559)	(1126,842)	(1283,670)	(1116,619)

(continued)

Table 1 (continued)

Instance	SPEA	CMOEA	MODE-ms1	MODE-ms2	MODE-ms3	MODE-ms4	MODE-ms5
ORB5	(1002,1)	(988,23)	(922,386)	(915,411)	(907,644)	(929,742)	(916,612)
	(1012,5)	(989,45)	(928,247)	(927,367)	(908,643)	(942,713)	(927,558)
		(994,18)	(939,198)	(934,328)	(911,597)	(955,709)	(929,367)
			(953,170)	(938,298)	(912,567)	(959,655)	(932,241)
			(962,146)	(947,203)	(922,551)	(972,583)	(955,143)
LA26	(1405,4436)	(1366,3539)	(1272,3814)	(1296,3614)	(1327,4804)	(1375,5003)	(1304,4045)
	(1428,4346)	(1375,3537)	(1282,3427)	(1316,3377)	(1330,4571)	(1379,4973)	(1306,3073)
		(1394,4063)	(1304,3344)	(1325,3347)	(1341,4443)		
			(1306,3286)	(1341,3205)	(1344,3888)		
			(1320,3098)	(1348,3132)	(1384,3833)		
LA27	(1451,2960.5)	(1451,2968.5)	(1347,4487.5)	(1332,3116)	(1362,4258.5)	(1421,4766.5)	(1322,3850.5)
	(1452,2512.5)	(1452,2611.5)	(1350,3395)	(1372,3115.5)	(1366,3806.5)	(1441,3978.5)	(1336,3649.5)
			(1353,3340)		(1416,3752.5)		(1337,3614.5)
			(1367,3339.5)				(1370,3571.5)
LA28	(1400,3111)	(1398,3553)	(1316,3766)	(1321,3489)	(1338,4455)	(1367,4586)	(1303,3775)
	(1414,2926)	(1410,3339)	(1318,3520.5)	(1324,3482)	(1342,4247)	(1417,4508.5)	(1326,3539)
			(1319,3431)	(1343,2600)	(1350,3489)	(1433,4116)	(1341,3445)
			(1348,2422)	(1451,2543.5)	(1356,3134)	(1477,3964)	(1344,2888)
			(1368,2365)			(1541,3955)	

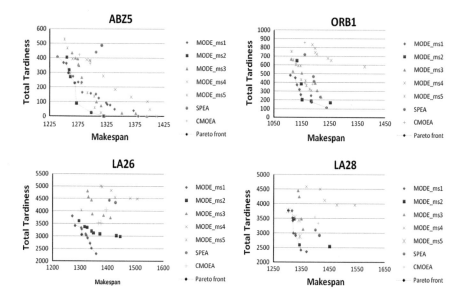

Fig. 2 Comparison of non-dominated solutions

CMOEA. In addition, MODE-ms1, MODE-ms-2, MODE-ms3, and MODE-ms5 clearly outperform SPEA and CMOEA since the majority of the solutions obtained by SPEA and CMOEA are dominated. However, the performance of MODE-ms4 is generally inferior to other MODE strategies, SPEA, and CMOEA since the solutions obtained from MODE-ms4 is mostly dominated by others. Figure 2 illustrates the comparison of non-dominated solutions generated from SPEA, CMOEA and all MODE strategies. It is important to note that the Pareto Front is obtained from the best non-dominated solutions found by all algorithms.

It can be easily seen from Fig. 2 that the non-dominated solutions obtained from SPEA and CMOEA are dominated by those from MODE-ms1, MODE-ms2, MODE-ms3, and MODE-ms5. Most of solutions on the Pareto front are from MODE-ms1, MODE-ms2, MODE-ms3, and MODE-ms5. MODE-ms4 again shows its poor performances since the solutions are very far from the Pareto front and mostly dominated by other algorithms.

5 Conclusions

This study presents an implementation of the MODE algorithm for solving multi-objective job shop scheduling problems with the objective to simultaneously minimize makespan and total tardiness of jobs. The MODE framework uses Elite group to store solutions and utilizes those solutions as the guidance of the vectors. Five mutation strategies with different search behaviors are used in order to search for the Pareto front. The performances of MODE are evaluated on a set of benchmark JSP problems and compared with results from SPEA and CMOEA. The results demonstrate that the MODE algorithm is a competitive approach and capable of finding a set of diverse and high quality non-dominated solutions on Pareto front.

References

Abbass HA, Sarker R, Newton C (2001) PDE: a Pareto-frontier differential evolution approach for multi-objective optimization problems. In: Proceedings of the 2001 congress on evolutionary computation, vol 2, pp 971–978
Bean JC (1994) Genetic algorithms and random keys for sequencing and optimization. ORSA J Comput 6(2):154–160
Bierwirth C (1995) A generalized permutation approach to job shop scheduling with genetic algorithms. OR Spectrum 17(2–3):87–92
Cheng R, Gen M, Tsujimura Y (1996) A tutorial survey of job-shop scheduling problems using genetic algorithms: I. representation. Com Ind Eng 30(4):983–997
Chiang TC, Fu LC (2006) Multiobjective job shop scheduling using genetic algorithm with cyclic fitness assignment. In: Proceedings of 2006 IEEE congress on evolutionary computation. Vancouver, Canada. 16–21 July 2006, pp 3266–3273
Deb K, Pratap A, Agarwal S, Meyarivan T (2002) A fast and elitist multiobjective genetic algorithm: NSGA-ii. IEEE Trans Evol Comput 6(2):182–197
Lei D (2008) A Pareto archive particle swarm optimization for multi-objective job shop scheduling. Com Ind Eng 54(4):960–971
Lei D, Wu Z (2006) Crowding–measure-based multiobjective evolutionary algorithm for job shop scheduling. Int J Adv Manuf Tech 30(1–2):112–117
Lei DM, Xiong HJ (2007) An efficient evolutionary algorithm for multi-objective stochastic job shop scheduling. In: Proceedings of the 6th international conference on machine learning and, cybernetics. 19–22 August 2007, pp 867–872
Madavan NK (2002) Multiobjective optimization using a Pareto differential evolution approach. In: Proceedings of the, evolutionary computation. 12–17 May 2002, pp 1145–1150
Ripon KSN, Tsang CH, Kwong S (2006) Multi-objective evolutionary job-shop scheduling using jumping genes genetic algorithm. In: Proceeding of international joint conference on neural networks. 16–21 July 2006, pp 3100–3107
Storn R, Price K (1995) Differential evolution: a simple and efficient adaptive scheme for global optimization over continuous spaces. Technical Report TR-95-012, Berkeley, CA: International Computer Science
Wisittipanich W, Kachitvichyanukul V (2010) Two enhanced differential evolution algorithms for job shop scheduling problems. Int J Prod Res 50(10):2757–2773
Wisittipanich W, Kachitvichyanukul V (2012) Mutation strategies toward Pareto front for multi-objective differential evolution algorithm. Int J Oper Res (article in press)
Zitzler E, Thiele L (1999) Multiobjective evolutionary algorithms: a comparative case study and the strength Pareto approach. IEEE Trans Evol Comput 3(4):257–271

Smart Grid and Emergency Power Supply on Systems with Renewable Energy and Batteries: An Recovery Planning for EAST JAPAN Disaster Area

Takuya Taguchi and Kenji Tanaka

Abstract This paper describes the design method of smart grid energy systems based on the simulation for introducing renewable energy (RE) and secondary batteries. The emergency power is also available. Managing RE systems and battery storage can minimize the cost of electricity by optimization balance between supply and demand as well as can reduce environmental impacts. Also the systems enable us to use emergency power supply for accidents and disasters, and to enhance robustness of electric power systems. To actualize the electricity management systems, the object-oriented and time-marching energy management simulator is developed which simulates power production, transmission and distribution, charge and discharge, and consumption of electricity every half hour. By using that simulator, peak and amount of power supply can be decreased. As a case study, we applied the systems with photovoltaic power (PV) generation and secondary batteries to reconstructed elementary schools that were damaged by Tohoku earthquake and tsunami on March 11, 2011, and consequently the validity of the analysis was obtained. Furthermore, it was verified the case that the systems were applied to business, which involves more profits than public facilities, and we examined requirements for business.

Keywords Smart grid · Operations management · Renewable energy · Battery · Earthquake disaster revival

---

T. Taguchi (✉)
Department of System Innovation, University of Tokyo, Tokyo, Japan
e-mail: taguchi@triton.naoe.t.u-tokyo.ac.jp

K. Tanaka
Graduate School of Engineering, University of Tokyo, Tokyo, Japan
e-mail: kenji_tanaka@sys.t.u-tokyo.ac.jp

1 Introduction

Nowadays, the expectancy for Renewable Energy (RE) is increased. Many countries implemented Feed-in tariff policies in order to encourage investments in RE, and consequently that led to significant amounts of RE generation in the world. However, Excess of applications for RE is likely to lower efficiency and robustness in grid because the output power of RE fluctuate much. Therefore, storage battery systems have been used to manage the power fluctuation. Nonetheless, optimal sizes of RE and storage batteries depend on power demand of consumers and climate in that area. Secondary batteries have also used for emergency power supply. It is required to design the size of RE appliances and secondary battery systems and emergency management. The goal of this paper is to introduce power management systems, which reduce power supply from grid and supply power in the case of blackout, with feasible capacities of PV and batteries in reconstructed schools in EAST JAPAN disaster area.

2 Literature Review

Kanchev et al. (2011) studies energy management systems in RE generation and storage batteries and discuss algorithms of charge/discharge of batteries. However, they did not describe technique about the optimization of sizes of RE and storage batteries. Shibata (2011) proposed how to optimize those systems on the basis of power demand and climate data in the past. Nevertheless, little is known about emergency power supply in his paper.

Considering these problems from the studies above, the purpose of this study is to design and optimize the sizes of RE and batteries in the power, economical, and emergency assessments. In addition, as a case study, we applied the systems with RE and secondary batteries to reconstructed elementary schools that were damaged by Tohoku earthquake and tsunami on March 11, 2011, and consequently the validity of the analysis was examined.

3 Model

3.1 Overview of the Model

In order to design optimal sizes of RE (or PV) and battery, we have developed a time-marching energy management simulator that simulates production, transmission and distribution, charge and discharge, and consumption of electricity every half hour. In authors' simulation system, changes of power demand and RE along temporal axes, the amounts of RE and battery are set as input data. And then

Smart Grid and Emergency Power

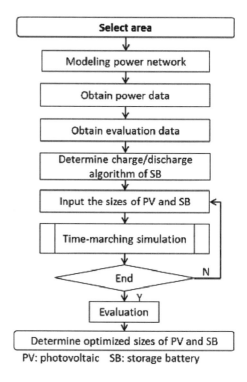

Fig. 1 Flow of the simulation

the effect of each scenario, in which different sizes of RE and battery are introduced, is obtained as output data by conducting time-marching simulation.

The steps of the model which design optimal sizes of RE and battery are shown in Fig. 1. First, the target area of the simulation is selected. Second, power network model is developed. Third, power demand data and RE data are obtained. Demand data and RE data are expressed in the same format. In this paper, each power data is for 365 days, 48 steps in a day. Fourth, data for evaluation such as unit price of RE, battery is obtained. Fifth, the algorithm that operates storage battery, which manages when and how much electric energy is charged or discharged, is selected. The sixth flow, combinations of the sizes of RE and battery are read into the time-marching energy management simulator. After the results of all the combinations are calculated, the calculation outputs are evaluated based on evaluation criteria. Finally, optimal sizes of RE and battery are designed.

3.2 Algorithm of Charge/Discharge of Storage Battery

Planning charge/discharge storage on the basis of power demand and generated power by RE needs algorithm of charge/discharge of storage battery. We should develop and choose optimal algorithm of them according to the purpose of that

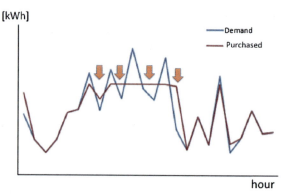

Fig. 2 The conceptual graph of peak-cut algorithm

project. On this paper, we developed "Peak-Cut Algorithm", which goal is to reduce maximum of purchased power supply in simulation period. The main reason why we use this algorithm is to minimize electricity charge cost. Figure 2 is conceptual graph of Peak-Cut Algorithm.

3.3 Output Data and Assessment Criteria

Each scenario is evaluated for the following three assessments; Power assessment, Economical assessment, and Emergency assessment (Table 1). Power assessment is to evaluate the amount and maximum of electricity. "Peak-cut" means how much maximum of power is reduced. Economical assessment is the evaluation concerned about profit. Emergency assessment is to evaluate the effect of disaster measure in each scenario.

3.3.1 Emergency Assessment

Design for emergency power supply in this paper means that the sizes of RE generation and storage batteries are optimized for the purpose of supplying electricity in case of disaster. We used reduction power demand which makes "Zero emission time ratio" 100 % as emergency assessment criteria. "Zero emission time ratio" is defined the time proportion when that area can be independent of the grid and supply power only by the systems with RE and battery. The calculate formula is shown in (1).

$$P = T(d)/T(0) \tag{1}$$

P　　Zero emission time ratios

Table 1 Assessment criteria

Power assessment	Energy sufficiency
	Peak-cut
	Power loss
Economical assessment	Reduction of electricity charge
	Payout period
Emergency assessment	The ratio of zero emission hours

T(d)   The number of day when power is not purchased from the grid through out the day, provided "d" % of power demand is reduced

4 Case Study

4.1 Simulation Setup

We applied that model and simulation to the elementary school in Ofunato city, Iwate prefecture, which was damaged by Tohoku earthquake and tsunami on March 11, 2011. Besides, PV power was used as RE generator in this case study. The amount of power demand in this school is 119 MWh and the power cost is 2,250,000 JPY (about 22,500 USD) in fiscal 2010. Power demand data was made from two factors, monthly power demand data of the school from April 2010 to March 2011, and daily fluctuation data of faculty of economics of the university of Tokyo. PV power data was made from solar radiation and temperature in Ofunato city in Iwate prefecture. Figure 3 displays daily power demand and PV power data.

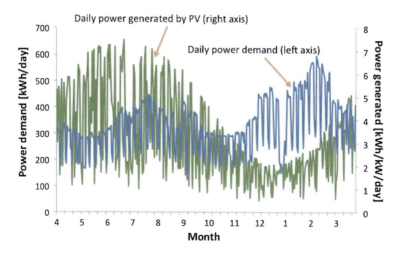

Fig. 3 Power demand and PV supply trend throughout a year

Table 2 PV and battery capacity in each scenario

	PV (kW)	Storage battery (kWh)
Scenario 1	15	50
Scenario 2	20	50
Scenario 3	30	100

Table 3 Economical criteria

Unit price of PV		300,000 (JPY/kW)
Unit price of lithium battery		100,000 (JPY/kWh)
Basic charge of power from the grid		1,348 (JPY/kW)
Commodity charge of power from the grid	Daytime	14 (JPY/kWh)
	Nighttime	9 (JPY/kWh)

We demonstrated three scenarios shown at Table 2. We evaluated these scenarios at three assessment criteria indicated at Table 1. Economical criteria such as the unit price of PV and battery are shown at Table 3. Also, service life of the systems of PV and batteries is given as 15 years. In addition, state of charge (SOC) of battery is set as 20–90 % of its capacity.

4.2 Simulation Results

4.2.1 Power Assessment

Table 4 shows the outputs of demonstration concerned with electricity power in each scenario. Peak of purchased power for a year is reduced to 29.4 % in scenario 3. Although, the size of PV in scenario 2 (15 kW) is larger than that of scenario 1 (20 kW), peak-cut ratio in both scenarios have little differences. It means that the effect of peak-cut depends more upon the PV size than the storage size. Besides, energy sufficiency in any scenario increases roughly in proportion of PV size because power loss is small compared with PV generated power.

Figures 4 and 5 display power indexes in winter and spring about scenario 3 (PV: 30 kW, Battery: 100 kWh). Power generated by PV, purchased power, power demand and power loss are shown by left axis. State of Charge (SOC) of secondary battery is shown by right axis. In winter, when power demand is high but power generated is low, purchased power is saved to about 60–70 %. Nevertheless, the

Table 4 Results of power assessment

Criteria	Scenario 1 (%)	Scenario 2 (%)	Scenario 3 (%)
Energy sufficiency	14.6	19.3	28.8
Peak-cut	16.7	17.8	29.4
Power loss compared with PV generated	0.1	0.7	1.2

Smart Grid and Emergency Power

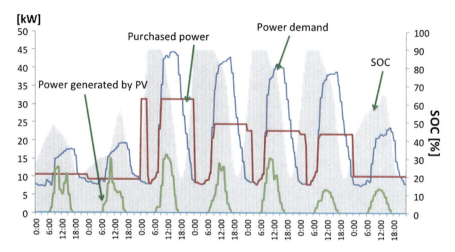

Fig. 4 Power indexes in the week of winter (2.10–2.16)

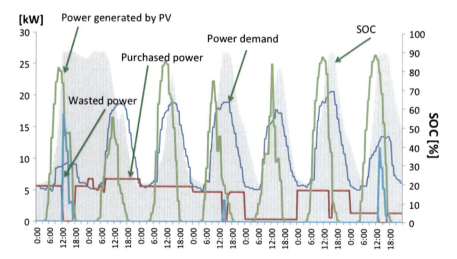

Fig. 5 Power indexes in the week of spring (5.18–5.24)

size of PV and battery is too small to smooth purchased power line. On the other hand, in spring, when power demand is small but power generated is high, peak at lunchtime is removed. In addition, power loss occurs on weekends in spring because power demand on weekends is lower than that on weekdays and SOC is high limited. However, the amount of power loss is negligibly small.

4.2.2 Economical Assessment

Table 5 shows the outputs of simulation concerned with economical evaluation. Similar to peak-cut ratio, reduction of basic charge in scenario 1 and scenario 2 have little differences because basic charge depends on maximum electricity power. Payout period was higher than service life of the systems with PV and battery in any scenario. Therefore, its introduction leads to loss in total. Public area such as school emphasizes facility for disaster and does not ask for profit, and accordingly is likely to introduce the system with PV and battery. By contrast, it is difficult for private company, which profit is important for, to introduce it in present economical conditions. We examined requirements for business by calculating IRR (Internal Rate of Return). Table 6 indicates IRR each PV unit price and service life when rising electricity to 120 % charge and using cheaper lead battery (20,000 JPY/kWh), which conditions are feasible, are presumed. Well-timed introduction that system for each company depends on how important

Table 5 Results of economical assessment

	Unit	Scenario 1	Scenario 2	Scenario 3
Reduction of electricity charge		17.7 %	21.4 %	32.5 %
Reduction of commodity charge		18.1 %	23.1 %	33.9 %
Reduction of basic charge		16.7 %	17.8 %	29.4 %
Implementation cost	JPY	9.5E+06	1.1E+07	1.9E+07
Annual income	JPY/year	4.0E+05	4.8E+05	7.3E+05
Payout period	year	23.9	22.8	25.9

Table 6 IRR based on each PV price and service life

		Service life (year)										
		15 (%)	16 (%)	17 (%)	18 (%)	19 (%)	20 (%)	21 (%)	22 (%)	23 (%)	24 (%)	25 (%)
The price of Pv modules (K JPY/ kW)	300	−2.7	−1.8	−1.1	−0.4	0.1	0.6	1.0	1.4	1.8	2.1	2.3
	290	−2.3	−1.4	−0.7	−0.1	0.5	0.9	1.4	1.7	2.1	2.4	2.6
	280	−1.9	−1.0	−0.3	0.3	0.8	1.3	1.7	2.1	2.4	2.7	2.9
	270	−1.5	−0.6	0.1	0.7	1.2	1.7	2.1	2.4	2.7	3.0	3.3
	260	−1.0	−0.2	0.5	1.1	1.6	2.0	24	2.8	3.1	3.4	3.6
	250	−0.6	0.2	0.9	1.5	2.0	24	2.8	3.2	3.5	3.7	4.0
	240	−0.1	0.7	1.3	1.9	2.4	2.8	3.2	3.5	3.8	4.1	4.3
	230	0.4	1.1	1.8	2.3	2.8	3.2	3.6	3.9	4.2	4.5	4.7
	220	0.9	1.6	2.3	2.8	3.3	3.7	4.0	4.4	4.6	4.9	5.1
	210	1.4	2.1	2.8	3.3	3.7	4.1	4.5	4.8	5.1	5.3	5.5
	200	2.0	2.7	3.3	3.8	4.2	4.6	4.9	5.2	5.5	5.7	5.9
	190	2.5	3.2	3.8	4.3	4.7	5.1	5.4	5.7	6.0	6.2	6.4
	180	3.1	3.8	4.4	4.9	5.3	5.6	5.9	6.2	6.5	6.7	6.8
	170	3.8	4.4	5.0	5.4	5.8	6.2	6.5	6.7	7.0	7.2	7.3
	160	4.4	5.1	5.6	6.0	6.4	6.8	7.1	7.3	7.5	7.7	7.9
	150	5.1	5.7	6.3	6.7	7.1	7.4	7.7	7.9	8.1	8.3	8.4

Fig. 6 Relationship between power saving and zero emission time

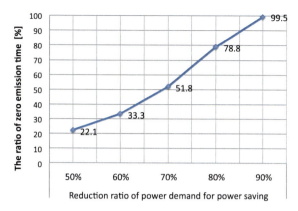

disaster measure is for the company. To take an example, small office, which is not used as refugee shelter in emergency situation, introduces that system with PV and battery in the condition that high IRR is expected.

4.2.3 Emergency Assessment

The graph in Fig. 6 illustrates the ratio of zero emission time for each reduction ratio of power demand at scenario 3. When power demand is saved 50 and 90 % of regular case, zero emission time reached 22.1 and 99.5 % respectively. If that area saves 90 % of electricity demand in the case of disaster, the facility accomplishes about 100 % self-sufficiency.

5 Conclusions

Conclusions of this paper describe below.

- We proposed the design of electricity management systems where sizes of RE and storage batteries is optimized by means of power demand and weather data in an area.
- As a case study, we design the electricity management system in a reconstructed elementary school, which were damaged by Tohoku earthquake and tsunami on March 11, 2011, and consequently the validity of the analysis was obtained.

Acknowledgments This work is partially supported by Social System Design Co. (http://www.socialsystemdesign.co.jp/) and its general manager Hideaki Miyata. We would like to express our deep appreciation for their kindness.

References

Kanchev H, Lu D, Colas F, Lazarow V, Francois B (2011) Energy management and operational planning of a microgrid with a PV-based active generator for smart grid applications. IEEE Ind Electron Mag 58(10):4583–4592

Shibata K (2011) Study on optimal design in smart house and power management with renewable energy system controlled by secondary batteries. Dissertation, University of Tokyo

Establishing Interaction Specifications for Online-to-Offline (O2O) Service Systems

Cheng-Jhe Lin, Tsai-Ting Lee, Chiuhsiang Lin, Yu-Chieh Huang and Jing-Ming Chiu

Abstract Information technology products such as smart phones, tablet PCs, eBooks and the intelligent-interactive digital signage have evolved and perfectly merged with the service systems to provide user-friendly user-interfaces and innovative service patterns. Online-to-offline (O2O) service model is one of the newest developments in the service systems where users in the physical world can interact with service providers in the cyberspace through various devices. Although traditional HCI studies have provided various research frameworks to describe interfaces and activities involved, there is a lack of interaction specifications which can clearly describe HCI in the realm of O2O service systems. This study developed a formal language that facilitates establishment of HCI specifications for O2O applications in proximity commerce based on interaction styles consisting of 4 interaction types represented in an interaction diagram. The formal language thus provides a common ground for service provider and service implementer to communicate and develop a concrete prototype effectively.

C.-J. Lin (✉) · T.-T. Lee · C. Lin
National Taiwan University of Science and Technology, Taipei, Taiwan, Republic of China
e-mail: Robert_cjlin@mail.ntust.edu.tw

T.-T. Lee
e-mail: charlenelee0912@gmail.com

C. Lin
e-mail: cjoelin@mail.ntust.edu.tw

Y.-C. Huang · J.-M. Chiu
Institute for Information Industry, The Innovative DigiTech-Enabled Applications and Services Institute, Taipei, Taiwan, Republic of China
e-mail: ychuang@iii.org.tw

J.-M. Chiu
e-mail: jmchiu@iii.org.tw

Keywords: O2O commerce · Human–computer interaction · Interaction specification

1 Introduction

In recent years, information technology products such as smart phones, tablet PCs, eBooks and the intelligent-interactive digital signage have evolved and perfectly merged with the service systems to provide user-friendly user-interfaces and innovative service patterns. A pattern is a solution to a problem in a context (Sharp et al. 2002). Different service patterns for a service system re therefore different ways to deliver desired user experience by interaction design. One intention of creating such service patterns are to communicate with each other in a design team based on the name of the pattern. Another intention is to produce a literature that documents user experience in a compelling form. A service pattern can be formed by a series of interactive steps, and the structure of those steps in this paper is called an "interaction style." The interaction styles consist of various types of interaction, which describe how a single step of interaction can be completed by a user with control or display devices. Since every service pattern is unique based on how steps of interaction are constructed (interaction styles) and how interaction is actually performed in each step (interaction types), the description of the service pattern can be used as a specification for human–computer interaction (HCI) to precisely describe the requirements and identify the components in HCI.

In the process of creating a service pattern, one may begin with working out how to design the physical interface (prototyping) or deciding what interaction styles to use (envisioning). Several prototyping approaches were commonly used including holistic design, sketching techniques or using scenarios and story boards. Holistic design approaches focused on how the interface should look in terms of their appearance and representation. Designers used cards with pictures of possible screens shots which users can manipulate to show how interaction can work (Ehn and Sjøgren 1991). Sketching techniques used explicit interface metaphors to create a cardboard as an analogy of the system and those metaphors help visually brainstorming best alternatives for the interface design (Verplank and Kim 1987). A scenario is a personalized story with characters, events, products and environments to help designer from concrete ideas of situations and a storyboard consisting of sequences of snapshots of visualized interaction is often used with a scenario to investigate possible interface problems (Nielsen 2002). Although abovementioned prototyping methodologies were often used in designing the interface, designers and service providers need repetitive communication and time-consuming discussion to minimize the gaps between ideas in their mind and then reach a common ground of concepts. The ineffectiveness of using these methodologies to work out a concrete design is partially attributable to their ambiguity in specifications and a lack of systematic description for interaction styles.

The term "interaction style" was used to describe ways in which users communicate or interact with computer systems. Five common interaction styles in HCI literature were command entry, menus, dialogues, form-fills, and direct manipulations (Preece et al. 1994). Commands can be single characters, short abbreviations or whole words which express instruction to the computer directly. Menus are sets of options displayed on the screen where the selection of one option results in executing the corresponding command. Dialogues can refer to rudimentary type of menus where the expected input is limited and the interaction takes up a form of a series of choices and response from the user. Form-fills are usually used when several different categories of data are fed into the system. In direct manipulation, there are usually icons representing objects and they can be manipulated to achieve desired commands. Those interaction styles are sufficient for traditional human–computer interaction when desktop or laptop computers were used. However, modern online-to-offline (O2O) applications of mobile computing devices in proximity commerce require non-traditional descriptions for innovative interaction styles. An O2O application in proximity commerce generally incorporates online service providers and resources (such as Cloud computing servers) to create an agent in the cyber space. When interacting with the online agent, users in the real world can experience the provided service through electronic portals (such as smart phones or intelligent digital signage) and obtain physical products. For example, Quick Response (QR) code is frequently used in O2O applications to enable communications between users and computers. By scanning QR codes with a smart phone, the user can be directed to a designated cyberspace, interact with online agent and obtain a coupon based on which a physical product can be redeemed. This interaction is beyond traditional interaction styles in that it cannot be classified into any of the abovementioned categories. Another example is using gyro sensors in the mobile computing devices to detect motions of the user as a way of interaction so that an instruction can be given accordingly. Versatile applications of various types of sensors in proximity commerce should be considered in the new systematic language to describe human–computer interaction, and the language should be able to give clear specifications for service providers to choose and designers to follow, i.e. a common ground for both sides in a design team to communicate.

The purpose of this paper is to propose such a systematic language that can be used as specifications of human–computer interaction to describe service patterns, especially for O2O application in proximity commerce. The language includes 4 basic interaction types, and an example is made to demonstrate how those interaction types can be used to describe an interaction style through a diagram representation. The contribution this language and future study directions are briefly discussed.

2 The Interaction Specification

2.1 Interaction Types

Four basic interaction types are developed to systematically describe a single step that may be included in interaction. The former 3 types (button, cursor and sensing) represent mostly controlling components that can be used to receive commands from users and give instructions to the service system. The display type mainly provides feedback to users regarding users' input and the system's state. In some circumstances, the display type can also be used to inform users necessary information for interaction. A summary of 4 types of interaction is listed in Table 1.

The button type is the most fundamental type of interactive patterns. Any interaction through either a physical button, a pseudo button on a touchscreen or an icon that signifies a change of status can qualify as a buttoning interaction. The activation of the button can be triggered by a single-click or a double-click, and the activation of the button can take its effect at the time when it is pressed (land-on), or at the time when it is released (lift-off). The physical buttons usually require single-clicks while icons usually need double-clicks. Land-on activation is common for the physical buttons while lift-off activation is believed to benefit accuracy for touch buttons (Colle and Hiszem 2004). Once a button is activated, it may maintain its status until it is touched again (self-sustaining) or recover its original state immediately after a trigger is sent (pulse). For button interaction, fingers or hand-arms are the most frequently used body part by users.

The cursor type interaction aims to acquire a single position or an entire path of a movement. The tracking of the cursor's position can be either discrete or

Table 1 O2O interaction types

Style	Parameter	Values
Button	Form	Physical/touch/Icon
	Activation	Single-click/double-click/land-on/lift-off
	Status	Self-sustaining/pulse
	Body part	Finger/hand-arm
Cursor	Target	Position/path
	Tracking	Discrete/continuous
	Activation	External/contact
	Body part	Head/finger/hand-arm/upper body/whole body
Sensing	Sensor	Video camera/depth camera/infrared (IR) sensor/electromagnetic sensor/gyro sensor
	Target	Physical objects/gesture/posture/motion
	Activation	Presence/identification/recognition
	Body part	Finger/hand-arm/upper body/whole body
Display	Channel	Visual/auditory
	Presentation	Light/text/graphics/sound/voice/video

continuous, depending on applications. For example, the position of a cursor can be continuously tracked to acquire a single target by averaging all positions obtained during a period of time (e.g. in a balance game, Fig. 1 on the left). In contrast, a path can be formed by several discrete positions (e.g. in a drawing application, Fig. 1 on the right). The activation of cursor tracking can be triggered either by an external signal (e.g. a button pressed or a target sensed) or by contact with a target area. For example, the slide to unlock function found on most mobile phones is a cursor type interaction where the tracking is activated when the finger contacts the slide area and the position of the cursor (the slide brick) is continuously tracked until it reached the end of the slide area (Fig. 2). The user can control the cursor either by his/her head, fingers, hand-arms, upper body or even the whole body.

The sensing type is the most advanced type of interaction in which a sensor is used to detect whether a target is present and whether the condition of the target meets the requirement for activation. Depending on the target, different types of sensors can be used for interaction. For presence or identification of a target, infrared (IR) sensor, electromagnetic sensor or video camera can detect the presence of various physical objects respectively such as a person with body temperature, a passive proximity card or a tag with Quick Response (QR) codes. The difference between presence detection and identification is that the former one needs no further confirmation of information or identity carried by the target, while the latter one requires recognition of identity or acquisition of information. If gestures, postures or motions performed by users are used to interact, depth camera

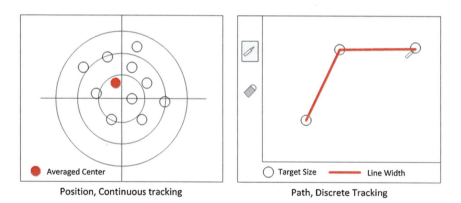

Fig. 1 Cursor interaction of different targets and tracking

Fig. 2 Slide to unlock

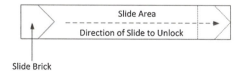

or gyro sensor should be used. The activation is thus based on recognition of spatial (e.g. distance or angle), geometrical (e.g. orientation or shape) or kinematic (speed or acceleration) features. To interact with sensors, different body parts may be used to move physical objects and perform gestures, postures or motions themselves.

Interaction can also take place without explicit motions from the user. The display type of interaction defined in Table 1 specifies such interaction. Information can be received through visual and/or auditory channels in various formats, including light, text, graphics (for visual channels only), sound, voice (for auditory channels only) and video (for both visual and auditory channels). The display interaction mainly provides instructions and feedback to users during the interaction so that the user may know the current status of the interaction and what to do next.

2.2 An Exemplified O2O Application in Proximity Commerce

"Make a Wish!" is one O2O application (Fig. 3) developed by Innovative Digi-tech-Enabled Applications & Services Institute (IDEAS) of the Institute for Information Industry (III), Taiwan. The application served as a promotion activity and was constructed on a system consisting of an intelligent-interactive digital signage, a smart phone and a user. Using the smart phone, the user firstly scanned a QR code appeared on the digital signage which automated a download of an application software (APP) designed specifically for the application. After installing and executing the APP, the user was asked to input his/her wish in text columns provided. Once a submit button was clicked, another QR code containing the user's wish would then appear on the screen of his/her smart phone. At the next

Fig. 3 Make a Wish application (the user is letting the camera installed beside the digital signage scan the QR code on his cell phone's screen)

step the user should move his/her mobile phone toward the digital signage and let another camera installed on the digital signage scan the second QR code on the smart phone's screen. Since the camera was camouflaged as a mailbox, the user's approaching his/her own smart phone simulated a mailing motion of a post card. Finally a greeting card would appear on the digital signage with the words of the user's best wishes.

2.3 Representing an Interaction Style in a Diagram

To define an interaction style specified above, an interaction diagram is used. Firstly, the user's entering texts into text columns were composed of two types: button interaction and cursor interaction. Once touched, the text column became active and a visible cursor indicated the current position for text entry. Therefore, this is cursor interaction by fingers based on positions, with discrete tracking and activation by contact. However, the typing of words should not be ignored. It was implemented by a software keyboard consisting of buttons. So the typing is an iteration of touch button interaction with single-click and lift-off activation. The buttons were operated by fingers also and sent out one letter a time, i.e. pulse signals. The end of the text entry was also a button interaction similar to those in typing. After the text message was transformed into a QR code, a sensing type interaction took place. The user moved the screen of his/her cell phone where the QR code was shown by hand and the QR code was identified by a video camera on the digital signage. The QR code contained the message the user just entered so that the digital signage was able to decode the wish and showed the message in the form of a Christmas card on its screen. The display interaction occurred here using visual channel in video format to provide the user enjoyable experience and feedback regarding information received by the digital signage. Afterward a product coupon was shown on the user's cell phone screen by which the user can redeem a free cup of coffee.

The interaction of Make a Wish can be therefore represented in a flowchart-like diagram (Fig. 4). In the Fig. 4 interaction points are defined (circled numbers): message entry, message submission, QR code scanning and message display. Beneath the interaction sequence there is a table defining interaction patterns and parameter used in each of the interaction points. Using this interaction diagram, a service provider can easily communicate with hardware and software engineers/ designers regarding its needs of interaction patterns. Hardware and software engineers/designers are also benefitted from such a diagram where specifications for interfaces can be easily seen and extracted.

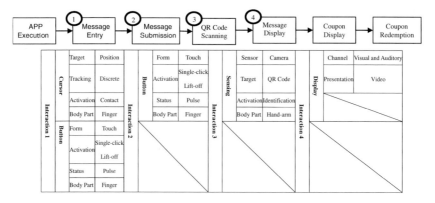

Fig. 4 The interaction diagram for Make a Wish

3 Conclusion and Future Directions

This study describes interaction specifications in terms of interaction styles represented by an interaction diagram consisting of 4 possible interaction types and their parameters which provide details in hardware selection, software implementation and criteria for interaction. The formal language used in the interaction specifications can be used for service provider and service implementer to communicate and develop a concrete prototype effectively because the process of interaction, the requirements for devices and how devices communicates with user are clearly and precisely defined. The specifications, however, are not comprehensive as technology advances and service pattern evolves. Future studies will focus on improving the specification to include innovative interaction design.

Acknowledgments This study is conducted under the "Digital Convergence Service Open Platform" of the Institute for Information Industry which is subsidized by the Ministry of Economy Affairs of the Republic of China.

References

Colle HA, Hiszem KJ (2004) Standing at a kiosk: effects of key size and spacing on touch screen numeric keypad performance and user preference. Ergonomics 47(13):1406–1423

Ehn P, Sjøgren D (1991) From system descriptions to scripts for action. In: Greenbaum JM, Kyng M (eds) Design work: cooperative design of computer systems. Routledge, London, pp 241–268

Nielsen L (2002) From user to character: an investigation into user-descriptions in scenarios. Paper presented at the proceedings of the 4th conference on designing interactive systems: processes, practices, methods, and techniques, London, England

Preece J, Carey T, Rogers Y, Holland S, Sharp H, Benyo D (1994) Interaction styles. In: Dix A (ed) Human-computer interaction. Addison-Wesley Publishing Company, Essex, pp 237–255

Sharp H, Rogers Y, Preece J (2002) Design, ptototyping and construction. In: Rogers Y et al (eds) Interaction design: beyond human-computer interaction. Wiley, NY, pp 239–275

Verplank B, Kim S (1987) Graphic invention for user interfaces: an experimental course in user-interface design. SIGCHI Bull 18(3):50–66. doi:10.1145/25281.25284

Investigation of Learning Remission in Manual Work Given that Similar Work is Performed During the Work Contract Break

Josefa Angelie D. Revilla and Iris Ann G. Martinez

Abstract According to literature, when a worker performs a task repeatedly, the time it takes to do the task decreases. This is based on the concept of learning curve. When the worker spends time away from work, there is usually an observed time decrement, as described by learning remission. A question may be "in case the worker does not completely stop working during the work break but performs 'similar work', how much remission can be expected?" Similar work may be work that share operations with the work prior to the break, but not completely the same. To investigate on learning remission, this research worked with a semiconductor and a hockey glove sewing business. The objective was to find the amount of similar work needed to be done during the work break to possibly reduce or avoid learning remission. The finding for the test case is, when at least 21 % of the supposed work break is spent doing the previous work, learning remission may not result. It is possible that for other industries, researchers may find similar results. This will help organizations transferring workers from one kind of work to another, not have increase in work time as a result of learning remission.

Keywords Learning curve · Learning remission · Similar work · Work break

J. A. D. Revilla (✉)
Assistant Professor of the Department of Industrial Engineering,
University of the Philippines, Los Baños, Laguna, Philippines
e-mail: angelie.revilla@gmail.com

I. A. G. Martinez
Associate Professor of the Department of Industrial Engineering and Operations Research,
University of the Philippines, Diliman, Quezon City, Philippines

1 Introduction

Work interruption is commonly experienced by workers. Causes of work interruptions include:

1. Transfer to other lines within the organization because of seasonality of demand for the different lines;
2. Transfer to another organization because of end of contract with the previous one;
3. Complete stop while looking for employment because of end of contract with the previous employer.

Whichever the reason may be among the aforementioned, work interruption may cause learning loss or decline in work performance. This is known in literature as "forgetting" or "learning remission".

1.1 Learning Remission and Output

Learning remission describes decrement or decline in performance as a result of being away from work for some period of time. Learning remission a longer time to do the task or effectively, lower number of output for the same time period. This effect of lower output for the same time period is a concern of some manufacturing organizations. Specifically, for example, in manufacturing organizations that compensate their workers on piece-rate method, there is a minimum number of output required per worker per work day. The minimum number of output is set based on the performance of the standard worker. Typically, all workers are rated based on their performance compared with the standard. When workers take time away from work after a period of time, i.e., six-month contract period, it is often a question whether learning remission would really set in. In effect, a lower output should be expected from the worker from his resumption of work compared to his output prior to the work break.

1.2 Possibility of Avoiding Learning Remission

Extending the aforementioned question, a challenge can be stated as "would it be possible to avoid learning remission by performing related activities during the *supposed* work break?" Furthermore, if that will be possible, what related activities can be done? These questions are pressing concerns of countries where *contract work* is practiced.

1.3 Definition of Work Break

As mentioned above, in this research, the work break is "supposed". Strictly speaking, the work break should be complete stoppage from work. However, since most workers don't really stop working after they complete their work contract but instead work on an organization within the same or similar industry where their skills are required, work that is related to the work that they performed during Stint 1 may be done by the worker during the "supposed" work break.

2 The Objectives of This Research

It is the aim of this research to determine the level of output of a worker upon resumption from a work break. Furthermore, this research has the objective of identifying factors that affect the rate of output upon resumption. Knowledge of the factors will enable organizations institute programs and measures to ensure same or similar level of output upon resumption of work (i.e., Stint 2), as with the output prior to the work break (i.e., Stint 1). Please see Fig. 1 for an illustration of the objective of this research.

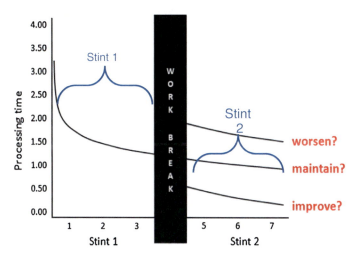

Fig. 1 The objective of this research

2.1 Specific Objectives

The specific objectives of this research are as follows:

1. To determine whether the output at the last point during Stint 1 varies significantly from the output at the first point during Stint 2 given work break of less than six (6) months.
2. To identify the work break-related factors that will significantly affect the level of output during Stint 2. Among the numerous work-break related factors, the following are considered in this paper:

    (a) Length of the work break (i.e., one day to six months)
    (b) Number of unique operations during the "supposed" work break
    (c) Frequency of performing during the "supposed" work break, the operations that are common with Stint 1.

3 Related Studies on Learning Curve and Forgetting

Learning curve traces its background from 1885 when Ebbinghaus first described it for memorizing syllables. Later, a mathematical model of the learning curve was proposed following work describing the effect of learning on production costs in the aircraft industry (Wright 1936). From then on, various research activities have been undertaken about the learning curve. Some of these are on the application of machine learning such as that of Principe and Euliano (2000).

While the concept of learning has been much explored, studies on forgetting, as a result of time away from work, have not been numerous. One of these few studies claims that an interruption in work (i.e., work break in this research) warrants the return of the worker to the top of the learning curve. Another study concluded with two findings: (1) Forgetting (i.e., defined as the excess of actual time over learning curve-predicted time) may be negligible for "continuous control" tasks but significant for procedural task and (2) Forgetting is a function of the amount learned and the passage of time, not of the learning rate or other variables. Similar to these conclusions, still another research stated that the degree of forgetting is a function of the work break length and the level of experience before the break.

4 The Problem and Hypothesis of This Research

The specific problem that this study seeks to address is "How can forgetting be minimized or avoided?"

The hypothesis of this research is as follows: "Given a 'time away' of six months or less from the work of Stint 1, forgetting can be avoided if similar operations as with those of Stint 1 are done during the supposed work break or time away from work." This research will consider "forgetting" to have been avoided if the output at the last data point in Stint 1 is at least equal to the output at the last data point of Stint 2.

5 Methodology: Case Study of a Semiconductor Company and a Hockey Glove Sewing Company

Figure 2 presents the methodology of this research:

As shown, the study initially took 15 samples of workers of a semiconductor company and compared the output rates for Stint 1 and Stint 2. To pursue the study further, this study took 79 samples of workers of a hockey glove manufacturing company again to compare the output rates for Stint 1 and Stint 2. Note that for each of the two companies, the observations made were done on workers who indeed have two stints. It was not difficult to perform the observations of Stint 1 and Stint of the same worker because it was typical for the organizations in the case study to re-hire workers after the workers have spent a work break of about six months after their contracts are completed.

Furthermore, Table 1 shows more detailed description of the observed workers and operations.

Fig. 2 Methodology of this research

1. Take 15 samples of inspection operation and 79 samples of sewing operation.
2. Compute for average processing time for Stint 1 and average processing time for Stint 2.
3. Compare:
   (a) average processing time Stint 1 versus average processing time Stint 2
   (b) output of last point of Stint 1 versus output of last point of Stint 2

4. Investigate work break-related factors and their correlation with the average processing time of Stint 2 and with the output of the first point of Stint 2.

Table 1 Description of the observed workers and operations

Case	Product	General process	Kind of work	Use of contractual operators
1	Actuator	1. Machining 2. Washing 3. Inspection 1 4. Plating 5. Inspection 2 6. Packing	Inspection process • 10 % machine-paced and 90 % worker-paced	70 % are contractual operators • Stint 1 = 15-16 days • Work break = 10 days • Stint 2 = 16-17 days
2	Baseball gloves	1. Cutting 2. Stamping 3. Skiving 4. Matching color 5. Inspection 1 6. Sewing (accessories) 7. Hot hand 8. Linings 9. Sewing (fingers) 10. Lacing 11. Inspection 2 12. Packing	Sewing process • 20 % machine-paced and 80 % worker-paced	80 % are contractual operators • Stint 1 = 2-3 months • Work break = 3 months • Stint 2 = 2-3 months

6 Results and Discussions

Both companies have partially machine-paced and partially operator-paced operations. Case study 1, done on a semiconductor company, considered 15 operators and a shorter work break duration. From this, as initial study, an interesting result is found. It shows that, aside from the work break length, the frequency of performing similar operation during the *supposed* work break seems to affect the average processing time on stint 2. It is seen that a 30 % increase in average processing time on stint 2 seems to be an effect of a work break wherein similar operations are performed. Thus, to further analyze the initial investigation, another case study is conducted.

In case study 2, a sufficient sample size and a longer work break length is considered. Furthermore, factors that might affect the output at Stint 2 were investigated.

The work break-related factors that were investigated included:

1. Length of the work break
2. Number of unique operations performed during the work break
3. Frequency of doing the same operations during the supposed work break as with those of Stint 1

Correlation analysis showed that among the factors considered, the frequency of performing similar operations during the supposed work break, i.e., common operations as those done during Stint 1, can significantly reduce or even eliminate forgetting. This study was not able to establish correlation between Stint 2 performance and the other work break-related factors investigated. This study found a negative moderate correlation between the output or processing time on Stint 1 and the frequency of doing similar operations during the supposed work break. This means that when the frequency of doing similar operations becomes higher, the processing time on Stint 2 becomes shorter. This finding coincides with the principle of the learning curve. However, the further question is "how frequent should the similar operations be done to not experience forgetting?"

This research has found that for the cases considered, the processing time at the start of Stint 2 is always higher than the processing time at the end of Stint 1 before the work break. However, even if the processing time at the start of Stint 2 is higher, the decline of processing time will be steeper if more than 40 % of the operations done in Stint 1 are continued during the supposed work break. This eventual steeper decline in processing time will result to a lower average processing time for Stint 2. Otherwise, when 40 % or less of the operations done in Stint 1 are continued during the supposed work break, not only will the processing time during the start of Stint 2 will be higher than the processing time at the end of Stint 1, but also the eventual decline of processing time during Stint 2 as a result of learning will not be enough to cause the average processing time of Stint 2 to be lower than the average processing time of Stint 1 (Fig. 3).

The finding about 40 % degree of similarity of operations between Stint 1 and the work during the supposed work break has been an interesting result for this study. The decrease in average processing time implies that forgetting may be avoided by making the worker perform similar tasks (i.e., at least 40 % similar) so

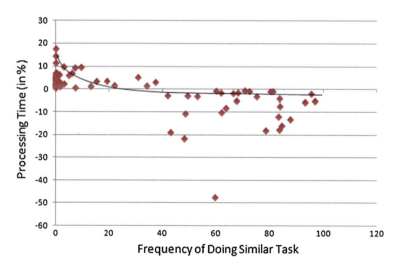

Fig. 3 Trend analysis between frequency of doing similar operation and average processing time on Stint 2

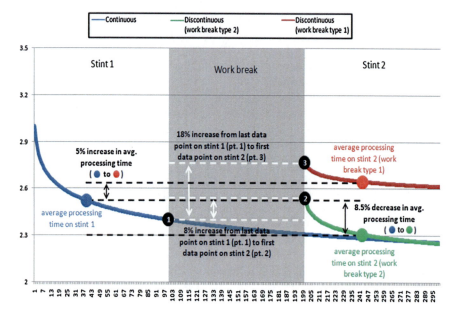

Fig. 4 Learning and forgetting when similar work is done during the supposed work break

that the same output can be expected on him when he goes back to the same work as performed during Stint 1. This finding may perhaps be also observed in other similar "procedural" work and may prove useful as a basis for computing production quotas, wage and productivity-related measures.

7 Conclusions

Work interruption is a typical occurrence. When work is interrupted, there is usually an associated forgetting by the worker. This causes a decline in performance that will be observed when the worker resumes the work on a future time. Several studies have been conducted on worker learning and forgetting. This study, in particular, was concerned with knowing if forgetting can be empirically observed for work breaks of less than six months and if forgetting can be avoided by instituting measures such as requiring the worker to perform similar work during the supposed work break. This study pursued the research by conducting two case studies: one on a semiconductor company and the other on a hockey glove manufacturing company (Fig. 4).

It was interesting for this study to find that forgetting could be observed for both case studies. An increase in the processing time was always observed between the end of Stint 1 and the start of Stint 2. However, when more than 40 % of operations were done similarly during the work break as with Stint 1, the learning during Stint 2 can be said to be better as shown by steeper decline in the eventual performance time of the task. In this case, the average processing time is observed to be smaller than the average processing time during Stint 1. This phenomenon was not observed when 40 % or less of the operations were common between those done in Stint 1 and during the supposed work break. The findings of this research can be helpful when organizations would like to hire and re-hire, assign and re-assign workers into tasks with work breaks in between "stints" and where forgetting needs to be minimized or avoided.

Acknowledgments This work is partially supported by the Engineering Research and Development for Technology program of the Department of Science and Technology of the Philippines.

References

Principe JC, Euliano NR, Lefebvre WC (2000) Neural and adaptive systems: fundamentals through simulations. Wiley, NY

Wright TP (1936) Factors affecting the cost of airplanes. J Aeronaut Sci 3(4):122–128

A Hidden Markov Model for Tool Wear Management

Chen-Ju Lin and Chun-Hung Chien

Abstract Determining the best time of tool replacement is critical to balancing production quality and tool utilization. A machining process would gradually produce defective parts as a tool wears out. To avoid additional production costs, replacing a tool before the yield drops below a minimum requirement is essential. On the other hand, frequent tool replacement would cause additional setup and tool costs. This paper proposes a hidden Markov model (HMM) to study the unknown nature of a tool wear progress by monitoring the quality characteristic of products. With the constructed model, the state of tool wear is diagnosed by using the Viterbi Algorithm. Then, a decision rule that evaluates the yield of the next machining part is proposed to determine the initiation of tool replacement. The simulation analysis shows that the proposed method could accurately estimate the model and the status of tool wear. The proposed decision rule can also make good use of tools whereas controlling yield.

Keywords Tool replacement · Tool wear · Yield

1 Introduction

Online monitoring of tool conditions is important to many machining processes. As tools wear out, machining processes would produce more and more defective workpieces. Tool replacement should be initiated once production yield drops

C.-J. Lin (✉) · C.-H. Chien
Department of Industrial Engineering and Management, Yuan Ze University,
No. 135 Yuan Tung Rd, Chung-Li, Taiwan
e-mail: chenju.lin@saturn.yzu.edu.tw

C.-H. Chien
e-mail: s995434@mail.yzu.edu.tw

below a minimal requirement to avoid additional production cost. On the other hand, premature tool replacement would incur extra tool costs and increase machine downtime. Accordingly, determining the best time of tool replacement is important to balance between production quality and tool utilization.

Recent advances in sensor technology and computational capabilities have facilitated in-process monitoring of tool conditions. Condition-based maintenance (CBM) that suggests maintenance decision depending on the information collected through condition monitoring is commonly applied in practice (Jardine et al. 2006). However, a critical challenge of CBM is high installation costs. Manufacturers may not be able to afford expensive monitoring equipment but afford general inspection instruments that measure outer appearance or features of the workpieces machined. Instead of determining the time of tool replacement based on the indicators like temperature, vibration data, X-ray images or others collected during condition monitoring, this paper directly focuses on the quality characteristic of workpieces.

2 Tool Wear Diagnostics Support

To diagnose tool wear, the time series approach was applied to analyze data variance to monitor tool condition (Kumar et al. 1997). The task of clustering tool condition becomes more difficult as the machining processes become more complicated. The collected data for analyzing tool wear condition are versatile. Cutting forces were investigated to monitor tool condition under the structure of neural network (Saglam and Unuvar 2003). However, neural network requires abundant training data which may not be available in practical problems. Fuzzy system was studied instead (Devillez and Dudzinski 2007). Other researchers completely reviewed the methods that diagnose tool condition by analyzing vibration and wavelet data (Heyns 2007; Zhu et al. 2009a, b).

Hidden Markov model (HMM) is appropriate to explain the tool wear problem. An HMM for tool condition consists of two stochastic processes: (1) A Markov chain with finite but unknown states describing an underlying process of tool wear. The number of states could be either known or unknown. (2) An observable process depending on the hidden state. The observations could be the condition monitoring data studied in the CBM domain or quality characteristics which will be introduced in the next section. When fixing number of states, the Baum-Welch algorithm (Baum 1972) was applied to estimate the parameters of HMM based on training data (Heck and Mcclellan 1991). The tool state of a drilling process is then determined by maximizing the conditional probability of state sequence given the estimated model and observation sequence. Similar approaches were applied to different machining process as follows, drilling: (Ertunc et al. 2001) and (Baruah and Chinnam 2005); rotating: (Wang et al. 2002) and (Li et al. 2005); computer numerical control (CNC): (Vallejo et al. 2005); milling: (Zhu et al. 2009a, b);

cutting: (Zhang and Yue 2009). The condition probabilities used for state diagnostics are slightly different in these papers.

When the number of states is unknown, a similarity index was used to determine the best number of states for a CNC machining process (Kumar et al. 2008). An adaptive process that increases the number of states by one whenever k consecutive observations cannot be categorized into the existing states was proposed for analyzing online data (Lee et al. 2010). The concept was close to statistical process control techniques.

3 Tool Wear Monitoring System

To effectively monitor tool condition and determine the time of tool replacement, this paper proposes a monitoring system containing three main phases: (1) model construction—estimating the HMM parameters and the number of states of tool wear, (2) state estimation—estimating tool wear status, and (3) tool replacement determination—an yield-based decision rule.

3.1 Modeling Tool Wear

Let the observation sequence $X = (x_1, x_2,..., x_{T(h)})$ be the quality characteristic of workpieces collected from training data. A workpiece with the quality characteristic beyond its upper specification limit (USL) and lower specification limit (LSL) is considered as a defect. To construct the HMM of tool wear, a brand new tool is used in the machining process until producing h defective workpiece, $h > 1$. The collected quality characteristics of workpieces form a training data set for model construction.

After data acquisition, apply the Baum-Welch algorithm to estimate the parameter $\lambda = (\pi, A, B)$ of the HMM. π is the initial probabilities of the states, A is the transition probability matrix, and B is the distribution parameters of the quality characteristic under each tool wear state. Here we assume the distribution is Normal. Since the number of states is unknown, we applied the Bayesian information criterion (BIC) to determine the best model. The way of HMM construction is:

Step 1. Start the machinery process and collect the quality characteristic of each workpiece until the hth defect occurs, $X = (x_1, x_2, ..., x_{T(h)})$. $T(h)$ is the hth time such that $x_{T(h)} \notin$ [LSL, USL].
Step 2. Set the number of states $k = 2$. Apply the Baum-Welch algorithm to estimate $\hat{\lambda} = (\hat{\pi}, \hat{A}, \hat{B})$.

Step 3. Calculate BIC (Yu 2010) under the estimated HMM with k states.

$$\mathrm{BIC}(k) = -2\ln L + v(k)\ln T(h) \qquad (1)$$

where L is the likelihood function, $L = f(o_1, o_2, \ldots, o_T | \hat{\lambda}) = \prod_{i=1}^{T} f(o_i | \hat{\lambda})$, and $v(k)$ is the degree of freedom, $v(k) = k^2 + 2k - 1$.

Step 4. Set $k = k + 1$. Estimate $\hat{\lambda} = (\hat{\pi}, \hat{A}, \hat{B})$ and calculated $\mathrm{BIC}(k)$ again.
Step 5. If $\mathrm{BIC}(k) < \mathrm{BIC}(k-1)$, return to Step 4. Otherwise, stop. The final HMM has parameters $\hat{\lambda} = (\hat{\pi}, \hat{A}, \hat{B})$ estimated under $k*$ states, $k* = k - 1$.

3.2 Diagnosing Tool Wear Conditions

Suppose that all of the tools have the same performance as the one used in model construction. To diagnose the tool condition of in-process data, sequentially apply the Viterbi algorithm (Viterbi 1967) as follows. T would increase as a machining process continues.

Initialization:

$$\delta_1(i) = \pi_i b_i(o_1), \quad 1 \leq i \leq k^* \qquad (2)$$

$$\psi_1(i) = 0 \qquad (3)$$

Recursive:

$$\delta_t(j) = \max_{1 \leq i \leq k^*} \{\delta_{t-1}(i) a_{ij}\} b_j(o_t), \quad 2 \leq t \leq T, \ 1 \leq j \leq k^* \qquad (4)$$

$$\psi_t(j) = \arg\max_{1 \leq i \leq k^*} \{\delta_{t-1}(i) a_{ij}\}, \quad 2 \leq t \leq T, \ 1 \leq j \leq k^* \qquad (5)$$

Termination:

$$P^* = \max_{1 \leq i \leq k^*} \{\delta_T(i)\} \qquad (6)$$

$$q_T^* = \arg\max_{1 \leq i \leq k^*} \{\delta_T(i)\} \qquad (7)$$

Optimal path:

$$q_t^* = \psi_{t+1}(q_{t+1}^*), \quad t = T-1, T-2, \ldots, 1 \qquad (8)$$

o_t is the quality characteristic of the workpiece at time t. $b_t(o_t)$ is the probability of observing o_t at state i. π_i is the initial probabilities of state i. a_{ij} is the transition probability from state i to state j. q_t^* is the best state of tool wear at time t.

3.3 Diagnosing Tool Wear Conditions

Severity level of tool wear directly affects the quality of workpieces machined. Effectively monitor and control machinery health is critical to produce qualified workpieces. We proposed a yield-based decision rule to initiate tool replacement. With the estimated HMM explained in Sect. 3.1, we are able to evaluate how likely the next workpiece will be qualified given the currently observed sequence O_t, $O_t = \{o_1, o_2, \ldots, o_t\}$. A tool is considered as capable if such probability is high. On the other hand, a tool is considered as worn out if such probability cannot meet a minimum requirement of yield. The proposed yield-based rule is as follows.

Decision rule:

Initiate tool replacement when $P(o_{t+1} \in [LSL, USL] | O_t) < R$, where R is the pre-specified minimum requirement of yield. Otherwise, continue machining.

$P(o_{t+1} \in [LSL, USL] | O_t)$ can be calculated as

$$\sum_{j=1}^{k^*} P\left(o_{t+1} \in [LSL, USL] \,|\, q_{t+1} = S_j\right) \times \left(\sum_{i=1}^{k^*} \hat{a}_{ij} \frac{\hat{\alpha}_t(i)}{P(O_t | \hat{\lambda})}\right) \quad (9)$$

where

$$\hat{\alpha}_1(i) = \hat{\pi}_i \hat{b}_i(o_1), \quad 1 \leq i \leq k^* \quad (10)$$

$$\hat{\alpha}_{s+1}(j) = \hat{b}_j(o_{s+1}) \sum_{i=1}^{k^*} \hat{\alpha}_s(i) \hat{a}_{ij}, \quad 1 \leq s \leq t-1, 1 \leq j \leq k^* \quad (11)$$

$$P(O_t | \hat{\lambda}) = \sum_{i=1}^{k^*} \hat{\alpha}_t(i) \quad (12)$$

4 Simulation Analysis

In the following simulation, a drilling process is used to demonstrate the efficacy of the proposed system. The machining process drills one hole on each workpiece. The depth of a hole drilled is the quality characteristic measured in this example. As a tool wears out, the depth becomes shallow. The USL and LSL of the depth are set to 11 and 9 mm, respectively. Once there are three defects, the system stops collecting the training data, $h = 3$. As for the testing stage, the drilling process would stop and initiate tool replacement when the conditional yield of the next workpiece drops below 90 %, $R = 90$ %.

Assume that the tool wear condition of the drilling process is a three-state left to right Markov chain. From a brand new tool to worn out, the three states are the slight, moderate, and severe wear states. The distributions of the depths are normally distributed with N(10.4, 0.04), N(10, 0.04), and N(9.4, 0.09) under the slight, moderate, and severe wear states. The initial probabilities of the states are $\pi = [1\ 0\ 0]$. The transition probability matrix is

$$A = \begin{bmatrix} 0.97 & 0.03 & 0 \\ 0 & 0.97 & 0.03 \\ 0 & 0 & 1 \end{bmatrix} \qquad (13)$$

In the simulation, the 81st, 82nd, and 84th workpieces are defective. Thus, the training phase stops collecting data at time $T(3) = 84$. The Baum-Welch algorithm is applied to estimate the parameters of the model. The minimum BIC value occurs when setting the number of states to three. The final estimated model is a three-state HMM with parameters $\pi = [1\ 0\ 0]$,

$$A = \begin{bmatrix} 0.973 & 0.027 & 0 \\ 0 & 0.972 & 0.028 \\ 0 & 0 & 1 \end{bmatrix} \qquad (14)$$

N(10.4, 0.03), N(10, 0.11), and N(9.5, 0.23), which are highly close to the preset model. This estimated model is then applied to determine the time of tool replacement.

100 tools are then tested in the same drilling process to demonstrate the performance of the proposed system, including the accuracy in diagnosing tool condition and the efficiency in tool replacement. Suppose that all of the 100 tools start from brand new and have the same features as the one applied to construct the model of tool wear. The analysis shows promising performance of the Viterbi algorithm in evaluating the status of tool wear. For one tool, 93.02 % of tool condition can be accurately diagnosed before replacing the tool on average. The final state of tool condition is often considered as the worn out status when a tool must be immediately replaced. However, tool wear is a continuous progress. A tool may still be capable of machining a few more qualified workpieces at the beginning of the final state. Comparing to the naïve method that replaces a tool whenever the machining process enters the final state, Table 1 shows that the proposed system can prolong the usage of a tool whereas producing more qualified workpieces. The beneficial result is due to the proposed yield-based decision rule that is designed to balance between tool utilization and quality.

The stopping rule h of collecting training data would change the data set of training data and might influence model estimation. To further analyze the sensitivity of h to model estimation, we tested different rules that stop collecting

Table 1 Performance comparison between the proposed system and the naïve method

	Proposed system	Naïve method
Average time to replace tools	47.65	41.61
Average number of qualified products	47.36	41.50

training data when having 2–5 defects. The results suggests that parameter estimation of HMM is rather robust to h. The stopping rule has more influence on the parameters of the final state than the other states. The phenomenon may due to the number of observations taken in each state.

5 Conclusions

Monitoring the status of tool wear and scheduling tool replacement are critical tasks to many machining process due to high requirement in quality and cost control. This paper proposes a HMM to diagnose tool conditions and a yield index to determine tool replacement. The simulation results show that applying Baum-Welch algorithm along with BIC can effectively model the status of tool wear. The Viterbi algorithm also performs well in estimating the condition of tool wear with accuracy rates higher than 90 %. Comparing to the naïve policy that replaces tools when a machining process enters the final state of tool life, the proposed policy based on the yield index can lead to a few more qualified workpieces whereas maintaining yield requirement. The promising results suggest a rather balance between quality and tool utilization.

The analysis in this paper shows that the estimated HMM depends on the training data. It would be worthwhile to investigate the optimum stopping rule h of collecting the training data in the future. We are also working on building the HMM for tool wear based on the tool life of multiple tools rather than one, which could be biased. With a better estimated model, the proposed system could even better estimate the condition of tool wear and justify the time of tool replacement.

References

Baruah P, Chinnam RB (2005) HMMs for diagnostics and prognostics in machining processes. Int J Prod Res 43(6):1275–1293

Baum LE (1972) An inequality and associated maximization technique in statistical estimation for probabilistic functions of markov processes. Inequalities 3:1–8

Devillez A, Dudzinski D (2007) Tool vibration detection with eddy current sensors in machining process and computation of stability lobes using fuzzy classifiers. Mech Syst Signal Pr 21(1):441–456

Ertunc HM, Loparo KA, Ocak H (2001) Tool wear condition monitoring in drilling operations using hidden markov models (HMMs). Int J Mach Tool Manu 41(9):1363–1384

Heck LP, McClellan JH (1991) Mechanical system monitoring using hidden markov models. Paper presented at the IEEE international conference on acoustics, speech and signal processing 3, Toronto, Canada, pp 697–1700, 14–17 Apr 1991

Heyns PS (2007) Tool condition monitoring using vibration measurements a review. Insight 49(8):447–450

Jardine A, Lin D, Banjevic D (2006) A review on machinery diagnostics and prognostics implementing condition-based maintenance. Mech Syst Signal Pr 20(7):1483–1510

Kumar SA, Ravindra HV, Srinivasa YG (1997) In-process tool wear monitoring through time series modeling and pattern recognition. Int J Prod Res 35(3):739–751

Kumar A, Tseng F, Guo Y, Chinnam RB (2008) Hidden-markov model based sequential clustering for autonomous diagnostics. Proceedings of the International Joint Conference on Neural Networks, Hong Kong, China, 3345-3351, 1-8 June 2008

Lee S, Li L, Ni J (2010) Online degradation assessment and adaptive fault detection using modified hidden markov model. J Manuf Sci E-T ASME 132(2):021010-1–021010-11

Li Z, Wu Z, He Y, Chu F (2005) Hidden Markov model-based fault diagnostics method in speed-up and speed-down process for rotating machinery. Mech Syst Signal Pr 19(2):329–339

Saglam H, Unuvar A (2003) Tool condition monitoring in milling based on cutting forces by a neural network. Int J Prod Res 41(7):1519–1532

Vallejo AJ, Nolazco-Flores JA, Morales-Menendez R, Sucar LE, Rodrıguez CA (2005) Tool-wear monitoring based on continuous hidden markov models. Lect Notes Comput Sc 3773:880–890

Viterbi J (1967) Error bounds for convolutional codes and an asymptotically optimal decoding algorithm. IEEE T Inform Theory 13(2):260–269

Wang L, Mehrabi MG, Kannatey-Asibu E (2002) Hidden markov model-based tool wear monitoring in turning. J Manuf Sci E-T ASME 124(3):651–658

Yu J (2010) Hidden markov models combining local and global information for nonlinear and multimodal process monitoring. J Process Contr 20(3):344–359

Zhang C, Yue X, Zhang X (2009) Cutting chatter monitoring using hidden markov models. Paper presented at the international conference on control, automation and systems engineering, Zhangjiajie, China, pp 504–507, 11–12 July 2009

Zhu K, Wong YS, Hong GS (2009a) Multi-category micro-milling tool wear monitoring with continuos hidden markov models. Mech Syst Signal Pr 23(2):547–560

Zhu K, Wong YS, Hong GS (2009b) Wavelet analysis of sensor signals for tool condition monitoring: a review and some new results. Int J Mach Tool Manu 49(7):537–553

Energy Management Using Storage Batteries in Large Commercial Facilities Based on Projection of Power Demand

Kentaro Kaji, Jing Zhang and Kenji Tanaka

Abstract This study provides three methods for projection of power demand of large commercial facilities planned for construction, for the operation algorithm of storage batteries to manage energy and minimize power costs, and for derivation of optimal storage battery size for different amounts of power demand and building use. The projection of power demand is derived based on statistics of building power demand and floor area. The algorithm for operating storage batteries determines the amount of purchased electricity on an hourly timescale. The algorithm aims to minimize the cost of power through two approaches: first by reducing the basic rate determined by the peak of power demand, and second by utilizing the power purchased and charged at nighttime when the price of power is lower. Optimization of storage battery size is determined by calculating internal rate of return, which is derived by considering the profit from energy management, cost, and storage battery lifetime. The authors applied these methods to commercial facilities in Tokyo. The methods successfully helped the facility owners to determine appropriate storage battery size and to quantify the profit from their energy management system.

Keywords Energy management · Storage battery · Power demand projection

---

K. Kaji (✉) · J. Zhang
Department of Technology Management for Innovation, Graduate School of Engineering, The University of Tokyo, Tokyo, Japan
e-mail: kaji@triton.naoe.t.u-tokyo.ac.jp

J. Zhang
e-mail: chou@sys.t.u-tokyo.ac.jp

K. Tanaka
Department of System Innovation, Graduate School of Engineering, The University of Tokyo, Tokyo, Japan
e-mail: tanaka@triton.naoe.t.u-tokyo.ac.jp

1 Introduction

After the Great East Japan Earthquake on March 11, 2011, utilization of stationary batteries for demand-side electricity management has gathered social interest, and many companies such as real estate and electronics companies have been developing energy management services for offices, commercial buildings, and condominiums using energy storage systems (NEC press release 2012). Behind this trend are two energy-related anxieties shared among companies. The first is securing electricity in the event of blackout, and the second is energy cost, which has been rising because power companies cannot run their nuclear power plants since safety guidelines have not yet been established following the accident at the Fukushima Daiichi Nuclear Power Plant (Tokyo Electric Power Co. 2011).

Breakthroughs in large-scale storage batteries are also enabling utilization of storage batteries for stationary use. Batteries with capacities of 50 kWh to 2 MWh are being used in demonstration experiments conducted by the Japanese government and some electronics companies in communities in Japan.

By introducing stationary batteries, four effects can be expected:

1. Reduction of the basic rate, which is determined by the peak of power demand
2. Utilization of the power purchased and charged at nighttime when the price of power is cheaper than daytime
3. Enhancement of energy durability in the event of blackout
4. Energy supply service for adjacent buildings

However, methodologies have not yet been established for specification design of storage batteries optimized for energy consumption patterns, for the amount of power demand, or for operation of storage batteries to minimize energy cost.

There are some studies on the optimization of storage batteries and renewable power generators (Borowy and Salameh 1995; Protogeropoulos et al. 1998; Koutroulis et al. 2006; Yang et al. 2007); however, those studies do not consider the optimization of storage battery size from both technical and economical aspects. There are no studies that consider algorithms to decide the amount of purchased electricity on an hourly timescale while minimizing energy cost and pursuing the above effects. The authors have therefore developed an algorithm to pursue two objectives: minimizing energy cost and deriving appropriate storage battery size. A method to project power demand was also developed, because the proposed models are intended for planning building construction.

2 Model

In this section, methods for projection of power demand and a storage battery operation algorithm are discussed. To derive optimized storage batteries and evaluate the effect, power demand on hourly scale is projected, and then the appropriate storage battery size is decided by a time-marching energy management simulation using the algorithm. An evaluation method for the simulation results is also discussed in this section.

2.1 Projection of Power Demand

In this part, the method of power demand projection is discussed. The projection method consists of two essentials: "daily fluctuation," which means the fluctuation of the amount of power demand per day during the year, and "hourly fluctuation," which means the fluctuation of the amount of power demand per 30 min during the day.

Figure 1 shows the process of power demand projection. A daily fluctuation pattern is first derived by the process shown in Fig. 2.

First, the floor size of each target building usage is obtained. Usage is defined in this study as commercial, office, hotel, or other use. Second, power demand by floor size of each usage is obtained from statistics data or other existing data. Third, derive the amount of average daily power demand of each floor usage by multiplying the floor size with the power demand of each floor usage. Fourth, the average daily power demand of the whole building is derived by adding the amount of power demand of each usage and multiplying by a weighting factor k, as shown in Eq. (1). Fifth, the daily fluctuation pattern is derived by multiplying average power demand with a daily fluctuation coefficient, which is defined as a weighting factor on each day of the year, derived by dividing the amount of each

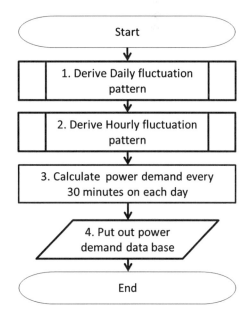

Fig. 1 Flow of power demand projection

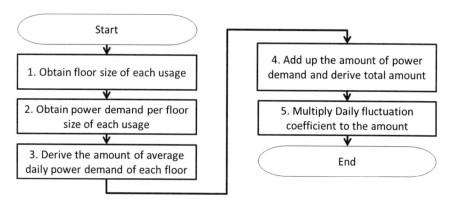

Fig. 2 Flow of deriving daily fluctuation pattern

daily power demand by the average amount of power demand of existing statistics data. The computation of the fifth step is shown as Eq. (2).

$$Average\ Power\ Demand = k * \sum Floor\ Size * Power\ Demand\ of\ Each\ Usage \quad (1)$$

$$Daily\ Fluctuation\ Pattern\ (n) = DFC(n) * Average\ Power\ Demand$$
$$DFC(n) : Daily\ Fluctuation\ Coefficient\ of\ the\ "n"th\ day\ (n = 1, 2, \ldots, 365) \quad (2)$$

The second step of the process of power demand projection in Fig. 1 is shown in detail in Fig. 3.

The first step in deriving the hourly fluctuation pattern is to obtain power demand at 5:00 and 17:00 of an existing building whose usage is the same as the target building. This aims to obtain the maximum and minimum power demand of an existing building. According to our previous case studies of existing buildings,

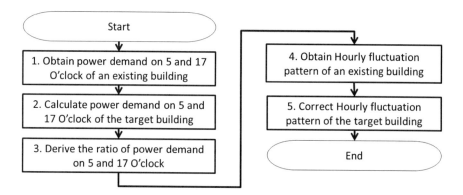

Fig. 3 Flow of deriving hourly fluctuation pattern

power demand is minimal at 5:00 and maximal at 17:00. Second, calculate power demand at 5:00 and 17:00 for the target building by adding up power demand of each floor in use at 5:00 or 17:00. The approach to deriving power demand of each usage floor is the same as in Eq. (1). The process of the second step is shown as Eq. (3). Third, derive the ratio of power demand by dividing power demand at 17:00 by that at 5:00. Fourth, obtain the hourly fluctuation pattern of the building whose power demand at 5:00 and 17:00 is obtained in the first step. Fifth, the hourly fluctuation pattern of the target building is derived by correcting the pattern of the existing building according to Eq. (4).

$$Power\ Demand(t) = \sum PDEUF(t)$$
$$Power\ Demand(t) : Power\ demand\ at\ t : 00 \tag{3}$$
$$PDEUF(t) : Power\ demand\ of\ each\ usage\ floor\ on\ t : 00\ (t = 5, 17)$$

$$D(k) = E(5) + 0.5 * \frac{\alpha_{target}-1}{\alpha_{existing}-1} * \{E(17) - E(5)\} * \left[1 - \cos\left\{\frac{E(k)-E(5)}{E(17)-E(5)} * \pi\right\}\right]$$
$D(k)$: *Power demand at* $k : 00$ *of the target building (normalized form)*
$E(k)$: *Power demand at* $k : 00$ *of the existing building (normalized form)*
α_{target} : *Ratio of power demand at* $5 : 00$ *and* $17 : 00$ *of the target building*
$\alpha_{existing}$: *Ratio of power demand at* $5 : 00$ *and* $17 : 00$ *of the existing building*

$$\tag{4}$$

The third step of the process of power demand projection in Fig. 1 is to calculate power demand every 30 min by dividing the amount of daily power demand into 48 steps according to hourly fluctuation pattern. Then in the fourth step, a database of projected power demand of the target building is attained in 48 steps a day, over 365 days.

2.2 Algorithm for Operating Storage Batteries

The purpose of the algorithm is to decide the minimum value of the maximum amount of power per 30 min, purchased from the grid. By deciding an appropriate minimum peak, the amount of power charged in the storage battery can be discharged effectively during the peak time, so the basic electricity fee ratio can be minimized.

Another purpose is to minimize the minimum amount of power purchased from the grid, to prevent excessive charging of the storage battery. By deciding the minimum amount of power purchased, ill effects on purchase planning such as a sudden increase of power purchasing at the beginning of nighttime can be prevented. Figure 4 shows the flow of the algorithm. After obtaining and sorting projected power demand data in the first and second steps, the maximum amount of purchased power is decided in the third and fourth steps. In the fourth step the maximum amount of power is derived in between line $[i - 1]$ and line $[i]$, using the bisection method. The minimum amount of purchased power is decided in the fifth and sixth steps in the same way as deciding maximum purchased power.

2.3 Evaluation Method for the Result of Energy Management Simulation

The feasibility of introducing storage batteries is evaluated based on the unit price of storage batteries and maintenance cost, and on the profit considering the amount of reduced basic electricity fees and commodity electricity fees. In calculating the profit and the cost, device lifetimes and profit discount ratios are also considered, as are the price of electricity and storage batteries.

3 Case Study

3.1 Projection of Power Demand

We applied the model to commercial buildings planned for construction in Tokyo, Japan. Table 1 shows an example of statistical data on power demand and the result for the projected amount of power demand in the target building.

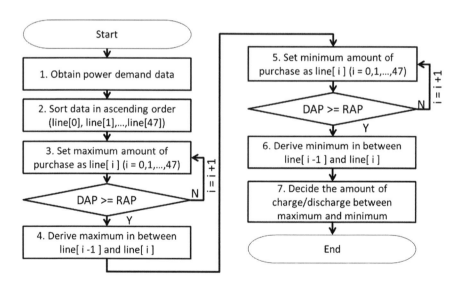

DAP: Dischargeable Amount of Power
RAP: Requested Amount of Power

Fig. 4 Algorithm for operating storage battery

Energy Management Using Storage Batteries 1171

Table 1 Projected power demand for the target building

Statistics data		Target building		
Purpose	Power demand (kWh/day/m^2)	Purpose	Floor size (m^2)	Power demand (MWh/day)
Office	1.4	Office	69,000	72.5
Commerce	2.6	Commerce	29,000	56.6
Other	0.38	Other	72,000	20.5
–		Total	170,000	149.5

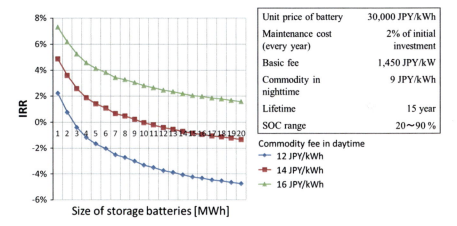

Fig. 5 Internal rate of return versus battery size for three daytime commodity fees

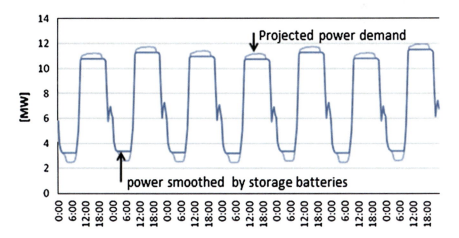

Fig. 6 Change in power purchased from the grid (July 3–9)

3.2 Evaluation of Simulation Result

We simulated energy management based on the projection, changing the storage battery size; then the internal rate of return of each case was calculated as shown in Fig. 5. Calculating the commodity fee in daytime as 16 JPY/kWh, the internal rate of return for all storage battery sizes up to 20 MWh is positive during the battery lifetimes. The size is limited when the commodity fee in daytime is 12 or 14 JPY/kWh. When deciding the storage battery size, the effect of energy durability in the event of blackout, and battery space limitations are also considered, as is profitability. We confirmed that the algorithm reduced projected peak by the storage batteries shown in Fig. 6, which shows change in purchased power after introducing 5 MWh storage batteries.

4 Conclusions

This study developed a projection method for power demand of buildings planned for construction and an algorithm for operating storage batteries, which can pursue profit by reducing the basic electricity fee and by utilizing cheaper power purchased in nighttime. As a case study, the models were applied to commercial buildings planned for construction, and the effect of the models and the profit brought by storage batteries were verified.

Acknowledgments This work was partially supported by Social System Design, Co. and its president Hideaki Miyata, to whom we express our gratitude. The authors would also like to express our deep appreciation for the reviewers' helpful comments and advice, which greatly improved this paper.

References

Borowy B, Salameh Z (1995) Methodology for optimally sizing the combination of a battery bank and PV array in a wind/PV hybrid system. IEEE Trans Energy Conver 11:367–375

NEC press release (2012) Tokyo Electric Power Co. press release. http://www.tepco.co.jp/e-rates/individual/kaitei2012/

Protogeropoulos C, Brinkworth BJ, Marshall RH (1998) Sizing and techno-economical optimization for hybrid solar photovoltaic/wind power systems with battery storage. Int J Energy Res 21:465–479

Koutroulis E, Kolokotsa D, Potirakis A, Kalaitzakis K (2006) Methodology for optimal sizing of stand-alone photovoltaic/wind-generator systems using genetic algorithms. Sol Energy 80:1072–1088

Yang H, Lu L, Zhou W (2007) A novel optimization sizing model for hybrid solar-wind power generation system. Sol Energy 81:76–84

The Optimal Parameters Design of Multiple Quality Characteristics for the Welding Thick Plate of Aerospace Aluminum Alloy

Jhy-Ping Jhang

Abstract The welding of different metal materials such as aerospace aluminum alloy has superior mechanical characteristics, but the feasible setting for the welding parameters of the TIG has many difficulties due to some hard and crisp inter-metallic compounds created within the weld line. Normally, the setting for welding parameters does not have a formula to follow; it usually depends on experts' past knowledge and experiences. Once exceeding the rule of thumb, it becomes impossible to set up feasibly the optimal parameters, and the past researches focus on thin plate. This research proposes an economic and effective experimental design method of multiple characteristics to deal with the parameter design problem with many continuous parameters and levels for aerospace aluminum alloy thick plate. It uses TOPSIS (Technique for Order Preference by Similarity to Ideal Solution) and Artificial Neural Network (ANN) to train the optimal function framework of parameter design for the thick plate weldment of aerospace aluminum alloy. To improve previous experimental methods for multiple characteristics, this research method employs ANN and all combinations to search the optimal parameter such that the potential parameter can be evaluated more completely and objectively. Additionally, the model can learn the relationship between the welding parameters and the quality responses of different aluminum alloy materials to facilitate the future applications in the decision-making of parameter settings for automatic welding equipment. The research results can be presented to the industries as a reference, and improve the product quality and welding efficiency to relevant welding industries.

Keywords TIG · TOPSIS · ANN · Aerospace aluminum alloy · Taguchi method

---

J.-P. Jhang (✉)
Department of Industrial Engineering and Management Information, Hua Fan University, Taiwan, Republic of China
e-mail: jpjhang@huafan.hfu.edu.tw

1 Introduction

The welding of different metal materials has superior mechanical characteristics, but the feasible setting for the welding parameters of the TIG has many difficulties due to some hard and crisp inter-metallic compounds created within the weld line. Normally, the setting for welding parameters does not have a formula to follow; it usually depends on experts' past knowledge and experiences. Once exceeding the rule of thumb, it becomes impossible to set up feasibly the optimal parameters, and the past researches focus on thin plate. This research proposes an economic and effective experimental design method of multiple characteristics to deal with the parameter design problem with many continuous parameters and levels for the aerospace aluminum alloy thick plate.

It is difficult to solve the optimization problem of multiple parameters by analytical method. The search algorithm is easy to fall into local optimal but not global optimal.

Jhang and Chan (2001) applied Taguchi Method with orthogonal table of L18 and quality characteristic of smaller-the-better to improve the process yield rate for air cleaners in Toyota Corona.

Tong and Wang (2000) propose the algorithm of Grey relational analysis and TOPSIS for multiple quality characteristics.

Tong and Su (1997a, b; Tong and Wang 2000) propose multi-response robust design by principal component analysis and by Fuzzy multiple attribute decision making.

Su et al. (2000) use Soft Computing to overcome the limitations of practical applications for Taguchi method. The methods used the ANN (Artificial Neural Network), GA (Simulated Anneal) and SA (Genetic Algorithm), to compare and find the global optimal solution for multiple quality characteristics.

Juang and Tarng (2002) find that the factors of welding current and welding torch drift speed are important factors for the quality of welding.

Chan et al. (2006) propose a new method for the propagation system evaluation in wireless network by neural networks and genetic algorithm.

Chang (2006) the proposed approach employs a BPN to construct the response model of the dynamic multi-response system by training the experimental data. The response model is then used to predict all possible multi-responses of the system by presenting full parameter combinations.

Chi and Hsu (2001) propose a Fuzzy Taguchi experimental method for problems with multi-attribute quality characteristics and its application on plasma arc welding.

Lin and Lin (2002) propose the use of the orthogonal array with grey relational analysis to optimize the electrical discharge machining process with multiple performance characteristics.

In order to be efficient for solving optimal parameters problems, our research uses TOPSIS (Technique for Order Preference by Similarity to Ideal Solution) and ANN to find the global optimal function framework of parameter design for the thick plate weldment of aerospace aluminum alloy.

2 Methodologies

2.1 Structure

This research collects the data of welding Taguchi experiments. There are non-destructive quality characteristics such as weld width, thickness, the ratio of melting into the deep, and the destructive quality characteristics such as tensility, shock. We compute *S/N* ratios, response graph, response table, the optimal combination of factor levels, ANOVA, contribution rate for multiple quality characteristics, which are compiled into a Cross Table to find the integrated optimal combinations. We use TOPSIS method to integrate all *S/N* ratios of multiple quality characteristics into C_i. The factors level and C_i values are training by ANN to find the optimal frame which associates all combinations to find global optimal solution. Finally, the global optimal is obtained by the confirmation experiment of different optimal solutions with respect to different methods.

2.2 Topsis

Hwang and Yoon (1981) have developed multiple criteria evaluation method called TOPSIS, taking into account the basic concept that are the distances from each program to the ideal solution and negative ideal solution, so the selected program is near ideal solution and far from the negative ideal solution. The analysis steps are as follows:

Step 1. Create the performance matrix with respect to the evaluation criterion.
Step 2. The performance values are standardized. As follows:

$$r_{ij} = \frac{x_{ij}}{\sqrt{\sum_{i=1}^{m} x_{ij}^2}} \quad (1)$$

where x_{ij} is i program under j evaluation criteria.

Step 3. The performance matrix is multiplied by the weight of each criterion.
Step 4. To calculate the distance of ideal solution (S_i^+) and the distance of the negative ideal solution (S_i^-).

$$S_i^+ = \sqrt{\sum_{j=1}^{n}(v_{ij} - v_j^+)^2} \quad \text{where, } v_j^+ = \max_i[v_{ij}], \tag{2}$$

$$S_i^- = \sqrt{\sum_{j=1}^{n}(v_{ij} - v_j^-)^2} \quad \text{where, } v_j^- = \min_i[v_{ij}], \tag{3}$$

Step 5. Arrange the priorities of the programs.

$$C_i = \frac{S_i^-}{S_i^+ + S_i^-} \tag{4}$$

where C_i is between 0 and 1, the priority of the ith program is higher when C_i is closer to 1.

Figure captions should be below the figures; table names and table captions should be above the tables. Use the abbreviation "Fig." even at the beginning of a sentence.

3 Experimental Planning

3.1 Experimental Allocation

In this study, we use the welding material is the aerospace aluminum alloy (7075) thick plate(8 mm), size is 80 × 60 × 8 mm, the welding diagram is showed in Fig. 1', 5 sets of control factors are considered; each control factor has 3 levels.

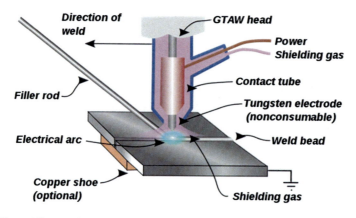

Fig. 1 The welding graph

Table 1 Experimental factors and levels

Control factor	I	II	III	Unit
1. Electric current	170	180	190	A
2. Moving speed	15	16	17	cm/min
3. Welding gap	1.5	1.7	1.9	mm
4. Striking Tungsten length	5	8	11	mm
5. Gas flow rate	11.5	13.5	15	l/min
Noise factor		3 different welding operators A, B, C		

Please refer to Table 1 for the experimental factor and its level. The noise factor is 3 different welding operators. This research adopts the orthogonal Table of L_{27}.

There are five quality characteristics as follows.

1. Welding thickness and width
   In the welding track of aluminum alloy plates, from left to right we measure the welding thickness and width for the five points of 20, 25, 30, 35, 40 mm.

2. The ratio of melting into the deep
   The ratio is welding length in the front side over the reverse side.

3. Tensile strength and shock value
   Tensile test specimens conform CNS 2112 G2014, and in accordance with the specimen 13B. Shock test is the specimen compliance CNS 3033 G2022, and in accordance with V-concave regulations. The formula of energy shock is

$$E = Wh_1 - Wh_2 = WR(COS\beta - COS\alpha) \quad (5)$$

where, α: initial angle, 143°; β: shocking angle; W: weight, 26.63 kgf; R: radius, 0.635 m

3.2 Analysis of Individual Quality Characteristic

In this study, the quality characteristics of welding thickness, tensile strength and shock value are all considered as larger-the-better, but the quality characteristics of welding width and the ratio of melting into the deep are considered as nominal-the-best.

$$\text{Nominal-the-best } S/N = 10 \times \log\left[\frac{S_m - V_e}{n \times V_e}\right] \text{ where, } S_m = \frac{(\sum y_i)^2}{n}, V_e$$

$$= \frac{1}{n-1}\left(\sum y_i^2 - S_m\right) \quad (6)$$

$$\text{Larger-the-better } S/N = -10 \times \log\left[\frac{1}{n}\sum_{i=1}^{n}\frac{1}{y_i^2}\right] \quad (7)$$

3.3 Analysis of Multiple Quality Characteristics

We compute the S/N ratios, the optimal combination of factor levels, ANOVA, contribution rate of multiple quality characteristics respectively, which are compiled into a Cross Table to find the optimal combinations.

We also use S/N ratios of multiple quality characteristics to transform into the Ci value of TOPSIS. The value of the five levels of control factors as input, the Ci value of TOPSIS as output, use BNN to build models. In this study, we select the marginal value (the maximum and minimum) and the median value of 27 groups of samples as the test samples, and the remaining samples for training, the criteria of decision-making is according to the MSE values of ANN, the MSE of training samples and test samples are the more smaller the more better. The optimal frame which associates all combinations finds global optimal solution.

4 Results Analysis

4.1 The Optimal Combinations of Cross Table

We compute the S/N ratios, the combination of factor levels, ANOVA, contribution rate of multiple quality characteristics, which are compiled into a Cross Table to find the optimal combinations, as shown in Table 2.

4.2 The Optimal Combinations of TOPSIS

We use TOPSIS method to integrate all S/N ratios of multiple quality characteristics into Ci and to find the optimal combination as shown in Table 3.

4.3 Confirmation Experiment

The 95 % Confidence interval of Ci for the confirmation experiment is [0.44, 1.07].

4.4 Results and Discussions

From Table 4, the Ci of ANN and all combinations is larger than Ci of Cross table, and it falls into the 95 % confidence interval of Ci for the confirmation experiment. So the optimal combination ANN and all combinations is the total optimal welding parameters design of aerospace Aluminum alloy thick plate.

The significant factors are welding gap and striking Tungsten length.

The Optimal Parameters Design of Multiple Quality

Table 2 Cross table

Factor	A	B	C	D	E	Factor	A	B	C	D	E
Welding thickness (10 %)						**Shock value (15 %)**					
Optimal combination	A3	B1	C1	D1	E3	Optimal combination	A3	B1	C3	D2	E3
Significant of S/N	*			*	*	S/N Significant		*	*	*	*
Contribution rate (%)	18 %	2 %	4 %	43 %	12 %	Contribution rate (%)	4 %	4 %	4 %	4 %	4 %
Welding width (10 %)						**Tensile strength (40 %)**					
Optimal combination	A2	B2	C3	D1	E2	Optimal combination	A2	B2	C3	D2	E3
S/N Significant	*	*	*		*	S/N Significant			*	*	*
Contribution rate (%)	0 %	3 %	5 %	26 %	0 %	Contribution rate (%)	6 %	0 %	23 %	15 %	6 %
The ratio of Melting into the deep (25 %)						**Optimal parameters levels**	A2	B2	C3	D2	E3
Optimal combination	A2	B2	C1	D3	E2						
S/N Significant		*		*							
Contribution rate (%)	5 %	2 %	3 %	37 %	10 %						

Table 3 Response table of C_i

Factor		A	B	C	D	E	Average
C_i (TOPSIS)	Level1	0.45	0.47	0.46	0.34	0.45	0.43
	Level2	0.56	0.55	0.43	0.57	0.53	0.53
	Level3	0.49	0.48	0.61	0.58	0.54	0.54
	Comparison	0.10	0.08	0.18	0.24	0.14	0.14
	Best Level	A2	B2	C3	D3	E2	
	Rank	3	4	2	1	5	
	Significant			*	*		

Table 4 The comparison C_i of confirmation experiment

Optimal combinations	Cross table A2B2C3D2E3	TOPSIS A2B2C3D3E2	ANN and all combinations
C_i value	0.77	0.90	0.93

5 Conclusions

The conclusions are summarized in the following:

1. The ANN and all combinations method used in this case are better than others. So the optimal combination of ANN and all combinations is the total optimal welding parameters design of aerospace Aluminum alloy thick plate.
2. The significant factors are welding gap and striking Tungsten length in this case.
3. In the future, we can consider using the ANN, GA and SA to find the optimal solution for multiple quality characteristics. We can also consider other welding techniques, such as CO_2 welding, GMAW (Gas Tungsten Arc Welding) and LAFSW.

References

Chan HL, Liang SK, Lien CT (2006) A new method for the propagation system evaluation in wireless network by neural networks and genetic algorithm. Int J Inf Syst Logistics Manag 2(1):27–34

Chang HH (2006) Dynamic multi-response experiments by back propagation networks and desirability functions. J Chin Inst Ind Eng 23(4):280–288

Chi SC, Hsu LC (2001) A fuzzy Taguchi experimental method for problems with multi-attribute quality characteristics and its application on plasma arc welding. J Chin Inst Ind Eng 18(4):97–110

Hwang GL, Yoon K (1981) Multiple attributes decision making methods and applications. Springer, New York

Jhang JP, Chan HL (2001) Application of the Taguchi method to improve the process yield rate for air cleaners in Toyota Corona vehicles. Int J Reliab Qual Saf Eng 89(3):219–231

Juang SC, Tarng YS (2002) Process parameter selection for optimizing the weld pool geometry in the tungsten inert gas welding of stainless steel. J Mater Process Tech 122:33–37

Lin JL, Lin CL (2002) The use of the orthogonal array with grey relational analysis to optimize the electrical discharge machining process with multiple performance characteristics. Int'l J Mach Tools Manuf 42(2):237–244

Su CT, Chiu CC, Chang HH (2000) Optimal parameter design via neural network and genetic algorithm. Int J Ind Eng 7(3):224–231

Tong LI, Wang CH (2000) Optimizing multi-response problems in a dynamic system by grey relational analysis. J Chin Inst Ind Eng 17(2):147–156

Tong LI, Su CT (1997a) Optimizing multi-response problems in the Taguchi method by Fuzzy multiple attribute decision making. Qual Reliab Eng Int'l 13:25–34

Tong LI, Su CT (1997b) Multi-response robust design by principal component analysis. Total Qual Manag 8(6):409–416

Synergizing Both Universal Design Principles and Su-Field Analysis to an Innovative Product Design Process

Chun-Ming Yang, Ching-Han Kao, Thu-Hua Liu, Ting Lin and Yi-Wun Chen

Abstract To promote developing more usable and accessed, daily-used products that could meet the rigorous requirements from diverse consumers at the present time, this research proposed an innovative product design process by synergizing both universal design (UD) principles and TRIZ tools. This newly developed process started with stating the design problems via a UD evaluation, followed by PDMT analysis to develop the preliminary design directions. The directions were then analyzed by using Su-Field models in order to locate the potential resolutions from TRIZ's 76 Standard Solutions. Finally, a case study was conducted to demonstrate how this innovative design process works. Study result shows that this approach can help identify the core of the problem and locate the improved product concepts effectively, resulting in generating more creative and usable product design.

Keywords TRIZ · Su-field analysis · Universal design · PDMT

C.-M. Yang (✉) · C.-H. Kao · T.-H. Liu · T. Lin · Y.-W. Chen
Department of Industrial Design, Ming Chi University of Technology, 84 Gungjuan Road, Taishan Distric, New Taipei City, Taiwan
e-mail: cmyang@mail.mcut.edu.tw

C.-H. Kao
e-mail: kaoch@mail.mcut.edu.tw

T.-H. Liu
e-mail: thliu@mail.mcut.edu.tw

T. Lin
e-mail: linting20@gmail.com

Y.-W. Chen
e-mail: lisa60832@gmail.com

1 Introduction

Declining birth rates are leading to an increase in the proportion of aged people in the population. For this reason, more emphasis is being placed on universal design. However, difficulties often occur in the design of universally applicable products because the principles offered are too general. Effectively using the principles of universal design requires a tool to provide direction in the design process. TRIZ is an instrument capable of dealing with a lack of inspiration and provides solutions to creativity-related problems. This study integrated universal design with TRIZ to establish a process for the systematic development of products, capable of guiding designers toward innovative solutions.

2 Literature Review

2.1 Universal Design

After World War II, the medical treatment for the injured soldiers and social turbulence led to the concern over the issue of "Barrier-free Design". During the implementation of this concept, related issues were expanded to a broader scope, not only covering individuals with physical and mental disabilities, but also the broad user population. Thus, it evolves into the concept of universal design in modern time (Duncan 2007). In 1990, the U.S. approved ADA. Although laws are passed to protect the individuals with disabilities, they still face much inconvenience in use of space or products. Therefore, The Center of Universal Design led by Ronald L. Mace, based on ADA, treated universal design as "all products and the built environment to be aesthetic and usable to the greatest extent possible by everyone, regardless of their age, ability, or status in life", and advocated "the design of products and environments to be usable by all people, to the greatest extent possible, without the need for adaptation or specialized design" (Duncan 2007; The Center for universal design 1997).

Scholars have presented a variety of definition for the principles of universal design. The Center for Universal Design (1997) proposed seven principles: (1) equitable use, (2) flexibility in use, (3) simplicity and intuitive operation, (4) perceptible information, (5) tolerance for error, (6) low physical effort, and (7) size and space for approach and use. These can be used to evaluate whether the design of a product is universal and direct the actions of designers accordingly. Nakagawa (2006) pointed out the shortcomings of the seven universal design principles and outlined three additional attributes: durability and production economics, quality and aesthetics, and health and natural environment. These were combined with 37 sub-principles to create an evaluation form for universal design, known as the Product Performance Program (PPP).

Universal design principles have been widely mentioned and applied by researchers. According to Preiser and Ostroff (2001), universal design refers to the planning and design regardless of the user. They also claimed that universal design involves a sense of space, which means that knowledge databases related to ergonomics can also be used in universal design. Preiser (2008) categorized the literature dealing with universal design as relating to industrial design, product design, fashion design, interior design, architecture, urban design and planning, information technology, health facility planners, administration, facility managers, and environmental psychologists. He also suggested that PPPs be designed to fit the domain. Muller (1997) provided background descriptions, definitions, and case analysis based on the seven principles outlined by the Center for Universal Design.

2.2 TRIZ

The Theory of Inventive Problem Solving (TRIZ) was created by Genrich Altshuller (1926–1998), who began investigating solutions to the problems of invention in 1946. During his patent research, Altshuller found that among hundreds of thousands of patents, only approximately 40,000 (2 %) were actual pioneering inventions (Altshuller et al. 1997). Altshuller (1999) observed three types of obstacles to thought processes during the evolution of innovation and invention: psychological inertia, limited domain knowledge, and trial and error method. Overcoming these obstacles and avoiding being led astray require a theory of innovation and invention and TRIZ is an instrument capable of dealing with multiple obstacles to thought processes.

TRIZ provides systematic solutions to problems, including scenario analysis, contradiction analysis, substance-field (su-field) analysis, the ARIZ problem-solving system, 40 inventive principles, and 76 standard solutions (Terninko et al. 1998). Despite the many techniques in TRIZ, the primary goal and problem-solving techniques still focus on identifying contradictions and ideal solutions (Ideation International Inc. 2006), and in the event that a system requires improvements, su-field analysis can be used for prediction.

2.3 Substance-Field Analysis

Su-Field Analysis was proposed by Altshuller in 1979. He developed models that described structural problems within systems and used symbolic language to clearly express the functions of technical systems (subsystems), thereby accurately describing the constituent elements of these systems and the relationships among them (Altshuller 1984). The elements necessary to construct and define the technical model of a system include two substances (S) and a field (F) (Fig. 1). A triangular model presents the design relationships among them. The substances can

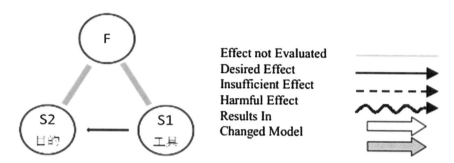

Fig. 1 Triangular model of substance-field analysis and legend of triangular model

be any objects. The symbol for substance is S, and S1 generally represents a passive role, S2 denotes an active role, and S3 indicates the introduced substance. The field displays the manner in which the substances interact. In physics, there are only four types of fields: gravity, electromagnetism, the nuclear force field, and the particle field (Savransky 2000; Mann 2007). The type of field determines the process of actions, and different relationships between the substances and the field correspond to different standard solutions. In the constructed model, the structure of the interactions among the three often demonstrates the following effects: the desired effect, harmful effects, and insufficient effects. These effects are displayed using different lines (Fig. 1) (Mann 2007). These symbols clearly show how the problem is structured, thereby enabling users to clarify the problem.

In the framework of the su-field analysis model, a system generally includes a variety of functions, each of which requires a model. TRIZ divides functions into the following four categories (Terninko et al. 1998):

1. Effective complete system: This function possesses all three components, which are all effective and display the effects desired by the designers.
2. Incomplete system: A portion of the three components does not exist and therefore requires additional components to achieve the effective complete function or a whole new function as a replacement.
3. Ineffective complete system: All of the components that provide the function exist, but the effects desired by the designers are incomplete.
4. Harmful complete system: All of the components that provide the function exist, but the effects produced conflict with the ones that the designers desire. During the process of innovation, harmful functions must be eliminated.

With the exception of (a), all of the models have problems that require improvement. 76 Standard Solutions, developed by Altshuller in conjunction with substance-field, can be used to find solutions to problems in any given model. 76 Standard Solutions can be divided into five classes (Table 1) (Savransky 2000). Although su-field analysis and the 76 Standard Solutions can be adopted independently, combining them is generally more effective in deriving elegant solutions.

Table 1 76 standard solutions

Class	Description	Number of standard solutions
Class 1	Improving the system with no or little change	13 Standard solutions
Class 2	Improving the system by changing the system	23 Standard solutions
Class 3	System transitions	6 Standard solutions
Class 4	Detection and measurement	17 Standard solutions
Class 5	Strategies for simplification and improvement	17 Standard solutions

2.4 Development of Systematic Universal Design Procedure Based on Substance-Field Analysis

Striving to follow all the principles of universal design can easily result in products that lack creativity. Thus, this study integrated universal design to detect problems and evaluate products, and then used the problem-solving techniques of TRIZ to derive solutions. This approach results in design concepts featuring the best of both worlds. This study utilized universal design principles as the main framework and integrated TRIZ theory with su-field analysis models to establish an innovative procedure for the implementation of universal design. We first investigated problems that did not conform to universal design principles using the PPP questionnaire. We then defined the preliminary direction for problem-solving using the PDMT approach before analyzing the problem using a su-field analysis model. Adopting a suitable solution from the 76 Standard Solutions of TRIZ enabled the generation of innovative solutions based on universal design.

Participants were required to actually operate the existing product before answering the 37 question items in the PPP questionnaire regarding their satisfaction towards the design of the product. Scoring was based on a Likert 5-point Scale, ranging from "very satisfied" (40) to "very dissatisfied" (0). We categorized and calculated the mean scores from problems originating from the same principles and selected the universal design principle with the lowest score as the defect that was most in need of improvement. We assumed an optimal solution method based on the objective and used PDMT (Chen 2008) to analyze the primary problem. This study derived the purpose, directions, methods, and tools for PDMT analysis. The universal design principle with the lowest score in the PPP scale was set as the problem. Concept development was then based on the optimal solution method, and the solution was extended from the direction of concept development.

Using su-field analysis, we then constructed a triangular model for the preliminary direction of problem-solving. We substituted the action of the component or product into the triangular model as F, the solution field (solution), and the medium of the object as S2, the improvement tool; S1 denoted the product or component to be improved. Once completed, the triangular model clearly identifies the part of the system requiring improvement using the 76 Standard Solutions. We then analyzed the problem (Table 2) and determined whether the problem was

Table 2 Problem analysis using 76 standard solutions

Problem analysis using 76 standard solutions	YES/NO
1. Does the system possess two substances and one field?	
2. Is the problem related to measurement?	
3. Do harmful relationships exist in the system?	

related to measurement. If this were the case, Class 4 (detection and measurement) would be taken into consideration. If the problem was not related to measurement, we would seek principle solutions from the other four classes and develop concepts to derive the final solution for the designers.

3 Results

This study adopted the common pen as a case study. According to the questionnaire results, Principle 6 (low physical effort) was identified as the primary issue in the use of pens. We therefore set "no fatigue after long-term use" as the ultimate objective of our ideal solution. This objective provided a suitable problem-solving direction and feasible solutions as well as direction analysis. We then derived the characteristics of weight, shape, reminder, and method of use, from which we extended eight improvement methods: material, ergonomics, time reminder, sound, vibration, other external forces, cease in pen use, and contact with more than paper. Using su-field analysis, we constructed a model and selected a solution from those derived using PDMT to create a substance-field model.

Case 1: In this instance, weight function was considered. Once the su-field model was constructed (Fig. 2), we defined the system accordingly and considered the sequence of events in using the pen. In accordance with the answers in the table above, we had to consider all five classes. For the sake of conciseness, we simply outline the results of our search and comparison, 5.1.1.1 of introducing substances, which involves the use of "nothing", such as vacuum, air, bubbles, foam, voids, and gaps. These suggestions led us to produce a pen with a skeletal structure to reduce the weight of the pen and prevent fatigue after long-term use (Fig. 2).

Case 2: In this case, cease using the pen was considered. With constructing the substance-field model, we incorporated an unknown F, in the hopes of changing the way pens are generally used (Fig. 3). For this, we employed 2.4.1 of detection and measurement in the 76 Standard Solutions, adding ferromagnetic materials. We magnetized the ink and placed a magnetic pad before the fingers of the participants so that they could write effortlessly (Fig. 3).

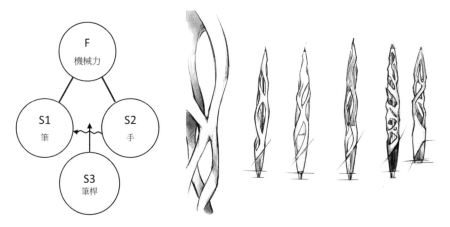

Fig. 2 Triangular model and concept development for case 1

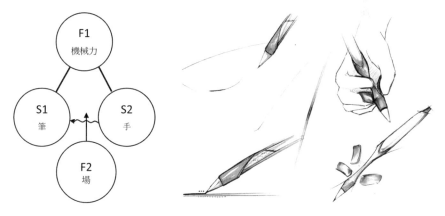

Fig. 3 Triangular model and concept development for case 2

4 Conclusion

Universal design is meant to satisfy the needs of the greater majority rather than a perspective of user issues from social concern. Universal design principles can provide reasonable suggestions for the development and design of products; however, designers often fail to make breakthroughs or are unsure of how to proceed. This study used TRIZ tools to provide innovative design procedure applicable to all domains. We also adopted su-field analysis to depict problems graphically in order to enable users to clarify and identify problems. Our results indicate that the integration of universal design principles and TRIZ tools provides a logical procedure for designers to follow to ensure that the resulting products are creatively realized and in line with the principles of universal design.

Acknowledgments The authors are grateful for support of the National Science Council, Taiwan under grant NSC101-2221-E-131-002-MY2.

References

Altshuller G (1984) Creativity as an exact science. Gordon and Breach, New York
Altshuller G, Shulyak L, Rodman S (1997). 40 principles: TRIZ keys to innovation (Vol. 1). Technical Innovation Center Inc
Altshuller G (1999) The innovation algorithm: TRIZ, systematic innovation and technical creativity. Technical innovation center, Worcester
Chen CH (2008) Introduction to TRIZ and CREAX. Workshop (Pitotech Co Ltd). 2 Oct 2008
Duncan R (2007) Universal design-clarification and development. A Report for the Ministry
Ideation International (2006) History of TRIZ and I-TRIZ. http://www.ideationtriz.com/history.asp. Accessed 21 Nov 2012
Mann D (2007) Hands-on systematic innovation. Tingmao, Taipei
Mueller J (1997) Case studies on universal design. Des Res Methods J 1(1)
Nakagawa S (2006) Textbook for universal design. Longsea press, Taipei
Preiser WFE (2008) Universal design: from policy to assessment research and practice. Archnet-IJAR, 2
Preiser W, Ostroff E (2001) Universal design handbook. New York
Savransky SD (2000) Engineering of creativity: introduction to TRIZ methodology of inventive problem solving. CRC Press, Boca Raton
Terninko J, Zusman A, Zoltin B (1998) Systematic innovation-introduction to TRIZ. CRC Press, Boca Raton
The Center of Universal Design (1997) http://www.ncsu.edu/project/design-projects/udi/. Accessed 3 Nov 2012

The Joint Determination of Optimum Process Mean, Economic Order Quantity, and Production Run Length

Chung-Ho Chen

Abstract In this study, the author proposes a modified Chen and Liu's model with quality loss and single sampling rectifying inspection plan. Assume that the retailer's order quantity is concerned with the manufacturer's product quality and the quality characteristic of product is normally distributed. Taguchi's symmetric quadratic quality loss function will be applied in evaluating the product quality. The optimal retailer's order quantity and the manufacturer's process mean and production run length will be jointly determined by maximizing the expected total profit of society including the manufacturer and the retailer.

Keywords Economic order quantity · Process mean · Production run length · Taguchi's quadratic quality loss function

1 Introduction

The supply chain system is a major topic for the manufacturing industries in order to obtain the maximum expected total profit of society including the manufacturer and the retailer. The manufacturer's objective needs to consider the sale revenue, the manufacturing cost, the inspection cost, and the inventory cost for having the maximum expected profit. The retailer's objective needs to consider the order quantity, the holding cost, the goodwill loss of cost, and the used cost of customer for having the maximum expected profit. How to get a trade-off between them should be available for further study. Chen and Liu (2007) presented the optimum profit model between the producers and the purchasers for the supply chain system with pure procurement policy from the regular supplier and mixed procurement

C.-H. Chen (✉)
Department of Management and Information Technology, Southern Taiwan University of Science and Technology, Tainan, Taiwan
e-mail: chench@mail.stust.edu.tw

policy from the regular supplier and the spot market. Chen and Liu (2008) further proposed an optimal consignment policy considering a fixed fee and a per-unit commission. Their model determines a higher manufacturer's profit than the traditional production system and coordinates the retailer to obtain a large supply chain profit.

In Chen and Liu's (2008) model with traditional production system, they neglected the effect of product quality on the retailer's order quantity and only considered the order quantity obeying the uniform distribution. In fact, the retailer's order quantity is concerned with product quality. Chen and Liu's (2008) model with simple manufacturing cost did not consider the used cost of customers in traditional production system. Hence, the modified Chen and Liu's (2008) model needs to be addressed for determining the optimum process parameters. Chen (2010) proposed a modified Chen and Liu's (2008) model with quality loss and single sampling plan based on the dependent assumption of the retailer's order quantity and manufacturer's product quality. However, Chen (2010) neglected manufacturer's inventory cost and the cost for the non-conforming products in the sample of accepted lot.

Hanna and Jobe (1996) discussed the quality characteristic of product and quality cost on the effect of lot size. They determine the optimal order quantity lot when the model with quality cost evaluation based on 100 % inspection, sampling inspection, and no inspection for products. Jaber et al. (2009) considered entropic order quantity model when product characteristic is not perfect. Their results suggested that larger quantities should be ordered than those of the classical economic order quantity model. Economic selection of process mean is an important problem for modern statistical process control. It will affect the expected profit/cost per item. Recently, many researchers have addressed this work. Both 100 % inspection and sampling inspection are considered for different models. Taguchi (1986) presented the quadratic quality loss function for redefining the product quality. Hence, the optimum product quality should be the quality characteristic with minimum bias and variance. Recently, his quality loss function has been successfully applied in the problem of optimum process mean setting.

In this paper, the work will propose a modified Chen and Liu's (2008) model with quality loss, manufacturer's inventory cost, and single sampling rectifying inspection plan. Assume that the retailer's order quantity is concerned with the manufacturer's product quality and the quality characteristic of product is normally distributed. The non-conforming products in the sample of accepted lot are replaced by conforming ones. If the lot is rejected, then all of the products are rectified and sold at the same price as the products of accepted lot. Taguchi's (1986) symmetric quadratic quality loss function will be applied in evaluating the product quality. The optimal retailer's order quantity and the manufacturer's process mean and production run length will be jointly determined by maximizing the expected total profit of society including the manufacturer and the retailer. The motivation behind this work stems from the fact that the neglect of the quality loss within the specification limits and manufacturer's inventory cost should have the overestimated expected total profit of society.

2 Modified Chen and Liu's (2008) Traditional System Model

Taguchi (1986) redefined the product quality as the loss of society when the product is sold to the customer for use. The used product for customer maybe occur a lot of costs including maintenance, safety, pollution, and sale service. Chen and Liu's (2008) model also did not consider the used cost of customers. The neglect of the quality loss within the specification limits should have the overestimated expected profit per item for the retailer.

Assume that the quality characteristic of Y is normally distributed with unknown mean μ_y and known standard deviation σ_y, i.e., $Y \sim N\left(\mu_y, \sigma_y^2\right)$ and $X|Y \sim N(\lambda_1 + \lambda_2 y, \sigma^2)$, where λ_1, λ_2, and σ^2 are constants. Hence, we have $X \sim N\left(\lambda_1 + \lambda_2 \mu_y, \lambda_2^2 \sigma_y^2 + \sigma^2\right)$ and $Y|X \sim N\left(\frac{\lambda_2 \sigma_y^2 (x - \lambda_1) + \mu_y \sigma^2}{\lambda_2^2 \sigma_y^2 + \sigma^2}, \frac{\sigma_y^2 \sigma^2}{\lambda_2^2 \sigma_y^2 + \sigma^2}\right)$.

Taguchi (1986) proposed the quadratic quality loss function for evaluating the product quality. If the product quality characteristic is on the target value, then it has the optimum output value. However, we need to input some different resource in the production process. Hence, the process control needs to obtain minimum bias and variance for output product. According to Taguchi's (1986) definition for product quality, the retailer's expected profit should subtract the used cost of customer for product in order to avoid overestimating retailer's expected profit. Hence, the author proposes the following modified Chen and Liu's (2008) model.

The retailer's profit is given by

$$\pi_{PS}^R = \begin{cases} RX - WQ - H(Q - X) - X \cdot Loss(Y), & X <; Q, \; -\infty <; Y <; \infty \\ RQ - WQ - S(X - Q) \cdot Loss(Y), & X \geq Q, \; -\infty <; Y <; \infty \end{cases} \quad (1)$$

where X is the consumer demand which is an uniform distribution, $X \sim U[\mu_x - (\sigma_x/2), \mu_x + (\sigma_x/2)]$, μ_x is the mean of X, σ_x is the variability of X, and $f(x)$ is the probability distribution of X; R is a retailer purchasing a finished product from a regular supplier and reselling it at this price to the end customer; C is the regular manufacturer produces each unit at this cost; W is the regular manufacturer and the retailer entering into a contract at this wholesale price; Q is the regular manufacturer setting the wholesale price to maximize his expected profit while offering the buyer this specific order quantity; S is a goodwill loss for the retailer when realized demand exceeds procurement quantity; H is a carrying cost for the retailer when realized demand is less than procurement quantity; Y is the normal quality characteristic of product, $Y \sim N\left(\mu_y, \sigma_y^2\right)$; μ_y is the unknown mean of Y; σ_y is the known standard deviation of Y; $Loss(Y)$ is Taguchi's (1986) quadratic quality loss function per unit, $Loss(Y) = k(Y - y_0)^2$; k is the quality loss coefficient; y_0 is the target value of product.

The retailer's expected profit includes the sale profit when the demand quantity of customer is less than order quantity, the sale profit when the demand quantity of customer is greater than order quantity, the carrying cost when the demand quantity of customer is less than order quantity, and the goodwill loss when the demand quantity of customer is greater than order quantity. From Chen (2010), we have the expected profit of retailer as follows:

$$E(\pi_{PS}^R) = E(\pi_1) + E(\pi_2) - E(\pi_3) - E(\pi_4) \tag{2}$$

where

$$E(\pi_1) = (R+H)\left\{\mu_k \Phi\left(\frac{Q-\mu_k}{\sigma_k}\right) - \sigma_k \phi\left(\frac{Q-\mu_k}{\sigma_k}\right)\right\} - (W+H)Q\Phi\left(\frac{Q-\mu_k}{\sigma_k}\right) \tag{3}$$

$$E(\pi_2) = (R-W+S)Q\left[1 - \Phi\left(\frac{Q-\mu_k}{\sigma_k}\right)\right]$$
$$- S\left\{\mu_k\left[1 - \Phi\left(\frac{Q-\mu_k}{\sigma_k}\right)\right] + \sigma_k \phi\left(\frac{Q-\mu_k}{\sigma_k}\right)\right\} \tag{4}$$

$$E(\pi_3) = kA^2\left\{\mu_k^3 \Phi\left(\frac{Q-\mu_k}{\sigma_k}\right) + 3\mu_k^2 \sigma_k\left[-\phi\left(\frac{Q-\mu_k}{\sigma_k}\right)\right]\right.$$
$$+ 3\mu_k \sigma_k^2\left[-\frac{Q-\mu_k}{\sigma_k} \cdot \phi\left(\frac{Q-\mu_k}{\sigma_k}\right) + \Phi\left(\frac{Q-\mu_k}{\sigma_k}\right)\right] + \sigma_k^3\left[-\left(\frac{Q-\mu_k}{\sigma_k}\right)^2 \cdot \phi\left(\frac{Q-\mu_k}{\sigma_k}\right)\right.$$
$$\left.-2\phi\left(\frac{Q-\mu_k}{\sigma_k}\right)\right\} + 2kAB\left\{\mu_k^2\left[\Phi\left(\frac{Q-\mu_k}{\sigma_k}\right)\right] - 2\mu_k \sigma_k \phi\left(\frac{Q-\mu_k}{\sigma_k}\right)\right.$$
$$\left.+ \sigma_k^2\{-\frac{Q-\mu_k}{\sigma_k}\phi(\frac{Q-\mu_k}{\sigma_k})\}\right\} + k(B^2+C_0)\left\{\mu_k \Phi\left(\frac{Q-\mu_k}{\sigma_k}\right) - \sigma_k \phi\left(\frac{Q-\mu_k}{\sigma_k}\right)\right\} \tag{5}$$

$$E(\pi_4) = kA^2 Q\left\{\mu_k^2\left[1 - \Phi\left(\frac{Q-\mu_k}{\sigma_k}\right)\right] + 2\mu_k \sigma_k \phi\left(\frac{Q-\mu_k}{\sigma_k}\right)\right.$$
$$\left.+ \sigma_k^2\left\{\left[\left(\frac{Q-\mu_k}{\sigma_k}\right)\phi\left(\frac{Q-\mu_k}{\sigma_k}\right)\right] + \left[1 - \Phi\left(\frac{Q-\mu_k}{\sigma_k}\right)\right]\right\}\right\}$$
$$+ 2kQAB\left\{\mu_k\left[1 - \Phi\left(\frac{Q-\mu_k}{\sigma_k}\right)\right] + \sigma_k \phi\left(\frac{Q-\mu_k}{\sigma_k}\right)\right\}$$
$$+ kQ(B^2+C_0)\left\{1 - \Phi\left(\frac{Q-\mu_k}{\sigma_k}\right)\right\} \tag{6}$$

where $\mu_k = \lambda_1 + \lambda_2 \cdot \mu_y$; $\sigma_k = \sqrt{\lambda_2^2 \sigma_y^2 + \sigma^2}$; $A = \frac{\lambda_2^2 \sigma_y^2}{\lambda_2^2 \sigma_y^2 + \sigma^2}$; $B = \frac{\mu_y \sigma^2 - \lambda_1 \lambda_2 \sigma_y^2}{\lambda_2^2 \sigma_y^2 + \sigma^2} - y_0$; $C_0 = \frac{\sigma_y^2 \sigma^2}{\lambda_2^2 \sigma_y^2 + \sigma^2}$; $\Phi(\cdot)$ is the cumulative distribution function of standard normal random variable; $\phi(\cdot)$ is the probability density function of standard normal random variable.

Assume that the retailer's order quantity is equal to the lot size of single sampling rectifying inspection plan. If the lot is accepted, then the selling price of product per unit is W. The non-conforming products in the sample of accepted lot are replaced by conforming ones. Let R_I denote the cost of replacing a defective item by an acceptable item in the accepted lot. If the lot is rejected, then all of the products are rectified and sold at a price W. Let R_L denote the expected cost of replacing all rejected items found in a rejected lot. Hence, the manufacturer's profit under adopting single rectifying inspection plan for determining the quality of product lot is given by

$$\pi_{PS}^S = \begin{cases} WQ - ni - DR_I - Qc\mu_y, & D \leq d_0 \\ WQ - Qi - R_L - Qc\mu_y, & D > d_0 \end{cases} \quad (7)$$

where n is the sample size; c is the variable production cost per unit; i is the inspection cost per unit; d_0 is the acceptance number; D is the number of non-conformance in the sample; c is the cost of processing per unit; i is the inspection cost per unit; $R_L = R_I \cdot d_{rl}$; d_{rl} is the expected number of defective items in a rejected lot (= the expected number of defectives found in the sample, given that the lot was rejected + the expected number of defectives in the non-sample portion of the lot),

$$d_{rl} = E(D|D > d_0) + p(Q - n) \quad E(D|D > d_0) = \frac{np\left[1 - \sum_{x=0}^{d_0-1} \frac{e^{-np}(np)^x}{x!}\right]}{1 - \sum_{x=0}^{d_0} \frac{e^{-np}(np)^x}{x!}}; p \text{ is the prob-}$$

ability of a defective item $\left(= 1 - \left[\Phi\left(\frac{U-\mu_y}{\sigma_y}\right) - \Phi\left(\frac{L-\mu_y}{\sigma_y}\right)\right]\right)$; L is the lower specification limit of product; U is the upper specification limit of product; $\Phi(\cdot)$ is the cumulative distribution function of the standard normal random variable.

The manufacturer's expected profit for the product lot is

$$\begin{aligned} E_1(\pi_{PS}^S) &= (WQ - ni - DR_I - Qc\mu_y)P_1 + (WQ - Qi - R_L - Qc\mu_y)(1 - P_1) \\ &= [R_L + (Q - n)i]P_1 - R_I n p P_0 + (WQ - R_L - Qi - Qc\mu_y) \end{aligned} \quad (8)$$

where

$$P_1 = \sum_{d=0}^{d_0} \frac{e^{-np} \cdot (np)^d}{d!} \quad (9)$$

$$P_0 = \sum_{d=0}^{d_0-1} \frac{e^{-np} \cdot (np)^d}{d!} \quad (10)$$

The manufacturer should consider the inventory cost if the product is produced and unsold before the retailer's order. Hence, the expected total profit for the manufacturer with imperfect quality of product is that the expected total profit for the product lot subtracts the total inventory cost including the set-up cost and the holding cost as follows:

$$E\left(\pi_{PS}^S\right) = E_1\left(\pi_{PS}^S\right) - S_1 \cdot \frac{Q}{I_1 T} - \frac{B_1(I_1 - O_1)T}{2} \quad (11)$$

where Q is the order quantity from the retailer; O_1 is the demand quantity in units per unit time; S_1 is the set-up cost for each production run; I_1 is the production quantity in units per unit time; B_1 is the holding cost per unit item per unit time; T is the production run length per unit time.

The expected total profit of society including the retailer and the manufacturer is

$$ETP(Q, \mu_y, T) = E\left(\pi_{PS}^R\right) + E\left(\pi_{PS}^S\right) \quad (12)$$

In Chen and Liu's (2008) model, the retailer determines the order quantity and the manufacturer sequentially determine the wholesale price for maximizing respective objective function. Their solution is based on independence between order quantity and wholesale price. However, the dependence exists in the modified Chen and Liu's (2008) model because the order quantity is related with the manufacturer's product quality characteristic. Hence, we need to solve Eq. (13) to simultaneously obtain the optimal retailer's order quantity (Q^*), the optimal manufacturer's process mean $\left(\mu_y^*\right)$, and the optimal production run length (T^*) with the maximum expected profit for the retailer and the manufacturer.

It is difficult to show that Hessian's matrix is a negative definite matrix for Eq. (13). One cannot obtain a closed-form solution. To decrease decision variables in solving the optimization problem, we consider maximizing the expected total profit of society, partially differentiating Eq. (13) with respect to T and equaling to zero:

$$\frac{\partial ETP(Q, \mu_y, T)}{\partial T} = \frac{S_1 Q}{I_1 T^2} - \frac{B_1(I_1 - O_1)}{2} = 0 \quad (13)$$

From Eq. (13), we get an explicit expression of T in terms of order quantity Q:

$$T = \sqrt{\frac{2S_1 Q}{I_1(I_1 - O_1)B_1}} \quad (14)$$

The heuristic solution procedure for the above model (13) is as follows:

Step 1. Set maximum $Q = Q_{\max}$.
Step 2. Let $Q = 1$
Step 3. Compute $T = \sqrt{\frac{2S_1 Q}{I_1(I_1 - O_1)B_1}}$.
Step 4. Let $L < \mu_y < U$. One can adopt direct search method for obtaining the optimal μ_y^* with the maximum expected total profit of society for Eq. (13) with the given order quantity Q and production run length T.

Step 5. Let $Q = Q + 1$. Repeat Steps 3–4 until $Q = Q_{\max}$. The combination $\left(Q^*, \mu_y^*, T^*\right)$ with maximum expected total profit of society is the optimal solution.

3 Numerical Example and Sensitivity Analysis

Assume that some parameters are as follows: $R = 100$, $W = 40$, $S = 3$, $H = 2$, $\lambda_1 = 100$, $\lambda_2 = 0.8$, $n = 16$, $d_0 = 1$, $\sigma = 2$, $y_0 = 10$, $\sigma_y = 0.5$, $i = 0.05$, $k = 50$, $c = 0.5$, $L = 8$, $U = 12$, $R_I = 1$, $I_1 = 10$, $O_1 = 8$, $S_1 = 2$, and $B_1 = 4$. By solving Eq. (13), one obtains the optimal process mean $\mu_y^* = 10.08$, the optimal order quantity $Q^* = 112$, and the optimal production run length $T^* = 2.37$ with retailer's expected profit $E\left(\pi_{PS}^R\right) = 5029.82$, manufacturer's expected profit $E\left(\pi_{PS}^S\right) = 3895.79$, and expected total profit of society $ETP(Q, \mu_y) = 8925.61$.

We do the sensitivity analysis of some parameters. From Table 1, we have the following observations:

1. The order quantity, the process mean, and the production run length almost is constant as the sale price per unit (R) increases. The retailer's expected profit, the manufacturer's expected profit, and the expected total profit of society increase as the sale price per unit increases. The sale price per unit has a have a major effect on the retailers' expected profit and the expected total profit of society.
2. The order quantity increases, the process mean is constant, and the production run length increases as the intercept of mean demand of customer (λ_1) increases. The retailer's expected profit, the manufacturer's expected profit, and the expected total profit of society increase as the intercept of mean demand of customer increases. The intercept of mean demand of customer has a have a major effect on the retailers' expected profit, manufacturer's expected profit, and the expected total profit of society.

Table 1 The effect of parameters for optimal solution

R	Q	μ_y	T	$E\left(\pi_{PS}^R\right)$	$E\left(\pi_{PS}^S\right)$	$ETP(Q, \mu_y, T)$	Per
80	111	10.08	2.36	2904.38	3860.91	6765.29	−24.20
120	112	10.08	2.37	7190.67	3895.79	11086.47	24.21
λ_1	Q	μ_y	T	$E\left(\pi_{PS}^R\right)$	$E\left(\pi_{PS}^S\right)$	$ETP(Q, \mu_y, T)$	Per
80	92	10.08	2.14	4067.35	3198.36	7265.71	−18.60
120	132	10.08	2.57	5992.24	4593.37	10585.60	18.60

Note Per $= \frac{ETP(Q, \mu_y, T) - 8925.61}{8925.61} \cdot 100\%$

4 Conclusions

In this paper, the author has presented a modified Chen and Liu's (2008) traditional system model with quality loss of product. Assume that the retailer's order quantity is concerned with the manufacturer's product quality and the quality characteristic of product is normally distributed. The quality of lot for manufacturer is decided by adopting a single sampling rectifying inspection plan. The process mean of quality characteristic, the production run length of product, and the order quantity of retailer are simultaneously determined in the modified model. From the above numerical results, one has the following conclusion: The sale price per unit has a have a major effect on the retailers' expected profit and the expected total profit of society and the intercept of mean demand of customer has a have a major effect on the retailers' expected profit, manufacturer's expected profit, and the expected total profit of society. Hence, one needs to have an exact estimation on these two parameters in order to obtain the exact decision values. The extension to integrated model with 100 % inspection may be left for further study.

References

Chen CH (2010) The joint determination of optimum process mean and economic order quantity. Paper presented at the 2010 international conference in management sciences and decision making, Tamsui, Taiwan, pp 285–292

Chen SL, Liu CL (2007) Procurement strategies in the presence of the spot market-an analytical framework. Prod Plann Control 18:297–309

Chen SL, Liu CL (2008) The optimal consignment policy for the manufacturer under supply chain coordination. Int J Prod Res 46:5121–5143

Hanna MD, Jobe JM (1996) Including quality costs in the lot-sizing decision. Int J Qual Reliab Manag 13:8–17

Jaber MY, Bonney M, Rosen MA, Moualek I (2009) Entropic order quantity (EnOQ) model for deteriorating items. Appl Math Model 33:564–578

Taguchi G (1986) Introduction to quality engineering. Asian Productivity Organization, Tokyo

Developing Customer Information System Using Fuzzy Query and Cluster Analysis

Chui-Yu Chiu, Ho-Chun Ku, I-Ting Kuo and Po-Chou Shih

Abstract Customer information is critical to customer relationship management. The goal of this research is to improve the efficiency of customer relationship management through developing a customer information system. Fuzzy terms with linguistic variables can help specific queries to be more versatile and user friendly for customer data mining. In this paper, we propose a method integrating cluster analysis with linguistic variables in the context of fuzzy query logistics. Based on the proposed method, we constructed a customer information system that can offer the user useful information as regards with strategy, decision making, and better resource allocation methods. We expect to decrease total execution time and to increase the practicability with the feature of customer information cluster analysis.

Keywords Fuzzy query · Cluster analysis · Linguistic variable · Relational database

1 Introduction

Along with the rapid development within and peripheral to the information industry, database systems are increasingly being utilized. However, information is often vague and ambiguous to the user. We can divide problematically imprecise

C.-Y. Chiu (✉) · H.-C. Ku · I.-T. Kuo · P.-C. Shih
Department of Industrial Engineering and Management, National Taipei University of Technology, Taipei 106, Taiwan, Republic of China
e-mail: cychiu@ntut.edu.tw

H.-C. Ku
e-mail: ruokku@yahoo.com.tw

I.-T. Kuo
e-mail: 298177@gmail.com

P.-C. Shih
e-mail: pojo0701@hotmail.com

Y.-K. Lin et al. (eds.), *Proceedings of the Institute of Industrial Engineers Asian Conference 2013*, DOI: 10.1007/978-981-4451-98-7_142,
© Springer Science+Business Media Singapore 2013

information into two types in this context: the first considers the possibility or similarity of imprecise queries within the classic database; the second concerns the storage problem as regards the imprecise information and the database system. However, no matter what type the imprecise information, it is important to acquire various types of information for making optimal decisions.

The fuzzy theory (Zadeh 1965), which has been developed for many years has been applied to indefinite data with regard to a diverse range of applications in a diverse range of industries and has been integral to the effective solutions to numerous problems. The applications in fuzzy theory can mainly divide into two categories within relational database systems: the translation of fuzzy queries and fuzzy relational databases.

However, user demand is also changeable and often simultaneous. Many have attempted to further improve the effect of processing uncertainties and the field of research has been developing over many years. Fuzzy sets were first used for querying conventional databases firstly in 1977 (Tahani 1977) where there were imprecise conditions inside queries. SQLf was a fuzzy extension to SQL (Bosc and Pivert 1995). This was proposed to represent a synthesis method among flexible queries in relational databases. FSQL, developed in 2005, is an extension of the SQL language. It contains Data Manipulation Language (DML) of FSQL, including SELEC, INSERT, DELETE, and UPDATE instructions.

Cluster analysis is one of the more useful methods/techniques used to glean knowledge from a dataset. Cluster analysis is based on the similarity of clusters and it can detect high heterogeneity among clusters and high homogeneity within a cluster.

We will use fuzzy theory to translate the uncertain demand in the context of clear query language. We are able to use the result of cluster analysis to create a new relational table within the relational database and divide the data set into several sub sets. According to the query, the search will select the subset according to a non-global search. A decrease in the sizes of searches will improve the efficiency (Chen and Lin 2002). Furthermore, the query based on the integration with the features of cluster analysis and linguistic variables are expected to be more practical.

2 Literature Review

2.1 Development of Relational Database

Databases are used in numerous business contexts, i.e., banking, airlines, universities, sales, online retailers, manufacturing, human resources, etc. Notably, the relational database, which is based on a relational model, is one of the more commonly used databases in business. The relational model was first proposed by Dr. E.F. Codd of IBM in 1970.

2.2 Fuzzy Query

Fuzzy theory was proposed by Zaden (1965); the method involves processing something uncertain in order to estimate the value exactly. In the past, we have received some information which has been well approximated as regards value but not is precise. This has yielded, in some cases, incorrect information causing executors to make poor decisions more often. Hence, fuzzy set theory is applied to problems in engineering, business, medicine, and natural sciences (Guiffrida and Nagi 1998).

Applying fuzzy set theory to relational databases has matured and diversified. However, there are two main directions of this type of research. The first direction is fuzzy query translation for relational database systems. Users who search for information from databases can input conditions and the system checks the data to satisfy the query conditions of the user. Sometimes the value of fuzzy terms is not as same as the conditions but this data is still relevant to the user. In order to improve upon this, we might translate the common nature language into SQL first and then process it through relational databases. The other direction involves fuzzy relational databases (Qiang et al. 2008). To improve the restrictions within the traditional relational databases, one can construct the database model so that it has an extended type of data.

2.3 Cluster Analysis

Clustering is an exploratory method used to help solve classification problems and is a useful technique for the elucidation of knowledge from a dataset. Its use is germane when little or nothing is known about the category structure within a body of data. The objective of clustering is to sort a sample of cases under consideration into groups such that the degree of association is high between members of the same group and low between members of different groups. A cluster is comprised of set of entities which are alike, while entities from different clusters are not alike. Clustering is also called unsupervised classification, where no predefined classes are assigned. Some general applications of clustering includes pattern recognition, spatial data analysis, imaging processing, multimedia computing, bioinformatics, biometrics, economics, WWW, and so on.

Aldenderfer and Blashfield decided on five basic steps that characterized all clustering studies (Aldenderfer and Blashfield 1984).

Clustering of data is broadly based on two approaches: hierarchy and partition (Jain and Dubes 1988). Over the last several decades, due to the development of artificial intelligence and soft computing, clustering methods based on other theories or techniques has advanced.

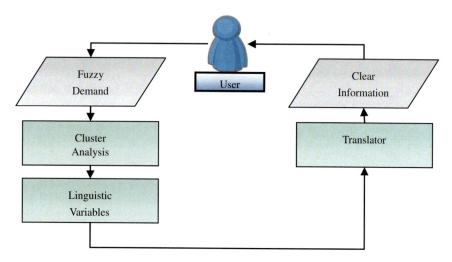

Fig. 1 The research architecture

3 Methodology

3.1 Framework

In this study, we propose a method for integrating cluster analysis and fuzzy terms in linguistic variables to improve the effect of fuzzy queries and, as such, this is expected to shorten the execution time for the query. Further, the practicability of the query could advance according to the feature of cluster analysis simultaneously. Finally, our method translates the results of the process into clear query language within the relational database and, in doing so, assists queries using an RFM analysis. We discuss how this could benefit an enterprise or firm in making decisions or allocating resources as regards different customers. The research architecture is shown as Fig. 1. However, we design the query systems architecture as shown in Figs. 2 and 3.

3.2 Cluster Analysis

This study uses the results of cluster analysis from the former experiment (Chiu et al. 2008). The steps of the method are:

- Step 1: build the relational table to record the cluster analysis. This is called the table of cluster analysis. There are records of information about features of each cluster such as measures of central tendency, measures of dispersion tendency, and measures of shape tendency.

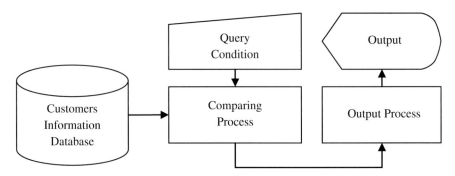

Fig. 2 Architecture of query system in global search

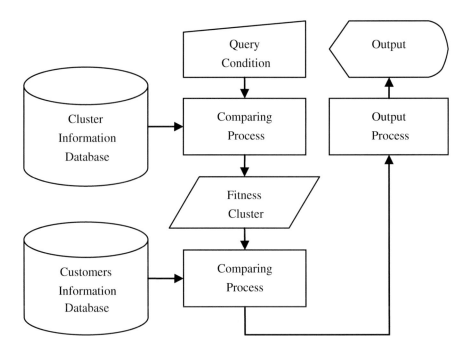

Fig. 3 Architecture of query system integrating the features of cluster analysis

- Step 2: divide the source database into several classes by its clusters. This will help to reduce the size of larger data sets and the execution times; this is amenable for administering the data at the same time.
- Step 3: according to the query conditions set by the user, count the distance to each centroid of clusters. When the distance is the minimum among these, its cluster will be the data set to compare with the query conditions.
- Step 4: adjust the parameter of function or the factor of query by its features of cluster analysis.

Table 1 The relational table of cluster analysis

G_index	RMean	FMean	MMean	Intra	Number
1	0.547866	0.556169	0.594098	0.216385	21
2	0.639573	0.205863	0.367448	0.268222	13
3	0.844350	0.680196	0.719214	0.293066	14
4	0.209397	0.278638	0.435888	0.239071	33
5	0.150691	0.073703	0.222419	0.208986	29
6	0.232697	0.662472	0.817353	0.310540	10

- Step 5: compare with the data in the above cluster whose centroid has the minimum distance to the query conditions and count the membership value to see if it satisfies as regards the threshold.

After these steps, the system can adjust the process of querying by the features of cluster analysis easily. This makes it more efficient to retrieve the data from database.

4 Experiment Process and Result Analysis

4.1 System Platform

- The programming language: PHP 4.3.11
- The database management system: MySQL 4.0.24:

The relational table of cluster analysis is shown in Table 1. It records the cluster information including the measurement of central tendency, measures of dispersion, measures of dispersion tendency, and so on.

The relational table of each cluster is shown in Table 2. This is used to count the membership value. Therefore, the value of tuples in the records is in relation to normalization, not the true value in reality.

Table 2 The relational table of cluster1

Index	NormalR	NormalF	NormalM
1	0.536763	0.581925	0.784784
7	0.568915	0.441738	0.440312
11	0.587959	0.691451	0.786775
31	0.703922	0.404430	0.636108

Table 3 The feature of simulation

	Mean			Standard Deviation		
	R	F	M	R	F	M
Cluster 1	19.80	139.47	1986197.0	8.77	114.54	2486626.0
Cluster 2	12.20	10.03	179139.0	9.51	8.99	270543.0
Cluster 3	2.37	552.40	10437012.0	2.62	719.80	12171818.0
Cluster 4	237.87	17.87	246326.2	172.96	14.18	306763.4
Cluster 5	326.30	2.73	15977.0	197.19	1.78	14074.0
Cluster 6	174.13	538.33	28255299.4	98.65	713.78	34120605.8

Table 4 The result of efficiency in both query systems

	Global	Cluster analysis
Cluster 1	0.022102	0.006931
Cluster 2	0.022723	0.005465
Cluster 3	0.021669	0.005613
Cluster 4	0.022655	0.007806
Cluster 5	0.022462	0.007514
Cluster 6	0.021418	0.005054

4.2 The Improvement of Query

4.2.1 System Simulation

In order to evaluate the performance of system, we construct the set of test data to simulate the query. The set of test data is generated by random numbers in normal distribution. There are 30 queries in each cluster. Table 3 summarizes the set of test data in each cluster. Each query will get all membership values which correspond to the total data in the same cluster.

4.2.2 The Evaluation of Efficiency

As regards the experiment in the set of test data, the results of efficiency with both query system architectures are shown in Table 4. We found that the efficiency of query system integrating cluster analysis is better than that of query system in a global search. However, in the query system integrating cluster analysis, the size of each cluster still affects the execution time within each cluster.

5 Conclusion

According to the above research and experiments, the effect of querying has undergone improvements. Not only has there been shown to be a decrease in execution time but the practicality of queries has also been improved by the integration of cluster analysis.

The query system offers accurate customer information. Therefore, this method can further help to track customers with the pertinent customer information including individual experiences and purchasing records. The market feature helps enterprises segment customers to decide upon the degrees of importance regarding demographics. We conclude that either of these benefits will optimize an enterprise's ability to glean important types of quality information to better allow them to assess and improve their own market strategy.

References

Aldenderfer MS, Blashfield RK (1984) Cluster analysis: quatitative applications in the social sciences. Sage Publication, Beverly Hills

Bosc P, Pivert O (1995) SQLf: a relational database language for fuzzy querying. IEEE Trans Fuzzy Syst 3(1):1–17

Chen SM, Lin YS (2002) A new method for fuzzy query processing inrelational database systems. Cybern Sys Int J 33:447–482

Chiu CY, Chen YF, Kuo IT, Ku HC (2008) An intelligent market segmentation system using k-means and particle swarm optimization. Expert Systems with Applications

Guiffrida AL, Nagi R (1998) Fuzzy set theory application in production management research: a literature survey. J Intell Manuf 9(1):39–56

Jain AK, Dubes RC (1988) Algorithms for clustering data. Prentice-Hall, Englewood Cliffs

Qiang L, Yang W, Yi D (2008) Kernel shapes of fuzzy sets in fuzzy systems for function approximation. Inf Sci 178:836–857

Tahani V (1977) A conceptual framework for fuzzy querying processing: a step toward very intelligent databases systems. Inf Process Manage 13:289–303

Zadeh LA (1965) Fuzzy sets. Inf Control 8:338–353

Automatic Clustering Combining Differential Evolution Algorithm and *k*-Means Algorithm

R. J. Kuo, Erma Suryani and Achmad Yasid

Abstract One of the most challenging problems in data clustering is to determine the number of clusters. This study intends to propose an improved differential evolution algorithm which integrates automatic clustering based differential evolution (ACDE) algorithm and *k*-means (ACDE-*k*-*means*) algorithm. It requires no prior knowledge about number of clusters. *k*-means algorithm is employed to tune cluster centroids in order to improve the performance of DE algorithm. To validate the performance of the proposed algorithm, two well-known data sets, Iris and Wine, are employed. The computational results indicate that the proposed ACDE-*k*-means algorithm is superior to classical DE algorithm.

Keywords Automatic clustering · Differential evolution · *k*-means

1 Introduction

Clustering is an unsupervised data classification for observations, data items, or features vectors based on similarity (Jain et al. 1999). There are numerous scientific fields and applications utilized clustering techniques such as data mining, image segmentation (Frigui and krishnapuram 1999), bioinformatics (Yeung et al. 2001), documents clustering (Cai et al. 2005), market segmentation (Kuo et al. 2012a, b).

R. J. Kuo (✉) · A. Yasid
Department of Industrial Management, National Taiwan University of Science and Technology, 43 Keelung Road, Section 4, Taipei 106, Taiwan, Republic of China
e-mail: rjkuo@mail.ntust.edu.tw

E. Suryani
Department of Information Systems, Institut Teknologi Sepuluh Nopember, Jl. Raya ITS, Surabaya 60111, Indonesia
e-mail: erma@is.its.ac.id

Clustering algorithms can be broadly divided into two groups: *hierarchical* and *partitional*. Hierarchical algorithm finds nested cluster either in agglomerative mode (begin with treat each data point as a cluster and merge the nearest two clusters iteratively until stopping criterion met) or divisive mode (begin with all data points grouped in one cluster and recursively dividing each cluster into smaller clusters until stopping criterion met). *Partitional* algorithm, on the other hand, attempts to find all clusters simultaneously. The most well-known hierarchical clustering algorithms are single-link and complete link; while in the *partitional* clustering, *k*-means algorithm is the most popular and simplest algorithm.

Clustering algorithm can also divided in two categories: crisp (hard) and fuzzy (soft) algorithm. In crisp clustering, any data point may only belong to one cluster. While in fuzzy clustering, data point may belong to all clusters with a specific degree of membership. The work described in this study is concerned crisp clustering algorithms.

However, most of these methods require the user to subjectively define number of clusters. Thus, one of the most challenging problems in clustering is how to find the k cluster number automatically. Recently, using the evolutionary computation to get cluster number has been widely applied such as genetic algorithm (Bandyopadhyay and Maulik 2002), particle swarm optimization (Paterlini and Krink 2006), and differential evolution (Das et al. 2008).

Differential evolution algorithm (Storn and Price 1997) is a novel evolutionary algorithm (EA) for global optimization. It uses floating point to encode for its population and has been successfully applied in clustering problem. Compared with PSO algorithm and GA, DE algorithm outperforms PSO algorithm and GA over *partitional* clustering (Paterlini and Krink 2006). In order to improve DE algorithm, Kwedlo (2011) combined DE algorithm with *k*-means algorithm. Das et al. (2008) also proposed automatic clustering based DE (ACDE) algorithm. The ACDE algorithm modifies scale factor F in original DE algorithm with random manner in the range [1, 0.5] and linearly decreases the crossover rate along with iterations. Therefore, this study will propose a novel automatic clustering approach based on DE algorithm combining with *k*-means algorithm for crisp clustering (ACDE-*k-means*). Combination of DE algorithm and *k*-means algorithm can balance exploration and exploitation processes.

The rest of this paper is organized as follows. Section 2 briefly discusses literature study about data clustering and DE algorithm, while the proposed method is described in Sect. 3. Section 4 presents the computational results. Finally, the concluding remakes are made in Sect. 5.

2 Literature Study

2.1 Clustering

Given a data set $X = \{x_1, x_2,..., x_N\}$ contains N data points in d dimension. The dataset can be grouped into K number of clusters $C = \{C_1, C_2,..., C_K\}$. Hard clustering problems have three properties that should maintain, i.e., (1) each cluster should have at least one data point assigned; (2) no data point common in two different clusters; and (3) each pattern must be attached into a cluster. The related equations are illustrated as follows.

$$C_i \neq \emptyset, \quad \forall i \in \{1, 2, ..., K\} \tag{1}$$

$$C_i \cap C_j = \emptyset, \quad \forall i \neq j \text{ and } i,j \in \{1, 2, ..., K\} \tag{2}$$

$$\cup_{i=1}^{K} C_i = X \tag{3}$$

2.1.1 k-Means

k-means algorithm is a well-known clustering algorithm. The basic step of k-means algorithm (MacQueen 1967) is as follows:

1. Randomly select k initial data points as cluster centroids,
2. Assign each data point to the closest centroid,
3. Recalculate the centroid of each cluster using average method,
4. Repeat steps 2 and 3 until stopping criterion is met.

2.1.2 Cluster Validity Index

The clustering result is evaluated using cluster validity index on a quantitative basis. Cluster validity index serves two purposes: determining number of clusters and finding the best partition. Thus, two aspects of partitioning namely *cohesion* and *separation* should be considered. There are many validity measurements proposed previously, yet this study will only discusses two validity measurements which will be employed in the proposed automatic clustering algorithm.

1. *VI Index*

This index calculates ratio between *intra* and *inter* dissimilarity (Kuo et al. 2012a, b). Let *VI* be the fitness function to minimize the value. The formula is shown as follows:

$$VI = (c \times N(0, 1) + 1) \times \frac{intra}{inter}, \qquad (4)$$

where $(c \times N(0, 1) + 1)$ is a punishment value to avoid having too few clusters, c is a constant value and set to 30. The $N(0,1)$ is Gaussian function of cluster numbers. In this experiment, $N(0,1)$ is adopted for Iris and Wine datasets since the numbers of clusters for these data sets are small.

Intra is the average distance between centroid m_k to the data point x in a cluster. Calculate the Euclidian distance of data point to the centroid of cluster, sum up all the shortest distance of each data point to the centroid of cluster, and then divide it by the total number of data tuples, N_p. The formula of *intra* is as follows:

$$intra = \frac{1}{N_p} \sum_{k=1}^{K} \sum_{x \in C_k} \|x - m_k\|^2 \qquad (5)$$

Inter as illustrated in Eq. (6) is a minimum centroid distance among clusters. The distance from each cluster centroid to another cluster centroid is calculated to get the minimum value.

$$inter = \min\{d(\vec{m}_k, \vec{m}_{kk})\}$$
$$\forall k = 1, 2, \ldots, K - 1 \text{ and } kk = k + 1, \ldots, K. \qquad (6)$$

2. *CS Measure*

CS measure is a simple clustering measurement index that can assign more cluster centroids to the area with low-density data than conventional clustering algorithms (Chou et al. 2004). First, the cluster centroid is average of all the data points in the cluster as shown in Eq. (7). Then, the CS measure can be formulated by Eq. (8), where $d(\vec{x}_i, \vec{x}_q)$ is distance metric between two data points, $\vec{x}_i$ and $\vec{x}_q$.

$$\vec{m}_i = \frac{1}{N_i} \sum_{x_j \in C_i} \vec{x}_j \qquad (7)$$

$$CS(K) = \frac{\frac{1}{K} \sum_{i=1}^{K} \left[\frac{1}{N_i} \sum_{\vec{x}_i \in C_i} \max_{\vec{x}_q \in C_i} \{d(\vec{x}_i, \vec{x}_q)\} \right]}{\frac{1}{K} \sum_{i=1}^{K} \left[\min_{j \in K, j \neq i} \{d(\vec{m}_i, \vec{m}_j)\} \right]}$$
$$= \frac{\sum_{i=1}^{K} \left[\frac{1}{N_i} \sum_{\vec{x}_i \in C_i} \max_{\vec{x}_q \in C_i} \{d(\vec{x}_i, \vec{x}_q)\} \right]}{\sum_{i=1}^{K} \left[\min_{j \in K, j \neq i} \{d(\vec{m}_i, \vec{m}_j)\} \right]} \qquad (8)$$

2.2 Differential Evolution Algorithm

DE algorithm is an evolution-based algorithm proposed by Storn and Price. Like other evolutionary algorithms, the initial population $V_{i,d}(t)$, is randomly generated. The ith individual vector (chromosome) of population at time step (generation) t has d components (dimensions) as shown in Eq. (9). Then, it will evolve using mutation and crossover.

$$V_{i,d}(t) = v_{i,1}(t), v_{i,2}(t), \ldots, v_{i,d}(t) \tag{9}$$

Mutation is a process to generate new parameter by adding weighted difference between two population vectors to a third vector. Mutant vector $Z_{i,d}(t+1)$ as illustrated in Eq. (10) is generated by three random vectors, i.e., $V_{j,d}(t)$, $V_{k,d}(t)$, and $V_{l,d}(t)$, from the same generation (for distinct $i, j, k,$ and l) according to scale factor F.

$$Z_{i,d}(t+1) = V_{j,d}(t) + F(V_{k,d}(t) - V_{l,d}(t)) \tag{10}$$

In order to increase the diversity of perturbed parameters vectors, trial vector $U_{ji,d}(t+1)$ as shown in Eq. (11) is created using crossover operator.

$$U_{ji,d}(t+1) = \begin{cases} Z_{i,d}(t+1) & \text{if } rand_j(0,1) \leq CR \text{ or } j = rand(d) \\ V_{i,d}(t), & \text{if } rand_j(0,1) > CR \text{ or } j \neq rand(d) \end{cases} \tag{11}$$

Furthermore, selection process is executed. In order to decide whether offspring can become a member of new generation, the trial vector is compared to the fitness function. If the new offspring $U_i(t+1)$ yields a better value (minimum) of the objective function $f(V_i(t))$, it replaces its parent in the next generation. Otherwise the parent $V_i(t)$ is retained to be parent for the next generation $V_i(t+1)$. The decision rule is as follows:

$$V_i(t+1) = \begin{cases} U_i(t+1), & \text{if } f(U_i(t+1)) > f(V_i(t)) \\ V_i(t), & \text{if } f(U_i(t+1)) \leq f(V_i(t)) \end{cases} \tag{12}$$

3 Automatic Clustering Based DE Algorithm

The algorithm of the proposed ACDE-k-means is as follows:

step 1. Initialize each chromosome to contain k (randomly generate) number of randomly selected cluster centers and k randomly activation threshold [0,1].
step 2. Find out the active centorids $v_{i,k}T_k$ in each chromosome using rule as shown in Eq. (13).
step 3. For $t = 1$ to t_{max} do

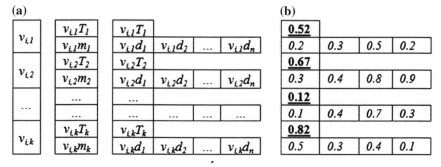

Fig. 1 **a** Detail of chromosome schema, **b** An individual vector with four threshold (*bold underlined*) and four centroid of each cluster (*italic*)

a. Calculate distance of each data vector to all actives centroids of the *i*th chromosome.
b. Assign each data vector to a cluster with shortest distance.
c. Change the population member based on DE algorithm. Use fitness function to select better population
d. Check whether the vector chromosome is met hard clustering properties.
e. Apply *k*-means algorithm to adjust centroids of *i*th active chromosome. Use the active cluster number as input of *k*-means algorithm.
step 4. Output the global best chromosome (the minimum fitness) as final result.

In this proposed method, the chromosome representation is based on (Das et al. 2008). Every chromosome $v_i(t)$ is a real number vector. It comprises of activation threshold $v_{i,k}T_k + (v_{i,k} \times d_n)$ dimensions. Each of activation threshold $v_{i,k}T_k$ is random number in the range of [0,1] that act as control parameter to determine whether the cluster is active or inactive. The rule to specify the cluster is active or inactive is as follows:

$$\text{IF } v_{i,k}T_k > 0.5, \text{ THEN the kth cluster center } v_{i,k}m_k \text{ is ACTIVE} \\ \text{ELSE } v_{i,k}m_k \text{ is INACTIVE} \quad (13)$$

As an example, (0.52), (0.67), (0.12) and (0.82) are activation threshold, while (0.2, 0.3, 0.5, 0.2), (0.3, 0.4, 0.8, 0.9), (0.1, 0.4, 0.7, 0.3) and (0.5, 0.3, 0.4, 0.1) are centroid of vector chromosome $v_{i,k}m_k$ (Fig. 1).

4 Computational Results

The proposed ACDE-*k*-*means* algorithm is validated using two data sets namely iris and wine. To judge the accuracy of ACDE-*k*-means, we let each algorithm to run for 30 times. Two fitness functions, CS measure and VI measure are applied. Simulations were executed using C++ on a PC Intel Core2 Quad at 2.4 GHz with

Table 1 Real life datasets

Data set	N	Number of dimensions	Number of clusters	Composition of each cluster
Iris	150	4	3	50, 50, 50
Wine	178	13	3	59, 71, 48

Note better result is printed in bold

Table 2 Tunning parameters

Parameter	Value
Population size	$10x$ dimension
CR_{min}; CR_{max}	0.5; 1
F (ACDE-k-means)	[0.5,1.0]
F (Classical DE)	0.5
k_{min}; k_{max}	2; $\sqrt{Npop}$

Note better result is printed in bold

Table 3 Number of clusters

Dataset	Algorithm	Number of clusters (average ± st. dev) CS	Number of clusters (average ± st. dev) VI	CS value	VI value
Iris	ACDE-k-means	**2.8 ± 0.407**	**3.2 ± 0.4068**	**0.605 ± 0.013**	**0.238 ± 0.013**
	Classical DE	2.83 ± 0.531	2.5 ± 0.7768	0.6791 ± 0.013	0.117 ± 0.101
Wine	ACDE-k-means	**3.2 ± 0.761**	**3.77 ± 0.504**	**0.947 ± 0.052**	**0.525 ± 0.036**
	Classical DE	3.4 ± 0.724	2.33 ± 0.547	0.852 ± 0.054	0.135 ± 0.123

Table 4 Accuracy

Dataset	Algorithm	Accuracy (average ± st. dev) CS	Accuracy (average ± st. dev) VI
Iris	ACDE-k-means	**0.8267 ± 0.0813**	**0.756 ± 0.0153**
	Classical DE	0.6791 ± 0.0033	0.4391 ± 0.1242
Wine	ACDE-k-mean	**0.8484 ± 0.1175**	**0.8291 ± 0.0654**
	Classical DE	0.4071 ± 0.0619	0.4721 ± 0.1097

2 GB of RAM environment. The tuning parameters are set based on Das et al. (2008). The only difference is that the initial number of clusters (k_{max}) is determined by using $\sqrt{Npop}$. The results are compared with those of classical DE algorithm (Tables 1 and 2).

Table 3 gives number of clusters obtained from 30-time independence runs where each run uses 100 iterations. The result shows that average cluster number found by ACDE-k-means is more similar to the actual value than classical DE algorithm both in CS and VI measures. For Iris data set, the average cluster numbers obtained from ACDE-k-means for CS and VI are 2.8 and 3.2, respectively. Table 4 summarizes accuracy obtained by both algorithms. It reveals that ACDE-k-means algorithm has better accuracy than classical DE algorithm for both data sets. Basically, no matter CS or VI, ACDE-k-means algorithm outperforms classical DE algorithm.

5 Conclusions

In this paper, a differential evolution algorithm based for automatic clustering has been proposed. ACDE algorithm is combined with k-means algorithm to improve the performance of DE algorithm. ACDE-k-means algorithm owns the ability to find the number of clusters automatically. Moreover, the proposed method is also able to balance the evolution process of DE algorithm so that it can achieve a better partition compared with classical DE algorithm. Results from different fitness measures (CS and VI) and two different data sets (Iris and Wine) indicate that ACDE-k-means performs well in terms of both number of clusters and clustering results.

References

Bandyopadhyay S, Maulik U (2002) Genetic clustering for automatic evolution of cluster and application to image classification. Pattern Recogn 35:1197–1208

Cai D, Xiaofei H, Han J (2005) Documents clustering using locality preserving indexing. Knowl Data Eng IEEE Tran 17(12):1624–1637

Chou C-H, Su M-C, Lai E (2004) A new cluster validity measure and its application to image compression. Patter Anal Applic 7:205–220

Das S, Abraham A, Komar A (2008) Automatic clustering using improved differential evolution algorithm. IEEE Trans Syst Man Cybern Part A 38:218–237

Frigui H, Krishnapuram R (1999) A robust competitive clustering algorithm with applications in computer vision. IEEE Trans Pattern Anal Mach Intell 21:450–465

Jain AK, Murty MN, Flynn PJ (1999) Data clustering: a review. ACM Comput Surv (CSUR) 31(3):264–323

Kuo RJ, Akbaria K, Subroto B (2012a) Application of particle swarm optimization and perceptual map to tourist market segmentation. Expert Syst App 39:8726–8735

Kuo RJ, Syu YJ, Chen ZY, Tien FC (2012b) Integration of particle swarm optimization and genetic algorithm for dynamic clustering. Inf Sci 195:124–140

Kwedlo W (2011) A clustering method combining differential evolution with the k-means algorithms. Pattern Recognit Lett 32:1613–1621

Macqueen J (1967) Some methods for classification and analysis of multivariate observations. In: Proceedings of the fifth Berkeley Symposium on mathematical statistics and probability, vol 1, pp 281–297

Paterlini S, Krink T (2006) Differential evolution and particle swarm optimization in partitional clustering. Comput Stat Data Anal 50:1220–1247

Storn R, Price K (1997) Differential evolution—a simple and efficient heuristic for global optimization over continuous spaces. J Global Optim 11:341–359

Yeung KY, Fraley C, Murua A, Raftery AE, Ruzzo WL (2001) Model-based clustering and data transformations for gene expression data. Bioinformatics 17:977–987

Application of Two-Stage Clustering on the Attitude and Behavioral of the Nursing Staff: A Case Study of Medical Center in Taiwan

Farn-Shing Chen, Shih-Wei Hsu, Chia-An Tu and Wen-Tsann Lin

Abstract The purpose of this study was to probe into and organizing personnel representing attitude and behavior that the function should possess, a case study of medical center nursing categories of employees for the study, due to the professionalism of nursing staff for the establishment of a good nurse-patient relationship as well as to demonstrated the attitude of an important condition, in this study was with Situational judgment test the collect a nursing staff in the true working situational, the view on the function importance. Testing the materials is obtained by the database of this case hospital, carry on Two-stage clustering method hiving off laws (Self-Organizing Maps and K-means) secondary analysis, will be importance of "situational judgment test functions questionnaire" that the information collected, analyze attitude and behavior that nursing staff should possess. According the analysis results found that personnel in different posts and ranks between two groups of 'staff' and 'executive', to the attitude and behavioral cognition, think 'responsible for seriously' with 'quality leading' the most important, cultivation of the future nursing staff, should strengthen the cultivation of professional attitude, and implement the rigorous of standard operation procedure, period of to reduce the gap between of health professional education and clinical/nursing practice.

Keywords Attitude and behavior · Competency · Situational judgment test · Two-step cluster method

---

F.-S. Chen · S.-W. Hsu (✉) · C.-A. Tu
Department of Industrial Education and Technology, National Changhua University of Education, Changhua 500, Taiwan, Republic of China
e-mail: kb80284@gmail.com

F.-S. Chen
e-mail: iefchen@cc.ncue.edu.tw

W.-T. Lin
Department of Industrial Engineering and Management, National Chin-Yi University of Technology, Taichung 411, Taiwan, Republic of China
e-mail: lin505@ncut.edu.tw

1 Introduction

Spencer and Spencer (1993) proposed the iceberg model theory, which divides competency into explicit components, such as knowledge, skills, and abilities, and implicit components including traits and values. They also indicated that competency is the integration of explicit and implicit employee abilities required for a specific position in a business or organization. In this study, nurses were selected as the primary subjects because they are a group who constitutes the majority of employees in health care institutions and who are closest to patients. More than 80 % of health care treatment behavior and care activities are conducted by nurses. This indicates that nurses occupy an important role in ensuring patient safety. Therefore, the professionalism and behavior of nurses must be cultivated to establish excellent nurse-patient relationships and patient service attitudes. Consequently, the objectives of this study are as follows:

1. Explore the problem-solving and coping measures related to health care services that are adopted by nurses in the case hospital, as well as their attitudes and behaviors in the work environment, using the situational judgment test (SJT) model.
2. To examines the nursing professionalism exhibited by nurses delivering health care services, and recommends attitude and behavior.

2 Literature Review

2.1 Attitude and Behavior

Regarding the relationship between attitude and behavior, most early studies that examined the factors influencing individual behavior investigated the influence that attitudes toward a specific item had on behavior. Therefore, Thomas and Znaniecki (1918) were the first scholars to use attitude to explain social behavior. They contended that attitude is a psychological process in which individual people determine their real behaviors and latent reactions. After the 1960s, the concept of attitude shifted toward being a type of behavioral tendency (Taylor et al. 1997), and with appropriate or accurate knowledge, resulting attitudes could influence behaviors (Ben-Ari 1996). Chien et al. (2006) indicated that the more positive people's attitudes toward a given behavior are, the higher their level of cognition and tendency to adopt a certain type of behavior becomes. Consequently, Askariana et al. (2004) explored the relationship between knowledge (i.e., cognition), attitude, and practice (i.e., behavior) and reported that a significant correlation existed between increasing knowledge and expected behavioral changes. However, Lin et al. (2009) indicated that merely increasing knowledge does not necessarily induce behavioral changes directly, and attitude can serve as a media that provides a continuous emotional input or investment. Based on this concept, this study explores the influence that nurses' attitudes have on their health care service provision behavior.

2.2 Competency

Competency refers to the ability and willingness to employ existing knowledge and skills to perform work tasks. In other words, competency is a general term for behavior, motivation, and knowledge related to work success (Burgoyne 1993; Byham and Moyer 1996; Fletcher 2001). According to the description provided by the American Society for Training and Development, competency is the crucial ability required to perform a specific task. Furthermore, the European Centre for the Development of Vocational Training defined competency as the unique knowledge, skills, attitude, and motivations required to complete a specific task. Spencer and Spencer (1993) contended that competency represents the underlying characteristics of a person, which not only relate to the person's work position, but also facilitate understanding the person's expected or real reactions and expressions influencing behavior and performance.

2.3 Situational Judgment Tests

Health care behavior involves high complexity and uncertainty, when nurses perform health care behavior-based work, they must possess highly professional nursing knowledge and skills to ensure the provision of safe patient care services. Furthermore, nurses primarily and directly participate in health care behaviors that occur during the treatment of patients by physicians. Therefore, using the SJT method, this study collected nurses' opinions regarding the importance of competency for real work situations in the case hospital. McDaniel et al. (2007) divided the answering models for SJT into the knowledge format (What is the best answer?) and the behavioral tendency format (What is the action you are most likely to adopt?).

3 Methods

3.1 Validation of Sources for Secondary Analysis

The objective of this study was to explore the influence that nurses' attitudes and behaviors had on expressions or demonstrations of competency in the case hospital. To acquire expected results more appropriately, this study used the database established through the case hospital's Competency Importance Questionnaire to conduct secondary analysis. The case hospital distributed the questionnaires to understand the self-expectations of employees and supervisors and their cognitions or perceptions regarding the attitudes and behaviors they should possess for their positions. Therefore, by conducting in-depth analysis of existing data, novel conclusions and explanations could be derived. Furthermore, only by screening and accessing adequate data from the database could this study achieve its research goals.

3.2 Participant Descriptions

The case hospital examined in this study was selected from 144 health care institutions accredited on the "2009–2012 list of contracted National Health Insurance teaching hospitals or institutions at higher levels" by the Department of Health, Executive Yuan, R.O.C. (Taiwan). The selected case hospital also attained accreditation as a medical center of excellent rank in 2011 (valid from January 1, 2012, to December 31, 2015). Thus, this medical center was selected for the case study, and nurses employed at the center were recruited as participants.

3.3 Explanations Regarding Administration and Completion of the Questionnaire

To select participants from the case hospital to complete the Competency Importance Questionnaire, hospital personnel were grouped according to profession type using professional and work attributes. Because the greater complexity of the physician profession hinders profession-based grouping, non-physician personnel of the case hospital were recruited as participants for this type of grouping. Subsequently, after the Human Resources Department of the case hospital met with the supervisors of each hospital unit and reached a consensus, the various professions in the case hospital were divided into 21 categories. Because this study only adopted nurses of the case hospital as participants.

3.4 Analysis and Implementation

This study used two-stage clustering to conduct a confirmatory comparison. This process was performed to classify data into numerous clusters and ensure high similarity among data in a cluster to understand nurses' perceptions regarding the various profession categories in the case hospital. The consistencies and differences between the participants were compared, and the results of the two-stage clustering analysis were organized in sequential order. We employed the 80/20 theory to extract the attitude and behavior items that influence the case hospital nurses' displays of competency. Chang et al. (2011) examined medical unit cost analyses and medical service procedural problems using quality control circle activities. They also applied the 80/20 theory to identify the primary reasons or factors requiring imperative improvement, allowing for enhancement of unit costs and medical procedures and enabling the hospital to increase income, reduce expenses, and achieve sustainable development. Consequently, this study determined that 80 % of the medical services delivered by nurses were influenced by 20 % of the attitude and behavior items affecting competency displays.

4 Data Analysis and Discussion

4.1 Cluster Comparison and Analysis of Two-Stage Clustering

Employing the SPSS Clementine 10.1 software package for two-stage clustering analysis, this study found an initial solution using the unsupervised self-organizing map (SOM) network in the first stage.

Subsequently, using the two-stage clustering analysis results of SOM and K-means, we obtained the distribution of each cluster composition. To facilitate clear examination of the distribution, we observed every cluster when all contained 100 data rows, using color normalization to explore the proportion of each cluster in relation to all clusters. Thus, normalization was used to further clarify and observe the relevance between each cluster and the percentage of each cluster in relation to the whole, as shown in Fig. 1a, b.

Figure 1a, b show the cluster distribution, where the relevance between the cluster size and personnel cognition of attitude and behavior in each profession category is clear and specific. Therefore, all nurses serving in the case hospital

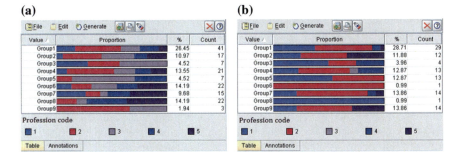

Fig. 1 a Employee self-evaluations. b Supervisor evaluations

possess a consensus that attitude and behavior can affect representations or demonstrations of competency. Regarding employee self-evaluations, the percentage sum of Clusters 1, 2, 4, 6, and 7 was 74.84 %, comprising all categories of nurse professions. Regarding supervisor evaluations, the percentage sum of Clusters 1, 2, 4, 7, and 9 was 81.18 %, which shows a similar level to that for employee self-evaluations. However, the cluster distribution graph only demonstrates the percentages of each cluster. To further examine the attitude and behavior items that influence the case hospital nurses' competency displays, we extracted the attitude and behavior items that influence competency based on the 80/20 rule. The extracted attitude and behavior items can provide a reference for nurses of the case hospital for improving medical services offered by the hospital.

4.2 Extracting the Attitude and Behavior Items that Influence Displays of Competency Based on the 80/20 Rule

This study divided case-hospital nursing respondents into employees and supervisors according to their position rank. For the questionnaire administration methods mentioned previously, multiple-response items were designed. Consequently, this study used the SPSS Statistics 17.0 software package to conduct multiple-response item analysis and rank the attitude and behavior items extracted from corresponding clusters of the employee self-evaluations and supervisor evaluations using the 80/20 rule. The results showed that in different clusters, the cognitive or perceived importance of various attitude and behavior items differed between employees and supervisors. Thus, to further understand importance rankings for attitude and behavior items in employee self-evaluations and supervisor evaluations, we organized the overall importance rankings into Table 1.

4.3 Discussion

In summation, the two-stage clustering analysis results were first used to divide employee self-evaluations and supervisor evaluations into nine clusters, and clustering analysis results were then adopted to plot a distribution map or diagram to explore the composition and relationship between each profession category. The objective of this study was to explore the influence of attitude and behavior items. Therefore, clusters comprising all categories of nurses were screened to conduct in-depth analysis and comparisons of the relationship between employee self-evaluations and supervisor evaluations. The 80/20 rule was used to extract the attitude and behavior items that influence competency displays. Because the Competency Importance Questionnaire comprised 35 items, the initial 20 %

Table 1 Importance rankings of the various attitude and behavior items

Employee self-evaluations			Supervisor evaluations		
Ranking	Attitude and behavior items	%	Ranking	Attitude and behavior items	%
1	04_Earnest and responsible	15.71	1	02_Quality-oriented	17.14
2	07_Teamwork	12.86	2	04_Earnest and responsible	16.43
3	08_Communication and coordination	12.86	3	10_Customer service	15.00
4	35_Crisis management	12.86	4	07_Teamwork	12.86
5	06_Proactive	8.57	5	03_Work management	8.57
6	03_Work management	7.86	6	30_Ability to bear stress	8.57
7	02_Quality-oriented	7.14	7	06_Proactive	7.86
8	31_Care and empathy	5.71	8	08_Communication and coordination	5.00
9	13_Adaptability	5.00	9	13_Adaptability	4.29
10	30_Ability to bear stress	3.57	10	11_Integrity	1.43
11	10_Customer service	2.14	11	12_Self-development	1.43
12	05_Work vitality	1.43	12	01_Pursuit of excellence	0.71
13	11_Integrity	1.43	13	17_Customer-oriented	0.71
14	24_Work guidance	1.43			
15	29_Problem-solving	1.43			

(i.e., the first seven items in the sequential order or ranking of each cluster) were extracted. Furthermore, to understand the importance rankings of attitude and behavior items obtained from employee self-evaluations and supervisor evaluations, we organized the overall importance rankings to concentrate and consolidate the results. According to the ranking results, this study can propose perceptions found in employee self-evaluations and supervisor evaluations regarding commonalities and differences for attitude and behavior items and can identify the critical attitude and behavior items included in these evaluations.

5 Conclusions

The primary conclusions for the research objectives of this study were below:

Through the SJT model, we determined that 10 common cognitive items considered to influence workplace situations existed between employees and supervisors in the case hospital. Based on this result, employees believed that they must work earnestly to validly complete their tasks and demonstrate responsibility. By contrast, supervisors believed that quality must be ensured before completing tasks to implement safe medical services.

Regarding the notion that attitude and behavior affect demonstrations of competency, we aim to enhance health care institutions' emphasis on developing attitudes and behaviors that can influence competency displays. The percentage of common cognition between employees and supervisors was 77.14 and 97.14 %, respectively. This result shows that supervisors already sufficiently comprehend the attitude and behavior items that influence competency displays, whereas employees should enhance this cognition.

Regarding the provision of vocational nursing education that cultivates and inspires appropriate attitudes, behaviors, and nursing concepts among students, we examined the current situation of case hospital nurses using the attitude and behavior perceptions of nursing personnel who perform practical health care services and by applying scientific methods for analysis. We anticipate that by adopting a combination of methodology, theory, and practice using existing data, the study results can balance theory and practice and provide referential value for schools and the industry when training competent and adequate nurses for health care institutions.

References

Askariana M, Honarvara B, Tabatabaeeb HR, Assadianc O (2004) Knowledge, practice and attitude towards standard isolation precautions in Iranian medical students. J Hosp Infect 58:292–296

Ben-Ari A (1996) Israeli professionals' knowledge of attitudes towards AIDS. Soc Work Health Care 22(4):35–52

Burgoyne JG (1993) The competence movement: issues, stakeholders and prospects. Pers Rev 22(6):6–13

Byham WC, Moyer RP (1996) Using competencies to build a successful organization. Development Dimensions International Inc, USA

Chang WD, Lai PT, Tsai CT, Chan HC (2011) Using quality control circle to dispose of the cost problems in a medical unit: an experience of rehabilitation center. J Health Manage 9(1):17–27

Chien CW, Lee YC, Lee CK, Lin YC, An BY (2006) The influence of inpatients' attitude on recognition of patient safety: a theory of planned behavior perspective. Cheng Vhing Med J 2(4):18–25

Fletcher S (2001) NVQS, standards and competence: a practical guide for employers, managers and trainers. Kogan Page Ltd, London

Lin CC, Huang ST, Chuang CM, Lee HM (2009) A study of knowledge attitude and behavior of infection control of nurses. Bull Hungkuang Inst Technol 55:1–16

McDaniel MA, Hartman NS, Whetzel DL, Grubb WL (2007) Situational judgment tests, response instructions, and validity: a meta-analysis. Pers Psychol 60:63–91

Spencer LM, Spencer SM (1993) Competence at work: models for superior performance. Wiley, New Jersey

Taylor SE, Peplau LA, Sears DO (1997) Social psychology. Prentice Hall, New Jersey

Thomas WI, Znaniecki F (1918) The polish peasant in Europe and America. Badger, Boston

The Effects of Music Training on the Cognitive Ability and Auditory Memory

Min-Sheng Chen, Chan-Ming Hsu and Tien-Ju Chiang

Abstract Previous research indicated a link between music training and cognitive ability. Despite some positive evidence, there is a different point about which abilities are improved. In this study, one experiment was conducted to investigate the effect of music training on memory retention. Two input modalities (visual and auditory) and delay time (0, 4, 6, 8, and 10 s) were manipulated. Participants were asked to finish prime task (memory retention) and distraction task (press the direction key to match the word or arrow). The result showed that music trained group performed better than the non-music trained one. For music trained group, their performance showed no significant difference between visual and auditory modalities. Participants without music training performed much better on visual than on auditory modality. This result of this experiment may support the idea that the music training influences participant's performance on memory retention. This research shows that music training is an important part of adolescent education and can help improve children's cognitive abilities.

Keywords Music training · Retention ability · Auditory memory · Cognitive ability · Children

---

M.-S. Chen (✉) · C.-M. Hsu · T.-J. Chiang
Department of Industrial Engineering and Management, National Yunlin University of Science and Technology, Douliu, Taiwan, Republic of China
e-mail: chens@yuntech.edu.tw

C.-M. Hsu
e-mail: k21973@hotmail.com

T.-J. Chiang
e-mail: g9921801@yuntech.edu.tw

1 Introduction

There is a cultural belief in Taiwan that children who play music will not misbehave, and many Taiwanese families provide their children with music training. Yet, we know little about the impact the experience of learning a musical instrument has on children's cognitive behaviors. This paper investigates the relationship between cognition and musical training among young children in Taiwan.

Previous research has shown that music training not only provides learners with more experience and background knowledge, but also improves the cognitive ability of the brain to process musical information. There is some evidence, hypothesized that non-music learners would distinguish lyrics according to their surface features, such as tone and rhythm, while music learners would use structural features, such as keys and chords (Gabriel et al. 1995). Study categorized participants into two groups according to their musical abilities in order to understand the correlation between music ability and phonological awareness. The results showed that children with better music abilities also displayed better phonological awareness. The authors concluded that both language and music experiences aided children's auditory development and were beneficial to their auditory cognitive behavior (Zehra et al. 2002), and explores the impacts of music experience on intervals and melody outlines (Dowling 1978).

Baddeley and Hitch's (1974) propose a working memory model, which includes an attention controller, the central executive, which is assisted by two subsidiary systems, the phonological loop, and the visual-spatial sketchpad. Baddeley et al. (1998) tested children with a non-word span task, which was more accurate than the digit span task for measuring phonological loop span. Children relied completely on the operation of the phonological loop store when asked to pronounce an unfamiliar word during this test.

Korenman and Peynircioglu (2007) explored the effect of different learning methods, including visual and audio styles, on music learning and memory. The study experiment tested participants' memories of written words and a melody. Results showed that the visual style learner performed better when words or tunes were presented visually. Likewise, Moore's (1990) result showed that visual style learners demonstrated a higher learning efficiency and precision in music notation learning experiments. Ho et al. (2003) also explored the effect of music training on children's memory. This study also showed that the group with music training had better auditory memory and memory retention.

Comparisons of music training and non-music training participants represent experiments that have spatial–temporal (Rauscher et al. 1997; Gromko and Poorman 1998), spatial abilities (Husain et al. 2002), memory (Chan et al. 1998; Jakobson et al. 2003, 2008; Lee et al. 2007), mental abilities (Brandler and Rammsayer 2003; Helmbold et al. 2005), modularity (Peretz and Coltheart 2003) and transfer (Schellenberg 2005; Schellenberg and Peretz 2008).

The goal is to investigate whether differences in learning styles and preferences play a role in this relationship, and determine the impact of music training on children's working memory and memory retention capability.

2 Experiment

2.1 Participants and Design

At the stage 1, seventy one participants joined this experiment. They were asked to finish Barsch learning style inventory (BLSI) test were further divided into two equal groups according to the learning preference (visual or auditory).

At the stage 2, forty eight participants were selected from the BLSI test. There were 24 participants who had music training (7 males and 17 females), with a mean age of 11.67 years (SD = 0.87), and had an average music learning time of 4.17 years (SD = 1.84, longest was 7). In this group, there were 12 visual learners (3 males and 9 females) with an average age of 12 years (SD = 1.04) and 12 auditory learners (4 males and 8 females) with an average age of 11.33 years (SD = 0.49). Additionally, there were 24 participants who did not have music training (14 males and 10 females), with a mean age of 11.88 years (SD = 0.85), and had an average music learning time of 0.46 years (SD = 0.57 year). Eleven participants had brief music training experience, with longest of 1.5 years, and the rest had no experience. In this group, there were 12 visual learners (6 males and 6 females), with an average age of 11.92 years (SD = 1.08), and 12 auditory learners (8 males and 4 females), with an average age of 11.83 years (SD = 0.58). See Table 1 for a description of these participants.

The design of this study had two between-subject variables: music training group (music vs. non-music) and learning preference (visual vs. auditory), and two within-subject variables: modality (visual vs. auditory presentation) and delay time (0, 2, 4, 6, 8, and 10 s). Delay time = 0 refers to an immediate repetition with no intervening items. The dependent variable was accuracy.

2.2 Materials and Procedure

Tasks were completed with E-Prime 2.0, using a 17-inch LCD screen and 2.1 channel speakers. Numbers between one and nine were used in this digit memory task. No single digit was repeated and digits were not in consecutive order. For the

Table 1 Sample descriptions of participants in experiment

Group	Average music learning time	Learning preferences	Number	Average age
Music	4.167 (max = 7)	Visual/auditory	12/12	12.000/11.333
Non-music	0.458 (max = 1.5)	Visual/auditory	12/12	11.917/11.833

visual tasks, images were drawn using Photoshop 6.0 and presented at the size of 17 by 17 cm. For the auditory tasks, audio files were recorded with male voice at one second per digit. Participants sat approximately 50 cm from the computer screen.

This experiment involved a visual and auditory memory retention task. At the beginning of the experiment, participants were given a practical trial to ensure that they could understand and be familiar with the procedure. Two presentation conditions were manipulated including visual presentation and auditory presentation. Participants were instructed that they would see or hear two to nine digits for a visual or auditory presentation test, participants was requested to answer immediately or later. The tasks were repeated twice with random arrangement. Participants were presented with 198 trials (six practice trials and 192 test trials).

The word "start" was displayed at the center of the screen for the duration of the initially trial appear for a second and then a blank screen would appear for a second. Stimuli were presented in either visual or auditory presentation and the set size of digits ranged from two to nine. Amount of time required depended on the number of digits tested. Each digit appeared for 1000 ms.

When participants chose to answer questions immediately, 'please answer' would first appear on the screen and participants were required to press the space key to proceed to the next question (upper part of Fig. 1). Delay tie between prime task and distraction task with different time period. Distraction task when the sign '←' was presented or the word 'left' was spoken, participants needed to press the left key on the keyboard and vice versa. Participants were asked to write down the digits in the order presented, and a break of 5 min during the task (without feedback) was given.

2.3 Results

A repeated measures analysis of variance (ANOVA), the independent variables were music training group, learning preference, modality and delay time. The results revealed significant main effects for between and within variables: for groups ($p < 0.001$), for modality ($p < 0.001$) and for delay time ($p = 0.019$), but no significant for learning preference ($p = 0.469$). These analyses indicated that the mean accuracy was greater in music trained ($m = 0.88$, $SD = 0.14$) than in non-music trained ($m = 0.73$, $SD = 0.19$), and that the mean accuracy was better in visual presentation ($m = 0.88$, $SD = 0.13$) than in auditory presentation ($m = 0.73$, $SD = 0.20$). The results revealed significant two-factor interactions: for group × modality ($p = 0.002$), and for modality × delay time ($p = 0.000$). There was no significant difference in three-factor interactions.

A Tukey-HSD post hoc test revealed that participants performed better on visual tasks compared to auditory tasks for both the music trained ($p = 0.033$) and non-music trained ($p = 0.000$) groups. For music trained participants, they showed better performance on visual and auditory tasks, and visual task greater

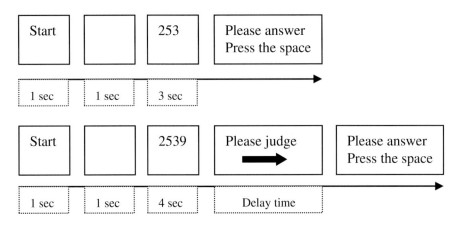

Fig. 1 Example display of experimental procedure for two presentation conditions

than auditory task. However, no obvious difference was found with visual task in the non-music trained group and auditory task in the music trained group ($p = 0.991$), see Fig. 2.

A Tukey-HSD post hoc test, with delay time an advantage for visual task was observed ($p < 0.016$), see Fig. 2. The visual task had no significantly response to changes in delay time ($p > 0.05$). Participants performed better on auditory task in 0 s delay than in 4, 8, and 10 s delay time ($p < 0.013$), see Fig. 3.

3 Discussion

Results of the experiment showed that the music trained group performed better than the non-music trained one on memory retention tasks. These results are similar to the findings reported in Ho et al. (2003). The difference between learning

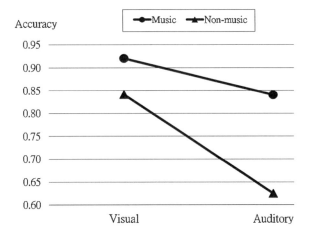

Fig. 2 Accuracy for two groups on visual and auditory task music training demonstrated overall higher accuracy ($p = 0.002$) by the presence of two tasks

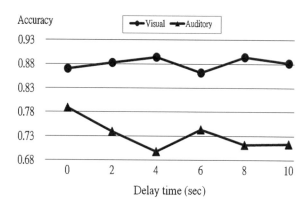

Fig. 3 Accuracy for different delay period on visual and auditory task, the interaction between modality presentation and delay time was significant (p < 0.016). The accuracy at each delay period was expected to be better for the visual presentation than for the auditory one

styles was not obvious, therefore we speculated that learning preference does not have an effect on memory retention.

Performance was better overall for visual tasks compared to auditory ones. Similar results were reported by Moore et al. Additionally, greater learning efficiency was observed among the non-music trained group for visual tasks. That is, in the music trained group, accuracy was 92.06 % for auditory tasks and 84.07 % for visual tasks, with difference of 7.99 %. In the non-music trained group, accuracy was 84.12 % for visual tasks and 62.46 % for auditory tasks, with difference of 21.66 %. Thus, a greater difference was observed in non-music trained group. The reason for this finding likely has to do with the constant flow of auditory information during music training. As a result, the music trained group might perceive less of a difference between visual and auditory cognition during memory retention tasks. On the other hand, children without music training likely rely mainly on visual perception. As a result, these children perceive a greater difference between the visual and auditory memory retention tasks.

Results also showed differences between the two modalities in terms of the time delay tasks. In this case, participants performed better with the visual tasks compared to the auditory tasks. The presence of a time delay did not affect performance on visual tasks, which had an accuracy of memory retention of 88.09 %. However, for auditory tasks, the accuracy decreased from 78.91 % with a 0 s time delay to 69.79 % with a 4 s delay time.

The results from this experiment were inconsistent with Jensen (1971). In our case, memory retention was greater for visual tasks. The inconsistency is likely different experimental methods and approaches in this experiment.

Participants with music training performed better than the group without music training. In terms of memory retention, the music trained group retained memory better than the non-music trained group. Overall, both groups performed better on visual tasks. A great study demonstrating this modality effect was reported by Mayer and Moreno (1998). The authors also conducted the experimental procedure for the two presentation conditions (visual text condition and auditory text condition). The results demonstrated that learning outcomes text presentation was higher for auditory. Given the results of our study, in terms of learning styles, auditory learners out-performed visual learners on the visual memory tasks.

Given the results of our study, participants' ability on memory retention was influenced by music training on visual and auditory memory retention. These findings suggest that children's visual-spatial abilities could be improved through music training. More curriculums should be included such as general music training and musical instrument lessons. This study shows that increased engagement in music and other interactive activities could improve children's overall cognitive abilities and enhance their learning experiences.

4 Conclusions

Our findings suggest that general associations between music training and modalities cognitive abilities from individual differences in memory retention. In terms of memory retention, music training retained memory better than the non-music training group. Both groups performed better when tasks were visually presented. No significant difference between difference style learners was found in memory retention task.

References

Baddeley AD, Hitch GJ (1974) Working memory. In: Broadbent DE (ed) Functional aspects of human memory. The Royal Society, London, pp 73–86

Baddeley A, Gathercole S, Papagno C (1998) The phonological loop as a language learning device. Psychol Rev 105:158–173

Brandler S, Rammsayer TH (2003) Differences in mental abilities between musicians and non-musicians. Psychol Music 31:123–138

Chan AS, Ho YC, Cheung MC (1998) Music training improves verbal memory. Nature 396(6707):128

Dowling WJ (1978) Scale and contour: two components of a theory of memory for melodies. Psychol Rev 85:341–354

Gabriel AR, Kevin JF, Julie AS (1995) Timbre reliance in nonmusicians' and musicians' memory for melodies. Music Perception 13:127–140

Gromko JE, Poorman AS (1998) The effect of music training on preschoolers' spatial-temporal task performance. J Res Music Educ 46:173–181

Helmbold N, Rammsayer T, Altenmüller E (2005) Differences in primary mental abilities between musicians and nonmusicians. J Individ Differ 26:74–85

Ho YC, Cheung MC, Chan AS (2003) Music training improves verbal but not visual memory: cross-sectional and longitudinal explorations in children. Neuropsychology 17(3):439–450

Husain G, Thompson WF, Schellenberg EG (2002) Effects of musical tempo and mode on arousal, mood, and spatial abilities. Music Perception 20:151–171

Jakobson LS, Cuddy LL, Kilgour AR (2003) Time tagging: a key to musicians' superior memory. Music Perception 20:307–313

Jakobson LS, Lewycky ST, Kilgour AR, Stoesz BM (2008) Memory for verbal and visual material in highly trained musicians. Music Perception 26:41–55

Jensen AR (1971) Individual differences in visual and auditory memory. J Educ Psychol 62:123–131

Korenman LM, Peynircioglu ZF (2007) Individual differences in learning and remembering music: auditory versus visual presentation. J Res Music Educ 55(1):48–64

Lee Y-S, Lu M-J, Ko H-P (2007) Effects of skill training on working memory capacity. Learn Instr 17:336–344

Mayer RE, Moreno R (1998) A split-attention effect in multimedia learning: evidence for dual processing systems in working memory. J Educ Psychol 90:312–320

Moore BR (1990) The relationship between curriculum and learner: music composition and learning style. J Res Music Educ 38:24–38

Peretz I, Coltheart M (2003) Modularity of music processing. Nat Neurosci 6:688–691

Rauscher FH, Shaw GL, Levine LJ, Wright EL, Dennis WR, Newcomb RL (1997) Music training causes long-term enhancement of preschool children's spatial-temporal reasoning. Neurol Res 19:2–8

Schellenberg EG (2005) Music and cognitive abilities. Curr Dir Psychol Sci 14:317–320

Schellenberg EG, Peretz I (2008) Music, language, and cognition: unresolved issues. Trends Cogn Sci 12:45–46

Zehra FP, Aydyn YD, Banu ÖK (2002) Phonological awareness and musical aptitude. J Res Reading 25(1):68–80

Control Scheme for the Service Quality

Ling Yang

Abstract This research constructs a novel quality control scheme for monitoring the service quality. Providing high-quality services can enhance company's productivity and strengthen its competitiveness. Due to the diversity of service operation, measuring and monitoring the service quality becomes very difficult. PZB's SERVQUAL is a commonly used scale to measure the service quality. The SERVQUAL scale has been shown to measure five underlying dimensions with 22 quality elements. After a questionnaire investigation, the collected information of service quality is often not monitored continually. If the service quality has variation, there will be no way for immediate correction. Precise instruments for measuring quality and accomplishing quality control have been developed and widely used in the manufacturing sector. The quality control chart is one of the commonly used tools of statistical process control for on-line control. Applying the control chart in the service quality can improve the control effect. In this work, the Ridit analysis is used to transform the collected data, which are mostly in Likert-scale, and to find the priority of quality elements. Some more important elements can be selected for the construction of control chart.

Keywords Control chart · Service quality · PZB quality model · SERVQUAL · Ridit analysis

1 Introduction

Facing global competition, the company should provide high-quality services and products to meet customers' needs. The diversity of service operation makes the measurement of service quality challenging. After many studies about service

L. Yang (✉)
Department of Marketing and Logistics Management, St. John's University, 499, Sec. 4, Tamking Road, Tamsui District, New Taipei City 25135, Taiwan, Republic of China
e-mail: lgyang@mail.sju.edu.tw

quality, Parasuraman et al. (called PZB for short) (1985) provided the famous PZB service quality model. PZB's SERVQUAL scale and their various extensions (Parasuraman et al. 1988, 1991, 1993, 1994; Zeithaml et al. 1993) indicated the factors that influence customer expectations and customer perceptions, and tried to quantify customer satisfaction using service performance gaps. The main gaps with SERVQUAL scale are: the knowledge gap, the standards gap, the delivery gap, the communications gap, and the overall gap. In SERVQUAL, the service quality is divided into five dimensions with 22 quality items. After SERVQUAL, there are a lot of studies for service quality scales proposed. SERVQUAL is still the most widely used.

SERVQUAL scale is based on the gap theory of service quality. In SERVQUAL's questionnaire, the twenty-two items of service quality are asked in "expected quality" sector and "perceived quality" sector, respectively. For each quality item, the service quality (SQ) is computed as the gap score of the perceived quality (P) and the expected quality (E), i.e.,

$$SQ = P - E \qquad (1)$$

If $SQ > 0$, it means that the customer is satisfied, and the higher the positive number, the more satisfaction. If $SQ < 0$, it means that the customer is not satisfied, and the higher the negative number, the more dissatisfaction.

Most of the service quality studies conduct a questionnaire survey in a specific period of time, but little research had been carried out for the monitoring of usual service quality. There are some firms engaged in regular surveys. Due to lack of appropriate quality control tool, the collected data of regular surveys cannot be used in a continuous quality monitoring. In order to ensure the product or service is in control, the application of on-line quality control is essential. The aim of this study is to construct a novel control scheme for monitoring the service quality in a firm's daily services.

2 Previous Research

2.1 Ordered Samples Control Chart

Franceschini et al. (2005) proposed an ordered samples control chart for ordinal scale sample data. Their study was based on the process sampling of manufacturing sector, and only one quality characteristic was considered. Such a quality control scheme cannot display all of the information of a firm's service quality.

2.2 Expectation-Perception Control Chart

Donthu (1991) constructed an expectation-perception control chart which used the SERVQUAL scales to generate quantitative data. Based on five dimensions of service quality, their collected data were analyzed in expectation analysis and perception analysis, respectively. And, the control limits of the expectation-perception control chart were computed by all respondent data. Unlike the concept of online control, the samples of the expectation-perception control scheme were selected randomly from the respondent data in the same survey. Their sampling method did not comply with the spirit of process control.

3 The Proposed Service Quality Control Scheme

The purpose of this study is to provide a control scheme for monitoring the service quality. Quality control chart is one of the commonly used tools in on-line quality control. The control chart is a graph used to study how a process changes over time with data plotted in time order. The construction of control chart includes the control center line, the upper control limit, and the lower control limit.

Most of the general service quality scales have about 20 items, or even more. If all of the quality items are put to control individually, it seems difficult to interpret these data. If the quality items can be sorted by the degree of importance, and select the more important items to control, the work of service quality control can be more focused. In questionnaire surveys, most of the answers of respondents are measured with Likert-scale. Likert-scale assumes that each quality item has the same value, i.e., each answer is considered as an interval scale. However, Schwarz et al. (1991) found that the respondents tended to the center-right options.

For many respondents, a Likert-scale from 1 (meaning "strongly disagree") to 5 (meaning "strongly agree") is not equidistant between 1 and 5 of each option, but more like ordinal scale to express their views. Using the average of interval scale to determine the importance sequence of the quality items seems inappropriate. Therefore, the collected data should be converted appropriately before construct a service quality control chart.

Some studies (Yager 1993; Wu 2007) tried to convert Likert scale. Ridit (Relative to an Identified Distribution Unit) analysis (Fielding 1993) is one of the commonly used methods in the context of ordered categorical data. Chatterjee and Chatterjee (2005) proposed a prioritization methodology of the service quality items with Ridit analysis (Bross 1958; Agresti 2010). In this study, Ridit analysis is invoked to transform the collected data of the survey questionnaire, and prioritize the quality items according to the increasing order of service gap (as shown in 1). Then some mean control charts, such as the Shewhart-$\bar{X}$ control chart (Montgomery 2009) and the exponentially weighted moving average (EWMA)-$\bar{X}$ control chart (Roberts 1959) can be used to construct the service quality control chart.

4 Conclusions

The proposed service quality control scheme allows the firms of service sector to sustain the quality control. The use of Ridit analysis makes data conversion more reasonable and identifies the priority of quality items. The design of the control charts for more important items allows the quality control more focused. The proposed service quality control scheme has practical value.

Acknowledgments This work is partially supported by the National Science Council, Taiwan, ROC, under Contract no. NSC 101-2221- E-129-011.

References

Agresti A (2010) Analysis of ordinal categorical data. Wiley, New York
Bross ID (1958) How to use ridit analysis. Biometrics 14:18–38
Chatterjee S, Chatterjee A (2005) Prioritization of service quality parameters based on ordinal responses. Total Quali Manag 16:477–489
Donthu N (1991) Quality control in services industry. J Profl Serv Market 7:31–55
Fielding A (1993) Scoring functions for ordered classifications in statistical analysis. Qual Quant 27:1–17
Franceschini F, Galetto M, Varetto M (2005) Ordered samples control charts for ordinal variables. Qual Relia Engi Intl 21:177–195
Montgomery D (2009) Statistical quality control: a modern introduction, 6th edn. Wiley, New York
Parasuraman A, Zeithaml VA, Berry LL (1985) A conceptual model of service quality and its implications for future research. J Market 49:41–50
Parasuraman A, Zeithaml VA, Berry LL (1988) SERVQUAL: a multiple-item scale for measuring consumer perceptions of service quality. J Retail 64:12–40
Parasuraman A, Berry LL, Zeithaml VA (1991) Refinement and reassessment of the SERVQUAL scale. J Retail 67:420–450
Parasuraman A, Berry LL, Zeithaml VA (1993) More on improving service quality measurement. J Retail 69:140–147
Parasuraman A, Zeithaml VA, Berry LL (1994) Alternative scales for measuring service quality: a comparative assessment based on psychometric and diagnostic criteria. J Retail 70:201–230
Roberts SW (1959) Control chart tests based on geometric moving averages. Technometrics 1:239–250
Schwarz N, Knauper B, Hippler HJ et al (1991) Rating scales: numeric values may change the meaning of scale labels. Pub Opin Quart 55:570–582
Wu CH (2007) An empirical study on the transformation of Likert-scale data to numerical scores. Appl Math Sci 1:2851–2862
Yager R (1993) Non-numeric multi-criteria multi-person decision making. Group Deci Negot 2:81–93
Zeithaml V, Berry L, Parasuraman A (1993) The nature and determinant of customer expectation of service quality. J Acad Market Sci 21:1–12

Particle Swam Optimization for Multi-level Location Allocation Problem Under Supplier Evaluation

Anurak Chaiwichian and Rapeepan Pitakaso

Abstract The aim of this paper is to propose the method for solving multi-level location allocation problems under supplier evaluation. The proposed problem is solved by particle swarm optimization (PSO). Generally, the multi-level location allocation problem considers the suitable locations to service customers or to store inventory from the suppliers. The proposed problem is to determine the suitable location to store and produce the product from the selected suppliers and then delivers product to the customers. The selected suppliers are determined by the capability of them. The capability of the suppliers means the quality of the material delivered and the reliability of delivery date which gather from their past statistics. The problem solving can be divided into two steps: the first step is to evaluate each potential supplier using fuzzy approach and second step is to selects the locations in order to serve the customer demand with minimum cost using PSO by calculating the amount of material shipped to location by a specified supplier closeness coefficient (CC_h). As the results, the percentage error is between 0.89 and 16.90 % and the average runtime is 4.2 s.

Keywords Capacitated p-median problem · Fuzzy · Closeness coefficient (CC_h) · Particle swarm optimization (PSO)

A. Chaiwichian (✉) · R. Pitakaso
Department of Industrial Engineering, Ubonratchathani University, Ubonratchathani 34190, Thailand
e-mail: anuak.aun@gmail.com

R. Pitakaso
e-mail: enrapepi@mail2.UBU.AC.TH

Y.-K. Lin et al. (eds.), *Proceedings of the Institute of Industrial Engineers Asian Conference 2013*, DOI: 10.1007/978-981-4451-98-7_147,
© Springer Science+Business Media Singapore 2013

1 Introduction

The capacitated p-median problem (CPMP) was a well-known discrete location problem. We have to partition a set of customers, with a known demand, in p-facilities and consider between capacities that enough supply to a set of customers and minimized total distance. This problem has many researchers to create various new heuristic algorithms. Such Lorena and Senne (2003) used Lagrangean/surrogate relaxation which improves upper bounds applying local search heuristics to solutions. Similar problem, Mulvey and Beck (1984) examined the Lagrangean relaxation of assignment constraints in a 0–1 linear integer programming problem formulation. Diaz and Fernandez (2006) proposed heuristic algorithms for capacitated p-median problem. They used 3 methods for solving including Scatter Search, Path Relinking, and combining Scatter Search and Relinking together.

Various heuristic algorithms were applied to determine CPMP. This was modified for complicated problem, which was known to be NP-hard (Garey and Johnsoh 1979). Among the more recent works on application heuristics was that of Ahmadi and Osman (2005) propose a greedy random adaptive memory search method and later offer a guided construction search heuristic. Fleszar and Hindi (2008) presented Variable Neighbour hood Search for solving CPMP.

For the recent problem, CPMP was mixed with customer and matching routing supplier or probability customer in order to minimize total cost or minimize distance. Corresponding to this, it has many researches try to solve this. Albareda-Sambola et al. (2009) introduced a new problem called the Capacity and Distance Constrained Plant Location Problem which constraints include capacity of plants and total distance between plant and customers. Zhu et al. (2010) offers a new problem called the capacitated plant location problem with customer and supplier matching (CLSM). They merge a distribution trip and a supply trip into one triangular trip for saving allocation cost. In same line, Kuban Altinel et al. (2009) considered constraints under capacity with probabilistic customer locations. A review of previous works indicates that the methods for solving problems are pointed to the discrete optimization with minimization cost under capacitated constrain but it could not consider the supplier evaluation together. So, in this paper we present an approach for capacitated p-median problem under supplier evaluation with mixed discrete-continuous optimization. To determine this problem, we present the fuzzy algorithms for evaluating supplier and we use Lingo software and PSO method to solve problem.

In this paper, the next section presents the mathematical models and Fuzzy method. In Sect. 3, we propose methodology of this research. Section 4 describes the method of PSO and Sect. 5 explains the methodology. Finally, the experiments are illustrated by a table which is a result of testing in 3 cases.

2 Mathematical Models

2.1 Mathematics

We have divided problem into 3 cases and defined the meaning of supplier in each case.

For the case 1, the plants receive products only from the suppliers which are passed and all customer demands are enough to serve. Objective function and constrains are as below

$$\text{Min} = \sum_{i=1}^{m}\sum_{j=1}^{n} A_i C_{ij} X_{ij} + \sum_{h=1}^{k}\sum_{j=1}^{n} U_h Y_h W_{hj} Z_{hj} - \sum_{j=1}^{n}\sum_{h=1}^{k} \frac{U_h}{(1+U_h)} Y_h W_{hj} Z_{hj} \quad (1)$$

Subject to

$$\sum_{h=1}^{k} U_h Y_h Z_{hj} \leq P_j, \ \forall j \quad (2)$$

$$\sum_{h=1}^{k} U_h Y_h Z_{hj} \geq \sum_{i=1}^{m} A_i X_{ij}, \ \forall j \quad (3)$$

$$\sum_{i=1}^{m} X_{ij} \geq 1, \ \forall j \quad (4)$$

$$\sum_{j=1}^{n} X_{ij} = 1, \ \forall i \quad (5)$$

$$\sum_{h=1}^{k} Z_{hj} \geq 1, \ \forall j \quad (6)$$

$$\sum_{j=1}^{n} Z_{hj} \leq 1, \ \forall h \quad (7)$$

$$X_{ij} \in \{0, 1\} \quad (8)$$

$$Z_{hj} \in [0, 1] \quad (9)$$

$$U_h > 0 \quad (10)$$

where,
i   Customer
j   Plant

h	Supplier
A_i	Customer demand at i
C_{ij}	Cost per piece for delivering products from plant j to customer i
U_h	Percentage for ordering products from supplier h
Y_h	Capacity of supplier h
X_{ij}	$\begin{cases} \text{if customer i is served by plant j.} \\ 0 \text{ otherwise.} \end{cases}$
CC_h	The closeness coefficient of supplier h
Z_{hj}	Plant j is served by supplier h
P_j	Maximum capacity of plant j
W_{hj}	Penalty cost per piece for delivering products from supplier h to plant j

In the Eq. (2), it guarantees that quantities in each supplier h which sends to plant j are less than or equal to maximum capacity of plant j. Equation (3) shows that the quantities in each supplier h sending to plant j are greater than or equal to customer demand at i. Equation (4) indicates that plant can send products more than one customer. Equation (5) guarantees that customer can receive products one plant. For the Eq. (6), plant can receive products more than one supplier. In the Eqs. (7) and (9), Z_{hj} is a real number and less than or equal one.

For the case 2, the plants receive products from the passing suppliers but they can't serve all customers. So, plants have to receive the residual products from the passing suppliers in order to fulfill demand of the customers. Objective function and constrains are shown in below,

$$\text{Min} = \sum_{i=1}^{m}\sum_{j=1}^{n} A_i C_{ij} X_{ij} + \sum_{h=1}^{k}\sum_{j=1}^{n} U_h Y_h W_{hj} Z_{hj} + \sum_{j=1}^{n}\sum_{h=1}^{k} (1 - U_h) Y_h \left[W_{hj} + \lambda_{hj} \right] Z_{hj}$$

$$- \sum_{j=1}^{n}\sum_{h=1}^{k} \frac{U_h}{(1 + U_h)} Y_h W_{hj} Z_{hj} \qquad (11)$$

subject to

$$\sum_{h=1}^{k} U_h Y_h Z_{hj} + \sum_{h=1}^{k} (1 - U_h) Y_h Z_{hj} \leq P_j, \forall j \qquad (12)$$

$$\sum_{h=1}^{k} [U_h Y_h + (1 - U_h) Y_h] Z_{hj} \geq \sum_{i=1}^{m} A_i X_{ij}, \forall j \qquad (13)$$

$$\sum_{i=1}^{m} X_{ij} \geq 1, \forall j \qquad (14)$$

$$\sum_{j=1}^{n} X_{ij} = 1, \forall i \qquad (15)$$

$$\sum_{h=1}^{k} Z_{hj} \geq 1, \forall j \tag{16}$$

$$\sum_{j=1}^{n} Z_{hj} \leq 1, \forall h \tag{17}$$

$$X_{ij} \in \{0, 1\} \tag{18}$$

$$Z_{hj} \in [0, 1] \tag{19}$$

$$U_h > 0 \tag{20}$$

where
λ_{hj} Penalty cost per piece that sends products form supplier h to plant j.

In the Eq. (12), it assures that products from supplier h are less than or equal to maximum capacity of plant j. Equation (13) shows that the products of supplier h are greater than or equal to customer demand at i.

In the case 3, the plants accept the products from the passing suppliers but are not enough to serve customers. Therefore, the plants must be received from the other suppliers. Objective function and constrains are shown in below,

$$\begin{aligned} \text{Min} = &\sum_{i=1}^{m}\sum_{j=1}^{n} A_i C_{ij} X_{ij} + \sum_{h=1}^{k}\sum_{j=1}^{n} U_h Y_h W_{hj} Z_{hj} + \sum_{j=1}^{n}\sum_{h=1}^{k} (1-U_h) Y_h [W_{hj} + \lambda_{hj}] Z_{hj} \\ &+ \sum_{j=1}^{n}\sum_{h=1}^{k} Y_{h|U_h=0} [W_{hj} + \lambda_{hj}] Z_{hj} - \sum_{j=1}^{n}\sum_{h=1}^{k} \frac{U_h}{(1+U_h)} Y_h W_{hj} Z_{hj} \end{aligned} \tag{21}$$

subject to

$$\sum_{h=1}^{k} U_h Y_h Z_{hj} + \sum_{h=1}^{k} (1-U_h) Y_h Z_{hj} + \sum Y_{h|U_h=0} Z_{hj} \leq P_j, \forall j \tag{22}$$

$$\sum_{h=1}^{k} [U_h Y_h + (1-U_h) Y_h] Z_{hj} + \sum_{h=1}^{k} Y_{h|U_h=0} Z_{hj} \geq \sum_{i=1}^{m} A_i X_{ij}, \forall j \tag{23}$$

$$\sum_{i=1}^{m} X_{ij} \geq 1, \forall j \quad (24) \quad / \quad \sum_{j=1}^{n} X_{ij} = 1, \forall i \tag{25}$$

$$\sum_{h=1}^{k} Z_{hj} \geq 1, \forall j \quad (26) \quad / \quad \sum_{j=1}^{n} Z_{hj} \leq 1, \forall h \tag{27}$$

$$X_{ij} \in \{0, 1\} \quad (28) \qquad / \qquad Z_{hj} \in [0, 1] \quad (29)$$

$$U_h > 0 \quad (30)$$

In the Eq. (22), it guarantees that quantities include the other suppliers are less than or equal to maximum capacity of plant j. Equation (23) shows that the quantities in each supplier h sending to plant j are greater than or equal to customer demand at i.

3 Fuzzy Method

In this paper, we use 2 variables in term of linguistic variables. One is the importance weights of various criteria and the second is the ratings of qualitative criteria. All of them are considered as linguistic variables. We have 3 steps to proceeding in fuzzy methods. First step, we assign the linguistic variables to 7 criterions for assessments. The linguistic variables for importance weight are Very Low, Low, Medium Low, Medium, Medium High, High and Very High (Chen et al. 2006) and the importance weights in each criterion are represented as shown in Fig. 1.

For example, the linguistic variable "Medium High" can be represented as (0.5, 0.6, 0.7, 0.8) for the importance weight of each criterion. The second step, the linguistic variables for ratings of qualitative are as shown in Fig. 2.

In term of "Good" in ratings, the linguistic variables can be represented as (7, 8, 9) for the ratings of qualitative in each criterion. In this paper, there are 4 criterions including Supplier's product quality, Supplier's delivery/order fulfillment capability, Price/Cost reduction performance, and Supplier's post-sales service (Wang 2010). The third step is reviewed from (Chen et al. 2006). In this

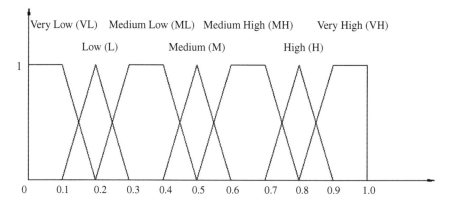

Fig. 1 Linguistic variable for importance weight of each criterion

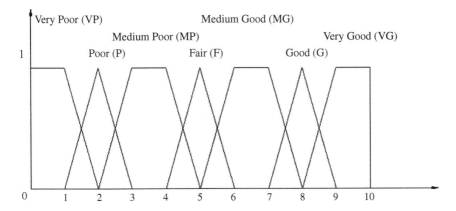

Fig. 2 Linguistic variable for ratings

research, we have 4 sets to group parameters for evaluating supplier. The sets describe as below

1. a set of K decision-makers called E = {$D_1, D_2, \ldots D_k$}
2. a set of m possible suppliers called A = {$A_1, A_2, \ldots, A_m$}
3. a set of n criteria, C = {$C_1, C_2, \ldots, C_n$}, with which supplier performances are measured
4. a set of performance ratings of Ai (i = 1,2, …,m) with respect to criteria C_j (j = 1,2, …,n), called X = {X_{ij}, i = 1,2, …,m, j = 1,2, …,n}.

The fuzzy ratings of each decision-maker D_k (k = 1,2,…,K) are represented as a positive trapezoidal fuzzy number $\widetilde{R}_k$ (k = 1,2,…,K) with membership function $\mu_{\widetilde{R}_k}(x)$ which translates the linguistic variable as below.

For example, the linguistic variable for importance weight in term of "Medium High (MH)" can be represented as (0.5, 0.6, 0.7, 0.8), the membership function of which is

$$\mu_{MediumHigh}(x) = \begin{cases} 0, & x < 0.5, \\ \frac{x-0.5}{0.6-0.5}, & 0.5 \leq x < 0.6, \\ 1, & 0.6 \leq x < 0.7, \\ \frac{x-0.8}{0.8-0.7}, & 0.7 \leq x < 0.8, \\ 0 & x \geq 0.8. \end{cases} \quad (31)$$

And the linguistic variable for ratings in term of "Very Good (VG)" can be represented as (8, 9, 9, 10), the membership function of which is

$$\mu_{VeryGood}(x) = \begin{cases} 0, & x < 8, \\ \frac{x}{9-8}, & 8 \leq x < 9, \\ 1, & 9 \leq x < 10, \end{cases} \quad (32)$$

The fuzzy ratings of all decision-makers is trapezoidal fuzzy numbers $\tilde{R}_k = (a_k, b_k, c_k, d_k)$, $k = 1, 2, \ldots, K$. so, the aggregated fuzzy rating is defined as $\tilde{R} = (a, b, c, d)$, $k = 1, 2, \ldots, K$
where

$$\left. \begin{array}{ll} a = \min_k \{a_k\} & b = \frac{1}{K} \sum_{k=1}^{K} b_k \\ c = \frac{1}{K} \sum_{k=1}^{K} c_k & d = \max_k \{d_k\} \end{array} \right\} \quad (33)$$

The fuzzy rating is $\tilde{x}_{ijk} = (a_{ijk}, b_{ijk}, c_{ijk}, d_{ijk})$ and importance weights is $\tilde{w}_{jk} = (w_{jk1}, w_{jk2}, w_{jk3}, w_{jk4})$ where $i = 1, 2, \ldots, m$; $j = 1, 2, \ldots, n$; $k =$ the kth decision maker. The aggregated fuzzy ratings $(\tilde{x}_{ij})$ with respect to each criterion can be calculated as $\tilde{x}_{ij} = (a_{ij}, b_{ij}, c_{ij}, d_{ij})$ where

$$\left. \begin{array}{ll} a_{ij} = \min_k \{a_{ijk}\} & b_{ij} = \frac{1}{K} \sum_{k=1}^{K} b_{ijk} \\ c_{ij} = \frac{1}{K} \sum_{k=1}^{K} c_{ijk} & d_{ij} = \max_k \{d_{ijk}\} \end{array} \right\} \quad (34)$$

The aggregate fuzzy weights $(\tilde{w}_j)$ of each criterion can be calculated as $\tilde{w}_j = (w_{j1}, w_{j2}, w_{j3}, w_{j4})$ where

$$\left. \begin{array}{ll} w_{j1} = \min_k \{w_{jk1}\} & w_{j2} = \frac{1}{K} \sum_{k=1}^{K} w_{jk2} \\ w_{j3} = \frac{1}{K} \sum_{k=1}^{K} w_{jk3} & w_{j4} = \max_k \{w_{jk4}\} \end{array} \right\} \quad (35)$$

A supplier-selection can be expressed in matrix format as below:

$$\left. \tilde{D} = \begin{bmatrix} \tilde{x}_{11} & \tilde{x}_{12} & \cdots & \tilde{x}_{1n} \\ \tilde{x}_{21} & \tilde{x}_{22} & \cdots & \tilde{x}_{2n} \\ \cdots & \cdots & \cdots & \cdots \\ \tilde{x}_{m1} & \tilde{x}_{m2} & \cdots & \tilde{x}_{mn} \end{bmatrix} \right\} \quad (36)$$
$$\tilde{w} = [\tilde{w}_1, \tilde{w}_2, \ldots, \tilde{w}_n]$$

where

$$\tilde{x}_{ij} = (a_{ij}, b_{ij}, c_{ij}, d_{ij}) \text{ and } \tilde{w}_j = (w_{j1}, w_{j2}, w_{j3}, w_{j4});$$

$i = 1, 2, \ldots, m$, $j = 1, 2, \ldots, n$. can be approximated by positive trapezoidal fuzzy numbers. We use linear scale transformation to reduce mathematical operations in a decision process. The set of criteria can be divided into benefit criteria (using the larger the rating, the greater the preference: B) and cost criteria (using the smaller the rating, the greater the preference: C). The normalized fuzzy-decision matrix can be shown as

$$\tilde{R} = [\tilde{r}_{ij}]_{m \times n} \tag{37}$$

where B and C are the set of benefit criteria and cost criteria, respectively

$$\tilde{r}_{ij} = \left(\frac{a_{ij}}{d_j^*}, \frac{b_{ij}}{d_j^*}, \frac{c_{ij}}{d_j^*}, \frac{d_{ij}}{d_j^*}\right), \quad j \in B \tag{38}$$

$$\tilde{r}_{ij} = \left(\frac{a_{\bar{j}}}{d_{ij}}, \frac{a_{\bar{j}}}{c_{ij}}, \frac{a_{\bar{j}}}{b_{ij}}, \frac{a_{\bar{j}}}{a_{ij}}\right), \quad j \in C, \tag{39}$$

$$d_j^* = \max_i d_{ij}, \quad j \in B \tag{40}$$

$$a_{\bar{j}} = \min_i a_{ij}, \quad j \in C. \tag{41}$$

The weighted normalized fuzzy-decision matrix is arranged as

$$\tilde{V} = [\tilde{v}_{ij}]_{m \times n}, \quad i = 1, 2, \ldots, m, \quad j = 1, 2, \ldots, n \tag{42}$$

where

$$\tilde{v}_{ij} = \tilde{r}_{ij}(\circ)\tilde{w}_j \tag{43}$$

The fuzzy positive-ideal solution (FPIS, A*) and fuzzy negative-ideal solution (FNIS, A⁻) can be described as

$$A^* = (\tilde{v}_1^*, \tilde{v}_2^*, \ldots, \tilde{v}_n^*), \quad A- = (\tilde{v}_{\bar{1}}, \tilde{v}_{\bar{2}}, \ldots, \tilde{v}_{\bar{n}}), \tag{44}$$

where

$$\tilde{v}_j^* = \max_i [v_{ij4}] \text{ and } \tilde{v}_{\bar{j}} = \min_i [v_{ij1}] \tag{45}$$

The distance of each alternative A* and A⁻ is as below

$$d_i^* = \sum_{j=1}^{n} d_v(\tilde{v}_{ij}, \tilde{v}_j^*), \quad i = 1, 2, \ldots, m. \tag{46}$$

$$d_{\bar{i}} = \sum_{j=1}^{n} d_v(\tilde{v}_{ij}, \tilde{v}_{\bar{j}}), \quad i = 1, 2, \ldots, m. \tag{47}$$

Where $d_v(\bullet, \bullet)$ is the distance measurement between two fuzzy numbers which are calculated by using the vertex method as Chen et al. (2006).

$$d_v(\tilde{m}, \tilde{n}) = \sqrt{\frac{1}{4}[(m_1 - n_1)^2 + (m_2 - n_2)^2 + \cdots + (m_m - n_n)^2]} \tag{48}$$

Final step of fuzzy method, a closeness coefficient (CC_i) is described to calculate the ranking order of all suppliers. It has an equation as

Table 1 Percentage of U_h

Closeness coefficient $(CC_h)^*$	Assessment status**	U_h (%)
$CC_h \in [0, 0.2)$	Do not recommend.	0
$CC_h \in [0.2, 0.4)$	Recommend with high risk.	0
$CC_h \in [0.4, 0.6)$	Recommend with low risk.	70
$CC_h \in [0.6, 0.8)$	Approved.	85
$CC_h \in [0.8, 1]$	Approved and preferred	100

*,** Chen-Tung Chen

$$CC_i = \frac{d_i^-}{d_i^* + d_i^-}, \quad i = 1, 2, \ldots, m. \tag{49}$$

The closeness coefficient is corresponded to Table 1.

4 Particle Swarm Optimization

The various steps involved in Particle Swarm Optimization Algorithm are as follows

step 1: The velocity and position of all particles are randomly set to within pre-defined ranges.

step 2: Velocity updating—At each iteration, the velocities of all particles are updated according to,

$$v_i^{t+1} = w_i v_i^t + c_1 R_1 (p_{i,best}^t - p_i^t) + c_2 R_2 (g_{i,best}^t - p_i^t). \tag{50}$$

step 3: Position updating—The positions of all particles are updated according to,

$$p_i^{t+1} = p_i^t + v_i^{t+1} \tag{51}$$

$$w_i = \frac{[(w_w - 0.4)(\text{no.iter} - \text{iter}_i)]}{(\text{no.iter} + 0.4)} \tag{52}$$

step 4: Memory updating—Update $p_{i,best}$ and $g_{i,best}$ when condition is met,

Fig. 3 Flowchart of process PSO

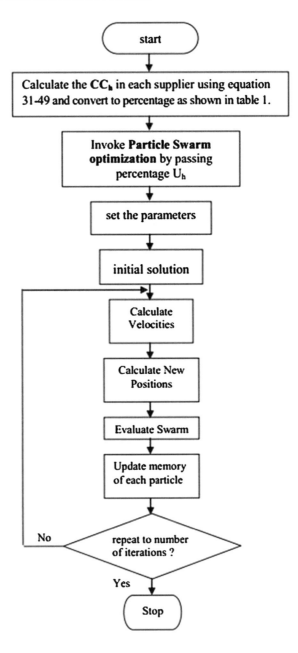

$$p_{i,best} = p_i \text{ if } f(p_i) > f(p_{i,best})$$
$$g_{i,best} = g_i \text{ if } f(g_i) > f(g_{i,best})$$
(53)

Table 2 Results of the experiments

Case	Optimum	Objective function of PSO	CPU time (min: sec: ms) Lingo	CPU time (min: sec: ms) PSO	% Error
1	3625090	3901156	01:28:00	00:04:22	7.62
1	3602470	4103373	01:28:00	00:04:20	13.90
1	3621240	3980837	01:28:00	00:04:18	9.93
1	3622100	3976914	01:28:00	00:04:20	9.80
1	3610880	4221071	01:28:00	00:04:20	16.90
2	10372902	11277106	>2 h	00:04:24	8.72
2	10560304	10412461	02:58:00	00:04:00	1.40
2	9983255	9339444	23:28:00	00:04:00	6.45
2	10043602	9445484	>2 h	00:04:06	5.96
2	11182905	9855059	>2 h	00:04:08	11.87
3	39872924	36894452	12:33:00	00:04:12	7.47
3	34572025	35201078	03:53:00	00:04:10	1.82
3	29506023	25449149	01:33:00	00:04:06	13.75
3	24641804	24421456	01:34:00	00:04:04	0.89
3	19433021	18476195	04:01:00	00:04:08	4.92
Average			1,197.33 s	4.12 s	8.09

where $f(x)$ is the objective function to be optimized and w_i is the inertia of the velocity

5 Methodology

In this study, PSO is set by the parameters as $c_1 = 3$, $c_2 = 1.5$, $R_1 = 4$, $R_2 = 2$, $w_w = 1$ and number of particles = 10, iterations = 10. This problems set the number of supplier (h), plant (j) and customer (i) as 2,000, 50 and 4,000 consequently. The process of PSO is shown as flowchart in Fig. 3.

6 Experimental Results

In this study, we run on a notebook with the configuration of 2.10 GHz CPU and 4.0 GB memory. We solved problems and compared with optimal solution to test the performance of our algorithms. The %error are reported in column 10–11 and calculated as

$$\frac{f - f_{opt}}{f_{opt}} \times 100, \tag{54}$$

where f denotes the best solution found and f_{opt} is the optimal value in objective function.

The results of experiments are shown in Table 2. As the results, the maximum %error in PSO is 16.90 and 0.89 % for minimum %error, and average %error of PSO is 8.09 %. The average runtime of PSO is 4.12 s while average runtime for founding optimal solutions is 1,197.33 s. In no. 6, 9 and 10 of Table 2, they run more than 2 h to find a solution but PSO runs time about 4.13 s and have %error is 8.72, 5.96 and 11.87 respectively.

7 Conclusion

A new problem of location allocation problem has proposed in this study. It has considered between quantity and quality for finding the optimal solutions. We use fuzzy method in this problem for evaluating the quality.

A particle swarm optimization is proposed as an alternative for solving this problem and it is compared to optimal solutions solved by Lingo software. The results shown that a heuristics PSO can generate good solutions to location allocation problem.

As the results, PSO outperforms Lingo when the computational time is limited and the optimal solutions are nearly the best solutions. Moreover, this PSO can be easily implemented by location allocation problem under supplier evaluation. Based on our computational test, we believe that PSO have potential to be useful heuristic for this problem.

For the future work, this problem can be applied to the modified PSO in order to find the exactly optimal solutions and run on big scale.

References

Ahmadi S, Osman IH (2005) Greedy random adaptive memory programming search for the capacitated clustering problem. Eur J Oper Res 162(1):30–44

Chen CT, Lin CT, Huang SF (2006) A fuzzy approach for supplier evaluation and selection in supply chain management. Int J Prod Econ 102:289–301

Kuban Altinel I et al (2009) A location-allocation heuristic for the capacitated multi-facility Weber problem with probabilistic customer locations. Eur J Oper Res 198:790–799

Diaz JA, Fernandez E (2006) Hybrid scatter search and path relinking for the capacitated p-median problem. Eur J Oper Res 169:570–585

Fleszar K, Hindi KS (2008) An effective VNS for the capacitated p-median problem. Eur J Oper Res 191:612–622

Lorena LAN, Senne ELF (2003) Local search heuristics for capacitated p-median problems. Netw Spat Econ 3:407–419

Garey MR, Johnsoh DS (1979) Computers and intractability : a guide to the theory of NP-completeness. W.H. Freeman and Co, New York

Albareda-Sambola M, Fernandez E, Laporte G (2009) The capacity and distance constrained plant location problem. Comput Oper Res 36:597–611

Mulvey JM, Beck MP (1984) Solving capacitated clustering problems. Eur J Oper Res 18:339–348

Wang WP (2010) A fuzzy linguistic computing approach to supplier evaluation. Appl Math Model 34:3130–3141

Zhu Z, Chu F, Sun L (2010) The capacitated plant location problem with customers and supplier matching. Transp Res Part E 46:469–480

Evaluation Model for Residual Performance of Lithium-Ion Battery

Takuya Shimamoto, Ryuta Tanaka and Kenji Tanaka

Abstract This study suggested the evaluation model for Lithium-ion battery life. Considering the trend that eco-system has become serious concern against the global environment, Lithium-ion battery is most promising represented by Electric Vehicle. However, estimation method for residual battery performance has not been established. Therefore, asset value and payout period are forced to be unsure, namely, the spread of Lithium-ion battery has been prevented. This evaluation model developed in this study can calculate the residual battery performance by degradation rate database and assumed battery use pattern. The degradation rate database was established using charge–discharge test of Lithium-ion battery cell. By applying the database, the residual battery performance can be calculated under any battery use scenario. This model was validated by comparing the experimental data with the simulation result.

Keywords Lithium-ion battery · Life cycle simulation · Evaluation model

1 Introduction

Today, the necessity of large-scale storage battery is growing for following the electric power supply and demand that changes complicatedly. This trend is made from expanding the use of renewable energy, the challenges of power generation equipment stemmed from the Great East Japan Earthquake. Moreover, expanding the use of various electric vehicles (xEVs: Electric Vehicles/Plug-In Hybrid Electric Vehicles/Hybrid Electric Vehicles) aiming at Greenhouse Gas emission

T. Shimamoto (✉) · R. Tanaka · K. Tanaka
Department of Systems Innovation, School of Engineering, The University of Tokyo,
7-3-1 Hongo Bunkyo-ku, Tokyo 113-8656, Japan
e-mail: shimamoto@m.sys.t.u-tokyo.ac.jp

reduction, and introducing smart grid are also increasing the importance of batteries.

Against this trend, Lithium-ion Battery (LiB) is most promising. LiB has a number of advantages compared with the conventional batteries, such as lead-acid storage battery and nickel-metal hydride battery. The points are high energy efficiency, charge and discharge energy density, and possibility of rapid charge and discharge. The market has so far been formed on small capacity articles such as a portable electric device or personal computer. Now, the research and development and use about the large capacity batteries are progressing rapidly adjusted to the above-mentioned purpose.

However, LiB has two major problems that the unit price and recycling cost per capacity are high, and degradation mechanisms are not fully understood (Vettera et al. 2005). These are factors inhibiting the spread of LiB.

With respect to these issues, on the user side of LiB, total life cost of LiB is tried to reduce through secondary use such as diversion xEVs use into stationary use. However, evaluation criteria for used LiB have not been established that is common to the various manufactures, storage capacity, and materials. In addition, usually, evaluation for the characterization of degradation of LiB is performed by the charge and discharge cycle test, the full charged preservation test, or the assumed pattern of LiB use.

This evaluation method is consistent for the purpose of the performance improvement, but it is not consistent for the purpose of the lifetime prediction for demand side expected complex usage patterns. Especially, the thing that a demand pattern becomes more complicated as storage batteries are introduced into community deeply, and the deviation with pre-supposed assessment conditions becomes large can be assumed easily. Moreover, preventive maintenance (PM) to prevent that the failure of battery triggered the hazard can be considered. From the above situations, there is an increasing necessity and importance of LiB evaluation method that can respond to all use patterns.

2 Existing Research

Abe et al. (2012) showed a method of separating LiB degradation into the "storage degradation" by the storage of LiB and the "charge–discharge degradation" by the charge and discharge cycles of LiB. Then, determined the approximate curve that represents each characteristic in order to estimate the degradation separately. Though this method can be set LiB operation freely, the use pattern across the State of Charge (SOC) range will be difficult to predict because the number of charge–discharge has been used in the degradation prediction formula.

Kaji (2012) build a degradation rate database by calculating the degradation rate based on the approximate curve through the initial state and the time of evaluation, however, it is difficult to predict in all conditions because it uses the number of charge–discharge as well. In addition, the effect of C-rate (=the amount of current/nominal full charge capacity) is not considered.

3 Aim and Approach

This study aimed the evaluation of LiB life with any use patterns. In order to achieve this purpose, the following approaches were taken.

- Define the battery performance.
- Establish the method to complement the experiment result in order to change the experiment result into database.
- Establish the method to estimate residual battery performance using database and assumed LiB use pattern.

Establishing the method to complement can reduce the number of required experiment. That means the time, money, and man-power required to the experiment can be saved. In addition, any use pattern of LiB will get to able to be dealt with.

It should be noted that the experiment was carried out by Japan Automobile Research Institute. It was storage test and charge–discharge cycle test for 18650-type LiB (2.4Ah, LCO-C).

4 Evaluation Method for Performance of LiB

4.1 Definition of Performance of LiB

In this study, the battery performance is defined from the user's point of view. We classified the battery performance into 2 factors: specific energy E (kWh/kg) and specific power P (kW/kg). As for EVs, each of the performance factors corresponds to the driving range and acceleration of the vehicle.

The equations of specific energy and specific power are shown as (1) and (2), where V represents the voltage, $V_{average}$ the average voltage during the discharge, Q the capacity, I the current, V_{OCV} the open circuit voltage, and R the internal resistance of the battery. As shown in the equations, the specific energy is a function of capacity (Q) and the specific power a function of internal resistance (R). Hence, the battery degradation can be attributed to the capacity fade and the increase in internal resistance. This indicates that it is sufficient to measure the capacity and internal resistance of the battery when evaluating the battery degradation.

$$E = \int P dt = \sum (V \cdot \Delta Q) = V_{average} \cdot Q. \quad (1)$$

$$P = I \cdot V = I \cdot (V_{OCV} - I \cdot R). \quad (2)$$

4.2 Method to Complement the Experimental Result and Develop the Degradation Rate Database

Complementing the experiment data is to deal with any use pattern of LiB. However, the method of interpolation has not been understood. This study established it by surveying the given graphs of the existing researches. This task was performed on 4 use-parameters (elapsed day, SOC, temperature, C-rate) that affect the degradation of battery. After determining each interpolation method, the experiment data was changed into database with that interpolation method. Figure 1 shows the procedure for establishing degradation rate database.

4.3 Method to Calculate the Residual Battery Performance

In this study, capacity degradation was supposed to have the additive property of the degradation due to storage ($\Delta Q_{storage}$) and the degradation due to current flows (ΔQ_{cycle}). As shown below, the total amount of degradation was calculated by

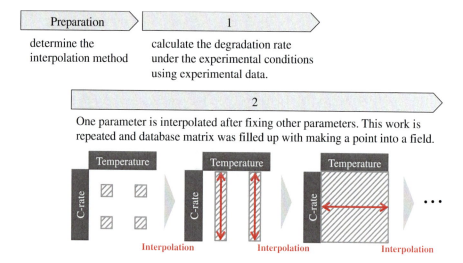

Fig. 1 Procedure for developing the degradation rate database

Evaluation Model for Residual Performance

integrating degradation rate and stay time in each condition. In this calculation model, giving assumed LiB use pattern, the residual battery performance can be calculated using degradation rate database.

$$\Delta Q(time) = \Delta Q_{storage}(time) + \Delta Q_{cycle}(time). \tag{3}$$

$$\Delta Q_{storage}(time) = \int_0^{Day} \int_{SOC_L}^{SOC_H} \int_{T_L}^{T_H} \int \frac{dQ_{storage}(Day, SOC, T)}{dt} dt\, dT\, dSOC\, dDay. \tag{4}$$

$$\Delta Q_{cycle}(time) = \int_0^{Day} \int_{SOC_L}^{SOC_H} \int_{T_L}^{T_H} \int_0^{C} \int \frac{dQ_{cycle}(Day, SOC, T)}{dt} dt\, dC\, dT\, dSOC\, dDay. \tag{5}$$

5 Verification of Interpolation Method

5.1 Establishment of Interpolation Method

In this chapter, verification of interpolation method that was mentioned in Sect. 4.2 is performed. Among the 4 parameters, C-rate is described as a representative.

Figure 2 shows the digitizing data from the experimental results of 5 existing researches (Liu 2007; Gao et al. 2008; Passerini et al. 2000; Wu 2005; He et al. 2008). Through this figure, the dependence of capacity degradation upon C-rate could be seen. In this figure, capacity is seemed to degrade linearly with respect to C-rate.

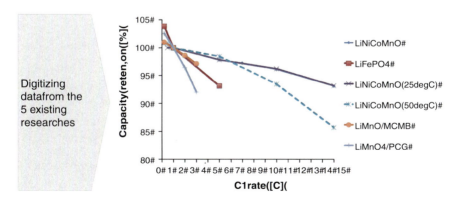

Fig. 2 Overall trend of capacity fade versus C-rate

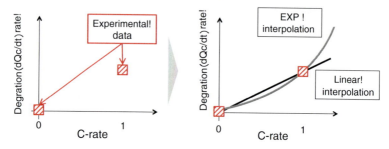

Fig. 3 Concept of C-rate interpolation

Through this survey, two interpolation methods that are the linear interpolation and the exponential interpolation were established as shown in Fig. 3. The exponential interpolation is for the possibility of accelerated degradation.

5.2 Establishment of Degradation Rate Database

These two interpolation methods were applied to the experimental result, then the degradation rate database was established as shown in Sect. 4.2. For other parameters (elapsed day, SOC, temperature), interpolation methods were fixed in order to verify the C-rate interpolation method. Each interpolation method was: elapsed day versus linear, SOC versus linear, temperature versus Arrhenius equation.

The experimental result that was changed into degradation rate database was performed by Japan Automobile Research Institute (JARI). This experiment was carried out for 18650-type LiB (2.4Ah, LCO-C). The test conditions were shown in Table 1.

5.3 Verification of C-Rate Interpolation Method

In order to verify the C-rate interpolation method, other experimental result that is shown in Table 2 was used. With the database based on Table 1 conditions and the use pattern based on Table 2 conditions, LiB degradation was calculated. The verification was performed by comparing the calculated result with the experimental result based on Table 2 conditions.

The outcomes are shown in Fig. 4. In this figure, plots mean the experimental result and lines mean the calculated result. In addition, the difference between the calculated result and the experimental result is shown in Table 3.

Evaluation Model for Residual Performance

Table 1 Test conditions for database

Items	SOC	C-Rate	Temp
Charge discharge cycle	5–25 %	1C (CC)	25 °C
	25–45 %		50 °C
	45–65 %		
	65–85 %		
	0–100 %		
	0–100 %	1C	
Storage	100 % (4.2 V)	(CCCV)	
	90 % (4.1 V)		
	78 % (4.0 V)		

Table 2 Test conditions for verification

Items	SOC (%)	C-rate	Temp
Charge discharge cycle	0–100	0.3C (CC)	45 °C
	0–100	2C (CC)	

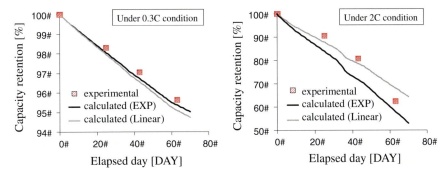

Fig. 4 Verification result under 0.3C/2C condition

Table 3 Differences between calculated result and experimental result

Elapsed day			0	25	43	63
0.3C	Difference (%)	EXP	0.0	−0.3	−0.3	−0.3
		Linear	0.0	−0.4	−0.5	−0.6
2C	Difference (%)	EXP	0.0	−7.4	−11.2	−7.3
		Linear	0.0	−3.2	−2.7	−9.0

Through these graphs and tables, linear interpolation was adopted as the C-rate interpolation method. The reason is that it could be estimated to within 0.5 % error at 0.3C condition.

In case of 2C condition, some care is needed. In Fig. 4, the right-most plot has fallen strongly. This is considered to be an exceptional degradation derived from very severe conditions of battery use. 2C condition means the battery usage

Table 4 High reproducible interpolation method

	Interpolation method
Elapsed day	Power
SOC	Exponential
Temperature	Arrhenius equation
C-rate	Linear

running out of full charge capacity in 30 min. Continuing this situation 24 h a day for 2 months brought about serious damage to the battery. Therefore, by regarding the rightmost plot as abnormal degradation and considering as non-existent, linear interpolation could be estimated the residual battery capacity with high accuracy.

From the above discussion, better interpolation method for C-rate is determined as linear one. Similar work was performed for other 3 parameters. Then it was found that the interpolation methods shown in Table 4 have high reproducibility for the experimental results.

6 Conclusion

The conclusion of this study is mentioned below.

- By converting the experimental data into degradation rate database, a system that can estimate the LiB residual performance under any operational patterns was developed.
- In order to build the degradation rate database, interpolation methods of experimental result was established.
- This approach can also be applied to the internal resistance estimation.

Acknowledgments This study was performed at Tanaka laboratory in the University of Tokyo. Grateful thanks for a lot of helpful supports of the all team members.

References

Abe M et al (2012) Lifetime prediction of lithium-ion batteries for high-reliability system. Hitachi Rev 94:334–337

Gao F et al (2008) Kinetic behavior of LiFePO4/C cathode material for lithium-ion batteries. Electrochim Acta 53:5071–5075

He Y-B et al (2008) Preparation and characterization of 18650 Li($Ni_{1/3}Co_{1/3}Mn_{1/3}$)O_2/graphite high power batteries. J Power Sources 185:526–533

Kaji K (2012) Evaluation model for used lithium-ion battery life. Bachelor thesis, The University of Tokyo

Liu X (2007) A mixture of LiNi1/3Co1/3Mn1/3O2 and LiCoO2 as positive active material of LIB for power application. J Power Sources 174:1126–1130
Passerini S et al (2000) Lithium-ion batteries for hearing aid applications: I. Design and performance. J Power Sources 89:29–39
Vettera J et al (2005) Ageing mechanisms in lithium-ion batteries. J Power Sources 147:269–281
Wu H-C (2005) Study the fading mechanism of $LiMn_2O_4$ battery with spherical and flake type graphite as anode materials. J Power Sources 146:736–740

A Simulated Annealing Heuristic for the Green Vehicle Routing Problem

Moch Yasin and Vincent F. Yu

Abstract Nowadays, the encouragement of the use of green vehicle is greater than it previously has ever been. In the United States, transportation sector is responsible for 28 % of national greenhouse gas emissions in 2009. Therefore, there have been many studies devoted to the green supply chain management including the green vehicle routing problem (GVRP). GVRP plays a very important role in helping organizations with alternative fuel-powered vehicle fleets overcome obstacles resulted from limited vehicle driving range in conjunction with limited fuel infrastructure. The objective of GVRP is to minimize total distance traveled by the alternative fuel vehicle fleet. This study develops a mathematical model and a simulated annealing (SA) heuristic for the GVRP. Computational results indicate that the SA heuristic is capable of obtaining good GVRP solutions within a reasonable amount of time.

Keywords Alternative fuel vehicle · Green vehicle routing problem · Simulated annealing

1 Introduction

During recent years, many researchers have shown a high level of interest in developing green supply chain models. Incorporation of the ethical and environmental responsibilities into the core culture of today's business world is now

---

M. Yasin (✉) · V. F. Yu
Department of Industrial Management, National Taiwan University of Science and Technology, Taipei 106, Taiwan, Republic of China
e-mail: yasin@saya.me

V. F. Yu
e-mail: vincent@mail.ntust.edu.tw

greater than it previously has ever been. With the high level of competition they are about to face, companies find promoting green supply chain very attractive. Escalating deterioration of the environment, e.g. diminishing raw material resources, overflowing waste sites and increasing levels of pollution are the main causes of the importance of Green Supply Chain Management (GSCM) implementation (Srivastave 2007).

Likewise, logistics reliance on transportation modes such as trucks and airplanes using fossil burning fuels and the subsequent emission of carbon dioxide (CO_2) can pollute the living environment such as air, water, and ground. In the European Union, transport is the largest consumer of oil products and second largest emitter of CO_2; within the sector, road transport dominates in both regards. To reduce oil dependency and to make transport more sustainable, the European Commission set out the target to replace 10 % of conventional transport fuels with renewable alternatives, such as biofuel, hydrogen, and green electricity, by the year 2020 (European Union 2009).

Raley's Supermarkets (Raley's), a large retail Grocery Company based in Northern California, decided to take participation on utilizing heavy-duty trucks powered by liquefied natural gas (LNG) in 1997. It was found that the LNG trucks emits lower levels of oxides of nitrogen and particulate matter than the diesel trucks. California, is willing (through the local air quality management district) to pay as much as $12,000 per ton of measurable NO_x reduction through the "Carl Moyer Program." Based on a 10-year life, this gives an annualized cost of $4,550 per year. Overall, the potential cost effectiveness would be $3,730/ton of NO_x. This cost effectiveness-compared to the $12,000 per ton of NO_x reduction that the state is willing to pay for a given project is extremely favorable for the Raley's project (Chandler et al. 2000).

Until recently, AFVs have not been sufficiently developed to appear competitive in the market. One of the most important reasons is the availability of the facilities or stations providing alternative fuel for the vehicles. Therefore, this research focuses on the Green Vehicle Routing Problem (G-VRP) which considers the need of utilizing alternative fuel vehicles in traditional Vehicle Routing Problem (VRP). The G-VRP aims at finding at most m tours which start and end at the depot. Each tour is serviced by a vehicle and required to visit a subset of vertices including Alternative Fuel Stations (AFSs) when needed. The goal is to minimize the total distance traveled by vehicles. A predetermined tour duration limit, T_{max}, specifies vehicle driving range that is constrained by fuel tank capacity and driver working hours. Without loss of generality, to reflect real-world service area designs, it is assumed that all customers can be visited directly by a vehicle with at most one visit to an AFS. This does not preclude the possibility of choosing a tour that serves multiple customers and contains more than one visit to an AFS.

Erdogan and Miller-Hooks (2012) proposed the G-VRP and implemented two construction heuristics, the Modified Clarke and Wright Savings and the Density-Based Clustering Algorithm with a customized improvement technique. This research proposes a Simulated Annealing (SA) algorithm for G-VRP. SA has some attractive advantages such as its ability to deal with highly nonlinear models, chaotic and noisy data and many constraints. In addition, SA is also empowered by the flexibility and ability to approach global optimality. SA does not rely on any restrictive properties of the model and this makes this method very versatile. Parameter settings significantly impact the computational results of SA. The coefficient used to control speed of the cooling schedule, Boltzmann constant used in the probability function to determine the acceptance of worse solution and the number of iterations the search proceeds at a particular temperatures are some of those. These parameters need to be adjusted and numerous trials are required to make sure that SA can provide good results (Lin et al. 2011).

2 Problem Statement

The G-VRP problem consists of a customer set, a depot and a set of alternative fuel stations (AFSs). It is assumed that the depot can also be used as refueling station, meaning that once the vehicle returns to the depot, the fuel tank will be filled to its maximum capacity. All refueling stations have unlimited capacities and will always be able to serve the vehicle until its tank reaches the full capacity. Once the truck visits an AFS, it is assumed that it is served until its maximum fuel capacity is reached. Each truck has the capacity of $Q = 50$ gallons and each truck has the fuel consumption rate of 0.2 gallons per mile or 5 miles per gallon fuel efficiency with the vehicle speed of 40 miles per hour. Visiting a customer node will cost a service time of 30 min and visiting an AFS node will cost a time of 15 min. The objective of the problem is to find at most m tours, one for each vehicle, which starts and ends at depot. The vehicles are required to visit all customer nodes and AFS, if necessary. The goal is to find the minimum total travel distance of vehicles. The travel distances are calculated by employing Haversine formula (radius of earth $= 4{,}182.44949$ miles). The problem needs to be solved without violating tour duration constraint. Vehicle driving range constraints depend on fuel tank capacity limitations. The AFSs can be visited more than once or not at all.

3 Mathematical Formulation

The notations and mathematical formulation for G-VRP are adopted from Erdogan and Miller-Hooks (2012).

Notations
- I_0 Set of customer nodes (I) and depot (v_0), $I_0 = \{v_0\} \cup I$
- F_0 Set of AFS nodes (F) and depot, $F_0 = \{v_0\} \cup F$
- p_i Service time at node i (If $i \in I$ then p_i is the service time at the customer node, if $i \in F$ then p_i is the refueling time at the AFS node, which is assumed to be constant.)

r	Vehicle fuel consumption rate (gallons per mile)
Q	Vehicle fuel tank capacity
x_{ij}	Binary variables equal to 1 if a vehicle travels from node i to j and 0 otherwise
y_j	Fuel level variable specifying the remaining tank fuel level upon arrival at node j. It is reset to Q at each refueling station node i and the depot
τ_j	Time variable specifying the time of arrival of a vehicle at node j, initialized to zero upon departure from the depot.

Mathematical Formulation

$$\min \sum_{i,j \in V', i \neq j} d_{ij} x_{ij} \tag{1}$$

$$\sum_{j \in V', j \neq i} x_{ij} = 1, \quad \forall i \in I \tag{2}$$

$$\sum_{j \in V', j \neq i} x_{ij} \leq 1, \quad \forall i \in F_0 \tag{3}$$

$$\sum_{j \in V', j \neq i} x_{ij} - \sum_{j \in V', j \neq i} x_{ij} = 0, \quad \forall j \in V' \tag{4}$$

$$\sum_{j \in V' \setminus \{0\}} x_{0j} \leq m \tag{5}$$

$$\sum_{j \in V' \setminus \{0\}} x_{j0} \leq m \tag{6}$$

$$\tau_j \geq \tau_i + (\tau_{ij} - p_j) x_{ij} - T_{\max}(1 - x_{ij}), \quad i \in V', \forall j \in V' \setminus \{0\} \text{ and } i \neq j \tag{7}$$

$$0 \leq \tau_0 \leq T_{\max} \tag{8}$$

$$\tau_{0j} \leq \tau_j \leq T_{\max} - (t_{j0} + p_j), \quad \forall j \in V' \setminus \{0\} \tag{9}$$

$$y_j \leq y_i - r d_{ij} x_{ij} + Q(1 - x_{ij}), \quad \forall j \in I \text{ and } i \in V', i \neq j \tag{10}$$

$$y_j = Q, \quad \forall j \in F_0 \tag{11}$$

$$y_j \geq \min\{r d_{j0}, r(d_{jl} + d_{l0})\}, \quad \forall j \in I, \forall l \in F' \tag{12}$$

$$x_{i,j} \in \{0, 1\}, \quad \forall i, j \tag{13}$$

The objective (1) is to find the minimum total distance traveled by the AFV fleet in a day. Constraint (2) ensures that each customer node has only one successor. The successor can be a customer, AFS or depot. Constraint (3) ensures that each AFS node can only have at most one successor because it is not required to visit all AFSs. Constraint (4) is flow conservation constraint that requires the equality between the number of arrivals and the number of departures at each node. Constraint (5) limits the number of vehicle to be routed out of depot. The vehicles to be routed must not be more than available ones, which is denoted as m. Constraint (6) ensures that at most m vehicles return to the depot in a day. A copy of the depot is created in order to differentiate departure and arrival times at the depot. This will be important in tracking the time at each node and preventing

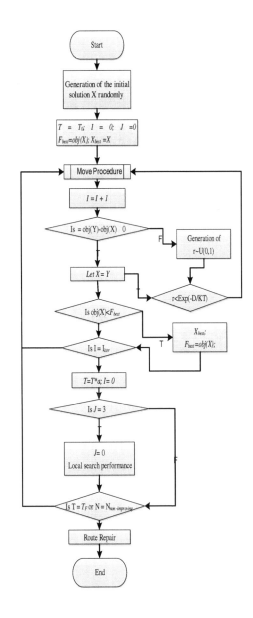

Fig. 1 Flowchart of the SA heuristic

sub-tours. The time of arrival at each node by each vehicle is tracked through constraint (7). Constraint (8) ensures that the vehicle will arrive and depart at depot without violating T_{max} constraint. Constraint (9) requires all vehicles to complete the service before T_{max} and no vehicles are allowed to perform service after T_{max}. Constraints (7), (8) and (9) ensure all vehicles to return at depot no later than T_{max}. Constraint (10) tracks a vehicle's fuel level based on node sequence and type. If

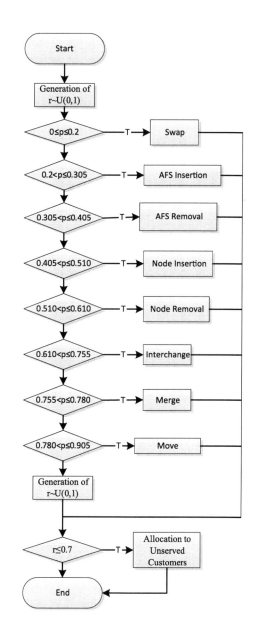

Fig. 2 Flowchart of move procedure

node j is visited right after vertex i ($x_{ij} = 1$) and vertex i is a customer node, the first term in constraint (10) reduces the fuel level upon arrival at node j based on the distance traveled from vertex i and the vehicle's fuel consumption rate. Time and fuel level tracking constraints, constraint (7) and (10), respectively, serve to eliminate the possibility of sub-tour formation. Constraint (11) reset the amount of fuel to the tank full capacity. This happens once the refueling is done at the AFS. Constraint (12) ensures that the fuel remaining on the vehicle will be enough for the vehicle to perform a route returning to the depot. Constraint (13) is about binary integrality.

4 Simulated Annealing Algorithm

The proposed SA algorithm begins by setting current temperature T to be T_o (3,500,000,000) and randomly generating an initial solution X. The current best solution X_{best} and the best objective function value obtained so far, denoted by F_{best}, are set to be X and $obj(X)$, respectively.

For each iteration, a new solution Y is generated from the neighborhood of the current solution X, $N(X)$, and its objective value is evaluated. Let $\Delta = obj(Y) - obj(X)$. If Δ is less than or equal to zero (Y is better than X), X is replaced with Y. Otherwise, the probability of replacing X with Y is $\exp(\Delta/KT)$. X_{best} and F_{best} record the current best solution and the best objective function value obtained so far, as the algorithm progresses. The current temperature T is decreased after I_{iter} (80) iterations after the previous temperature decrease, according to the formula $T = \alpha T$, where $\alpha = 0.99$. After every three temperature reductions, a local search procedure that sequentially performs swap, AFS insertion, AFS removal, Node insertion, Node removal, exchange, merge and move is conducted.

The algorithm is terminated when the current temperature T is lower than T_F (5) or the current best solution X_{best} has not improved for $N_{non\text{-}improving}$ (3,000) consecutive temperature decreases. Following the termination of SA procedure, the (near) optimal routing plan can be derived from X_{best}. Figures 1 and 2 illustrate the proposed SA heuristic the Move Procedure, respectively.

5 Computational Results

The proposed simulated annealing has proven to be very effective. From Tables 1 and 2, it can be seen that simulated annealing algorithm outperforms CPLEX on 8 out of 10 problems. The percentage improvement ranges from −6 to −15 %. Further, the proposed algorithm shows even higher improvement over MCWS and DBCA algorithms.

Table 1 Comparison between computational results of the proposed simulated annealing heuristic and CPLEX

Sample	CPLEX Exact solution (miles)	MCWS Lower bound	DBCA Lower bound	Simulated annealing Number of tours	Customer served	Total cost	Difference (%)
20c3sU1	1,797.51	1,818.35	1,797.51	6	20	1,269.34	−29
20c3sU2	1,574.82	1,614.15	1,613.53	7	20	1,427.14	−9
20c3sU3	1,765.9	1,969.64	1,964.57	6	20	1,361.38	−23
20c3sU4	1,482	1,508.41	1,487.15	8	20	1,635.21	10
20c3sU5	1,689.35	1,752.73	1,752.73	7	20	1,594.69	−6
20c3sU6	1,643.05	1,668.16	1,668.16	4	20	1,128.94	−31
20c3sU7	1,715.13	1,730.45	1,730.45	7	20	1,591.3	−7
20c3sU8	1,709.43	1,718.67	1,718.67	6	20	1,450.17	−15
20c3sU9	1,708.84	1,714.43	1,714.43	7	20	1,526.35	−11
20c3sU10	1,261.15	1,309.52	1,309.52	6	20	1,528.66	21
						Average	−10

Table 2 Comparison between computational results of the proposed simulated annealing heuristic and CPLEX

Sample	CPLEX Exact solution (miles)	MCWS Lower bound	DBCA Lower bound	Simulated annealing Number of tours	Customer served	Total cost	Difference (%)
S1_2i6 s	1,578.15	1,614.15	1,614.15	7	20	1,447.05	−8
S1_4i6 s	1,438.89	1,561.3	1,541.46	8	20	1,635.21	14
S1_6i6 s	1,571.28	1,616.2	1,616.2	6	20	1,365.18	−13
S1_8i6 s	1,692.34	1,902.51	1,882.54	6	20	1,423.66	−16
S1_10i6 s	1,253.32	1,309.52	1,309.52	7	20	1,611.76	29
S2_2i6 s	1,645.8	1,645.8	1,645.8	5	20	1,036	−37
S2_4i6 s	1,505.06	1,505.06	1,505.06	6	20	1,087.89	−28
S2_6i6 s	2,842.08	3,115.1	3,115.1	7	20	1,527.62	−46
S2_8i6 s	2,549.98	2,722.55	2,722.55	6	20	1,423.66	−44
S2_10i6 s	1,606.65	1,995.62	1,995.62	6	20	1,594.81	−1
						Average	−15

6 Conclusions

This research proposes a simulated annealing heuristic for G-VRP. Computational results indicated that the proposed SA outperforms two existing algorithms. This research can be extended by incorporating hybrid vehicles in the model. In addition, consideration of fuel prices variation, heterogeneous fleets, driving range variation and different sources of fuel will bring the model closer to reality.

References

Chandler K, Norton P, Clark N (2000) Raley's LNG truck fleet: final results. LNG Truck Utilization Report

Erdogan S, Miller-Hooks E (2012) A green vehicle routing problem. Transp Res Part E 48:100–114

European Union (2009) Directive 2009/28/EC of the European Parliament and of the Council of 23 April 2009 on the promotion of the use of energy from renewable sources and amending and subsequently repealing Directives 2001/77/EC and 2003/30/EC. Amendment, Strasbourg

Lin SW, Yu VF, Lu CC (2011) A simulated annealing heuristic for the truck and trailer routing problem with time windows. Expert Syst Appl 38:15244–15252

Srivastave SK (2007) Green supply-chain management: a state-of-the-art literature review. Int J Manage Rev, pp 53–80

Designing an Urban Sustainable Water Supply System Using System Dynamics

S. Zhao, J. Liu and X. Liu

Abstract This paper addresses the issue on designing an effective sustainable water supply system both in quantity and quality side, which is considered as a prerequisite for a sustainable development strategy. In order to achieve a sustainable water supply system, System Dynamics, which is an effective system analysis and development tool, is employed in modeling and simulating the system. The study presents a decision platform, where a quantified expected resilience model is built to measure water supply satisfaction rate, and to meet the requirements of sustainability indicators firstly. After determining the sustainability requirements, namely customer requirements, water supply system is constructed using system dynamics approach, through which the threats of the system such as demand boosting, pipeline aging, and other events that cause supply disruptions are illustrated subsequently. Further, prevention strategies will be taken into account to achieve the resilience ratio in the water system. Finally, a case study of water system in Shanghai is demonstrated to show the effectiveness of the proposed method.

Keywords Water supply · Sustainability · Resilience · System dynamics

S. Zhao (✉) · X. Liu
Department of Industry Engineering, Shanghai Jiao Tong University, 800 Dongchuan Road, Min-Hang District, Shanghai 200240, People's Republic of China
e-mail: Sixiang.zhao@hotmail.com

X. Liu
e-mail: X_liu@sjtu.edu.cn

J. Liu
Sino-US Global Logistic Institute, Shanghai Jiao Tong University, 1954 Huashan Road, Xu-hui District, Shanghai 200030, People's Republic of China
e-mail: Liujian007@sjtu.edu.cn

1 Introduction

Bulging population and rapid urbanization cause higher pressure on urban water supply system by deteriorating water quality and increasing water demand. As water is one of the most critical resources for human being, a sustainable urban water supply system (SUWSS) is needed to meet this challenge. Sustainable development is seen as an unending process—defined by an approach to creating change through continuous learning and adaptation (Mog 2004). The core indicators to measure the sustainability of water system can be concluded as water demand satisfaction rate and quality satisfaction rate (UN 2007; GCIF 2007). Thus, the ultimate goal of a sustainable urban water system is to meet the objective satisfaction rate of qualify water continuously. To achieve this goal, resilience in water system is the leading character (Milman and Short 2008). This paper defines resilience of SUWSS as joint ability of absorption and recovery to resist and recover from any disturbance, and ability of adaption to enhance the absorption and recovery capacity by self-adjustment (Walker et al. 2004; Bruneau et al. 2003; Ouyang et al. 2012).

Researches about sustainable water resource focus on assessment or indicators of sustainable water system (Lundin and Morrison 2002; Hedelin 2007; Milman and Short 2008), sustainability in Singapore water system (Xi and Kim 2013), sustainability in Yellow River (Xu et al. 2002). Few studies present a quantitative and resilient model in realizing a continuously satisfied demand, both long-term and short-term, in urban water system. To fill in this gap, this study combines resilience thinking and systematical thinking in realizing SUWSS. In this paper, we formulate a SUWSS model using system dynamics, which is an effective modeling and simulation tool. Then the case study of water system in Shanghai is presented, and several policy suggestions are given to achieve SUWSS.

2 Model

As mentioned earilier, the main indicators to achieve SUWSS is the continuity of both water quality and quantity satisfaction rate. These critical indicators are incorporated with other elements from both water quality/quantity demand side and water supply side in SD modeling. Combining resilience thinking, a casual loop digram (CLD) capturing the key elements of SUWSS is built as Fig. 1. Signs B represents negative feedback loop that means balance and equilibrium. Quality satisfaction rate and demand satisfaction rate, which are the most important aspects of SUWSS in the CLD, are quantified by Eqs. (1) and (2)

$$\text{Demand Satisfaction Rate} = \frac{\text{Water Supply}}{\text{Water Expected Demand}} \quad (1)$$

Designing an Urban Sustainable Water Supply System

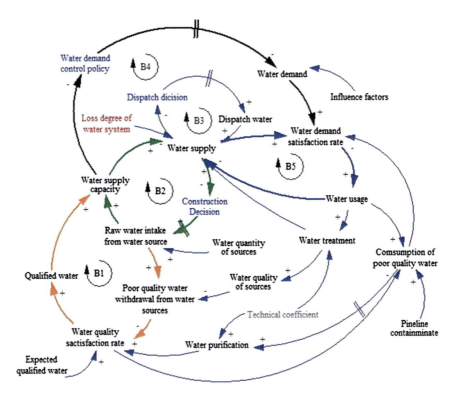

Fig. 1 Casual loop of the resilient SUWS

$$\text{Quality Satisfaction Rate} = \frac{\sum W_i \cdot Q_i}{\sum W_i} \quad (2)$$

where W_i is the total water withdrawn from water resource i and Q_i is the quality satisfaction rate in resource i. Incorporating the indicators of sustainability in urban water system, resilience is water satisfaction rate in this study. A metric of resilience is given as follows:

$$R = E\left[\frac{\int_0^T F(t)dt}{\int_0^T Tar(t)dt}\right] = \left[\frac{\int_0^T Tar(t)dt - \sum_{n=1}^{N(T)} AL_n(t_n)}{\int_0^T Tar(t)dt}\right] \quad (3)$$

where $F(t)$ represents the utility function of the system at a given time t; T is the time interval; $Tar(t)$ the target performance curve; $AL_n(t_n)$ is the area between the real performance curve and the target. In reality, loss of resilience can be caused by any disturbance of supply, such as demand boosting, pipeline aging etc. Therefore, for assurance of stable water supply and then to meet the target satisfaction rate, supply capacity redundancy and appropriate level of water reservoir are needed (B5). Besides, dispatching decision will be capable to recover the satisfaction rate when water emergency happens (B3). Meanwhile, construction decision will take effect to increase water supply when incapability is about to occur (B2). In addition, on the demand side, an infinite increasing demand can be control by demand-control policy to realize the continuity (B4). On the demand side, a balance loop is added to describe switching from poor quality resource to high quality will enhance quality satisfaction rate (B1).

3 Case Study

3.1 Model Development

Located at the estuary of the Yangtze River, Shanghai, with the population of 23.0 million, has been one of the largest city in Asia. The highly increasing urbanization rate and the poor water quality put great pressure on achieving SUWSS in Shanghai from both supply and quality side. Based on the CLD in Fig. 1, stock-flow diagram of water supply system in Shanghai is developed as Fig. 2. For the purpose of practicality and simplicity, some elements that exceed our research scope are omitted. Most of the input data are obtained from the Shanghai Statistics Bureau and Shanghai Water Authority (Shanghai Statistics Bureau 2002–2012; Shanghai Water Authority 2006–2010). Several assumptions have been made because of the limitation of available data. The first assumption is that the water quality of resources remains unchanged during the simulation horizen. The second assumption is that the maximum of total supply capacity is limited by the design withdrawal capacity of water reservoir, and there will be no new reservoir constructed during the simulation horizon.

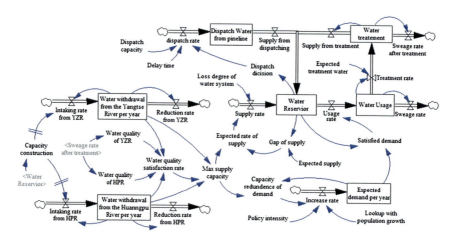

Fig. 2 Stock-flow diagram for Shanghai SUWS

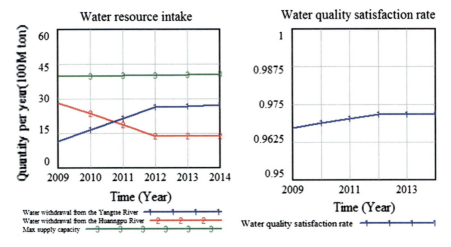

Fig. 3 Simulation results under switching resources policy

Yangtze river and Huangpu river are the two major water resources for Shanghai. In 2010, the total raw water withdrawal proportion from Yangtze and Huangpu river is 70 % and 30 % respectively. As the quality of Huangpu river is poor, the government raised the withdrawal proportion from Yangtze river to 60 % by 2012 through the construction of Qing Cao Sha reservoir. A simulation is run on the water quality satisfaction rate to illustrate this improvement after water resources switching (Fig. 3). Note that the satisfaction rate here is measured by water quality after purification. Though the improvement is slight, it shows a great releasing of purification pressure from poor quality water.

As the prediction of water demand, which consists of industrial, agricultural and residential demand, is a complicated mechanism, we perform a roughly prediction in total water consumption via second exponential smoothing method. Figure 4 illustrates the simulation results under a specific exploitation rate of existing capacity of water reservoir, which is derived by linear regression from the historical data. In 2033, the water supply capacity reaches its maximum without construction of new reservoirs, while the demand is still climbing. After 2045, the water demand will exceed the supply capacity, and 4 years later, water will be inadequate since the water of reservoir runs out. In our time horizon, the satisfaction rate, which is 97.55 %, can be easily calculated by Eq. 3. Obviously, the satisfaction rate will decrease as the time horizon expands.

3.2 Results and Discussions

Demand control policies are needed to limit the infinite increasing water demand to achieve SUWSS. Besides, to ensure stable water supply, capacity redundancy of 10 to 15 % is necessary as consensus of capacity redundancy in water supply

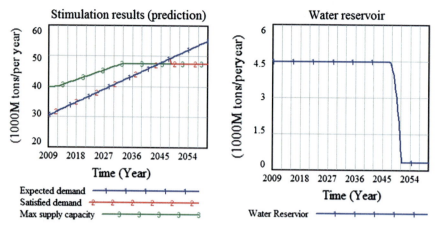

Fig. 4 Simulation results

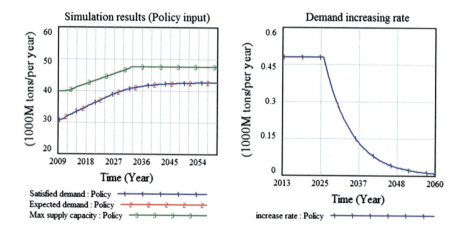

Fig. 5 Simulation results under policy control

industries. With an 11.3 % annually decline in water increasing rate, expected demand converge after demand control policy taking effect in 2026 (Fig. 5). The decline rate is determined by the policy intensity, and a high intensity policy will lead to a sharply decline in the increasing rate. A combination of certain policies such as stepped-pricing, control of immigration and industrial water saving polices will take effect in declining the demand increasing rate.

Simulation result under fluctuation is also presented by Fig. 6. If the expected demand soars by 20 % in a specific year every decade, water supply is also adequate under the coaction of both capacity redundancy and reservoir redundancy.

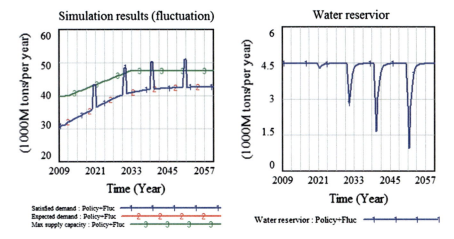

Fig. 6 Simulation results under demand fluctuation

Hereinbefore, we set a strict unchanged water quality assumption because of limitation of input data. However, poor water treatment of upriver plants will contaminate downriver water resource. Thus the treatment rate of sewage water of upriver plants must be improved to prevent the water resource from contamination.

4 Conclusions

This research has demonstrated how to achieve a sustainable urban water supply system with combination of systematical thinking and resilience thinking. A system dynamics framework was built and a quantified expected resilience model was proposed to measure the water satisfaction rate. In this framework, sustainable requirements are met by achieving continuously satisfied sustainability indicators. At the end of the article, the result of the simulation in Shanghai SUWSS suggested that the increasing rate of water demand should be controlled by certain policies. Meanwhile, a 10 % capacity redundancy was proved to be capable to withstand the disturbance of soaring demand.

However, this study is not without limitations. One is that the prediction of water demand by second exponential smoothing method is lack of precision. Another one is that this research doesn't present specific policies strategies to each water demand component. These limitations provide guidance to our further research.

Acknowledgments This research is partly supported by the National Research Foundation Singapore under its Campus for Research Excellence and Technological Enterprise (CREATE) and the NSFC (91024013, 91024131).

References

Bruneau M, Chang SE, Eguchi RT et al (2003) A framework to quantitatively assess and enhance the seismic resilience of communities. Earthquake Spectra 19(4):733–752

Global city indicators facility (2007) List of GCIF indicators. http://www.cityindicators.org/ProjectDeliverables.aspx. Accessed 12 Apr 2013

Hedelin B (2007) Criteria for the assessment of sustainable water management. Environ Manage 39:151–163

Lundin M, Morrison GM (2002) A life cycle assessment based procedure for development of environmental sustainability indicators for urban water systems. Urban Water 4:145–152

Milman A, Short A (2008) Incorporating resilience into sustainability indicators: an example for the urban water sector. Global Environ Change 18:758–767

Mog JM (2004) Struggling with sustainability: a comparative framework for evaluating sustainable development programs. World Dev 32:2139–2160

Ouyang M, Dueñas-Osorio L, Min X (2012) A three-stage resilience analysis framework for urban infrastructure systems. Struct Saf 36–37:23–31

Shanghai Statistics Bureau (2002–2012) The year book, online edition. Available: http://www.stats-sh.gov.cn/data/release.xhtml. Accessed 28 Apr 2013

Shanghai Water Authority (2006–2010) Water resources bulletin. Available: http://222.66.79.122/BMXX/default.htm?GroupName=%CB%AE%D7%CA%D4%B4%B9%AB%B1%A8. Accessed 28 Apr 2013

United Nations (2007) Indicators of sustainable development: guidelines and methodologies. http://sustainabledevelopment.un.org/index.php?page=view&type=400&nr=107&menu=35. Accessed 12 Apr 2013

Walker B, Holling CS, Carpenter SR, Kinzig A (2004) Resilience, adaptability and transformability in social–ecological systems. Ecol Soc 9(2):5

Xi X, Kim LP (2013) Using system dynamics for sustainable water resources management in Singapore. Paper presented at the conference on systems engineering research (CSER'13), Georgia Institute of Technology, Atlanta, GA, March 19–22, 2013

Xu ZX, Takeuchi K, Ishidaira H, Zhang XW (2002) Sustainability analysis for yellow river water resources using the system dynamics approach. Water Resour Manage 16:239–261

An Evaluation of LED Ceiling Lighting Design with Bi-CCT Layouts

Chinmei Chou, Jui-Feng Lin, Tsu-Yu Chen, Li-Chen Chen and YaHui Chiang

Abstract Light-emitting diodes (LEDs) became an important home-lighting device. Due to the property of high efficiency LED lighting sources, thus we expected to apply high efficiency LED lighting to improve or enhance our lighting environment. The purpose of this study is to design LED ceiling lightings layout based on evaluating human's physiological responses and subjective feelings, where the experiments were conducted in the office-like laboratory. We had four experimental combinations included two Correlated Color Temperature (CCT) and two different types of lighting sources (lower Blue-value and high-efficiency). Two different types of lighting sources, one was that the lower Blue-value lighting sources was equipped at the center of the device, the other was that high-efficiency equipped around the lower Blue-value lighting source. Six participants were recruited in this study to perform sheet, laptop-typing, and tablet-searching tasks under the four experimental combinations. In addition to working performance measures, heart rate, Galvanic Skin Response (GSR), eyes blink duration, blink time, and critical fusion frequency (CFF) values were measured as well. The results showed that CCT 4,000 K-high efficiency lightings design would affect human's physiological alert and stress. Thus we suggested the CCT 4,000 K-high efficiency participants had lower Tablet-searching error rate higher physiological alert and less eye fatigue.

Keywords LED · Correlated color temperature · Physiological responses · Ceiling lighting design

---

C. Chou (✉) · J.-F. Lin · T.-Y. Chen · L.-C. Chen
Department of Industrial Engineering and Management, Yuan Ze University, 135 Yuan-Tung Road, Chung-Li 32002 Taiwan, People's Republic of China
e-mail: kinmei@saturn.yzu.edu.tw

Y. Chiang
Industrial Technology Research Institute, 195, Sec. 4, Chung Hsing Road, Chutung, Hsinchu 31040 Taiwan, People's Republic of China

1 Introduction

The increase of green consciousness and energy saving. Light-emitting diode (LEDs) has become an important home-lighting device in the home or office lighting environment. Many research mentioned that the LED Correlated Color Temperature (CCT) would influence human's physiological responses. (Manav 2007) Investigated that CCT 4,000 Kwas preferred to CCT 2,700 K for impressions of 'comfort and spaciousness'. Navvab (2002) mentioned that between CCT 3,500 and 7,000 K, the higher CCT caused the higher reading corrective rate by participants. Thus we set two CCT 4,000 k and 6,500 K as our lighting CCT parameters. In this study we also discussed the difference of Blue-value in our lighting sources. Blue-value was a quantitative value, which concerns the ratio of the blue-ray in the visible lighting spectrum, and this value could be evaluated by the degree of human's blue light hazard. (Viola AU et al. 2008) mentioned that fitting blue-ray could increase human's working performance, higher blue value might increase the risk of blue hazard, which would affect human's health. To reduce the Blue Hazard we used low Blue-value lighting sources as our lighting element, however the low Blue-value lighting sources would consume more electricity. To solve this situation we design the high-efficiency lighting source that is expected to make a balance between decreasing the Blue hazard and saving more energy. The aim of this study was to design LED ceiling lightings layout base on evaluating human's physiological responses and subjective feelings where conducted in the office-like laboratory.

2 Methods

2.1 Experimental Setup and Equipment

This study was conducted in an office-like experimental room furnished with an office table, and interchangeable LED lighting devices. To measure physiological responses of the participants and eye movement, the equipment used in this study included Bluetooth wireless biofeedback system (NeXus-10), Eye Tracker (View Point Eye Tracker®), Flicker fusion apparatus (FLICKER), laptop and tablet. While conducting the experiment, the desktop illumination was set at 500 lx, room temperature was controlled in 25 °C, and the relative humidity was controlled at 50 % RH.

2.2 Participants

Six male participants, aged from 20 to 24 (mean age 21.33), were recruited in this study. All of them were familiar with basic computer and tablet usage (e.g., Windows office word processing tasks, and tablet article explore). All participants had normal or corrected-to-normal vision.

2.3 Parameters Design

In our study used two difference lighting sources and two CCT for our ceiling lighting parameter design. We set Lower Blue value and High-Efficiency lighting as our lighting sources and the CCT of lighting sources, which we set to 4,000 and 6,500 K. We designed four ceiling LED lighting devices which equipped lighting sources were mentioned above.

We divided the devices into two types first type of design were basic ceiling lighting design (as shown in Fig. 1a, b. They were all equipped the Lower Blue Value lighting source. Second types designed were equipped lower Blue-value and high-efficiency lighting sources (as shown in Fig. 1c, d. Which equipped the Lower Blue Value lighting source in the part of central and the higher-efficiency lighting sources was equipped around central which in the device.

These two kinds of lighting sources set CCT in 4,000 and 6,500 K which equipped in different types of lighting design as shown in Fig. 1.

2.4 Experimental Variables

The independent variable discussed in this study were the four ceiling lighting design included two different CCT which were 4,000 and 6,500 K, and two different type lighting sources. The dependent variables were four office-like task performance, objective physiological responses, and subjective questionnaire.

Task performance were measured the completion and error rates by evaluated participants' task performance. Physiological responses were measured as Heart Rate (HR), Galvanic Skin Response (GSR), eye Blink Duration (BD), Blink Time (BT) and Critical Fusion Frequency (CFF) values which could represent our emotion regulation and eye fatigue under the four LED lighting ceiling design.

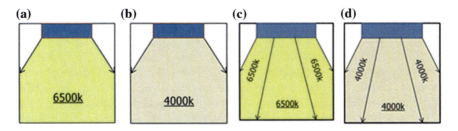

Fig. 1 Experimental ceiling lighting design

Table 1 Overview the Karolinska sleepiness scale (KSS)

Value	English rating
1	Extremely alert
2	Very alert
3	Alert
4	Rather alert
5	Neither alert nor sleepy
6	Some sign of sleepiness
7	Sleepy, no effort to stay awake
8	Sleepy, some effort to stay awake
9	Very sleepy, great effort to keep awake, fighting

The subjective questionnaires were measured as Karolinska Sleepiness Scale (KSS) as shown in Table 1. Linhart and Scartezzini (2011) KSS is a subjective rating scale which to state their actual alertness level on a 9-stage scale between "extremely alert" (1) to "very sleepy, great effort to keep awake, fighting sleep" (9).

2.5 Experimental Protocol

There were four tasks that were conducted in this study to simulate the office work. The first task was Psychomotor Vigilance Task, which executed at start and end of this protocol. It took five minutes for measuring average reaction time. The task was a laptop task. Participants had to enter the corresponded alphabet button on laptop keyboard to eliminate the alphabet, which was showed up on the laptop monitor. Each alphabet is shown randomly and only one alphabet is shown on the monitor during the task. Second task was Landolt Ring counting which was sheet task, took 50 min for measured the Completion and Error Rates as participants working performance. In this Task participants were gave A4 sheets randomly filled with four different directions of Landolt rings. Each sheet had 240 Landolt rings which diameter was 9 mm and gap width was 0.5 mm. Participants counted the number of Landolt rings for each gap direction of rings. Third Task was Three-digit Addition Laptop Typing Task, which took 50 min for measuring the Completion and Error Rates as participants working performance. Participants had to use laptop to calculate Three-digit Addition questions, the Three-digit Addition question of this task would randomly showed up the Three-digit numbers on laptop monitor. Fourth Task was Listening and Wrong Identify Tablet Searching Task, which took 50 min for measuring the Error Rates as participants' working performance. In this study participants had to listen the speech and identify the difference words between the speech and article. All of these tasks order were fixed. Between each Task participants have ten minutes to rest and write the Karolinska Sleepiness Scale which evaluated participants' subjective sleep feeling,

An Evaluation of LED Ceiling Lighting Design

and alertness. Critical Fusion Frequency values which measured at the beginning and end of this protocol.

2.6 Statistic

Analysis of variance general linear model and post-analysis Tukey comparisons were applied for analysis all physiological responses and the Karolinska Sleepiness. Critical Fusion Frequency values were applied Paired student T test for analysis.

3 Results

3.1 Task Performance

The results showed that the Tablet-Searching Task under the four lighting parameters did have significance effect on task performance. As shown in Fig. 2 we found the mean error rate which under the CCT 4,000 and 4,000 K-high efficiency. The mean error rates and standard deviation were 0.23 ± 0.07 and 0.22 ± 0.07 which had significance lower than the CCT 6,500 and 6,500 K-high efficiency lighting design.

3.2 Physiological Responses

Physiological responses including HR, GSR, BD, BT, and CFF value. The detail of significance effect under the four lighting parameters will be below. As shown in Fig. 3 shown participants' average HR was measured as the highest and the lowest (75.89 ± 8.65 Beats/min and 68.80 ± 5.37 Beats/min) when the HR at the set of CCT 6,500 K high-efficiency and CCT 4,000 K-high efficiency. As shown in

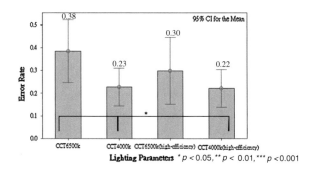

Fig. 2 The effect of lighting parameters on tablet-searching task performance

Fig. 3 The effect of lighting parameters on heart rate

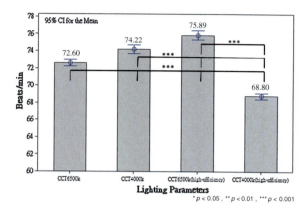

Fig. 4 The effect of lighting parameters on GSR

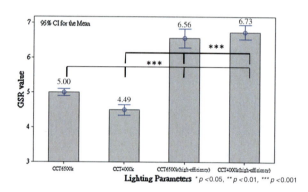

Fig. 4 shown participants' average GSR value was measured as the highest (6.56 ± 4.20 and 6.73 ± 3.15) when the Lighting Parameter was designed at the set of CCT 6,500 K high-efficiency and CCT 4,000 K high-efficiency. Figure 5 had shown under the high-efficiency lighting design included CCT 6,500 and 4,000 K the average Blink Duration was 12.42 ± 8.46 counts and 15.23 ± 11.14 counts. Participants had lower Blink Duration than original lighting design. Figure 6 had shown under the CCT 4,000 K-high efficiency, participants had lowest Blink Time was measured as 36.68 ± 1.93 ms.

3.3 Subjective Questionnaire

The Subjective Questionnaire of this study was Karolinska Sleepiness Scale (KSS), which measured about eyes fatigue. In Fig. 7 we found under the CCT 6,500 K high-efficiency would make participants had more awake than CCT 4,000 K high-efficiency on the Landolt Ring counting task performance in the result of KSS.

An Evaluation of LED Ceiling Lighting Design

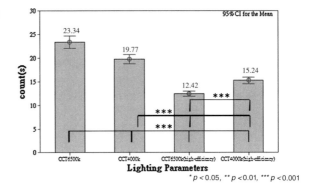

Fig. 5 The effect of lighting parameters on blink duration

$^*p<0.05, ^{**}p<0.01, ^{***}p<0.001$

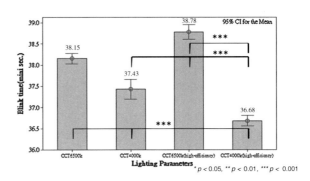

Fig. 6 The effect of lighting parameters on blink time

$^*p<0.05, ^{**}p<0.01, ^{***}p<0.001$

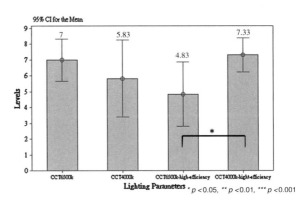

Fig. 7 The effect of lighting parameters on karolinska sleepiness

$^*p<0.05, ^{**}p<0.01, ^{***}p<0.001$

4 Discussion

In this study different Ceiling Lighting Design had significant affected on human's Tablet task performance, objective physiological responses and subjective Karolinska Sleepiness Scale (KSS). The reason which caused the significance might be due to the different CCT or the different Blue-Value lighting sources, these lightings would affect participants' emotion, however many research indicated different lighting conditions would affect human's psychological perception. According to (Riva et al. 2003), some specific physiological responses such as heart rate and GSR value while increasing the alert and exciting the heart and the conductivity of GSR would change, thus this research is focused in the heart rate and GSR value to represent human's psychological stress and awakens. In eye measurement (Schleicher et al. 2008) mentioned that Blink Duration and Blink Time could be a indicator to eye fatigue. Although in this study we had significance effect on our physiological responses, however in these results of our eye measurement we just only explained these results had a trend to become fatigue. We didn't have specific evidence to identify the eye fatigue.

5 Conclusion

In this study the effect of different LED ceiling lighting design on four types of office-like tasks Psychomotor Vigilance Task, Landolt Ring counting Sheet Writing Task, Three-digit Addition Laptop Typing Task, Listening and Wrong Identify Tablet Searching Task. Our results showed that participants working under our lighting parameters in CCT 4,000 K (included original and high-efficiency design) could lower their Tablet-Searching error rate. In physiological responses CCT 4,000 K high-efficiency would make participants feel comfort due to its lower stress (HR) and eye fatigue (Blink Time), however in the high-efficiency lighting design(included CCT 4,000 and 6,500 K) participants would feel alert and comfort due to the higher GSR value and less Blink Duration.

Acknowledgments This work is partially supported by Taiwan Industrial Technology Research Institute (ITRI).We would like to acknowledge the grant support from the Taiwan Industrial Technology Research Institute (ITRI).

References

Manav B (2007) An experimental study on the appraisal of the visual environment at offices in relation to colour temperature and illuminance. Building and Environment. doi:http://dx.doi.org/10.1016/j.buildenv.2005.10.022

Navvab M (2002) Visual acuity depends on the color temperature of the surround lighting. Illum Eng Soc 31:70–84

Viola AU JL, Schlangen LJM, Dijk D-J (2008) Blue-enriched white light in the workplace improves self-reported alertness, performance and sleep quality. Scand J Work Environ Health 34(34):297–306

Linhart F, Scartezzini J-L (2011) Evening office lighting e visual comfort vs. energy efficiency vs. performance? Build Environ 46:981–989

Riva G, Davide F, IJsselsteijn W (2003) 7 Measuring presence: subjective, behavioral and physiological methods. being there: concepts, effects and measurement of user presence in synthetic environments, pp 110–118

Schleicher R, Galley N, Briest S, Galley L (2008) Blinks and saccades as indicators of fatigue in sleepiness warnings: looking tired? Ergonomics. doi:10.1080/00140130701817062

Postponement Strategies in a Supply Chain Under the MTO Production Environment

Hsin Rau and Ching-Kuo Liu

Abstract Postponement strategies that decrease the impact of uncertainty of demand and improve customization have implemented for a long time in the supply chain. Postponement is to start to perform some operations until the customized order is received, and it does not go by forecast like the traditional approach. In modeling postponement, most researchers focus on manufacturing postponement and they do not consider production environment. This study considers the production environment of make to order to develop a supply chain network and to study the optimal combination of postponement operations in the supply chain, such as manufacturing, packaging, and logistics. We take notebook computer as an example to show the optimal combination of postponement strategies when the minimum total cost is reached. We believe that the results of this study would provide suggestions to managers for their supply chain operational decisions.

Keywords Supply chain · Make to order · Postponement

1 Introduction

The concept of postponement was introduced in the literature by Alderson (1950), but it only attracted more researches in recent years (Yang et al. 2004; Yang and Yang 2010). The concept of postponement is to redesign the product and process and delay the difference of product to the upmost, it doesn't complete the final operation until getting customer order, and it's a way for customization and could decrease many uncertain factor and operational risk (Lee and Billington 1994).

H. Rau (✉) · C.-K. Liu
Department of Industrial and Systems Engineering, Chung Yuan Christian University,
Chungli 320, Taiwan, Republic of China
e-mail: hsinrau@cycu.edu.tw

Lee et al. (1993) considered that the location of customization could affect the level of inventory and service, and the design of product and process also affects the operational process, for example, some products are suitable to be produced by the mother factory, some are by distributor, and the others are by final customer. The postponement of customization could improve the flexibility of demand and the inventory cost; on the other hand, the speed of delivery and the service level are likely to be decreased.

The classification of postponement strategies have been proposed by many scholars with different concepts. Zinn and Bowersox (1998) considered postponement is to delay the customer order de-coupling point to the downstream of the supply chain. They classified five postponement strategies: labeling, packaging, assembly, manufacturing and time. Cooper (1993) developed four postponed strategies, bundled manufacturing, deferred assembly, deferred packaging, and unicentric manufacturing if the specification of product and characteristic of the peripheral product had the same community of market. Bowersox and Closs (1996) mentioned time, space and form postponement strategies according to the concept of delaying the operational time, space and design of product. Pagh and Cooper (1998) integrated postponement and speculation (P/S) of manufacturing and logistics postponement into four general supply P/S strategies called P/S Matrix, included full speculation, manufacturing postponement, logistics postponement and full postponement strategy. Van Hoek (2001) proposed four postponement strategies with respect to the viewpoint of time and space by performing postponement, namely, time postponement, bundled manufacturing, deferred assembly and deferred packaging. Yang and Burns (2003) proposed six postponements, namely, design, purchasing, manufacturing, assembly, packaging/labeling, logistics and discussed with a continuum of customization and standardization, where includes make to forecast, shipment to order, packaging/labeling to order, final manufacturing/assembly to order, make to order, buy to order, and engineering to order. Yang et al. (2004) studied the role of postponement in the management of uncertainty; and the postponement includes product development, purchasing, production, logistics postponements. From the above literature review we could know that the classification of postponement may not be all the same, but their concept is very close. This study will use three kinds of postponements: manufacturing postponement, packaging postponement and logistics postponement to study the best postponement strategies for notebook computer in the supply chain under the make to order (MTO) production environment.

With respect to the mathematical model regarding postponement, Lee et al. (1993) considered the community of component with the expected benefits and product life cycle. They computed the difference of cost for performing postponement and considered the transshipment at a fixed service level. Lee and Tang (1997) developed a cost model for a multi-stage system of inventory and discussed the cost difference resulted from product difference. Ernst and Kamrad (2000) discussed modularization in product design and logistics postponement in light of processes of manufacturing, assembly, and packaging and developed four structures: rigid, modularized, postponed and flexible. They developed a total cost model for them and compared their cost. Rau and Liu (2006) developed a network of postponement for notebook computer industry and used a mathematical model to discuss the optimal combination of postponement with the minimum cost. From the above observation, we find that no research has developed a mathematical model to obtain the optimal postponement strategies under the production environment such as MTO.

The above literature review gives a motivation to this paper. This study considers the production environment of MTO (make to order) to develop a supply chain network and to study the optimal combination of postponement operations in the supply chain, such as manufacturing, packaging, and logistics. We take notebook computer as an example to show the optimal postponement strategies when the minimum total cost is reached.

2 Modeling the Optimal Combination of Postponement Strategies

2.1 Development of the Postponed Network of Supply Chain

This section focuses on developing the postponed operations in the supply chain. In modeling postponement, most researches pay more attentions on the description of theory and the discussion of benefits of performing postponement. Fewer researches discussed the implementation of postponement in the supply chain and most only focus the manufacturing postponement. This research extends the postponement to the whole supply chain and discusses the implementation of various postponements in the supply chain.

In order to describe our supply chain environment, we have the following assumptions:

- The finished product is composed of standard, crucial, differential components; semi-finished product 1 is composed of standard components; semi-finished product 2 is composed of standard and crucial components, which means it consists of semi-finished product 1 from product the component structure.
- The standard semi-finished product is made by the assembly of components after we purchase its components, and it is not from the purchase directly.
- Assume the product could perform any types of postponements and their combination.

Partners in our system have supplier, assembler, distributor, retailer and customer to form the supply chain network as shown in Fig. 1. This study would discuss manufacturing, packaging and logistics postponements and use notebook computer as an example.

Three postponements used in this study are described as follows:

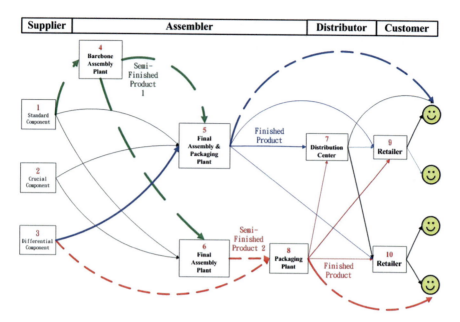

Fig. 1 Supply chain network

- Manufacturing postponement—this happens when standard components (1) are sent to a barebone assembly plant (4) to become semi-finished products 1, which are called as barebones in the notebook computer industry. These barebones can be postponed to become semi-finished products 2 according to the request from demand. If standard components (1) and crucial components (2) are assembled at a final assembly and packaging plant (5) or at a final assembly plant (6) to become semi-finished products 2, then we do not have manufacturing postponement.
- Packaging postponement—this happens when semi-finished products 2 stay at a final assembly plant (6) and not going to a packaging plant (8) right away to become finished products. If differential components (3) such as power supply, manual, etc. are sent to a final assembly and packaging plant (5) for becoming finished products, then there is no packaging postponement.
- Logistics postponement—this study defines logistics postponement as replacing de-centralized inventory with centralized inventory by direct distribution to final customer without passing the distribution center and retailer after the finished products have been completed. This happens when the finished products completed at a final assembly and packaging plant (5) or a packaging plant (8) in Fig. 1, they are delivered to a final customer directly without through a distribution center (7) and/or retailer (9 or 10). On the other hand, if the finished products are delivered through a distribution center and/or retailer, then there is no logistics postponement.

The MTO production environment for the notebook computer can be set by the decoupling point (DP) (Yang and Burns 2003) for the above three postponements as shown in Fig. 2.

2.2 Model Development

Due to space limitation, this paper only discusses the concept of mathematical model development and the mathematical model itself will be neglected. The goal of mathematical model in this study is to search the optimal combination of postponed strategies at the minimum cost and satisfying customer's lead time. The cost includes the purchasing cost, processing cost (including manufacturing, assembly, and packaging cost), inventory cost, transportation cost, penalty cost, over-production cost, and shortage cost. The penalty cost is for products not delivering at the lead time that a customer requests. In the consideration of time, it includes the processing time and transportation time.

3 Example Illustration

One of major Taiwan notebook computer OEM company is chosen. For its American market, the barebones are made in China, and then they are sent to the final assembly plant and final assembly and packaging plant in Houston of

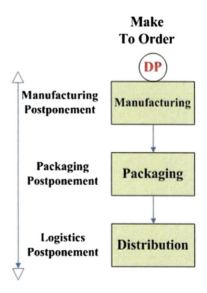

Fig. 2 The framework of postponement in MTO production environment

Table 1 Component classification

	Standard	Crucial	Differential
Components	Case	TFT-LCD	Keyboard
	Battery	CPU	Package and others
	Motherboard	DRAM	
	CD-ROM	HD	

Table 2 Sales data is used is from 2003 to 2006 (1000 units)

	2003	2004	2005	2006
Q1	360.44	401.80	545.6	915.02
Q2	337.74	430.49	697.66	
Q3	401.10	527.64	882.46	
Q4	489.98	599.98	1,010.06	
	Mean		Standard deviation	
	584.62		216.20	

The rate of market in the American about 17.6 %

Table 3 The optimal combination of postponed strategies

Postponement			Total cost (M US$)
Manufacturing	Packaging	Logistics	
0	0	1	8,528
0	0	2	9,460
0	0	3	10,515
0	1	1	8,708
0	1	2	9,696
0	1	3	10,755
1	0	1	6,735
1	0	2	**6,649**
1	0	3	7,704
1	1	1	6,916
1	1	2	6,886
1	1	3	7,944

American. Distribution center and packaging plant are located in the main market of American in order to process the delivery and postponed packaging.

The finished product is composed of standard, crucial, differential components; and the components used in this study are shown in Table 1. Sales data is used is from 2003 to 2006 as shown in Table 1. The software of Mathematica is used to solve the optimal combination of postponement strategies with the objective of minimum cost in the MTO production environment with consideration of operational lead time. The result of total cost for the optimal combination of postponed strategies is shown in Table 3, where 0 and 1 denote without and with postponement respectively for either manufacturing or packaging postponement. In logistics, 1 denotes with postponement (direct ship); 2 and 3 denote without postponement, and 2 is for two stage shipment and 3 is for three stage shipment (Table 2).

Table 3 shows that for all the manufacturing postponement cases are better than all the non-manufacturing postponement ones. This is because in the MTO environment, the lead time is very short, so we have to make barebones first that can reduce the lead time, increase the subsequent assembly efficiency to satisfy customer's service level. The optimal solution occurs at the situation with manufacturing postponement, without packaging postponement, and without logistics postponement; but with two stage shipment. This outcome balances between inventory cost and transportation cost. The result obtained here matches the real practice.

4 Conclusions

Postponement strategies can help reduce the effect of uncertainty of demand and improve customization in the supply chain operations. In modeling postponement, most researchers before focused on manufacturing postponement and they did not consider production environment. This study used notebook computer as an example to propose a supply chain model to study the optimal combination of postponement strategies including manufacturing, packaging, and logistics postponements for the MTO production environment using the concept of decoupling point.

The example chosen is one of major Taiwan notebook computer OEM company with its barebones manufacturing site in China and the final assembly plant and final assembly and packaging plant in Houston of American. Distribution center and packaging plant are located in the main market of American. The optimal solution occurs at the situation with manufacturing postponement, without packaging postponement, and without logistics postponement; but with two stage shipment. The result obtained in this study matches the real practice and give more insights.

Due to the space limitation, the mathematical modeling is not shown and various factors such as package size, inventory cost, and lead times are also not shown to explore the effects on postponement strategies selection.

Besides the MTO production environment, there are several other production environments such as packaging/labeling to order, final manufacturing/assembly to order, buy to order, and engineering to order to be able to study in the future.

Acknowledgments Although we only present a partial work in this paper due to space limitation, the full work is supported in part by National Science Council of Republic of China under the grants NSC 93-2213-E-033-042 and NSC 101-2221-E-033-037-MY3, and College of Electrical Engineering and Computer Science of Chung Yuan Christian University under the grants CYCU-EECS-10001.

References

Alderson W (1950) Marketing efficiency and the principle of postponement. Cost Profit Outlook 3:1–3

Bowersox DJ, Closs DJ (1996) Logistical management: the integrated supply chain process. McGraw-Hill, New York

Cooper JC (1993) Logistics strategies for global businesses. Int J Phys Distrib Logis Manag 23:12–23

Ernst R, Kamrad B (2000) Evaluation of supply chain structures through modularization and postponement. Eur J Oper Res 124:495–510

Lee HL, Billington C (1994) Designing products and processes for postponement. In: Dasu S, Eastman C (eds) Manage Des: Eng Manage Perspect. Kluwer Academic, Norwell, pp 105–122

Lee HL, Billington C, Carter B (1993) Hewlett-Packard gains control of inventory and service through design for localization. Interfaces 23:1–11

Lee HL, Tang SC (1997) Modelling the costs and benefits of delayed product differentiation. Manage Sci 43:40–53

Pagh JD, Cooper MC (1998) Supply chain postponement and speculation strategies: how to choose the right strategy. J Bus Logist 19:13–33

Rau H, Liu C-T (2006) The optimal conbination of postponement operations in a supply chain. J Chin Inst Ind Eng 23:253–261

Van Hoek RI (2001) The rediscovery of postponement a literature review and directions for research. J Oper Manage 19:161–184

Yang B, Burn N (2003) Implications of postponement for the supply chain. Int J Prod Res 41:2075–2090

Yang B, Burn N, Backhouse CJ (2004) Management of uncertainty through postponement. Int J Prod Res 42:1049–1064

Yang B, Yang Y (2010) Postponement in supply chain risk management: a complexity perspective. Int J Prod Res 48:1901–1912

Zinn W, Bowersox DJ (1998) Planning physical distribution with the principle of postponement. J Bus Logis 9:117–136

Consumer Value Assessment with Consideration of Environmental Impact

Hsin Rau, Sing-Ni Siang and Yi-Tse Fang

Abstract In response to climate change, environmental issues, such as greenhouse gas emissions and carbon footprint, become important. Products are required to take into account of life-cycle thinking in their product design phase. In the past, enterprises only pursued the quality improvement and technology development of green products, but they ignored the value of consumers in those eco-design products. This ignorance results in weak competition; however, it gives the motivation of this study. This study is based on the concept of eco-efficiency to develop a consumer value assessment model with consideration of environmental impact. Several different types of laptop computers are used as examples to illustrate the application of our assessment model. This model simultaneously considers the value of functional performance, the total cost of ownership, and the environmental impact for the product. The environmental impact includes the assessment of energy consumption, pollution and non-recyclability. We believe that our proposed model can be served as a guidance of product improvement or innovation for the designer, and a reference of purchasing products for consumers.

Keywords Consumer value · Environmental impact · Eco-design

1 Introduction

In recent decades, governments have legislated and formulated policies regarding to environmental issues in response to energy shortages, greenhouse gas emissions, climate change, and other serious environmental problems. As the environmental

H. Rau (✉) · S.-N. Siang · Y.-T. Fang
Department of Industrial and Systems Engineering, Chung Yuan Christian University,
Chungli 320, Taiwan, Republic of China
e-mail: hsinrau@cycu.edu.tw

S.-N. Siang
e-mail: g9602408@cycu.edu.tw

impact of a product is determined at the product development stage, the designers are supposed to have the responsibility for the sustainable product development through the Eco-design concept.

Eco-design, known as Design for Environment (DfE), can be referred to as green design focusing on the integration of environmental considerations and business oriented goal in product development. It attempts to include environmental variables at the same level of importance of functionality, costs, ergonomics, and efficiency (Platcheck et al. 2008). It incorporates the principles of the productive closed-loop supply chain, starting from the raw material selection, the production, the use, and the end-of-life of industrial products. The application of eco-design techniques have been done on various areas such as mini-compressor, mouse, and automotive electronics (Platcheck et al. 2008; Herrmann et al. 2005; Borchardt et al. 2009). Here, indicators such as factor X and eco-efficiency are needed as design criteria by product developers, as decision criteria by companies and as purchasing criteria by consumers.

Eco-efficiency analysis has been applied for several studies in different areas such as agricultural, furniture production, industrial system, paper mills, iron rod industry, and power plants (Reith and Guidry 2003; Michelsen et al. 2006; Zhang et al. 2008; Hua et al. 2007; Kharel and Charmondusit 2008; Korhonen and Luptacik 2004).

Based on a comprehensive literature review above, we found that most of the relevant literature concentrated on the implementation of eco-design. As we know, to date there are no reports discussing the value of consumer under environmental impact based on the concept of eco-design. Therefore, this study will take this advantage to introduce a new, simple, and practical methodology to both producers and consumers as a consideration in the decision making of producing or purchasing green products. Several different types of laptop computers were used as examples to illustrate the application of our proposed methodology.

2 Methodology

2.1 Model Development

This study develops a framework as a basis to calculate the value of consumer under environmental impact. The proposed method consists of two main steps, quantifying value of consumer and quantifying environmental impact. To quantify value of consumer, three steps are applied: quantifying value of product, determining lifespan of product, and quantifying total cost of ownership. We will discuss them as follows.

Zeithaml (1988) defines value of consumer (V_C) as consumer's overall assessment of the utility of a product based on perceptions of what is received and what is given, as shown in Eq. (1). Receipts (R_C) cover quality, functionality, personal value, and positive feelings while contributions (C_C) cover money, time or effort, and negative feelings.

$$V_C = \frac{R_C}{C_C} \quad (1)$$

This study defines the receipts of customer as a multiplication of the product value (V_p) and the lifespan (L_p) of a product obtained by consumer, as shown Eq. (2).

$$R_c = V_p \times L_p \quad (2)$$

1. To obtain the value of product (V_p), the calculation steps are the following:
2. Determine a reference product (R) and other products to compare or evaluate based on the reference product.
3. Identify key components of the product ($m = 1, 2, \ldots M$).
4. Choose the most significant attributes or functions for each key component, then find their performance ($Y_{m,n}$, $m = 1, 2, \ldots M$, $n = 1, 2, \ldots N(m)$, where $Y_{m,n}$ is the performance of attribute n of key component m of the evaluated product.
5. As the contribution for each component and attribute is different, then we can apply the weighting factor to express their importance.
6. Using factor X to find each attribute performance after factor X as shown in Eq. (3), where $R_{m,n}$ is the performance of attribute n of key component m of the reference product.
7. Use root mean square to obtain the mean performance of component m, as shown in Eq. (4).
8. Calculate the value of product V_p, as shown in Eq. (5).

$$X_{m,n} = \frac{Y_{m,n}}{R_{m,n}} \quad (3)$$

$$U_m = \sqrt{\frac{1}{N(m)} \sum_{n=1}^{N(m)} X_{m,n}^2} \quad (4)$$

$$V_p = \sqrt{\frac{1}{M} \sum_{m=1}^{M} Q_m^2} \quad (5)$$

where $Q_m = U_m \times W_m$, W_m is the weighting factor of component m.

We use the term TCO (Total Cost of Ownership) (Kirwin 1987) to represent C_c in Eq. (1). TCO includes total cost of acquisition and operating cost. The cost of product to the consumer includes purchasing cost, any costs related to use of the product and any costs that arise when the product is disposed (Bengtsson 2004). In this study, total cost of ownership is modelled as the sum of purchasing cost (C_P), use cost (C_{Use}), repair cost (C_R), and other cost (C_O), as shown in Eq. (6).

$$TCO = C_P + C_{Use} + C_R + C_O \quad (6)$$

Life Cycle Assessment (LCA) is a powerful tool that identifies and evaluates the life cycle impact that the product has on the environment. Eco-design products should be more energy-efficient, less polluting, and efficiently recycled than other products. Thus, indicators including energy consumption (E_c), pollution (P), and recyclability (N_r) are used to measure the environmental impact of a product over its life cycle (E_{lc}) as defined by Eq. (7).

$$E_{lc} = E_c + P + N_r \quad (7)$$

In this study, we collect their value from a questionnaire.

We can define the value of consumer under environmental impact based on the concept of eco-design as Eq. (8).

$$V_{CE} = \frac{V_C}{E_{lc}} \quad (8)$$

The value of consumer (V_C) can be described as the ratio of the value of product (V_p) during the lifespan (L_p) to Total Cost of Ownership (TCO). Substituting Eqs. (1), (2) and (6), then we can obtain the value of consumer under environmental impact as

$$V_{CE} = \frac{V_p \times L_P}{TCO \times E_{lc}} \quad (9)$$

From WBCSD (2000), eco-efficiency can be defined as the ratio of product value to environment impact, also following the above definitions, then we can define eco-efficiency as a multiplication of the product value (V_p) and the lifespan (L_p) of a product to its whole environmental impact (E_{lc}) based on Life Cycle Assessment (LCA), as shown in Eq. (10).

$$Eco - efficiency = \frac{V_p \times L_p}{E_{lc}} \quad (10)$$

Then the value of consumer under environmental impact based on the concept of eco-design can be defined as the ratio of eco-efficiency to TCO (customer contribution during life span).

$$V_{CE} = \frac{V_p \times L_P}{TCO \times E_{lc}} = \frac{Eco - efficiency}{TCO} \quad (11)$$

This study will use Eq. (11) to evaluate consumer value that involves product eco-efficiency and consumer contribution. However, the eco-efficiency considers product value under the environmental impact during the life cycle of the product;

therefore, the consumer value proposed here can be served as not only a purchasing reference for consumers but also a reference of product design improvement or innovation for producers with consideration of environment impact.

3 Example Illustration

3.1 Calculation of Product Value

In this section, we evaluated the applicability of our proposed methodology using laptop computers as examples. With regard to product type, there are three types of laptop computers are considered in this study as evaluated products. Each type has two computers and there are six computers (A to F) in total, as shown in Table 1. In order to compare with those six evaluated products, we determine a reference product as shown in Table 2 that is chosen as a reference product because its sales volume is number one among all types of laptop computer at that moment. Table 3 presents the value of product for each of the evaluated product. It is well known that the higher the functionality performance value, the greater the preference value of a product. Thus, it is not surprising that product F (notebook computer) and product B (CULV) got the higher score. Negative scores of product D and product C mean that the functional performance of those products (netbook computers) cannot compete with the reference product, where $V_P' = V_P/(V_P)_R$.

3.2 Calculation of Total Cost of Ownership

In order to quantify the total cost of ownership, purchasing cost, use cost, repair cost, and other cost should be defined first. In this study we used selling price as purchasing cost, electricity cost as use cost. The most important factors of electricity cost are energy consumption and lifespan of laptop computers. According to EuP Lot 3 (EuP 2005), different operational mode (off, sleep, and active mode) will lead to different total energy consumption. The percentage for each mode is off (49 %), sleep (37 %), and active (14 %). Lifespan is set 43800 h. Repair cost constant (F) = 85 USD.

3.3 Calculation of Value of Consumer

After calculating value of products and total cost of ownership, then we can find the value of consumer based on Eq. (1). Product D is in the first rank, and its

Table 1 Attributes of evaluated products

Type	A	B	C	D	E	F
	Consumer ultra-low voltage (CULV)		Netbook		Notebook	
CPU	Intel Core 2 Duo	Intel Core 2 Duo	Intel Atom N270	Celeron M	Intel Core i5 540 M	Core 2 Duo T4500
	Clock Speed(1.3 GHz)	Clock Speed(1.4 GHz)	Clock Speed(1.6 GHz)	Clock Speed(1.7 GHz)	Clock Speed(2.1 GHz)	Clock Speed(2.5 GHz)
	L2 Cache(3 MB)	L2 Cache(3 MB)	L2 Cache(1 MB)	L2 Cache(1 MB)	L2 Cache(2 MB)	L2 Cache(3 MB)
	Number of Cores(2)	Number of Cores(2)	Number of Cores(1)	Number of Cores(1)	Number of Cores(2)	Number of Cores(2)
	Bus Speed (800 MHz)	Bus Speed (1066 MHz)	Bus Speed (133 MHz)	Bus Speed (133 MHz)	Bus Speed (800 MHz)	Bus Speed (1066 MHz)
Memory	DDR2, 800 MHz, Max	DDR3, 1,066 MHz, Max	DDR2, 667 MHz, Max	DDR2, MHZ, Max 1GB	DDR3, 1,066 MHz, Max	DDR2, 800 MHz, Max
	4 GB	4 GB	1 GB		8 GB	8 GB
Display	12.1″	15.6″	8.9′	10′	14′	15″
	LED Backlight	LED Backlight	(1024 × 600)	(1024 × 600)	LED Backlight	LED Backlight (1366*
	(1366 × 768)	(1366 × 768)			(1366 × 768)	769)
Storage	320 GB, 5,400 rpm	500 GB, 5,400 rpm	12 GB, 4,800 rpm	80 GB, 4,800 rpm	640 GB, 5,400 rpm	320 GB, 5,401 rpm
Dimensions	296 × 210 × 25.1 mm	386 × 259 × 26.4 mm	225 × 175 × 39 mm	226 × 191 × 38 mm	349 × 238 × 36.5 mm	345 × 249 × 35.5 mm
Weight	1.5 kg	2.4 kg	1.1 kg	1.45 kg	2.2 kg	2.36 kg
Price (USD)	1,063	1,197	666	600	1,230	797

Table 2 Attributes of reference product

	Reference product
CPU	Intel Pentium T4500
Memory	DDR3 1,066 MHz 2 GB
Display	14″(1,366 × 768)
Hard disk	500 GB 5,400 rpm
Dimensions	239 × 342 × 38.6 mm (W × D × H)
Weight	2.36 kg

Table 3 Functional performance value

Reference product $(V_p)_R$	Type	Product value (V_p)	Improvement rate (V_p')	Rank
0.352	A	0.396	1.13	3
	B	0.405	1.15	2
	C	0.323	0.92	6
	D	0.347	0.99	5
	E	0.364	1.03	4
	F	0.418	1.19	1

product value is not the highest, but it has the lowest TCO. This phenomenon can explain why netbook computer was so popular a couple of years ago.

Figure 1 shows a comparison with the reference product in terms of product value and TCO, where $TCO' = TCO/TCO_R$. The value located in the second quadrant is the best and that has better performance for both V_p' and TCO'. However, none of the six laptop computers has this situation. Products D and C have better values in TCO' and others have better values in V_p'. Combining with these two factors, product D has the highest value in consumer value, followed by products F, C, A, C, and E.

3.4 Calculation of Value of Consumer under Environmental Impact

In this step, we quantify the environmental impact of those evaluated products based on LCA and use energy consumption, pollution, and recyclability as the indicators. We conducted a questionnaire to the computer company which owns

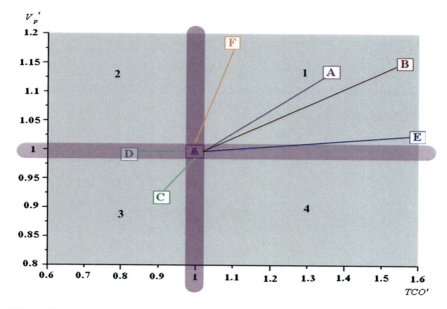

Fig. 1 Value of consumer

Table 4 Value of consumer under environmental impact

Type		Consumer value (V_C')	Environmental impact (E_{lc}')	V_{CE}'	Rank
CULV	A	0.81	1	0.81	4
	B	0.73	1	0.73	5
Netbook	C	1.00	1.01	0.98	3
	D	1.19	1.01	1.17	1
Notebook	E	0.64	1	0.64	6
	F	1.09	1	1.09	2

the six evaluated computer. The 58 questions are clustered into LCA processes: raw material extraction, production, transportation, use, and end-of-life aspects. Table 4 presents the environmental impact and the value of consumer under environmental impact of each evaluated products. From this indicator, product D is the most worthy product, followed by products F, C, A, B, and E. The rank order follows the order of consumer value, and this is due to there are no significant differences of environmental impacts among evaluated products.

When consumers demand for attributes such as compact, lightweight, easy to carry, and low price, netbook computer is the most recommended. Compared to product C, product D is better in the value of product and total cost of ownership. In the case of CULV, product A is better than product B. Product A has a lower value in product value than product B, but it has a lower value in total cost of ownership. As for notebook computer, product F is much better than product E as its value of product is much higher than product E and its total cost of ownership is also lower than product E.

4 Conclusions

The purpose of this study was to develop an assessment model that simultaneously considered the value of product based on functional performances, the total cost of ownership paid by consumer during life cycle, and the environmental impact complied by products. Six laptop computers ranging from CULV super light and thin, netbook, to main stream laptop computers were used to illustrate the applicability of our proposed assessment model. From this illustration, we can determine which laptop computer can be recommended to purchase for consumers in terms of not only functional performance and/or the total cost of ownership paid by consumer during life cycle, but also the environmental impact. At the same time, this model can be served as a reference of product design improvement or innovation for producers with consideration of environment impact. For the future study, we would suggest to extend our model to consider more data and develop software to make this assessment model in line with actual application.

Acknowledgments This work is supported in part by National Science Council of Republic of China under the grants NSC 98-2221-E-033-031-MY3 and 101-2221-E-033-037-MY3, and College of Electrical Engineering and Computer Science of Chung Yuan Christian University under the grants CYCU-EECS-10001.

References

Bengtsson S (2004) The BASF eco-efficiency analysis method: applied on environmental impact data from an LCA study of two colorants. Akzo Nobel Surface Chemistry. http://www.dantes.info/Publications/Publication-doc/. Accessed 1 May 2012

Borchardt M, Poltosi LAC, Sellitto MA, Pereira GM (2009) Adopting ecodesign practices: case study of a midsized automotive supplier. Environ Qual Manage 19:7–22

EuP (2005) Lot 3 personal computers (desktops and laptops) and computer monitors final report (Task 1-8). European Commission DG TREN http://extra.ivf.se/ecocomputer/downloads/Eup%20Lot%203%20Final%20Report%20070913%20published.pdf. Accessed 1 May 2012

Herrmann C, Stachura M, Yim HJ (2005) Methodic eco-design considering consumer needs and requirements: case study with computer mouse. In: Paper presented at the forth international symposium on environmental conscious design and inverse manufacturing. Tokyo, 12–14 Dec 2005

Hua ZS, Bian YW, Liang L (2007) Eco-efficiency analysis of paper mills along the Huai River: an extended DEA approach. Int J Manage Sci 35:578–587

Kharel GP, Charmondusit K (2008) Eco-efficiency evaluation of iron rod industry. J Cleaner Prod 16:1379–1387

Kirwin B (1987) End-user computing: measuring and managing change. Gartner Group Strategic Analysis Report, Stamford CN, USA

Korhonen P, Luptacik M (2004) Eco-efficiency analysis of power plants: an extension of data envelopment analysis. Eur J Oper Res 154:437–446

Michelsen O, Fet AM, Dahlsrud A (2006) Eco-efficiency in extended supply chains: a case study of furniture production. J Environ Manage 79:290–297

Platcheck ER, Schaeffer L, Kindlein W, Candido LHA (2008) EcoDesign: case of a mini compressor re-design. J Cleaner Prod 16:1526–1535

Reith CC, Guidry MJ (2003) Eco-efficiency analysis of an agricultural research. J Environ Manage 68:219–329

WBSCD (2000) Eco-efficiency: creating more value with less impact, World Business Council for Sustainable Development. ISBN 2-94-024017-5

Zeithaml VA (1988) Consumer perceptions of price, quality, and value: a means-end model and synthesis of evidence. J Mark 52:2–22

Zhang B, Bi J, Fan ZY, Yuan ZW, Ge JJ (2008) Eco-efficiency analysis of industrial system in China: a data envelopment analysis approach. Ecol Econ 68:216–306

A Study of Bi-Criteria Flexible Flow Lines Scheduling Problems with Queue Time Constraints

Chun-Lung Chen

Abstract This paper considers the scheduling problems in a flexible flow line (FFL) with queue time constraints. The objective of the scheduling problems is to minimize the primary criterion which is exceeding queue time constraint times and the secondary criterion which is makespan. The problem considered in the paper is a NP-hard in a strong sense. It requires much computation time to find the optimal solution; therefore, heuristics are an acceptable practice for finding good solutions. In this paper, a meta-heuristic is proposed to solve the candidate problems. In order to evaluate the performance of the proposed heuristics, a conventional tabu search algorithm is examined for comparison purposes. The results show the proposed meta-heuristic performs effective.

Keywords Flexible flow line · Queue time constraints · Bi-criteria · Dispatching rule · Meta-heuristic

1 Introduction

This research is aimed at developing heuristic algorithms to solve bi-criteria of flexible flow lines scheduling problems with queue time constraints. The objective of the scheduling problems is to minimize the primary criterion which is exceeding queue time constraint times and the secondary criterion which is makespan. The flexible flow lines scheduling problems are perceived as NP-Hard problems, therefore we intend to develop heuristic methods to solve the problems in this study. Heuristics are developed for the primary criterion which is exceeding queue time constraint times and the secondary criterion which is makespan.

C.-L. Chen (✉)
Department of Accounting Information, Takming University of Science and Technology, Taipei, Taiwan, Republic of China
e-mail: charleschen@takming.edu.tw

Flexible flow line (FFL) problems are also known as flexible flow shop (FFS), hybrid flow shop (HFS), or flow shop with multiple processors (FSMP) problems. A typical flexible flow line production system could be defined in this way: There are N number of jobs to pass through J number of stages, and every stage may contain one or more than one machine, and the waiting area between stages is assumed to be infinite, in other words, there is no job jammed in the waiting area. All jobs have the same routing over the stages of the shop and the flow of jobs through the shop moves in one direction from the first stage to the last.

It is common to control queue time in the production process in the making of many products. For example, queue time control is carried out in the assembly process of LCD (Cell), between wet etch and furnace in the fabrication of wafer in semiconductor industry, and in the procedure of wafer bumping in IC packaging industry. The queue-time constraint is defined as the time elapsed between two processes (Scholl and Domaschke 2000; Ono et al. 2006). In order to avoid quality defects due to long queue time in the production line, therefore these factories set up a maximum queue time between production processes so that no part needs to be reworked or discarded due to exceeding the queue time limit that causes production capacity reduction and economic loss (Su 2003).

Since queue time constraint phenomenon usually exists in FFL, they can be classified as flexible flow line with queue time constraint problems. There are some studies to FFL with no-wait or limited waiting time constraint. Wang et al. (2005) presents a heuristic algorithm to solve a two-stage flexible flowshop scheduling problem with no waiting time between two sequential operations of a job and no idle time between two consecutive processed jobs on machines of the second stage. Some studies relax the no-wait constraint that job must be processed continuously without waiting time between consecutive stages. Su (2003) presented a heuristic algorithm and a mixed integer program to solve a hybrid two-stage flowshop with limited waiting constraints. Chen and Yang (2006) proposed eight mixed binary integer programming models for the open-shop, job-shop, flow-shop, and permutation flow-shop scheduling problems with limited waiting time. The above two studies are to minimize makespan. Gicque et al. (2012) proposed a discrete time exact solution approach for a complex hybrid flow-shop scheduling problem with limited-wait constraints and considered the objective of minimizing the total weighted tardiness.

The mentioned above studies considered the jobs must be processed within every given queue time between each consecutive stages in the whole system. However, there are many situations usually occur in wafer fabrication and wafer bumping procedures, the mentioned above studies do not consider. For example, queue time constraints may be a time window set between two specified stages to prevent quality defects and the specified stages may not be consecutive. Figure 1 illustrates the physical relationship for queue time constraint between two specified stages. As shown in Fig. 1, there is a queue time among stage k to stage $k + b$, and the queue time is represented by $QT_{k,k+b}$, and among stage k to stage $k + b$ there is zero to b-1 stages. For convenience, the queue time constraint between two specified stages can be regarded as a queue time constraint block

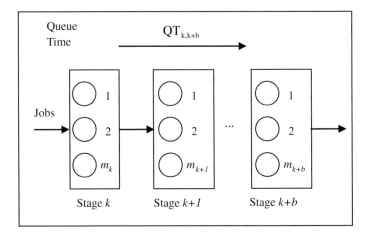

Fig. 1 Queue time constraint among stages

(QTCB). In addition, the QTCB may not include all stages in the whole FFL system. This means that there are many QTCBs in the FFL system controlled by several given queue times. For convenience, in this paper we named stage k as "the upstream specified process stage", and stage $k+1$ as "the downstream specified process stage". In addition, we name the stages in FFL except for stages k and $k+1$ as "general stage".

To the best of our knowledge, there are no studies that consider the bi-criteria of FFL scheduling problems with multiple QTCBs. The objective of the problem is to minimize the primary criterion which is exceeding queue time constraint times and the secondary criterion which is makespan. Hoogeveen et al. (1996) proved that minimizing makespan for two-stage FFL problems with stage 1 having one machine and stage 2 having two machines or with stage 1 having two machines and stage 2 having one machine is NP-hard. Therefore, the candidate FFL problem considered is at least NP-hard. Since it requires much computation time to find the optimal solution, heuristics are proposed to solve the candidate problems. A lot of test problems will be used to evaluate the performance of the proposed heuristics.

2 Description of the Considered Problem

The FFL with queue time constraints problems considered in this paper assumes that there are J stages and include at least a QTCB. We introduce the following characters which are considered in this paper.

There are m_j identical parallel machines in stage j and the number m_j may vary from stage to stage. Each stage has at least one machine, and at least one stage must have more than one machine. There are n jobs to be processed and each job has the same routing and must visit all the stages consecutively. Each job i has a positive processing time p_{ij}. A machine can process only one job at a time, and jobs cannot be preempted. There are unlimited buffers between stages. There are at least one QTCB existing in the system. There is no machine breakdown and all information is known in advance. If a job exceeds the queue time constraints, automatically return to the upstream specified process stage waiting re-work, and without taking into consideration the rejection. Bi-criteria are considered in the research, exceeding queue time constraint times and makespan. The objective is to find a schedule that optimize the primary criterion (exceeding queue time constraint times) followed by the optimization of the secondary criterion (makespan) subject to the primary objective value.

3 Proposed Heuristics

Due to the effect of queue time constraints, when a job is selected, it is important to pay attention to the queue time constraints lest the job to be reworked or discarded. Therefore, one needs to make several decisions and carry them out in the stage, and the first decision is whether to release a job from the waiting area to a upstream specified process stage, and secondly, to carry out the dispatching decision made under the queue time constraints at the downstream specified process stages.

The decision of upstream specified process stage is to determine whether the release of the job will collide with the queue time constraints. If it will, then the release of the job is not allowed. Due to the fact that once a job is selected at an upstream specified process stage, we have to make sure that the job won't collide with the queue time constraints at the downstream specified process stages. We develop a trial simulation method (TSM) that selects a job by the machine from the upstream specified process stage buffer, and computes the job whether it passes the queue-time constraint block.

In the downstream specified process stage, if a job stays in the waiting area for too long and exceeds the queue time limit, then it requires rework; therefore, we see the queue time limit as the most important decision criterion. In order to determine if the job exceeds the queue time limit of the various stages in the queue time area, the maximum queue time should be first established. The RQT rule is used to determine the job with the minimum remaining queue-time. The RQT can be expressed as (MQT_{ik}), where MQT_{ik} is the maximum queue time of job i could stay in the queue at stage k.

An extended iterated local search algorithm (EILS) is developed to solve the candidate problems. The workflow of the proposed EILS algorithm is shown as Fig. 2. The figure shows that an approach can be applied to generate a feasible initial solution, and a greedy local search is applied to improve the initial solution obtained. In addition, a shaking operator, Shake(), is constructed to perturb the incumbent solution and to increase diversification of the search in EILS. In this paper, the destruct-construct method (Ruiz and Stützle 2007) is applied to generate a set of input candidate solutions which are different from the incumbent solution. Then, a minimal solution from the input candidate solutions is selected to be the input solution in the next iteration.

4 Computational Experiments and Analysis of Results

In this paper, we developed a deterministic scheduling model to verify the performance of the proposed methods and the assumptions of the model are:

(1) There are 6 stages. The routing information and the related queue time information are shown in Table 1.
(2) The number of machines at each stage is generated from the uniform distribution U(1, 5).
(3) The number of jobs is 30 and 50.

Procedure *EILS*
Generate an initial solution S_0.
Set $S_b = S_0$. // S_b = the incumbent best solution
Set the value of exceeding queue time constraint times: $V_{qb} = f(S_0)$
Set the value of makespan: $V_{mb} = f(S_0)$
Repeat
　//apply a greedy local search and initial solution S_0 to obtain an improved solution S_k
　$S_k = \text{LS}(S_0)$
　Set V_{qk} = the value of exceeding queue time constraint times obtained by LS(S_0)
　Set V_{mk} = the value of makespan by LS(S_0)
　If $V_{qk} < V_{qb}$ **Then**
　　$V_{qb} = V_{qk}$
　　$S_b = S_k$
　ELSE IF $V_{qk} = V_{qb}$ **Then**
　　If $V_{mk} < V_{mb}$ **Then**
　　　$V_{mb} = V_{mk}$
　　　$S_b = S_k$
　　End If
　End If
　Apply Shake() function to generate a set of input candidate solutions which is different from the best solution
　S_0 = Select a minimal solution from the generated input candidate solutions
Until termination condition met
End

Fig. 2 The procedure of the proposed EILS algorithm

Table 1 Relevant data for job in multiple queue time constraints section

Routing number	Stage	Beginning stage with queue time constraints	Final stage with queue time constraints
N	Stage k		
N + 1	Stage $k + 1$	Stage $k + 1$	Stage $k + 3$
N + 2	Stage $k + 2$		
N + 3	Stage $k + 3$	Stage $k + 3$	Stage $k + 4$
N + 4	Stage $k + 4$		
N + 5	Stage $k + 5$		

(4) It is assumed that the processing time is generated from the uniform distribution U(10, 100) * M_j, where M_j means the number of machines at stage j.

(5) The queue time limit is generated from 100 (maximum processing time) * q_len * 5 (loose) and 100 * q_len * 4 (tight). q_len is the number of the downstream specified process stages within a QTBC.

(6) If a job exceeds the queue time constraints, automatically return to the upstream specified process stage waiting re-work, and without taking into consideration the rejection.

To verify the performance of the algorithms developed in this paper, we conducted computational experiments comparing the results between conventional tabu search and the proposed heuristics. The first-come first-served (FCFS) rule is used in the general and upstream specified process stages. RANDOM rule is applied to generate the input sequence of jobs. Six trials are conducted. In verifying the performance of the proposed methods, we considered the following performance measures: number of exceeding queue time constraint times, average makespan, relative deviation index (RDI), and number of best solutions (NBS). RDI is defined as:

$$RDI = \begin{cases} \frac{S_a - S_b}{S_w - S_b} & \text{if } (S_w - S_b) \neq 0, \\ 0 & \text{otherwise.} \end{cases}$$

S_a is the solution value obtained by method a, and S_b and S_w are, respectively, the best and worst values among the solutions obtained by the methods included in the comparison.

The results are presented in Table 2. In Table 2, we found that the proposed EILS with shaking 20 candidate solutions performs better than other algorithms. Furthermore, if the queue time limit is tight, the average of the RDI produced by EILS with shaking 20 candidate solutions will be 0; and the RDI of EILS with shaking one candidate solution will be 0.0776. This means that EILS with more candidate solutions performs better than EILS with one candidate solution. The proposed EILS with shaking 20 candidate solutions should be considered as the best choice for the candidate problem.

Table 2 Results for the computational experiments

Queuetime	Algorithms	Number of exceeding queue time constraint times	Average makespan	RDI	NBS
Loose	RANDOM	0	4,157	1.0000	0
	Tabu search	0	3,525	0.2528	0
	EILS with shaking one candidate solution	0	3,370	0.0363	3
	EILS with shaking 20 candidate solutions	0	3,346	0.0052	9
Tight	RANDOM	0	4,903	1.0000	0
	Tabu search	0	4,062	0.2388	0
	EILS with shaking one candidate solution	0	3,893	0.0776	0
	EILS with shaking 20 candidate solutions	0	3,810	0.0000	12

5 Conclusions

In this paper we propose a metaheuristic to solve the scheduling problems in a FFL with queue-time constraints. The objective of the scheduling problems is to minimize the primary criterion which is exceeding queue time constraint times and the secondary criterion which is makespan. To evaluate the performance of the proposed heuristic, we designed a lot of test scenarios to simulate practical shop-floor problems. For these candidate problems, we conducted computational experiments to compare the performance of the proposed heuristic. The proposed EILS with shaking 20 candidate solutions performs better than other algorithms. Our future research will consider other system characteristics, which are not included in this paper, such as machine availability and machine eligibility. Furthermore, the idea which generates a set of shaking candidate solutions by destruct-construct method can also be applied to solve other scheduling problems in the following studies.

Acknowledgments This paper was supported in part by the National Science Council, Taiwan, ROC, under the contract NSC 101-2221-E-147 -001.

References

Chen JS, Yang JS (2006) Model formulations for the machine scheduling problem with limited waiting time constraints. J Info Optim Sci 27(1):225–240

Gicquel C, Hege L, Minoux M, Canneyt W (2012) A discrete time exact solution approach for a complex hybrid flow-shop scheduling problem with limited-wait constraints. Comput Oper Res 39:629–636

Hoogeveen JA, Lenstra JK, Veltman B (1996) Preemption scheduling in a two-stage multiprocessor flowshop is NP hard. Eur J Oper Res 89:172–175

Ono A, Kitamura S, Mori K (2006) Risk based capacity planning method for semiconductor fab with queue time constraints. In: Paper presented at the 2006 IEEE international symposium on semiconductor manufacturing

Ruiz R, Stützle T (2007) A simple and effective iterated greedy algorithm for the permutation flowshop scheduling problem. Eur J Oper Res 177(3):2033–2049

Scholl W, Domaschke J (2000) Implementation of modeling and simulation in semiconductor wafer fabrication with time constraints between wet etch and furnace operation. IEEE Trans Semicond Manuf 13(3):273–277

Su LH (2003) A hybrid two-stage flowshop with limited waiting time constraints. Comput Ind Eng 44:409–424

Wang ZB, Xing WX, Bai FS (2005) No-wait flexible flowshop scheduling with no-idle machines. Oper Res Lett 33:609–614

Modeling the Dual-Domain Performance of a Large Infrastructure Project: The Case of Desalination

Vivek Sakhrani, Adnan AlSaati and Olivier de Weck

Abstract The performance of a large infrastructure project depends on not only technical design choices, but also contractual and other economic arrangements. These choices and arrangements interact in the context of uncertainty to result in the project's realized performance. Large infrastructure projects such as desalination plants are thus multi-dimensional design problems in which the dimensions can be broadly categorized into either the technical or institutional domains, creating the need for "dual-domain design". This paper describes the concept of dual-domain design for infrastructure in the context of desalination projects in the Kingdom of Saudi Arabia. It demonstrates the results of an analytical model that relates design choices along some technical and institutional design dimensions to plant economic performance. The analysis shows that plant design can be optimized subject to an uncertainty profile of water demand, and is sensitive to technology type, output capacity and potentially to price/contractual terms embedded in the delivery mode. The lens of dual-domain design thus provides a richer understanding of the relationship between project design and potential performance. Next steps can include multi-attribute assessments of performance (energy, environmental impact, etc.) as well as a greater variation in contractual forms in the institutional domain of design.

V. Sakhrani (✉)
Engineering Systems Division, Massachusetts Institute of Technology, 77 Massachusetts Avenue E40-252, Cambridge, MA 02139, USA
e-mail: sakhrani@mit.edu

A. AlSaati
KACST-MIT Center for Complex Engineering Systems, King Abdulaziz City for Science and Technology, P.O. Box 6086, Riyadh 11442, Kingdom of Saudi Arabia
e-mail: a.alsaati@cces-kacst-mit.org

O. de Weck
Aeronautics and Astronautics and Engineering Systems, Massachusetts Institute of Technology, 77 Massachusetts Avenue E40-261D/33-410, Cambridge, MA 02139, USA
e-mail: deweck@mit.edu

Y.-K. Lin et al. (eds.), *Proceedings of the Institute of Industrial Engineers Asian Conference 2013*, DOI: 10.1007/978-981-4451-98-7_155,
© Springer Science+Business Media Singapore 2013

Keywords Design · Uncertainty · Projects · Infrastructure · Desalination · Management

1 Introduction

This paper describes the concept of "dual-domain design" for infrastructure (Lessard et al. 2013) and links it to project performance. The performance of a large infrastructure project depends on not only technical design choices, but also contractual and other economic arrangements (Miller and Lessard 2001; Esty 2004; Merrow 2011; Lessard and Miller 2013), which interact in the context of uncertainty to result in the project's realized performance (de Weck et al. 2004; de Neufville and Scholtes 2011). Large infrastructure projects such as desalination plants are thus multi-dimensional dual-domain design problems. Considerations such as technology choice (reverse osmosis, multi-stage flash), output capacity, and load factor in the technical domain may interact with water tariff, delivery contract structure, or availability requirements in the institutional domain (Korn et al. 1999; Wolfs et al. 2002; Voutchkov 2012).

The paper first develops the notion of dual-domain design space, using desalination as a case example and data for the Kingdom of Saudi Arabia (Sect. 2), and then demonstrates how choices along those design dimensions affect performance under water demand uncertainty (Sect. 3). The paper concludes that project managers or sponsors can get a richer understanding of potential project performance through the lens of dual-domain design (Sect. 4).

2 Dual-Domain Attributes of Desalination in Saudi Arabia

To gain an understanding of desalination plant design choices and how they may affect plant performance, projects in the Kingdom of Saudi Arabia were analyzed for differences along design attributes or dimensions. The data source is Global Water Intelligence's (2013) IDA Plant Inventory online data repository. The database contains approximately 2,500 desalination projects in KSA. The final analysis retains about 1,400 plants that were commissioned in or after 1990, after accounting for plant closures, build cancellations, and unknown operating status.

The main independent variables affecting plant design is water demand from different end uses and the desired water quality for those uses. The main design dimensions are plant size or output capacity, desalination technology, and delivery mode. Choices along the design dimensions interact with the independent variables to result in performance, a multi-attribute dependent variable. While the database does not reflect either the aggregate demand (independent variable) or realized plant performance (dependent variable), the dual-domain design space can still be identified based on the plant design choices.

2.1 Technical Design

Plant capacity in KSA varies significantly with technology type, and to some extent, also with primary use for water. Figure 1 shows a comparison of plant capacities (using a log scale for ease of comparison) across main technology categories, and also by primary use within each technology category. The main technology categories are Multi-effect Distillation (MED), Multi-Stage Flash (MSF), Reverse Osmosis (RO), and Other (to denote experimental or less common technologies such as Electro Dialysis). Primary use is classified into Drinking (total dissolved solids, tds of 10– < 1,000 ppm), Industrial (tds < 10 ppm) and Other (which includes agriculture, tds > 1,000 ppm, and military/defense applications). MSF plants supplying drinking water are the largest in the Kingdom with a median plant size of around 13 log cubic meters/day (ln cu.m./d). Reverse Osmosis plants (RO) tend to be much smaller—around 6 ln cu.m./d—with a similar distribution of plant capacities across the drinking, industrial and other categories. MSF plants are commonly large capacity plants and more conducive to meeting drinking water and other mid-to-low water quality needs, whereas RO plants may be more versatile in meeting different types of end use, because of flexibility in sizing, water quality and modularity (Voutchkov 2012).

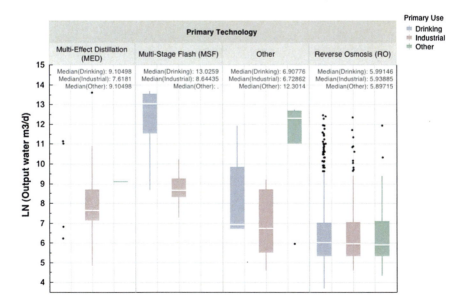

Fig. 1 Desalination plant capacity (log scale) in KSA by primary technology and use

2.2 Institutional Design

There are many 'delivery modes'—institutional-organizational approaches to delivering large projects such as desalination plants (Hallmans et al. 1999; Grimsey and Lewis 2007). The modes can be broadly classified into Asset Procurement (or Engineering-Procurement-Construction), Independent Water Producers (IWPs), or Service Provision (a hybrid category of water delivery contracts).

Plant capacity in KSA varies by delivery mode, and also by primary use. Figure 2 shows a comparison of plant capacities (using a log scale for ease of comparison) across delivery modes, and also by primary use within delivery mode. Plant capacities are much larger on average for Independent/Private plants. This arrangement is better suited to dedicated operations of industrial facilities, where a facility such as a refinery may contract with an Independent Water Producer. Service provision arrangements for drinking water also tend to make use of large plants, in comparison with the asset procurement (EPC) mode. These observations reflect potential economies of scale when water is delivered under independent or contractual arrangements (Hart 1996; Joskow 1988). While many technologies have been deployed under the EPC mode, large IWP plants have used the MED technology, whereas smaller service provision plants have used primarily RO technologies for supplying drinking water (Fig. 2).

The analysis in this section shows that there are many combinations of plant output capacity, technology types, and delivery modes for meeting different water end use needs. The next section describes an uncertainty-based analysis for how

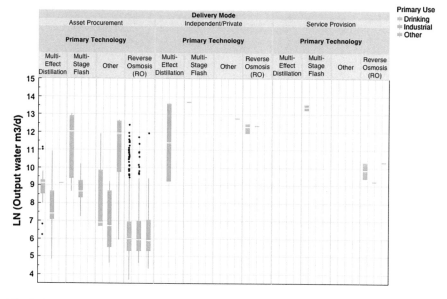

Fig. 2 Desalination plant capacity (log scale) in KSA by delivery mode, technology and use

project architects can evaluate project performance given different design choices in the fact of uncertain and volatile water demand.

3 Modeling Performance

Desalination project performance is evaluated using a stochastic simulation model in which uncertain water demand is the main independent variable, and Net Present Value is the main performance or dependent outcome variable. The analysis in this paper is limited to these variables, but other independent and dependent variables can easily be added.

3.1 Simulating Uncertain Water Demand

The Monte Carlo simulation generates 1,000 possible water demand scenarios, with a demand growth rate (drift) of 2 % per year and variability in growth rate (volatility) of 5 % per year, using Eq. (1) for water demand so that

$$D(t_i) = D(t_0) * e^{R(t_0, t_i)} \tag{1}$$

$$\text{where } R(t_0, t_n) = \sum_i R(t_i), \tag{2}$$

$$\text{and } R(t_i) = m + v * \tilde{\varepsilon}, \quad \text{where} \tag{3}$$

$R(t)$ Is the instantaneous return in demand or growth rate per time period (%/year);
m Is the drift, or expected growth rate per time period (%/year);
v Is the volatility, or standard deviation of growth rate per time period (%/year);
ε Is a standard normal random variable, $\sim$ N(0,1);
$D(t)$ Is the instantaneous water demand (million cu. m., MCM/year).

3.2 Simulating Plant Build and Operations

The design space for the desalination plant consists of technology type, output capacity and delivery mode.

Table 1 lists the possible values for the design vector and other key assumptions for the model. The design vector [Reverse Osmosis, 50 MCM/year, Asset Procurement/Fixed Tariff] fully specifies a plant design. The model iteratively tests 1,089 such plant designs.

3.3 Assessing Plant Performance

The main performance variable of is the Net Present Value (*NPV*) of plant cash flows. *NPV* is a useful indicator because it gives a much better idea of plant project value than capital build cost by including operating expenses, and adjusting for the time value of money by explicitly considering the timing of cash flows. The standard *NPV* formulation in Eq. (4) is

$$NPV = \frac{1}{N}\sum_{i}^{n}\frac{CF_t}{(1+r)^t} \quad (4)$$

$$\text{where } CF_t = B_t - (op_t + I_t) \text{ and} \quad (5)$$

NPV Is the Net Present Value (USD millions);
CF_t Is the cash flow in a given year (USD millions);
r Is the discount rate (%/year);
n Is the plant life, or horizon of the study (years)
B_t Is the revenue in that year (USD millions);
op_t Is the plant operating expense in that year;
I_t Is the capital investment in that year;
N Is the number of water demand scenarios.

The model first calculates the present value of cash flows for any plant design for each of the $N = 1,000$ water demand scenarios, then calculates *NPV* as the expected present value, and iterates over the 1089 possible designs. The discount rate for the analysis in this paper is 7.5 %/year. This approach gives both a

Table 1 Design vector choices and key assumptions

	Multi-stage flash (MSF)	Multi-effect distillation (MED)	Reverse osmosis (RO)
Capital cost ($/cu.m./day)	1,750	1,275	1,200
Operating cost ($/cu.m.)	0.45	0.50	0.55
Electricity consumption (kWh/cu.m.)	–	1.25	4.2
Plant capacity range (MCM/year)	1.5–50		
Delivery mode	Asset procurement/fixed tariff		
Water tariff ($/cu.m)	1		

probability distribution of present values for the plant and a single *NPV* number that is easier to compare across designs.

3.4 Model Results

Plant capacity for each technology type can be selected to maximize *NPV*, in effect truncating the downside (low or negative *NPV* scenarios) from the plant's distribution of value. Figure 3 (left) shows the plant's expected project value as a function of capacity for each technology type. The delivery mode is held constant as 'Asset Procurement/Fixed Tariff'. NPV first increases as the plant is able to meet an increasing share of demand with increased capacity, but then decreases as capacity is underutilized for large plant capacities. Subject to the assumptions, each technology peaks in roughly the same region (~ 29 MCM/year); however one technology may dominate the others in the "sub-optimal" capacity regions. Figure 3 (right) shows how selecting the optimal capacity (~ 29 MCM/year, thick lines) minimizes the chance of negative plant value, whereas oversizing (for ex, 45 MCM/year, faint lines) leaves 25–45 % chance across technology types that the plant value is negative.

Similar analyses can be conducted for the other delivery modes (not shown here), demonstrating that not only plant value (*NPV*), but also shares of the value accruing to the water customer (typically public sector) and plant owner or operator (typically private) are sensitive to the contract structure in place under different delivery modes (see Sakhrani and AlMisnad 2013).

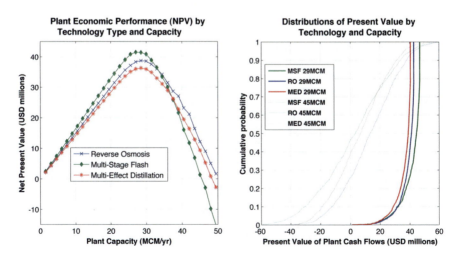

Fig. 3 Plant economic performance (net present value—USD millions), *left*, and distribution of present value (USD millions), *right*, by technology type and plant capacity (MCM)

4 Conclusions

This paper has described the concept of dual-domain design for infrastructure in the context of desalination projects in the Kingdom of Saudi Arabia. It has also demonstrated the results of an analytical model that relates design choices along some technical and institutional design dimensions to plant economic performance. The analysis shows that plant design can be optimized subject to an uncertainty profile of water demand, and is sensitive to technology type, output capacity and potentially to price/contractual terms embedded in the delivery mode. The lens of dual-domain design thus provides a richer understanding of the relationship between project design and potential performance. Next steps can include multi-attribute assessments of performance (energy, environmental impact, etc.) as well as a greater variation in contractual forms in the institutional domain of design.

Acknowledgments The authors are grateful for support from the Center for Complex Engineering Systems at KACST and MIT, and from the Sustainable Infrastructure Planning Systems project.

References

de Neufville R, Scholtes S (2011) Flexibility in engineering design. MIT Press, Cambridge
de Weck O, de Neufville R, Chaize M (2004) Staged deployment of communications satellite constellations in low earth orbit. J Aero Comp Info and Comm 1(3):119–136
Esty B (2004) Why study large projects? An introduction to research on project finance. Eur Fin Management 10(2):213–224
Global Water Intelligence (2013) IDA plant inventory. http://www.DesalData.com. Accessed 16 May 2013
Grimsey D, Lewis M (2007) Public private partnerships: the worldwide revolution in infrastructure provision and project finance. Edward Elgar, Cheltenham
Hallmans B, Stenberg C (1999) Introduction to BOOT. Desalination 123(2):109–114
Hart O (1996) Firms, contracts, and financial structure. Clarendon, Oxford
Joskow P (1988) Asset specificity and the structure of vertical relationships: empirical evidence. J Law Econ Org 4(1):95
Korn A, Bisanz M, Ludwig H (1999) Privatization of dual-purpose seawater desalination and power plants—structures, procedures and prospects for the future. Desalination 125(1):209–212
Lessard D, Miller R (2013) The shaping of large engineering projects. In: Priemus H (ed) International handbook on mega projects. Edward Elgar, Cheltenham (Forthcoming)
Lessard D, Sakhrani V, Miller R (2013) House of project complexity: understanding complexity in large infrastructure projects. In: Paper to be presented at the engineering project organizations conference, Winter Park, CO, USA, 9–11 July 2013
Merrow E (2011) Industrial megaprojects: concepts, strategies, and practices for success. Wiley, Hoboken
Miller R, Lessard D (2001) The strategic management of large engineering projects: shaping institutions, risks, and governance. MIT Press, Cambridge

Sakhrani V, AlMisnad A (2013) Evaluating contractual implications in the dual-domain design of a large infrastructure project. In: Paper submitted to the complex systems design and management conference, Paris, France, 4–6 December 2013

Voutchkov N (2012) Desalination engineering: planning and design. McGraw-Hill Professional, New York

Wolfs M, Woodroffe S (2002) Structuring and financing international BOO/BOT desalination projects. Desalination 142(2):101–106

Flexibility in Natural Resource Recovery Systems: A Practical Approach to the "Tragedy of the Commons"

S. B. von Helfenstein

Abstract "As overuse of resources reduces carrying capacity, ruin is inevitable" (Hardin 1998). In his controversial paper of 1968, Garrett Hardin introduced the concept of the *tragedy of the commons* in which the freedom of individuals to maximize their personal utility of common resources/goods (water, air, land, and so forth) leads to the destruction of those resources. While this problem is initially one of economic, socio-political, and ecological systems, in the end, it also an engineering problem since many current and future solutions depend on engineering systems design and implementation. Focusing on a single area of concern—construction and demolition (C&D) waste—I examine current U.S. practices in waste management and resource recovery. I then present a case study demonstrating the real-world use of flexibility in C&D natural resource recovery systems and explore its practical implications for a proposed paradigm shift that could resolve this one aspect of the *tragedy of the commons*.

Keywords Flexibility · Natural resource recovery systems · Construction and demolition waste

1 Introduction

The December 1968 issue of *Science* published a controversial paper by Garrett Hardin titled "The Tragedy of the Commons." Although the author was addressing the problem of human over-population, the phrase he coined, *the tragedy of the commons*, has since been used to describe a dilemma in which the

S. B. von Helfenstein (✉)
Value Analytics and Design, One Broadway—14th Floor, Cambridge, MA 02142, USA
e-mail: svonhelf@valueanalyticsanddesign.com

freedom of individuals to maximize their personal utility of common resources/goods (water, air, land, and so forth) leads to the destruction of those resources.

The locus of the *tragedy of the commons* initially resides in economic, sociopolitical, and ecological systems. But, in the end, it is also an engineering systems problem since many current and future solutions depend on engineering systems design and implementation. The *tragedy* takes its most poignant forms in the large-scale use/abuse of natural resource inputs and the overwhelming accumulation of waste outputs accompanying mass urbanization. Thus, it also becomes a critical consideration for new city design.

This paper is dedicated to the conceptual exploration of a single area of concern—construction and demolition (C&D) waste. It examines current waste management practice and proposes the use of flexibility to turn ineffective waste management systems into effective systems for resource recovery. A real-world case study is presented that demonstrates the attractiveness of flexibility in C&D resource recovery systems and the potential it offers to resolve this one major aspect of the *tragedy of the commons*.

2 Nature and Magnitude of the Challenge

As with other resource waste and recovery problems, the nature and magnitude of C&D waste involves both the use of non-renewable resource inputs and the generation of waste material outputs for which there is no current or future use. Consider rock aggregates as an example: "Aggregates play a major role in the construction industry as they are the major component of roadways, bridges, airport runways, concrete buildings, drainage systems, and many other constructed facilities. Because aggregates are the major component of much of the nation's infrastructure, their use is engineered to provide the necessary performance in place. For instance, concrete is approximately 75 % aggregate...

"In the past, when concrete structures reached the end of their service life or needed to be repaired or replaced, the resulting materials were considered waste and were disposed of in embankments and landfills. The costs of transporting these materials to waste areas were considered a necessary part of the replacement work. Likewise, the costs associated with the mining of new aggregates, the production of the replacement concrete, and the transportation and placement at the project were also considered a necessary part of the work" (CMRA 2012).

This is the description of an engineering, ecological, and economic input–output problem on a very large scale.

2.1 Urbanization as a Driver of C&D Waste

It is fair to say that the majority of C&D inputs and waste outputs are directly or indirectly generated to support urban growth and development.

While the definition of an *urban area* differs from country to country, the blistering pace of urbanization over the last 60 years and the expected pace of continued urbanization in the future is uncontested. Most sources state that over half of the world's population (e.g. 3 billion) now lives in cities, with at least 500 cities with populations of over 1 million. The forecast for 1015 is 50 megacities, 23 of which will boast a population over 10 million. CEIC Data Company Ltd, the U.N. Population Division, and *The Economist* forecast that by 2025, urbanization will be approximately 90 % in Western Europe, the U.S., and Brazil, 60 % in China, 50 % in Southeast Asia, and 38 % in India.

By way of contrast, in 1950, less than 30 % of the world's population lived in cities. Two hundred years ago, Peking was the only city with a population as large as 1 million.

Demographic shifts of the scale suggested above affect not only established cities. They also affect the surrounding suburbs/exurbs, rural areas, and the design of new cities in significant ways and for all related systems. Current economic, socio-political, ecological, and engineering systems are incapable of addressing such resource challenges. Instead urban systems around the globe are in gridlock.

2.2 Size of C&D Resource Use and Waste Streams

Further understanding of the magnitude of C&D resource use and waste streams generated by urbanized societies can be gained from the following—quite dated—U.S. statistics. It is notable that there appear to be no current updates indicating improvement in the status of these issues.

Landfill: In 1999, Staten Island, New York, contained what was the world's largest landfill. It received 26 million pounds of commercial and residential waste per day. In 1999, it contained 2.9 billion cubic feet (100 million tons) of trash. This was only 0.018 % of all the waste generated in the U.S. on a daily basis. Total annual waste in the U.S., excluding wastewater, exceeded 50 trillion pounds per year. Wastewater added another 200 trillion pounds. Less than 2 % of the total waste stream was being recycled (Hawken et al. 1999).

Construction and demolition waste: In the U.S., 40 % of all material flows were construction materials. 15–40 % of U.S. landfill space is taken by waste from these flows (Hawken et al. 1999). In 2006, un-recycled construction and demolition (C&D) waste in the U.S. was estimated at 325 million tons per year (Bouley 2006).

Roads: The U.S. has 3.9 million miles of public roads and an unknown number of miles of private roads. "The pervasiveness of roads and their cumulative effect on the environment are now of increasing concern..." (Deen 2003).

3 Current U.S. Waste Management and Resource Recovery Systems

3.1 Waste Management

Waste management is performed throughout the U.S. on organic and inorganic waste streams from municipalities, industry, healthcare, agriculture, and other sources. Current waste management systems are massive in scale, long-lived, capital intensive, and costly to run and replace. They are also centralized, stand-alone and single purpose, and toxic (even with best environmental efforts). With a far-reaching footprint, they are over-burdened with current use but lacking in scalability, and tightly bound to layers of conflicting public policy and funding—making them unresponsive to both planned change and unforeseen events. With regards to C&D waste, standard practice is to incinerate or landfill it all.

These are highly inflexible systems and they create further inflexibility in the economic, socio-political, and ecological systems that support them. Their entire function contributes to the *tragedy of the commons* because: (1) they utilize large amounts of common resources to operate; (2) they are themselves considered common goods by the populations they serve, and are not stewarded carefully; and (3) they produce further waste streams that are disposed of in the commons.

3.2 Resource Recovery

In the U.S., resource recovery is still a relatively narrowly-applied approach to waste management. Wikipedia defines *resource recovery* as "the selective extraction of disposed materials for a specific next use, such as recycling, composting or energy generation. The aim of resource recovery is to extract the maximum practical benefits from products, delay the consumption of virgin natural resources, and to generate the minimum amount of waste."

Further, "Resource recovery is the practice of reclaiming materials that were previously thought of as unusable. It is not managing waste, which is the standard for most garbage companies. Traditional waste companies collect and move wasted materials to large-scale, single-use sites such as landfills or incinerators. Unlike the management of waste, resource recovery recognizes that there is still value in those materials. The intention of resource recovery is always to make the best and highest use of all materials, and landfill only those materials for which there is no current use" (http://recology.com/).

At this time, most of the efforts of resource recovery practitioners are focused in a few areas: recycling of residential waste, wastewater treatment, and one or two others. Resource recovery systems are fragmented, heavily dependent on ever-changing, multi-level government regulation and funding, and conducted in industry/waste stream silos, much like their traditional waste management system peers. Unfortunately, they provide only a temporary relief for the *tragedy of the commons*, since their outputs end up back in traditional waste management systems, generally in less recoverable condition than they were the first time around.

3.3 State of the System

Environmental protection and sustainability have been a major topic of public discourse in the U.S. since the 1960s. The effects of waste streams on urban areas, rural land, flora and fauna, and humans are well and publicly documented. Decades of environmentalists have protested and proposed solutions. The four *eco-efficient* strategies—*reduce, reuse, recycle, and regulate* (McDonough and Braungart 2002)—have become mantras for green initiatives. Many eco-efficient solutions have been applied by government, commerce, and academia, codified in thousands of regulations, and become common practice.

"**But ultimately a regulation is a signal of design failure.** In fact, it is what we call a *license to harm*: a permit issued by a government to an industry so that it may dispense sickness, destruction, and death at an "acceptable" rate... [G]ood design can require no regulation at all...." (McDonough and Braungart 2002).

This paper proposes a paradigm shift involving system reconfiguration that will correct current design failure through incorporating flexibility.

4 Flexibility in Natural Resource Recovery Systems

To gain an appreciation of the nature of the paradigm shift represented by C&D Natural Resource Recovery Systems (NRRS), we must first define flexibility and develop a general description of a proposed flexible NRRS. The taxonomy used for system description is suggested by (Baldwin et al. 2011). This conceptual framework is then applied to a real-world C&D NRRS, built and operating in Maine, United States.

4.1 Flexibility

Flexibility is a term that describes a system's capacity for dealing dynamically with uncertainty. Flexible system design builds components into the system that provide for system change capacity, should it be desirable in the future. Not everything that might be needed is built into the system from the outset.

"The right kind of flexibility in design gives... three kinds of advantages. It can: (1) greatly increase the expected value of the project or products; (2) enable the system manager to control the risks, reducing downside exposure while increasing upside opportunities, thus making it possible for developers to shape the risk profile. This not only gives them greater confidence in the investment but may also reduce their risk premium and further increase value; and (3) often significantly reduce first costs of a project—a counterintuitive result due to the fact that the flexibility to expand means that many capital costs can easily be deferred until prospective needs can be confirmed" (de Neufville and Scholtes 2011).

4.2 General Description of the Proposed Flexible NRRS

In direct contrast to the rigid, massive, costly waste management system configurations currently in use, the proposed flexible NRRS is:

Local, exhibiting an intimate city-rural linkage: The proposed NRRS may be a city-based system. But, it receives from and contributes to its rural context rather than viewing that context as a dumping ground for its own waste. As a local system, it responds to local needs, "fits into the character of the land and its topography, soils, climate" and utilizes "locally available materials, regional construction techniques,... labor-saving functionality, and minimal cost to build, operate, and maintain" (Thorbeck 2012).

Small scale, decentralized, modular, and scalable: Unlike traditional systems, the proposed NRRS is built on a smaller scale and can be modular. It is a decentralized system, allowing it to be flexible and scalable—even mobile—to meet local needs and growth. Its scale and decentralization also allow it to change and innovate at a low cost, or to shut down at modest cost without leaving behind massive system remains.

Adaptive, amenable to innovation: Adaptive behavior is "the ability to alter one's own functions or goals... to adjust to environmental changes without significant changes to the system configuration" (Baldwin et al. 2011). The proposed NRRS's flexibility allows it to adjust and adapt to change with far less stress and disturbance to its context than its current peers. As the environment or technologies change, the system configuration can continue to absorb and incorporate change gracefully.

Small footprint: Since system inputs are large in scale, the proposed NRRS manages its technological and ecological footprint with care, reducing it wherever and however possible.

Closed loop-circular flow "waste is food" philosophy: "Our move toward sustainable cities will require an important shift in thinking of cities not as linear resource-extracting machines but as complex metabolic systems with flows and cycles, where, ideally, the things that have been traditionally viewed as negative outputs (e.g., solid waste, wastewater) are re-envisioned as productive inputs to satisfy other urban needs, including food, energy, and clean water" (Beatley 2011).

Simple, reasonably priced and cost-effective: Driven by private industry cost-benefit concerns, proposed NRRS components are as simple, accessible, and cost-effective as possible. Where components are initially more complex and costly, their flexible uses create operating efficiencies and short payback periods.

Self-organizing: Organizational development and management theory, the theory of free market systems, and many other disciplines attest to the success of self-organization in bringing about system creation, growth and change. Rather than being tightly overseen by government, the proposed NRRS is designed to be self-organized and self-regulated.

4.3 Case Example: CPRC Group, Scarborough, Maine

CPRC Group (Commercial Paving and Recycling http://www.cpcrs.com/) was founded in 1945 as a traditional asphalt paving company. In 1990, its original owner discovered that his cold-mix equipment, used to make asphalt, could turn contaminated soil into fully usable construction fill. Further experimentation led to uses of the same machinery for other C&D waste remanufacturing. Since 2004, current owners, John Adelman and Jim Hiltner, have turned CPRC into a leader in *conversion technology* and a classic example of flexibility in engineering design.

The company focuses on *making the turn*—i.e., taking in C&D (and other) waste materials and converting them into useful, saleable construction, landscaping, and agricultural product. One of its four divisions operates and manages the City of Portland's Riverside Recycling Facility. The others receive and convert: asphalt pavement, concrete, bricks, rock, ledge, and miscellaneous aggregate-based material; residential asphalt shingles of any size, shape, and color; asbestos-tested commercial asphalt roofing material; catch basin and sand-blast grit; stumps, branches, wooden pallets, demo woods, clean wood, leaves, brush, grass clippings; gypsum board that is free of paint, wallpaper, and contamination by wood, cans, paper or other debris; glass and porcelain materials (except containers that once held hazardous products, automotive headlights, and residential incandescent light bulbs); uncontaminated inert materials such as unscreened loam; soil containing heating oil, motor oil, and waste oil; and institutional food waste.

Once waste materials are sorted, CPRC remanufactures reusable components into an array of conversion products, such as: C&R gravel that is crushed, screened and blended from pavement, concrete, and rock materials, asphalt shingles, glass, and inert materials in various proportions, and then used to build roads, parking lots, bridge approach ramps, embankments, shoulders, construction project sites, and other heavy infrastructure projects; erosion control materials made from converted green waste; licensed inert fill dirt (made from converted petroleum-containing soil) that is highly compatible and uniform and is both structurally and environmentally sound; screened loam made from converted non-contaminated soils; biomass fuel made from demolition wood and clean wood; bark mulch for landscaping purposes; organic compost for agricultural uses.

Operations are designed to be highly flexible. Land, buildings, and technologies are multi-use. Employees are trained and incentivized to implement lean manufacturing methods. Customers are given valuable options that make using CPRC desirable. For instance, waste inputs can be transported to CPRC facilities by the customer, picked up by CPRC, or converted on-site by CPRC's mobile equipment. Remanufactured waste outputs can be picked up at CPRC facilities or delivered to the customer. In addition, outputs can be custom mixed, based on customer specifications.

CPRC is an excellent illustration of the proposed flexible NRRS system structure. Based in a rural suburb of Portland, Maine, it is local, but exhibits an intimate city-rural linkage through its use, conservation, and improvement of both urban and rural land and resources. Because CPRC is small scale, decentralized, modular, and scalable, it is also far more adaptive and amenable to innovation than larger traditional waste management systems.

While C&D remanufacturing facilities involve sizeable land allotments and large-scale technologies, CPRC manages its technological and ecological footprint with care. The company owns its land and intend to keep it pristine for a range of future uses. And, company inputs and outputs form a closed-loop circular flow in which waste inputs become "food," i.e. productive outputs to satisfy other needs.

Is the CPRC system simple, reasonably priced and cost-effective? It is, for the customer. In addition, although C&D conversion technologies are becoming increasingly sophisticated and costly, CPRC's use of these technologies demonstrates that older, simpler machinery can be re-engineered and used successfully where necessary. Even if new technologies must be purchased, they are long-lived, mobile, and can be redeployed for multiple uses with a minimum of retrofitting. This makes ongoing operations cost effective and efficient.

As for the attribute of self-organization, while all environmental activities in the U.S. are tightly enforced by regulation, CPRC and the industry of which it is a part exhibit a high degree of self-organization within the proscribed limits. Both industry and other literatures suggest that the complexity and arbitrariness of the regulatory environment and its slowness in accepting C&D conversion products currently present a hindrance to further beneficial contributions by this industry. But, hopes are high for a regulatory paradigm shift to match and support other system shifts.

5 System Valuation

Genichi Taguchi insisted that manufacturing waste created a significant cost to society. The same could be said about natural resource use/abuse and the *tragedy of the commons*. Thus, the economic, socio-political, ecological and engineering problems discussed in this paper are also problems of value. How do we value the commons? How do we value the systems that might contribute to its recovery and restoration?

Although the scope of this paper does not allow for a discussion of such valuation issues, valuation "plays an important role in any engineering field, primarily as an aid in making design decisions" (Kangas 2004). Therefore, we mention two areas in which progress has been made: (1) Ecological economics, a discipline that seeks "to reinvent economics with connections to ecology" (Kangas 2004); and (2) the work of de Neufville and Scholtes (2011) that suggests a portfolio of screening and valuation techniques allowing engineering systems designers to directly address uncertainty and flexibility.

There is more work to be done but these steps offer a way forward.

6 Extended Applications of Flexible C&D NRRS

While there are any number of potential applications of flexible C&D NRRS, two offer immediately appealing value propositions: landscape architecture/ecological engineering; and new city design and construction.

Landscape architecture and ecological engineering both concern themselves with natural resource use and restoration as well as innovation in urban fabric, infrastructure, and material technologies. The small-sample literature search performed for this paper indicates that there is little to no current use of remanufactured C&D waste in landscape architecture or ecological engineering projects. Yet, C&D conversion products seem like ideal candidates for such projects.

New city design and construction might also benefit from the use of C&D conversion products. Imagine bringing conversion equipment on site and using both the C&D waste from site preparation and the C&D waste generated during ongoing new construction to build out the baseline infrastructure of the city—a city that builds itself with greatly reduced utilization of virgin materials.

7 Conclusions

The conclusions to be drawn seem simple. In a world in which financial resources are becoming increasingly limited but natural resource use/abuse and waste increasingly prevalent and threatening, we can continue to design and build huge, costly, inflexible systems that exacerbate the very problems they purport to address and then impose these systems on stakeholders by regulatory diktat and tax schemes. Or, we can begin to explore and adopt flexible, smaller-scale, affordable systems that transform problems into benefits and self-organize through ingenious local capabilities and initiatives. If we choose the latter, we can look to C&D conversion technologies and the flexible NRRS they embody to show us a practical approach to reversing the *tragedy of the commons*.

References

Baldwin WC, Felder WN, Sauser BJ (2011) Taxonomy of increasingly complex systems. Int J Ind Sys Eng 9(3):298–316
Beatley T (2011) Biophilic cities: integrating nature into urban design and planning. Island Press, Washington
Bouley J (2006) Tearing up the road. Mainebiz 12(9)
Construction Materials Recycling Association (2012) Recycled concrete aggregate. White paper
de Neufville R, Scholtes S (2011) Flexibility in engineering design. The MIT Press, Cambridge
Deen TB (ed) (2003) Road Ecology: Science and Solutions. Island Press, Washington
Hardin G (1998) Extensions of the tragedy of the commons. Science 280(5364):682
Hawken P, Lovins A, Lovins LH (1999) Natural capitalism. Little, Brown and Company, New York
Kangas PC (2004) Ecological engineering: principles and practice. CRC Press LLC, London
McDonough W, Braungart M (2002) Cradle to Cradle. North Point Press, a division of Farrar, Strauss, and Giroux, New York
Thorbeck D (2012) Rural design: a new discipline. Routledge, New York

The Workload Assessment and Learning Effective Associated with Truck Driving Training Courses

Yuh-Chuan Shih, I-Sheng Sun and Chia-Fen Chi

Abstract Present study examined the workload and applied the theory of learning curve to evaluate the learning effective for training of driving courses. The trainees' workloads were assessed by the NASA-TLX twice, one on the 10th and the other on the last (28th) practice. Forty healthy male solders with an average age 23.2 years participated in this study, and a HINO 10.5T trunk was used for training in a standard training field. Five driving tasks evaluated were "going up and down a hill (up/down hill)", "three-point turn on a narrow road (3-point turn)", "moving forward and backward on an S curve (S-curve)", "reversing the car into a garage (reversing-into-garage)", and "parallel parking". For learning cures, the values of among 40 participants were averaged within each practice for each task, and the overall Wright's learning curves model for each driving task was fitted. Results showed all R^2s were significantly high with a range of 0.88–0.97. This implied that these learning curves of trunk driving tasks were able to be fitted by power function very well. Specifically, the learning rate was 0.9162 for up/down hill, 0.8912 for 3-point turn, 0.8802 for parallel parking, 0.8736 for reversing-into-garage, and 0.8698 for S-curve. For workload, the results indicated that the second measure (on 28th practice) was lower than the first measure (on 10th practice) for all evaluated tasks. This implied that practice was also able to reduce the overall workloads. Additionally for the task effect, S-curve task had the highest workload, 3-point-turn task had the lightest workload, and the rest three were not significantly different from each other. After practices, there were more

Y.-C. Shih (✉)
Department of Logistics Management, National Defense University, Taipei, Taiwan
e-mail: river.amy@msa.hinet.net

I.-S. Sun · C.-F. Chi
Department of Industrial Management, National Taiwan University of Science and Technology, Taipei, Taiwan
e-mail: s510678@yahoo.com.tw

C.-F. Chi
e-mail: chris@mail.ntust.edu.tw

reduction in workload for the tasks of S-curve, reversing-into-garage, and parallel parking.

Keywords Driving training · Learning curves · Workload · Trunk

1 Introduction

Trunks are one of the important and effective transporting tools for logistics. Anyone who wants to be a qualified trunk driver in Taiwan should first possess the license of small passenger vehicle for at least six months and, second, pass both written and driving tests in an examining agency. These requirements are the same for the military, in which solders possessing the license of small passenger vehicle were recruited and practiced in a training center for a period of 4 weeks. In these 4 weeks, they should accept an intensive training and have a formal driving examination on the last day. Five difficult and important driving tasks from the driving tests were evaluated, namely up/down hill, reversing into garage, parallel parking, 3-point turn, and moving forward and backward on a S-curve.

For decades the learning curves have been a valuable measurement tools to predict and monitor the performance of individuals, a group of individuals, and organization. They were widely applied in various sectors, such as manufacturing, healthcare, education, construction, and so on.

The most popular of all available models is one proposed by Wright in 1936. It is a power function of the number of practice, as shown in Eq. (1). Its popularity is attributed to its simple mathematics and to its ability to fit a wide range of data fairly well.

$$T_n = T_1 n^b \tag{1}$$

Where T_n is the time for the nth practice, and T_1 is the theoretical time to finish the first practice. n is the number of practice and b is the learning constant between 0 and -1. In addition, $\varphi = 2^b$ is called the learning rate between 0.5 and 1, and the smaller the learning rate, faster the learning.

Learning can be classified into two parts, cognitive learning and motor learning. The learning rate for purely cognitive learning task has been shown to be around 0.7, and around 0.9 for purely motor learning (Dar-el et al. 1995). Noticeably, most tasks usually involve both cognitive and motor learning. Konz and Johnson (2000) indicated that the learning rate was 0.74 for the machining and fitting of small castings, 0.83 for the assembly of a radio tube, and 0.89 for operating the punch press. Reid and Mirka (2007) used the learning curve to evaluate a patient lift-assist device and revealed that the learning rate was 0.83.

As to the applications of learning curves on driving simulator, authors concerned how to designed a practice scenario for adaptation for pedals (Sahami et al. 2009) and steering (Sahami and Sayed, 2010). Both revealed that a power function was suitable to model the adaptation for drivers to learn from the simulation system. Recently, Sahami and Sayed (2013) demonstrated that in driving simulator the adaptation time and leaning rate between male and female was not significantly different. More importantly, adaptation to a driving simulator was task-independent.

Driving training is, of course, to improve the driving skills for the novice drivers, even past studies used the learning curves to assess the adaptation time while a driving simulator was used, but in real driving tasks the information about the application the learning curves on the trainees during training is still deficient. This knowledge is helpful to how to allocation the training resources; for example, how many hours are needed for different training tasks.

2 Methods

2.1 Participants

Forty healthy male solders with an average age 23.2 years (s.d. = 3.4, range: 19–35 years.) participated in this study. All had valid license driving license of light vehicle for an average 30.5 months (s.d. = 23.7, range: 6–96 months).

2.2 Trunk and Driving Tasks

The trunk used for training was a HINO 10.5T trunk (KUOZUI MOTORS, LTD.), which was shown in Fig. 1 and its length, width, height, and wheelbase was 822.5, 217.5, 247.5, and 482 cm, respectively. Five driving tasks evaluated were "going up and down a hill" (denoted by up/down hill), "three-point turn on a narrow road" (denoted by 3-point-turn task), "moving forward and backward on an S curve" (denoted by S-curve task), "reversing the car into a garage" (denoted by reversing-into-garage task), and "parallel parking" (denoted by parallel-parking task). They are illustrated in Fig. 2 and all driving tasks were conducted in a standard training field.

2.3 Experimental Procedures and Data Acquisition

The training period was four weeks. During the first week the trainers instructed the basic knowledge about the trunk, and all trainees were allowed to familiarize with all training procedures and environment. The rest three weeks were for

Fig. 1 The HINO 10.5T trunk used in this study

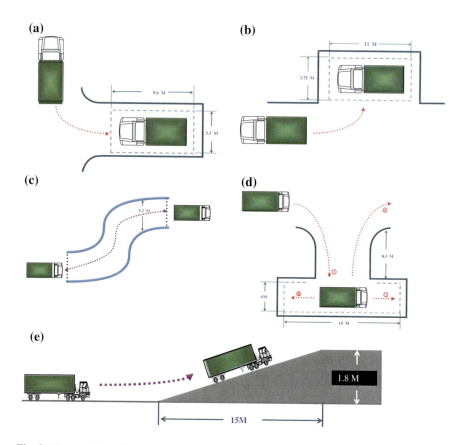

Fig. 2 The graphical illustration for five driving tasks. **a** Reversing-into garage. **b** Parallel parking. **c** S-curve. **d** 3-point turn **e** up/down hill

practice, five days per week, and the last day of the fourth week was for license examination. Each day participant had two times to practice, namely morning (09–12 o'clock) and afternoon (14–17 o'clock). Therefore, there were a total of 28 times for trainees to practice.

The allocation of driving field is a cycle with tasks in sequence of parallel parking, S-curve, 3-point turn, and reversing-into-garage. For each practice trainees should randomly start from one of these four tasks, and then completed them following the sequence. On the other hand, the site of up/down hill was located in the other side of training field, so it was the last to conduct for every practice. The finishing time associated with each practice was recoded. In addition, the NASA-TXL was used to assess the workload at the 10th practice (on the end of the second week) and 28th practice (the last one).

2.4 Experimental Design and Data Analysis

For the performance of practice, namely the time needed to complete each driving task, the effect of number of practice was examined. Additionally, the learning curve of each participant associated with each driving task was fitted according to the Wright's model, then the theoretical time to complete the first practice (T_1) and learning rate (ϕ) of each participant was adopted for ANOVA with the factor of driving tasks. For NASA-TLX, a factorial design was employed with two factors of driving task and stage (10th or 28th practice). In the all ANOVA models, the highest interaction order with participant was served as the error term to precisely test the influence of all main effect. The software Statistica 8.0 was used for data analysis, and a post hoc Newman-Keuls test was used to test paired differences for significant main effects and interactions. The level of significance (α) was set at 0.05.

3 Results

The ANOVA results indicated that there existed a significant effect of practice number on completing time for all driving tasks. Practice made the driving time gradually shorter, but this improvement became less and less as the times of practice increased. Generally speaking, there was not a significant improvement in the performance after 20th practices (during the fourth week).

Next, the learning curve of each participant associated with each driving task was fitted according to the Wright's model. The average coefficient of determination (R^2) was 0.76 (min–max: 0.51–0.95) for up/down hill task, 0.84 (min–max: 0.47–0.94) for 3-point-turn task, 0.86 (min–max: 0.69–0.95) for S-curve task, 0.79 (min–max: 0.23–0.93) for reversing-into-garage task, and 0.76 (min–max: 0.25–0.91) for parallel-parking task.

The theoretical time to complete the first practice (T_1) and learning rate (ϕ) of each participant was further adopted for ANOVA, which revealed that the effect of driving task was significant on both T_1 ($F(4,156) = 260.4$, $p < 0.001$) and ϕ ($F(4,156) = 12.2$, $p < 0.001$). The post hoc of Newman-Keuls test indicated that three groups could be classified for T1: up/down hill (mean = 91.3 s, s.d. = 11.2) and parallel-parking (mean = 94.8 s, s.d. = 14.7) tasks, following 3-point-turn (mean = 105.7 s, s.d. = 16.1) and reversing-into-garage (mean = 111.6 s, s.d. = 16.9) tasks, and the longest was S-curve task (mean = 222.7 s, s.d. = 37.2). As to the learning rate, the post hoc result according to the Newman-Keuls test demonstrated that the learning rates of S-curve task (mean = 0.8717, s.d. = 0.0331) was not significantly different from that of reversing-into-garage task (mean = 0.8752, s.d. = 0.0296), the learning rates of parallel-parking task (mean = 0.8815, s.d. = 0.0357) was not significantly different from that of 3-point-turn (mean = 0.8924, s.d. = 0.0273), and up/down-hill task (mean = 0.9156, s.d. = 0.0308) had the greatest learning rate.

Finally, the values of among 40 participants were averaged within each practice for each task, and the overall Wright's learning curves model for each driving task was fitted. All R^2s were significantly high with a range of 0.88–0.97. This implied that these learning curves of trunk driving tasks were able to be fitted by power function very well. Specifically, the learning rate was 0.9162 for up/down-hill, 0.8912 for 3-point-turn, 0.8802 for parallel-parking, 0.8736 for reversing-into-garage, and 0.8698 for S-curve.

The trainees' workloads were assessed by the NASA-TLX twice, one on the 10th and the other on the 28th practice. The scores were tested by means of ANOVA with factors of driving task and measured time. The results indicated that all main effects and two-factor interactions were significant on weighted NASA-TLX score. As the Fig. 3 shown, the second measure (on 28th practice) was significantly lower than the first measure (on 10th practice) for all evaluated tasks. This implied that practice was also able to reduce the overall workloads. Additionally for the task effect, averagely speaking, S-curve task had the highest

Fig. 3 The weighted NASA-TLX scores of five driving tasks for two measurements

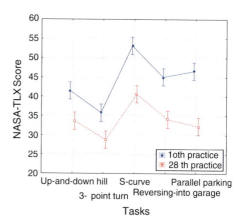

workload, 3-point-turn task had the lightest workload, and the rest three were not significantly different from each other. After practices, Fig. 3 also indicated that there were more reduction in workload for the tasks of S-curve, reversing-into-garage, and parallel-parking.

4 Discussions

The learning rate above 0.9 means pure motor skill, and the learning rate for a hybrid task containing cognitive and motor skill would be ranged 0.7–0.9. Learning can also be separated into two categories, cognitive learning and motor learning. Purely cognitive learning tasks have been shown to have a learning constant (learning rate) of ~ 0.70, while the learning constant for purely motor learning tasks is ~ 0.90 (Dar-el et al. 1995). Therefore, it implied that the task of up/down hill task was like a task with pure motor skill, and the rest with learning rates less than 0.9 needed more cognitive processes.

On the other hand, there was a negative correlation between T_1 and Φ for S-curve task ($r = -0.693$, $p < 0.01$), up/down-hill task ($r = -0.393$, $p < 0.05$), 3-point-turn task ($r = -0.812$, $p < 0.01$), reversing-into-garage task ($r = -0.854$, $p < 0.01$), and parallel-parking task ($r = -0.729$, $p < 0.01$).

As to the conclusion of Dar-El et al. (1995), a smaller learning rate seems to need more cognitive processes, which could spend more time. A new dual-phase model for learning industrial tasks is presented, based on the combined effects of cognitive and motor processes. The model proposes that cognitive elements dominate learning during the early cycles, whereas motor elements dominate the learning process as the number of repetitions becomes large. The implication is that the observed learning slope is a variable whose value gradually increases as experience is gained. Experimental studies are described whose results support the behavior of the dual-phase learning model.

It was found that longer the experience of holding the license of light vehicle led to a shorter T1 ($r = -0.324$, $p < 0.05$) of S-curve task. The learning rate for S-curve was 0.8698, the lowest among five tasks. The S-curve task also had the largest workload measured by NASA-TLX, say 53.2 and 40.6 for the first and the second measurement. The S-curve task could need more cognitive process due to the lower learning rate and the larger workload. Therefore, more experienced trainees were able to finish this task in shorter time.

On the other hand, older the trainee, smaller the learning rate for up/down hill task ($r = -0.393$, $p < 0.05$). The learning rate was 0.9162 for up/down hill, the largest among all five tasks. The workload measured by NASA-TLX for this task was 41.5 and 33.6 for the first and the second measurement, respectively. This workload among five tasks was in the middle place.

5 Conclusions

Practice made the driving time gradually shorter, but this improvement became less and less as the times of practice increased. The learning curve associated with each driving task was well fitted according to the Wright's model. This implied that these learning curves of trunk driving tasks were able to be fitted by power function very well. Specifically, the learning rate was 0.9162 for up/down-hill, 0.8912 for 3-point-turn, 0.8802 for parallel-parking, 0.8736 for reversing-into-garage, and 0.8698 for S-curve. Finally, the significantly lower second measurement of NASA-TLX implied that practice was also able to reduce the overall workloads.

References

Dar-el EM, Ayas K, Gilad I (1995) A dual-phase model for the individual learning process in industrial tasks. IIE Trans 27(3):265–271

Konz SA, Johnson S (2000) Work design: industrial ergonomics: Holcomb Hathaway

Reid SA, Mirka GA (2007) Learning curve analysis of a patient lift-assist device. Appl Ergon 38(6):765–771

Sahami S, Sayed T (2010) Insight into steering adaptation patterns in a driving simulator. Transp Res Rec J Trans Res Board 2185(1):33–39

Sahami S, Sayed T (2013) How drivers adapt to drive in driving simulator, and what is the impact of practice scenario on the research? Transp Res Part F Traffic Psychol Behav 16:41–52

Sahami S, Jenkins JM, Sayed T (2009) Methodology to analyze adaptation in driving simulators. Transp Res Rec J Transp Res Board 2138(1):94–101

Prognostics Based Design for Reliability Technique for Electronic Product Design

Yingche Chien, Yu-Xiu Huang and James Yu-Che Wang

Abstract Techniques that can effectively reduce failure rate, control life span and predict product life are highly expected in modern electronic products. Design for Reliability (DFR) technique introduced in this study, applies various robust design and system reliability modeling methods to evaluate whether reliability target can be met in product development stage. However, DFR technique often faces challenges mainly on insufficient accuracy of system reliability prediction. Prognostics and Health Management (PHM) technique applies failure precursors and their impact on product real failure to improve accuracy of reliability prediction in design phase. This study integrates DFR and PHM techniques for reliability prediction. Hard disk drive is selected as a case study for PHM application in design phase. A failure precursor of drive is selected and its statistical distribution of time-to-failure-precursor is established. Applying conditional reliability and residual mean-time-to-failure, remaining useful life (RUL) estimation is proposed. The prognostic based DFR developed in this study plays a key role in predicting product reliability during development stage as well as catastrophic failure prevention in maintenance stage.

Keywords Design for reliability · Prognostics and health management · Failure precursor · Remaining useful life

---

Y. Chien (✉) · Y.-X. Huang
Chung Yuan Christian University, Chung-Li, Taiwan, Republic of China
e-mail: cyc@cycu.edu.tw

Y.-X. Huang
e-mail: moumouhh@gmail.com

J. Y.-C. Wang
HP Taiwan, Taipei, Taiwan, Republic of China
e-mail: jimswung@gmail.com

1 Introduction

Design for Reliability techniques have been proved by many companies worldwide as a set of effective tools to achieve desired product functions and reliability objectives with low cost and short development cycle time.

DFR process offers a series of proactive actions on how to define a market winning reliability target, allocation of reliability target, prediction of reliability at early design phase by reliability modeling, critical components selection and qualification, derating review based on operating and environmental stresses, design Environmental Stress Tests (EST) for design verification, warranty and spares planning, integrated hardware and software reliability modeling for interface design and design optimization, etc.

With clear hand shake at each development process gates, implementation results of DFR process can be assured and design faults can be detected and removed to prevent expensive field failures. Enhanced DFR technique focuses on application of PHM methods for failure precursors identification and RUL prediction during product development phase. Thus, a fast and product oriented reliability prediction can be generated in development phase. The result can be compared with prediction results by parts count method generated early in development phase for prediction accuracy improvement. Investigation causes of failure precursors leads to find root causes of real failures early in design phase and improves product reliability.

2 Literature Review

In 1990, AT&T Bell Laboratories published a book, "Reliability by Design" which explores the approaches on reliability assurance in product life cycle and marked the beginning of new era of Design-for-Reliability. The reliability prediction methodology and system reliability model are two key techniques for the DFR. In 2006, *Telcordia Technologies* published the SR-332, Issue 2 document, "Reliability Prediction Procedure for Electronic Equipment". This document and other component failure rate standards based on company proprietary enable parts-count method for system reliability prediction in design phase.

Around year 2000, Prognostics and Health Management technology emerged as a promising approach for reliability prediction and logistic management. The technology is promoted by Dr. Michael Pecht and his research team, the Center for Advanced Life Cycle Engineering (CALCE), University of Maryland, USA.

PHM plays as a new paradigm in the fields of reliability, maintenance, and logistics. According to the introduction of the technology by CALCE, prognostics is a process of predicting a product's remaining useful life under expected future use by assessing the degradation or deviation of current health from expected state of health. PHM enables users, maintainers, and manufacturers to dynamically understand the state of product health and thereby helps them make informed and timely life cycle management decisions.

3 DFR and PHM Methodologies

3.1 Prognostics Based DFR Approach

The DFR process framework is built for effective reliability management. The framework can be represented by six-step process: (1) Identification of reliability requirements based on marketing and competition, (2) Product functionality design, (3) Component specification verification based on reliability requirements, (4) Design verification on schematic level for comparison with reliability objectives, (5) Design validation on system reliability by real product testing, and (6) Lessons learned from failures for design guidelines update and next generation product reliability improvement.

The DFR process consists inter-process gates for control and appraisal on project reliability status. For each gate, a development project team is responsible for complying to company's gate pass requirements and making pass/fail decision of the project as it proceed to each gate.

Prognostics based design-for-reliability starts from identifying failure precursor during product development phase. Test units with failure precursor to characterize their statistical distribution including mean-time-to-failure-precursor. Establish relationship of return rates between units with and without failure precursor for estimating MTTF of real failure after reveal of failure precursor. Prediction of new product reliability based on corresponding failure mode for the failure precursor can use sum of mean-time-to-failure-precursor and MTTF of real failure after precursor presented.

RUL of product can be estimated after the product health monitor on failure precursor during product life cycle using conditional reliability and residual MTTF based on distribution of failure precursor occurrence time. Result of product health monitor can be applied for preventive action on catastrophic failure.

3.2 Remaining Useful Life Estimate

Time-To-Failure-Precursor of a product can be analyzed for fitting some probability distribution. We define conditional probability as the probability of a product for successful operating t hours with no failure precursor detected after passing product health monitor for the precursor at time T_0. The conditional reliability can be expressed as,

$$R(T_0 \to T_0 + t | T_0) = R(T_0 + t | T_0) = \frac{R(T_0 + t)}{R(T_0)} = e^{-\int_{T_0}^{T_0+t} \lambda(t') dt'} \quad (1)$$

Furthermore, the first derivative of the conditional reliability (Eq. 2) shows that the conditional reliability is a decreasing function of T_0, if the failure rate is an increasing function. Similarly, if the failure rate is a decreasing function, the conditional reliability will be an increasing function of T_0. (Ebeling 2010)

$$\frac{dR(T_0 + t | T_0)}{dT_0} = R(T_0 + t | T_0)[\lambda(T_0) - \lambda(T_0 + t)] \quad (2)$$

$$MTTF(T_0) = \int_0^\infty R(T_0 + t | T_0) dt = \frac{1}{R(T_0)} \int_{T_0}^\infty R(t') dt' \quad (3)$$

where $t' = T_0 + t$

The residual *MTTF* denoted as *MTTF*(T_0), can be defined as Mean-Time-To-Failure-Precursor after passing product health monitor on failure precursor at time T_0. It can be derived from the conditional reliability function in Eq. (1). Calculation of residual *MTTF* is shown in Eq. (3). It is the expected time to reveal failure precursor after last product health check with no failure precursor found. The *MTTF*(T_0) is dependent on time T_0 and can be treated as remaining useful life of the product on failure precursor occurrence.

Our next step is to develop expected time for real failure occurrence given the product found with failure precursor. This expected time for real failure is denoted as $MTTF_2$ and will be used for taking preventive action to avoid unalarmed failure.

RUL of product after health monitor at time T_0 with no failure precursor found
$= MTTF(T_0) + MTTF_2$

$$(4)$$

At time $T_0 = 0$, product is ready for release, the *RUL* in Eq. (4) can be regarded as *MTTF* of product based on the failure precursor.

3.3 Relationship of MTTF of Product Carry No Failure Precursor and MTTF of Product Carry Failure Precursor

Considering to conduct a reliability test on two groups, each group consists of same product type. Units in the first group units have been checked before the test and found no failure precursor. Units in the second group have failure precursor before the test. It is reasonable to assume that the second group has K times higher return rate in h hours test period compared with that from the first group. For simplification, we assume that both groups are in their useful life period. The following relation hold.

$$1 - R_2(h) = K[1 - R_1(h)], \quad \text{where } h \text{ is test hours} \qquad (5)$$

$R_1(h) = $ reliability of product with no failure precursor $= e^{-\lambda_1 h}$

$R_2(h) = $ reliability of the product with failure precursor $= e^{-\lambda_2 h}$

where

$\lambda_1 = $ failure rate of product with no failure precursor

$\lambda_2 = $ failure rate of product with failure precursor

Solving Eq. (5) in terms of λ_1 and λ_2,

$$\lambda_2 = (-1/h) \, ln[1 - K(1 - e^{-\lambda_1 h})] \qquad (6)$$

Since $MTTF_1 = 1/\lambda_1$ and $MTTF_2 = 1/\lambda_2$,
the ratio of $MTTF_1$ and $MTTF_2$ the Accelerating Factor can be derived from Eq. (6).

$$\text{Accelerating Factor} = \frac{MTTF_1}{MTTF_2} = \frac{\lambda_2}{\lambda_1} = \frac{-1}{\lambda_1 h} \ln[1 - K(1 - e^{-\lambda_1 h})] \qquad (7)$$

It can be shown numerically, ratio of $MTTF_1/MTTF_2$ in Eq. (7) is greater than 1 with the scale depending on the value of K.

4 Case Study

A PHM study uses data from a group of hard disk drives (HDD) in desk top computers located in a computer lab. According to various studies on HDD, failure precursors of drives may include scan error, degrading data transmission rate, read/

write error, high internal temperature inside a drive, etc. Each of the precursor corresponds to their failure mode. The scan error is selected as the failure precursor for this study for its important role in HDD failure and its easy to identify from a built-in personal computer internal monitor program, Self-Monitoring Analysis and Reporting Technology (SMART). Scan error can be caused by bad sector(s) on hard disk or malfunction of magnetic head. For simplification, this research assumes that bad sector is a sole contributor to scan error. Bad sector refers to damage on some sectors in hard disk.

(A) Estimate of residual MTTF based on a failure precursor

Based on the investigation of 38 personal computers in a lab., time-to-failure-precursor of four (4) HDDs with scan error were obtained from the SMART monitoring record. The censored test hour records of the rest of 34 HDDs are also available.

Applying median rank method for regression analysis on the test records and using reliability software program, ReliaSoft, show that the scan error follows a two parameter Weibull distribution with shape parameter, β and scale parameter, η as

$$\beta = 1.62$$
$$\eta = 7418.5 \text{ h}$$
$$\gamma = 0$$

In general, random variable, T distributed as a 3—parameter Weibull,

$$f(T) = \frac{\beta}{\eta}(\frac{T-\gamma}{\eta})^{\beta-1} e^{-(\frac{T-\gamma}{\eta})^\beta}$$

$$R(T) = e^{-(\frac{T-\gamma}{\eta})^\beta} \qquad (8)$$

$$MTTF = \gamma + \eta\Gamma(\frac{1}{\beta}+1) \qquad (9)$$

From Eq. (3),

$$MTTF(T_0) = \frac{1}{R(T_0)} \int_{T_0}^{\infty} R(t')dt'$$

$$= \frac{1}{R(T_0)} \left[\int_0^{\infty} R(t')dt' - \int_0^{T_0} R(t')dt'\right] = \frac{1}{R(T_0)}\left[MTTF - \int_0^{T_0} e^{-(\frac{T-\gamma}{\eta})^\beta} dT\right] \qquad (10)$$

Assume that a drive has been scan tested by the SMART program at time T_0 of 0 h (new product), 2190 h (after 3 months use), 4380 h (after 6 months use) and 8760 h (after one year use) respectively, the residual MTTF can be calculated using Eq. (10) where the value of *MTTF* can be found from Eq. (9) and $\Gamma((1/\beta) + 1)$ is a Gamma function.

$$MTTF(0 \text{ h}) = MTTF = \frac{1}{R(0)} \left[7419 \cdot \Gamma\left(\frac{1}{1.62} + 1\right) \right] = 6644 \text{ h}$$

Using numerical method to find the integral value,

$$MTTF(2190 \text{ h}) = \frac{1}{R(2190)} \left[6644 - \int_0^{2190} e^{-\left(\frac{T}{7419}\right)^{1.62}} dT \right] = 5244 \text{ h}$$

Similarly,

$$MTTF(4380 \text{ h}) = 4429 \text{ h}$$

(B) Estimate of *MTTF* based real failure after failure precursor diagnosed

Pinheiro et al. (2007) investigated more than 100,000 units of HDD in servers installed at Google for impact of scan error on HDD reliability. The result showed that drives having one or more scan errors are 39 times more likely to fail within 60 days than drives with no scan errors. Field test result also showed HDD annual return rate is 6 % approximately.

Calculation shows that $MTTF_2$, the MTTF for drives with failure precursor (scan error) is 49 times shorter than $MTTF_1$, the MTTF for drive with no failure precursor. Notice that the failure defined here is real failure modes based on product specifications.

$$MTTF_1 = \frac{1}{\lambda_1} = 141,583 \text{ h} = 16.1 \text{ years}$$

$$MTTF_2 = \frac{1}{\lambda_2} = 2,889 \text{ h} = 0.33 \text{ years}$$

(C) Remaining Useful Life (RUL) based on PHM monitoring result

From eq. (4),

RUL of HDD based on scan error
$= MTTF(0) + MTTF_2 = 6644 + 2889 = 9533 \text{ h}$

The RUL based on scan error can be converted to failure rate of scan error for HDD. If more major failure modes in addition to scan error can be monitored and assumes that their occurrences are mutually exclusive, a total HDD failure rate

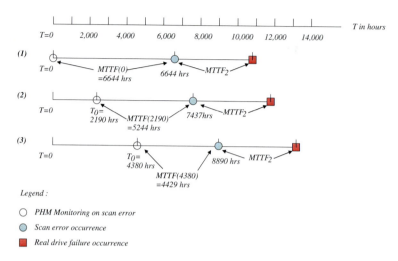

Fig. 1 Residual MTTF of failure precursor and MTTF of real failure for units carrying failure precursor

based on failure precursors can be obtained by summing up all failure rates of each failure mode.

The $MTTF_2$ can be regarded as an action window for taking necessary arrangements such as spare preparation and/or data duplication to prevent impact of unexpected real failure caused by the drive.

Figure 1 shows the relation for product health monitoring at age of (1) Start operation, (2) 3 months and (3) 6 months. It shows that the expected time for scan error occurrence is decreasing as the health monitoring time is increased (product getting older). The average time that a real HDD failure will occur after scan error revealed is $MTTF_2$.

5 Conclusion

This study provides method of application of PHM technique in reliability prediction during product development phase. Using failure precursor to predict product reliability establishes a product-oriented approach as compared with parts count method using generic approach. The failure precursor approach is also a sensitive way to detect product health status which leads to predict real failure occurrence.

Further study on establishing relation between RUL and damaging level for each failure mode can provide more precise basis for predicting product MTTF under different operational and environmental stresses.

Acknowledgments This work is supported by the National Science Council of Taiwan. The authors are also gratefully acknowledge the review from the reviewers.

References

Ebeling CE (2010) An introduction to reliability and maintainability engineering, 2nd edn. Waveland Press, Canada

Pinheiro E, Weber W-D, Barroso LA (2007) Failure Trends in a Large Disk Drive Population, 5th USENIX Conference on File and Storage Technologies, pp 17–29

A Case Study on Optimal Maintenance Interval and Spare Part Inventory Based on Reliability

Nani Kurniati, Ruey-Huei Yeh and Haridinuto

Abstract An engine fuel supply subsystem for particular type of aircraft plays an important role in providing, controlling, and distributing the fuel during engine operation. Failure on this subsystem will affect the readiness of the aircraft for operations. Therefore, for system experienced an aging characteristic or wear-out period, determining of the optimal preventive maintenance and optimal preventive replacement interval by considering the total cost of maintenance per unit time is important. In order to support the replacement activity, available number of spare parts required and must be well controlled to avoid either over stock or shortage. In this study, we attempt to determine the preventive maintenance interval and preventive replacement interval and its required inventory spare parts as well. We deliberate the cost structure, failure field data, and the reliability along the designing and managing the maintenance activity. We also examine the implication of the designed maintenance interval on reliability and availability of the system.

Keywords Reliability · Ware-out period · Maintenance interval · Replacement interval · Spare part inventory

---

N. Kurniati (✉) · R.-H. Yeh
National Taiwan University of Science and Technology, Taipei, Taiwan
e-mail: d10001804@mail.ntust.edu.tw; nanikur@ie.its.ac.id

R.-H. Yeh
e-mail: rhyeh@mail.ntust.edu.tw

N. Kurniati
Institute of Technology Sepuluh Nopember (ITS), Surabaya, Indonesia

Haridinuto
Department of Personnel Administration, Indonesian National Air Force,
Jakarta, Indonesia
e-mail: haridinuto@gmail.com

1 Introduction

Generally, there are several main systems in the aircraft include engine, hydraulics, air frame, and propeller. Among those systems, engine is the most important system responsible for operations as well as the main parameter determine whether an aircraft worth to fly or delayed due to failure. Inside the engine, there is important subsystem, called engine fuel supply, responsible to provide, control, and distribute the fuel to the engine during operations. The engine fuel supply is critical subsystem but has failure prone and needs to get more attention than the others.

For system experienced an aging characteristic or wear-out period as well as has potentially disastrous consequences of failure would suggest affording preventive maintenance (PM) rather than corrective maintenance (CM). In a particular aircraft type operated for more than 20 years, approaching to the end of life or wear-out periods, corrective action only is not an economical choice since the frequency of failure is high due to aging. It may also incurs cost of resulting in failure of the neighboring device, cost for more extensive repairs and maintenance overtime cost, cost for a large material inventory of repair parts required. These are costs we could minimize under different maintenance strategy rather than corrective action.

The PM's objective is to increase the reliability of the system over the long term by staffing off the aging effect (Lewis 1991). By simply expending the necessary resources to conduct maintenance activities intended by the equipment designer, equipment life is extended and its reliability is increased.

Maintenance activity can't be avoided from cost consequences (Ebeling 1997). If the maintenance activity done frequently with shorter interval period of time, the maintenance cost will increase however the cost due to failure will decrease. Conversely, the maintenance cost will decrease if we reduce the number of maintenance activities by a longer PM interval. Nevertheless, the cost of breakdowns may extremely increase for more complex failures. Therefore, minimizing the total cost for a certain period or the total cost per unit time is more preferable as maintenance decision basis.

Many studies on scheduling maintenance and spare-part problems have been done. Sherif (1982) studied a state-of-the-art review of the literature related to optimal inspection and maintenance schedules of failing systems. Huiskonen (2001) addressed the question of managing spare part logistics by discussing the basic principles affecting the strategic choices and related choices in different areas.

In this study, we attempt to determine the preventive maintenance interval and preventive replacement interval and its required inventory spare part as well. We deliberate the cost structure, failure field data, and the reliability along with the designing and managing the maintenance activity.

A Case Study on Optimal Maintenance Interval

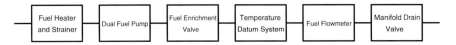

Fig. 1 Reliability block diagram for component of the engine fuel supply

2 Case study

In a particular aircraft, the engine fuel supply consists of several main components as provide in the diagram block. Examining the failure field data extensively for the following calculation, under following assumptions:

- Consider the failure time only, disregard the failure cause and consequences.
- The failure of items are independent and identical
- All items are repairable and replaceable (there is no discardable item)
- No delay due to resources, maintenance crew or equipment (Fig. 1)

Based on the collected failure data (time to failure, TTF) for several years, we determine the failure distribution of each component and consider the Weibull distribution of failure. Based on this failure distribution, we may determine the reliability function, the availability function, and further calculation to determine the preventive maintenance interval and preventive replacement. The failure distribution for consecutive components as block diagram provides are Weibull (4.42; 378.12), Weibull (4.12; 424.90), Weibull (4.81; 316.90), Weibull (3.46; 358.01), Weibull (2.95; 175.81), and Weibull (3.69; 348.13).

3 Determining the Preventive Maintenance Interval

The feature of the components considering maintenance time and numbers of maintenance labor both for corrective and preventive action. The maintenance labor cost is IDR 17,518 per hour person, and the cost due to failure or the consequence cost for 1 h of maintenance action refers to the operation lost for one flying hour with cost IDR 50,000,000 per hour. So we can get the maintenance cost by multiplying the summation of total labor cost and the cost due to failure with the time required to perform the maintenance action. The CM cost (C_f) is larger than the PM cost (C_p) since the time required to perform CM longer than for PM.

Define one cycle (t_p) is the length of interval between two consecutive PM. During time t_p, we may accommodate any corrective maintenance occur and one preventive maintenance action. The optimal PM interval is the length of interval t_p that minimize the total cost for one cycle. The total cost $C(t_p)$ for one cycle t_p given as (Jardine 1973):

Table 1 Feature and the length of PM interval for component of the engine fuel supply

Component	Corrective action		Preventive action		t_p (flying hour)
	CM time (h)	CM labor (prs)	PM time (h)	PM labor (prs)	
Temperature datum system	24	3	4	3	161
Fuel heater and strainer	8	2	1	2	177
Dual fuel pump	24	2	2	2	175
Fuel Enrichment Valve	4	2	1	2	177
Fuel flow meter	8	2	3	2	175
Manifold drain valve	4	2	1	2	178

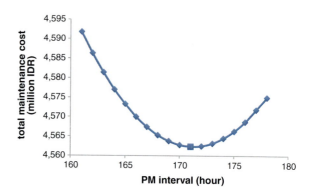

Fig. 2 Total maintenance cost profile under certain PM interval range

$$C(t_p) = \frac{\text{expected failure cost}}{\text{expected cycle length}} + \frac{\text{preventive cost}}{\text{cycle length}} = \frac{C_f(1 - R(t_p))}{\int_0^{t_p} R(t)dt} + \frac{C_p}{t_p} \quad (1)$$

where C_f and C_p are maintenance cost for failure or corrective action and preventive action, respectively. $R(t_p)$ is the reliability function at t_p (Table 1).

Adjustment is needed since the maintenance action for all components may take at the same flying hour. Therefore, we evaluate the total maintenance cost of all components along cycle length range from 161 to 178 flying hours which results in the lowest cost. The resulting PM interval for each component is 171 flying hours.

The comparison between this PM interval and the existing maintenance interval (200 flying hour, Koharmatau 2006) in terms of reliability (provide at Fig. 2) and availability shows that even though the calculated maintenance interval shorter than the existing, we can keep the reliability value better for every components as shown in the Fig. 3. The availability value also have identical pattern as well (Table 2).

A Case Study on Optimal Maintenance Interval

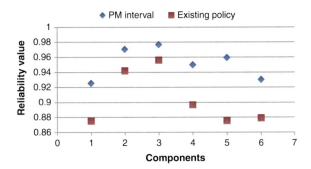

Fig. 3 The comparison of reliability value between the PM interval and the existing interval

Table 2 Analyzing the reliability importance of component of the engine fuel supply

Component	Method			
	Birbaum's measure	Criticality importance	Vessely fussel's	Improvement potential
Temperature datum system	0.6218	0.1701	0.2736	0.0775
Fuel heater and strainer	0.5777	0.0735	0.1271	0.0335
Dual fuel pump	0.5692	0.0547	0.0960	0.0249
Fuel enrichment valve	0.6070	0.1376	0.2267	0.0627
Fuel flow meter	0.6216	0.1697	0.2729	0.0773
Manifold drain valve	0.6191	0.1642	0.2652	0.0748

4 Determining the Preventive Replacement Interval

First, we focus on finding the critical component of the engine fuel supply based on the reliability value. There are several ways to analyze the reliability importance (Heley and Kumamoto Henly and Kumamoto 1992). As shown in Table 3, the most critical component of the engine fuel supply is temperature datum system. Its function is controlling the temperature of fuel before consume by the engine.

The collected time to failure data for several years is taken from maintenance component log book. The calculation of maintenance cost for sub-component is identical to that of maintenance cost for component except that account is taken of the spare part cost. The replacement policy is to perform a preventive replacement once the equipment has reached a specified age t_r plus failure replacement when necessary.

The optimal preventive replacement interval is the length of interval t_r that minimize the total expected replacement cost per unit time, denoted $C(t_r)$, is (Jardine 1973):

$$C(t_r) = \frac{C_p R(t_r) + C_f(1 - R(t_r))}{(t_r + T_p)R(t_r) + \int_0^{t_r} tf(t)dt + T_f(1 - R(t_r))} \quad (2)$$

Table 3 Feature and the length of PR interval for each sub-component

Sub-component	Failure distribution $f(t)$	Time T_p, T_f(h)	Labor (prs)	t_r (flying hour)
Fuel control unit	Weibull (9.49;6072.29)	8	3	4333
TD valve	Weibull (1.75;1615.01)	1	3	4398
Fuel nozzle	Weibull (2.08;2830.87)	3	3	4310
Thermocouple	Weibull (3.39;7422.16)	2	3	3988
TD amplifier	Weibull (1.39;2847.09)	1	2	4362

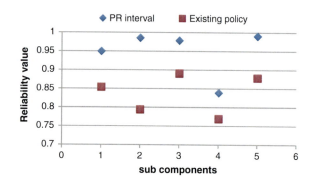

Fig. 4 The comparison of reliability value between the PR interval and the existing interval

where C_f and C_p are replacement costs for corrective and preventive replacement, respectively. T_f and T_p are times required for corrective replacement and preventive replacement, respectively.

In order to get beneficial meaning in operation, we attempt to find the multiple maintenance intervals closest to the replacement interval. The replacement interval, when we may replace all sub-components, is 4,446 flying hours. The comparison between this PR interval and the existing replacement interval (5,000 flying hour) shows the PR interval will keep the reliability better for every sub-component and availability as well (Fig. 4).

5 Determining the Spare Part Inventory Policy

In order to support the replacement plan, managing the inventory of the spare part is needed. Let s be the replacement interval, the component replaced by a new one after being used for s period or when it fails (Alkaff 1992). The mean time between replacements (MTBR) is $MTBR = \int_0^s R(t)dt$. The average number of spare parts required (N) both due to failure (N_f) and preventive replacement (N_p) during t is $N = t/MTBR$. The proportion of un-failed item after s period is $R(s)$, therefore $N_p = R(s)N$ and $N_f = (1 - R(s))N$ The average number of spare parts required is $N = [tR(s) + t(1 - R(s))] / \int_0^s R(t)dt$. Safety stock (ss) during interval s and order

A Case Study on Optimal Maintenance Interval

Table 4 Inventory policy for sub-component under different stock out risks

Sub-component	Inventory policy for $\alpha = 0\%$				Inventory policy for $\alpha = 5\%$			
	N	N_{max}	Ss	ROP	N	N_{max}	ss	ROP
Fuel control unit	2	4	3	3	2	2	1	1
TD valve	2	6	5	5	2	3	2	2
Fuel nozzle	7	10	4	4	7	6	0	0
Thermocouple	22	24	3	3	22	17	0	0
TD amplifier	2	5	4	5	2	2	1	2

lead time (L) is $ss = N_{max} - N$, where N_{max} is the maximum number of spare parts required that can be examined as follows. The probability of n times replacement at time t is defined as:

$$P(N(t) = n) = P_n(t) = (H^n(t)/n!)e^{-H(t)} \quad (3)$$

where $H^n(t) = \int_0^t \lambda(t)dt$ is the cumulative hazard rate. Under stock out risk α, N_{max} is the cumulative $P_n(t)$ for $n = 0$. N_{max} that can be found by iterated calculation from $\sum_{n=0}^{N_{max}} P_n(t) \geq 1 - \alpha$. Reorder point (ROP) is the number of stocks in which the reorder must be taken. The reorder point (ROP) defined as:

$$ROP = ss + D_L \quad (4)$$

where $D_L = L/MTBR$. Finally, the number of spare parts must order (OQ) which is defined as $OQ = N_{max} - ROP$. The inventory policies of the sub-component of temperature datum systems either for stock out risk at $\alpha = 0\%$ and $\alpha = 5\%$ are given in Table 4.

6 Conclusions

For a system experienced an aging characteristic or wear-out period, the evaluation for the interval of preventive maintenance or preventive replacement is necessary by examining the recent failure (field) data and its reliability value afterward. We determine the preventive maintenance interval based on minimizing the total expected cost per unit time by considering both preventive and corrective action cost, labor cost, and spare parts if needed. In our case, the comparison between the preventive maintenance interval and the existing interval in terms of reliability and availability shows that even though the calculated interval shorter than the existing, we can keep the reliability and availability better for every item.

To support the replacement action, managing the required spare part is important by considering both reliability and lead time. Providing the order quantity, safety stock, and the reorder point may keep the spare part available for both corrective and preventive replacement.

References

Alkaff A (1992) Teknik Keandalan Sistem. ITS press, Indonesia

Ebeling CE (1997) An introduction to reliability and maintainability engineering. University of Daytona. The Mc Graw Hill, USA

Henly EJ, Kumamoto H (1992) Probabilistic risk assessment reliability engineering design and analysis. IEEE Press, New York

Huiskonen J (2001) Maintenance spare part logistics: special characteristics and strategic choices. Intern J Prod Econ 71:125–133

Jardine AKS (1973) Maintenance, replacement, and reliability. Department of engineering production. University of Birmingham

Koharmatau (2006) Buku Petunjuk Pedoman Pemeliharaan Alutsista (BP3A) Pesawat C-130 Series. TNIAU, Indonesia

Lewis EE (1991) Introduction to reliability engineering. Department of Mechanical and Nulear Engineering Nortwestern University. John Wiley & Sons, USA

Sherif YS (1982) Reliability analysis: optimal inspection and maintenance schedules of failing systems. Microelectron Reliab 22:59–115

Developing Decision Models with Varying Machine Ratios in a Semiconductor Company

Rex Aurelius C. Robielos

Abstract In a semiconductor manufacturing, operators are usually faced with simultaneous activities and therefore it is a requirement that they should have adequate decision making skills. While their main responsibility is to ensure that the machines are continuously running, they are also expected to perform other activities during their assigned working hours. For cases of machine breakdown, one methodology being used is the recognition-primed decision model which is a pattern recognition problem diagnosis procedure. This methodology, however, is appropriate only for single machine breakdown. Thus, a revised decision model is developed to incorporate multiple decision points.

Keywords Recognition-primed decision model · Machine breakdown · Multiple decision points

1 Introduction

In a semiconductor manufacturing environment wherein the system is complex, operators may have difficulties in maintaining a concept of system performance and fitting that concept to the human role within the system. Since most of the activities in the manufacturing floor are assigned to operators, then it is imperative that operators should have the necessary decision making skills. In most cases, simultaneous tasks do occur and operator should have the capability to decide which task is more important and likewise be able to respond to this task appropriately.

R. A. C. Robielos (✉)
Mapua Institute of Technology, Manila, Philippines
e-mail: racrobielos@mapua.edu.ph

One methodology being used in a semiconductor industry is the recognition-primed decision model, a problem diagnosis procedure that is usually being done through pattern recognition. One such application of this model is during machine breakdown when there is a need to diagnose the cause of the breakdown. However, the model is appropriate only for single machine breakdown.

This study tries to analyze the different decision points of the operator when simultaneous activities occur due to varying machine ratios. The occurrence of simultaneous machine breakdown would also change the decision making process of the operator. Thus, a revised recognition-primed decision model is developed to include multiple decision points.

2 Man–Machine Task Assignments

Aside from ensuring that the machines are continuously running, operators are also expected to perform various tasks in the manufacturing floor. These activities are the Pre-Post activities, speed delay activities, internal activities, unplanned delay activities and other activities.

(1) **Pre-Post activities**

These are the activities being done by the operator before and after lot processing. There are currently 15 pre and post activities which are grouped according to their Promis Status (database nomenclature system) (Table 1).

Table 1 List of pre and post activities

Promis status	Pre and post activities
NVPACK	Count reconciliation (MIPS)
	Counting of finished lot/new lot
	Counting/checking of endorsed lot (from previous crew)
	Checking of summary versus actual
	Preparing for new lot/completion of fix's
	Printing of barcode and summary
	Labeling of tested units
PDCLEAN	Housekeeping
	For search of missing units
NVSET	Running SUV for new lot (same setup Type 1)
NVRESCR	Rescreen of rejects
NVOQA	Running ILS
	Running Rescr of failed ILS
	For endorsement to technician if still failed at rescr
PEDISPO	For TME's final dispo

(2) Speed Delay Activities

These are stoppages encountered during test processing of a lot that falls between the duration of greater than 30 s up to 10 min. These stoppages are commonly caused by 2 reasons:

(a) machine needing assistance because of jams or errors
(b) machine requiring operational assistance

(3) Internal Activities

These are activities being done by the operator while the machine is still running. There are 11 identified internal activities under the INUSE Promis Status (Table 2).

(4) Unplanned Delay Activities

These are the initial interventions provided by the operator before passing the job to the technician. Usually, these activities would require more than 10 min. Currently, there are 10 possible interventions that are being done by the operator (Table 3).

(5) Other activities

These are the activities that may be assigned by the supervisor from time to time (Table 4).

The decision making activities of operator becomes complex the moment that the machine assignment increases to more than one. Machine intensive manufacturing companies would always go into a higher machine ratio since it would bring the manpower expense at the minimum. But when more machines are assigned to operators, we are exposing them in situations that would test their decision making competencies. Thus, the intent of this study is to look at the major decision points and come up with the framework that will guide and help the

Table 2 List of internal activities

Promis status	Internal activities
INUSE	Preparing of untested units for load
	Loading of untested units for Test at the handler
	Unloading of tested units
	Stoppering at tested units
	Bundling of tested units
	Promis transaction (Promis mail viewing of other tester change status)
	Loading of clear tubes at the handler for tested good units
	Make clear lubes for tested good units
	Unloading of tested rejects
	Get stopper and tubes at the rack
	Unloading of used tubes at the handler (untested)

Table 3 List of unplanned delay activities

Promis status	Unplanned delay activities (Events encountered before UDWAIT)
UDHANDL	Handler Jamming
	3 or more time cleaning of jam
UNCNTCTR	Over rejection
UDSYS	Stop the handler
UDEQPT	Check device orientation
UDHANDL	
NVSET/UDSETUP	Failed SUV/SSUV
	Check device orientation
	Run another set of SUV
NVSET	System hang-up
UDSOFT	
UDHANDL	Handler hang-up
NVSET	Set-up of new device
NVSET/UDSETUP	Failed calibration
	Tester reset
NVSET	Unable to load program
	Tester reset
PDCLEAN	Missing unit
	For search of missing units
NVOQA	Failed ILS

Table 4 List of other activities

Promis status	Other activities
INUSE	Operator to get spare of fixtures (contactor, board, suv, etc.) if setup failed
	Operator to get summary report at supervisor room/bay if system encounter hang-ups
	Operator to get and return tubes at tube management if it is her turn

operator. Likewise, instituting a decision framework will definitely redound to the other bottom lines.

As illustrated in Fig. 1, there are 10 possible downtimes that may occur in any given point in time. Since the occurrence and the type of downtimes are random, operator will just have to wait for the machine to stop. Once it stops, operator has to diagnose the problem and then make the necessary course of action. This procedure is actually discussed in the paper by Patterson et al. (2009) wherein he used the recognition-primed decision making model. According to him, this model is composed of three components: one for matching, one for diagnosing, and one for simulating a course of action. Figure 2 depicts the matching and diagnosing components; the former is shown by the box labeled "pattern recognition" and the latter by the box labeled "clarify/diagnose (story building)."

In this model, an individual with expertise identifies a current problem situation as typical and familiar based upon a composite situation stored in memory and,

Developing Decision Models

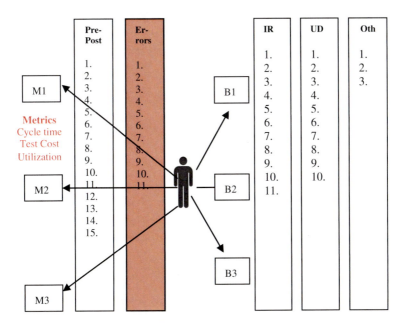

Fig. 1 Man-machine task assignments

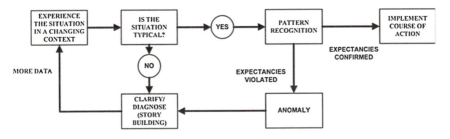

Fig. 2 Diagram of the recognition-primed decision model

with subsequent expectations confirmed, initiates an appropriate course of action, which is typically the first one considered. Klein (1997) also proposed that an individual may mentally simulate a course of action before actually implementing it. If the situation is unfamiliar, however, or if subsequent expectations are violated, then the individual will choose to diagnose and clarify the situation further, which may include story building. This section of the model is usually the case where the problem is beyond the capability of the operator. Thus, this is passed to the Line Technician for possible resolution since the problem is more complicated (Table 5).

If an individual experiences a situation as typical and familiar and recognizes the pattern, he or she decides to implement a course of action if expectancies are confirmed. If the situation is atypical and unfamiliar, or if expectancies are

Table 5 List of possible machine errors

Possible machine errors	Time to repair
Handler jamming	1–3 min
Over rejection	30 s
Failed SUV/SSUV	1 min
System hang-up	10–15 s
Handler hang-up	10–15 s
Set-up of new device	5–10 s
Failed Calibration	2–4 min
Unable to load program	3–5 min
Missing units	5–10 min
Failed ILS	1 min

violated, then the individual decides to clarify and diagnose the situation, which can involve story building.

Assuming that the man–machine ratio is one is to one, the priority of the operator is always the uptime of the machine. Therefore, she has to set aside her other activities whenever a machine breaks down. What complicate the tasks and decision making activity of the operator is when the assigned number of machines increases to more than one.

To illustrate, let us analyze the situation if the man–machine ratio increases to 1 is to 3. Let us start our analysis when the 3 machines are not yet running. Before the units are tested, the operator has to do the pre-activities such as count reconciliation, counting of new lot, preparing for new lot/completion of fixture, etc. After set-up, operator has to ensure that the machine is continuously running. After which, she has to start the pre-activities for Machine 2. There is a possibility that while she is doing the pre-activities for Machine 2, Machine 1 will stop. This is the time that the operator needs to decide which task is more important. Then another set of pre-activities has to be done for machine 3. While operator is doing pre-activities for Machine 3, there is a possibility that Machine 1 and Machine 2 will break down. There are actually many possible combinations that may occur considering that the downtimes of machine are random. Below are just some of the possible scenarios that may occur.

Case 1 Breakdown of Machines

    1. All 3 machines simultaneously broke down.
    2. 2 Machines broke down, 1 Machine is running

Case 2 Combination of Breakdown and Other Activities

    3. Machine 1 broke down, Machines 2 and 3 are running, Operator is doing internal activities
    4. Machines 1 and 2 broke down, Machine 3 is running, Operator is doing internal activities

5. Machines 1, 2 and 3 broke down,
   Operator is doing internal activities
6. Machine 1 broke down, Machines 2 and 3 are running,
   Operator is doing pre-post activities
7. Machines 1 and 2 broke down, Machine 3 is running,
   Operator is doing pre-post activities
8. Machines 1, 2 and 3 broke down,
   Operator is doing pre-post activities
9. Machine 1 broke down, Machines 2 and 3 are running,
   Operator is doing unplanned delay activities
10. Machines 1 and 2 broke down, Machine 3 is running,
    Operator is doing unplanned delay activities
11. Machines 1, 2 and 3 broke down,
    Operator is doing unplanned delay activities
12. Machine 1 broke down, Machines 2 and 3 are running,
    Operator is doing other activities
13. Machines 1 and 2 broke down, Machine 3 is running,
    Operator is doing other delay activities
14. Machines 1, 2 and 3 broke down,
    Operator is doing other delay activities

For Man–Machine Ratio of 1:3, here are the possible combinations of decision making activity.

Case 1 Breakdown of Machine

1. Since there are 11 possible downtimes, therefore there are $11 \times 11 \times 11 = 1{,}331$ combinations if 3 machines broke down
2. If 2 machines broke down, then there are $11 \times 11 = 121$ combinations

Case 2 Breakdown and Other Activities

3. There are 15 activities for the Pre-Post activities, 11 for internal Activities and 10 for Unplanned Delay, therefore the minimum combinations will be $15 \times 11 \times 10 = 1{,}650$

If the man–machine ratio increases to 1:4, the possible combinations of decision making activity of the operator will become 14,641.

Case 1 Breakdown of Machine

1. Since there are 11 possible downtimes, therefore there are $11 \times 1 \times 11 \times 11 = 14{,}641$ combinations if 4 machines broke down

The computations above are just estimate of the possible combinations of the decision making activities of the operator if the man–machine ratios are 1:3 and 1:4. The big leap in the values makes it very difficult for the operator to make a good judgment if no system or process is established.

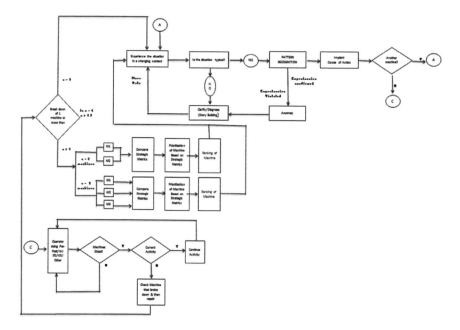

Fig. 3 Extended recognition-primed decision model

With all these activities expected from the operator, it is very important to determine the appropriate man–machine ratio and its corresponding implication on the operator's utilization. While the initial discussion concentrates on the various tasks and how it should be delivered based on the strategic metrics identified, the purpose of coming up with a man–machine ratio is to guide the company on what should be the optimal way of utilizing company's resources and at the same time still meet the various strategic metrics.

3 Conclusion

It is evident that the decision making activities of operators depend on machine assignment. As the number of machine increases, the decision making activity of the operator increases exponentially. Thus, operator should be provided with a framework in order to help them in their decision making process. An Extended Recognition-Primed Decision Model is developed to incorporate multiple machine assignment (Fig. 3).

References

Klein G (1997) The recognition-primed decision (RPD) model: looking back, looking forward. In: Zsambok CE, Klein G (eds) Naturalistic Decision Making. Lawrence Erlbaum Ass, Mahwah

Patterson R et al (2009) Modeling the dynamics of recognition primed decision model. Paper presented at the 9th International Conference on Naturalistic

New/Advanced Industrial Engineering Perspective: Leading Growth Through Customer Centricity

Inside Out to Outside In Through Expert Systems

Suresh Kumar Babbar

Abstract The expert-system-based automated process planning systems are prevalent in Manufacturing and become state of the art tool of successful Industrial Engineers. The author attempted to utilize the concepts of New/advanced Industrial Engineering to apply the capabilities of information technology to redesign business processes in educational institution to reduce the cost, time, and improve quality of its processes by embedding the knowledge of its best decision makers/experts in a "Teaching/learning expert system including scholarship authorization" as part of overall Director Academics Software. The author proposes a concept of software realization of an expert system by assembling experts experience in Personal Computer as knowledge base on the hypothesis that 'knowledge never dies', once we adopt the knowledge of some experts and use it in our system, this knowledge works more efficient than a simple work routine. Director Academics is a tool designed and created for Head of institutions. Here author explored his work in the field of Expert System development, especially what he experiences by working in the Institute and what he learns while he worked under professors. The proposed paper is based on the Rule Based and Case Based Reasoning. In the last author explored his and his senior's experiences.

Keywords Case based · Rule based · Reasoning academic · Expert system · New/advanced industrial engineering

---

S. K. Babbar (✉)
Bahra University, School of Mechanical Engineering, Solan,
Himachal Pradesh 173215, India
e-mail: sures193@hotmail.com

Y.-K. Lin et al. (eds.), *Proceedings of the Institute of Industrial Engineers Asian Conference 2013*, DOI: 10.1007/978-981-4451-98-7_161,
© Springer Science+Business Media Singapore 2013

1 Introduction

Teams striving to achieve excellence need to apply the proficiencies of information technology to redesign business processes (Yesser 2007) that should lead to growth especially in developing countries like India through customer centricity. Great problems in these countries like shortage of skilled manpower provide great opportunities for Industrial Engineers. Currently we are living in the age of dynamic evolution and in this situation the requirement of customer thereby industries/service organizations changes at a much faster speed. It becomes imperative for organizations to reorganize around customers rather than products for resilience (Gulati 2009). In the present scenario the organizations faces more complex problems for conventional approaches. To illustrate it, when we contrast Case Based Reasoning (CBR) (Watson 1995, 1999; Kolodner 1993) with Rule Based System, we see that the methodology for building and refining Knowledge Base (KB) is more sophisticated than the syntactic checks performed by the Rule Based method. The augmentation of Rule Based System with Case Based Reasoning allows us to handle exceptions gracefully, without making a rule set overly complicated.

A rule, by definition, is meant to capture generalization; it loses its power if it is heavily qualified. It attempts to handle the problems within the framework of mathematical logic has not yielded practical results. Case Based Reasoning provides a mechanism for domain-dependent inference that fills a gap in our proposed toolkit. In the proposed case, we would demonstrate the strength of CBR and its advantages to integrate the CBR with the present systems prevalent with industry/ service organizations and challenge the status quo that "Artificial Intelligence (Winston 1982) (AI) software is mostly platform dependent and implemented in environment that no one outside of AI communities uses". Further this paper presents the Case Based Reasoning, as an alternative approach to purely rule-based method, to build a decision support system. Software development for the given problem enables the system to tackle problems like high complexity, low experienced, new staff and changing industrial/service conditions and environment around them.

In the present scenario the purely rule-based method has its limitations like; requirement of explicit knowledge in detail of each domain hence takes years to build Knowledge Base (KB). Case Based Reasoning uses facts in the form of specific cases to solve a new problem, and the proposed solution is based on the similarities between the new problem and the available cases. In this paper we present a Case Based Reasoning which provides decision support for all domains unlike rule-based inference models which are highly domain knowledge specific. Experiments with real data clearly demonstrate the efficiency of the proposed method that has built-in capability to improve productivity of finding and implementing alternatives.

The proposed solutions given in this paper is based on Expert Systems (ES), to solve complicated practical problems of the world especially in developing countries like India that are becoming more and more widespread nowadays. Expert systems are being developed and deployed worldwide in innumerable applications, mainly because of their explanation capabilities.

2 New Industrial Engineering: Information Technology and Business Process Redesign

Business process redesign and information technology are predictable companions, yet industrial engineers have never fully utilized their relationship. The experts argue, in fact, that it has scarcely been exploited at all. But the organizations that have used IT to redesign boundary-crossing, customer-driven processes have benefited immensely. Industrial Engineer needs to change his/her approach from Inside-out to Outside-in to reorganize around customers. She/he must achieve resilience within the system for rapid reorganization for customer centricity that should lead to sustainable growth. That growth will take care of Industrial Engineer's core issues of productivity and economy to scale and will not adulterate the strategy of the organization.

Two new tools are renovating organizations for effectiveness to the degree that Taylors once did. These are information technology—the capabilities offered by computers, software applications, and telecommunications—and business process redesign with solving problems for customers—the analysis and design of work flows and processes within and between organizations along with corporate soul of developing with deep customer empathy. Working together, these tools have the potential to create a new industrial engineering (Davenport Thomas and Short James 1990) with focus on customer centricity with high potential to quickly reorganize for resilience. That can change the way the discipline (Industrial Engineering) is practiced and the skills necessary to practice the contemporary Industrial Engineering.

3 What is Case-Based Reasoning?

Case-based reasoning is used to solve problems by remembering a previous similar situation and by reusing information and knowledge of that situation. Let us illustrate this by following flow diagram (see Fig. 1).

A case-based reasoning (CBR) system generally refers to a computer programmed system that identifies a solution to a current problem by examining the descriptions of similar, previously encountered problems and their associated solutions. Matching the current problem with one or more similar previously encountered problems and using the associated solutions of the matching previously encountered problems to suggest a solution to the current problem. In

Fig. 1 Flow diagram for case-based reasoning

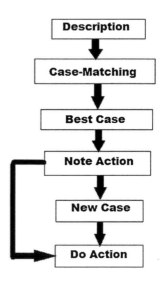

response to that the receipt of a description of a current problem, a conventional CBR system retrieves the closest matching cases from a case database using a search engine and iteratively asked the user for additional descriptive information until the retrieved case or cases identified by the search engine are sufficiently similar to the current problem to be considered as possible solutions. If a new solution (not previously stored in the case database) is subsequently validated, the validated solution can be stored into the case database and utilized to solve future problems.

4 The Case Base Reasoning Cycle-Interrelationship

CBR cycle-inter-relationship can be represented as the following comparative graph (see Fig. 2). When the number of cases is more, the effort for retrieval and ease of adoption is high or we can say that the effort for retrieval and ease of adoption is directly proportional to the number of cases. But when we see the proportionality between efforts for retrieval and ease of adoption it is reverse, when the ease of adoption is low effort of retrieval is high and vice versa.

5 Working of CBR Based Expert System

In the beginning CBR process number of cases is limited, most of the times we need to modify the solution but as the experience increases, the probability of similar or near similar case would increase. In normal times, a new problem is

Fig. 2 Proportionality of no. of cases with ease of adoption and effort for retrieval

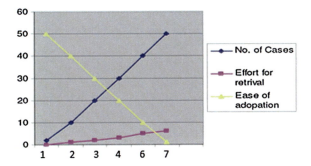

analyzed against cases in the Knowledge Base and one or more similar cases are retrieved. A solution suggested by the matching cases is then reused and tested for success. Unless the retrieved case is a close match, the solution will probably have to be revised producing a new case that can be retained.

The CBR cycle presented above (see Fig. 3) occurs without human intervention. For example many CBR tools act primarily as case retrieval and reuse systems. Case revision (i.e., adaptation) is often being undertaken by managers of the Knowledge Base. However, it should not be viewed as weakness of CBR that encourages human collaboration in decision support. The following sections will outline how each process in the cycle can be handled.

Fig. 3 CBR cycle

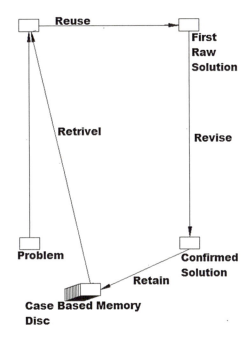

A Case is a piece of knowledge representing with experience. It contains the past solution that is the content of the case and the context in which the solution may be used. Typically a case comprises:

- The problem that describes the state of the world when the case occurred,
- The solution which states the derived solution to that problems, and/or,
- There is a lack of consensus within the CBR community as to exactly what information should be in a case. However, two pragmatic measures can be taken into account in deciding what should be represented in cases: the functionality and the ease of acquisition of the information represented in the case.

6 Advantages of Case Based Reasoning

- It RETRIEVES the most similar case from the galaxy of cases
- REUSE the problem cases to solve the new problem
- REVISE/MODIFY the proposed solution if required and
- RETAIN the new solution as a part of a new case.

7 Advantages of NEW/Advanced Industrial Engineering Approach

It provides quantum leap on the journey of disciplined integration and thus yields major competitive advantage. It prompts Industrial Engineer to shift 120-degree towards an outside-in perspective that otherwise is hardest advance to make.

References

Davenport Thomas H, Short James E (1990) The new industrial engineering: information technology and business process redesign. Sloan Manag Rev 31(4)
Gulati R (2009) Reorganize for resilience. Harvard Business School
Kolodner J (1993) Case-based reasoning. Morgan Kaufman, New York
Watson I (1995) Progress in case based reasoning. Lecture notes in artificial intelligence. Springer, Berlin, p 1020
Watson I (1999) Case-based reasoning is a methodology not a technology. Knowl-Based Syst 12:303–308
Winston PH (1982) Learning new principles from precedents and exercises. Artif Intell 19:321–350
Yesser (2007) Business process redesign methodology. The Saudi e-Government Program 1:1–26

Scheduling a Hybrid Flow-Shop Problem via Artificial Immune System

Tsui-Ping Chung and Ching-Jong Liao

Abstract This paper investigates a two-stage hybrid flowshop problem with a single batch processing machine in the first stage and a single machine in the second stage. In the problem, each job has an individual release time and the jobs are grouped into several batches. To be more practical in real applications, the waiting time between the batch machine and the single machine is restricted. Since the problem is NP-hard, an immunoglobulin-based artificial immune system (IAIS) algorithm is developed to find an optimal or near-optimal solution. To verify IAIS, comparisons with two lower bounds in the second problem are made. Computational results show that the proposed IAIS algorithm is quite stable and efficient.

Keywords Hybrid flowshop · Artificial immune system · Batch processing machine

1 Introduction

In this section, we consider a two-stage hybrid flowshop scheduling problem where there are n jobs to be processed first at stage one and then at stage two. The first stage contains a batch processing machine (BPM) and the second stage contains a single machine. The processing times of job j in stage one and two are p_{1j} and p_{2j}, respectively. Each job has a release time r_j and a corresponding size s_j. The capacity of each batching machine is up to B and the sum of the job sizes in a

---

T.-P. Chung (✉)
Department of Industrial Engineering, Jilin University, Changchun, China
e-mail: tpchung@jlu.edu.cn

C.-J. Liao
Department of Industrial Management, National Taiwan University of Science and Technology, Taipei, Taiwan
e-mail: cjliao@mail.ntust.edu.tw

batch must be less than or equal to B. The batch processing time is equal to the maximal processing time of the jobs in this batch, and all jobs of the same batch start and finish together. The waiting time between the BPM and the single machine is restricted by W. The single machine on stage two can process no more than one job at time. The proposed problems have been shown to be NP-hard by Lee and Uzsoy (1999) even in the case when the batch bounded is determined by job sizes in a single BPM.

There are also some research studies that focus on the hybrid flowshop problem with BPM. Xuan and Tang (2007) establish an integer programming model and propose a batch decoupling based Lagrangian relaxation algorithm for a hybrid flowshop problem with batch processing machine in the last stage. Oulamara (2007) deals with the problem of job scheduling in a no-wait flowshop with two batching machines. Bellanger and Oulamara (2009) develop several heuristics with the worst cases analysis for the hybrid flowshop problem where there are several batch processing machines in the second stage. Gong et al. (2010) consider a two-stage flowshop scheduling problem. They propose a priority rule and a worst case with a minimum value of 2. Su and Chen (2010) provide a two-machine flowshop where a batch processing machine is followed by a discrete machine. The objective functions in all the above literature are to minimize the makespan.

For the waiting time constraints in the hybrid flowshop problem with BPM, Su (2003) considers the same problem as ours but with identical job release times for all jobs. A heuristic algorithm and a mixed integer program are proposed. Fu et al. (2011) consider a flowshop scheduling problem with BPM and limited buffer to minimize the mean completion time. A lower bound and two heuristic algorithms are developed.

Mathirajan and Sivakumar (2006) provide a literature review of metaheuristics on the BPM scheduling in semiconductor. Genetic algorithm (GA) is applied by many authors for solving flowshop problem with the batch processing machine (Lu et al. 2008; Chiang et al. 2010; Malve and Uzsoy 2007; Chou et al. 2006). Liao and Huang (2010) propose a Tabu search algorithm for the two machine flowshop problem with batch processing machines. Xu et al. (2012) provide an ACO algorithm and a useful heuristic in the batch processing problem.

In the past few years, AIS has been successfully applied to a large number of combinational optimization problems (Engin and Döyen 2004; Bagheri et al. 2010; Woldemariam and Yen 2010). In this paper, we will improve an immunoglobulin-based AIS algorithm (Chung 2011) called IAIS, for the two BPM problems considered in this paper.

2 Lower Bounds

Two lower bounds, LB_1 and LB_2, are developed here. The final lower bound is determined by $LB = \max(LB_1, LB_2)$, which will be compared with the solution from the proposed algorithm for validation. Xu et al. (2012) provide a lower bound, and we extend it to a lower bound (LB_1). In additional, a new lower bound (LB_2) will also be proposed.

According to Xu et al. (2012), jobs are first indexed in the descending order of release times, so job n is the job with the longest release time. The lower bound based on Xu et al. (2012) is

$$LB_1 = \max_{j \in S_j}\{r_n + p_{1n} + p_{2n}, \min(r_j + p_{1j} + p_{2j} + p_{2n}, r_j + p_{1j} + p_{1n} + p_{2n})\} \quad (1)$$
$$\text{where } S_j = \{j | r_j + p_{1j} > r_n; s_j + s_n > B\}$$

Property 1 For the considered problem, there exists an optimal schedule where jobs in the same batch are consecutively processed on the machine without idle time in stage 2.

Proof According to Theorem 1 of Su (2003), the makespan of her problem is $\sum x_i + \sum p_{ij}$, which shows that processing jobs in the same batch consecutively at stage 2 can minimize the makespan. For our problem, the makespan is $r_{S_1} + \sum x_i + \sum p_{ij}$ because each batch has a start time and the batch start time is determined by the largest release time of the jobs in the same batch. Thus, Property 1 is also valid in our problem.

According to Property 1, if the single machine at stage 2 is a bottleneck machine, any feasible sequence is optimal. Thus, a lower bound LB_2 can be described as follows.

$$LB_2 = \min_{j \in n}\{r_j + p_{1j}\} + \sum\nolimits_j p_{2j} \quad (2)$$

3 The Proposed IAIS Algorithm

In this section, the structure of the proposed immunoglobulin-based AIS algorithm (IAIS) is presented. A new encoding and decoding method is proposed here.

In this study, possible schedules are represented by integer-valued strings of length n. The n elements of the strings are the batch numbers and each job belongs to a batch. Then, we sequence the batches by the Earliest Release Time (ERT) order on the BPM. Those strings are accepted as antibodies of the AIS. The algorithm goes up to solution by the evolution of these antibodies. This method is called the batch-based encoding and decoding method. affinity value of each schedule is calculated by function Affinity$(a) = 1/\text{makspan}(a)$, where a is the considered antibody. From this function, a lower makespan value creates a higher affinity value.

The method of population generation is based on the job's release time. Some notation is presented first. There is a random number $u \sim U(1, 100)$ and a fixed value F. n_b is the number of jobs in the current batch b.

Jobs in a non-increasing order of release times are arranged. The job at the head of the list is selected and placed in the first batch with three constraints as follows.

Constraint 1 The sum of the job sizes in the current batch must be less than or equal to B.

Constraint 2 The sum of the job processing time in stage two must be less than or equal to W.

Constraint 3 $u > F$.

If it does not satisfy any one of the three constraints, a new batch is created. The procedure is repeated until all jobs are assigned to a batch.

3.1 Isotypes Switching

There are four isotype immunoglobulins, IgM, IgG, IgE, and IgA. Each isotype has a different function. If the makespan of a new string in IgM is larger than that of the old one, three isotype immunoglobulins, IgG, IgE, and IgA are randomly selected to be the next mutation method.

Table 1 Comparison results of IAIS algorithm ($n = 20$)

Instance	LB	IAIS C_{max}	Gap	Time
J2S1O1-1	393	434	10.43	2.37
J2S1O1-2	355	391	10.14	2.12
J2S1O1-3	384	464	20.83	2.85
J2S1O1-4	366	414	13.11	2.43
J2S1O1-5	362	366	1.10	2.60
Ave.			11.13	2.47
J2S2O1-1	431	506	17.40	1.81
J2S2O1-2	530	585	10.38	1.83
J2S2O1-3	568	617	8.63	2.32
J2S2O1-4	544	566	4.04	1.86
J2S2O1-5	568	609	7.22	2.21
Ave.			9.53	2.01
J2S1O2-1	476	514	7.98	2.41
J2S1O2-2	476	489	2.73	2.20
J2S1O2-3	466	507	8.80	2.55
J2S1O2-4	410	442	7.80	2.30
J2S1O2-5	401	435	8.48	2.17
Ave.			7.64	2.32
J2S2O2-1	515	590	14.56	1.76
J2S2O2-2	576	596	3.47	1.87
J2S2O2-3	628	647	3.03	2.05
J2S2O2-4	529	572	8.13	2.13
J2S2O2-5	558	610	9.32	2.01
Ave.			7.70	1.96

In IgG, the pairwise mutation is used. For a sequence s, let i and j be randomly selected two positions in the sequence. A neighbor of s is obtained by swapping these two jobs. In IgA, an index value is given to each job calculated by r_j/b. The job with the smallest value is assigned to batch $b-1$ if $b>1$. The job with the largest value is assigned to batch $b+1$. In IgE, for an antibody, let i be a randomly selected position in the sequence. There is a random number $u \sim U(1, 100)$ and a fixed value G. If $u<G$, let i be assigned to batch $b+1$; otherwise, to batch $b-1$.

3.2 Elimination

According to the limited space of repertoire, the antibodies are deleted except the best antibody of the current generation. The new antibodies are generated randomly and are replaced with the deleted one.

Table 2 Comparison results of IAIS algorithm ($n = 50$)

Instance	LB	C_{max}	IAIS Gap	Time
J3S1O1-1	837	951	13.62	18.25
J3S1O1-2	650	740	13.85	27.07
J3S1O1-3	891	960	7.74	27.08
J3S1O1-4	717	853	18.97	18.88
J3S1O1-5	766	854	11.49	20.83
Ave.			13.13	22.42
J3S2O1-1	1,229	1,439	17.09	15.19
J3S2O1-2	1,257	1,291	2.70	14.97
J3S2O1-3	1,323	1,380	4.31	14.03
J3S2O1-4	1,389	1,456	4.82	12.32
J3S2O1-5	1,182	1,267	7.19	11.72
Ave.			7.22	13.65
J3S1O2-1	1,055	1,097	3.98	36.51
J3S1O2-2	975	999	2.46	15.03
J3S1O2-3	1,145	1,204	5.15	17.83
J3S1O2-4	1,007	1,065	5.76	14.87
J3S1O2-5	1,110	1,123	1.17	16.21
Ave.			3.71	20.09
J3S2O2-1	1,219	1,456	19.44	16.34
J3S2O2-2	1,297	1,414	9.02	17.00
J3S2O2-3	1,333	1,423	6.75	21.54
J3S2O2-4	1,380	1,468	6.38	13.29
J3S2O2-5	1,202	1,278	6.32	12.84
Ave.			9.58	16.20

4 Computational Results

In this section, we present the results of computational experiments. On an Intel Core 2 CPU (3.0 GHz) with 2.0 GB RAM, all the algorithms were coded in C^{++}.

The benchmark problems of Xu et al. (2012) are used as the data set in the first stage of the proposed problem. The waiting time of each job is limited by $W = 40$, and the processing times of jobs in the second stage are generated by the distributions of $O1 = U(1, 25)$ and $O2 = U(1, 40)$. The generated benchmark problems can be downloaded from http://web.ntust.edu.tw/~ie/index.html. Preliminary tests were conducted to find good parameter settings: $A = 10$, $D = \lfloor n/4 \rfloor$, $C = 10$ and $F = 10$.

The performance of the algorithms is calculated by Eq. (3). Each instance is run ten times to obtain the best C_{max} and the average CPU time for IAIS. The computational results for $n = 20, 50, 100$ are summarized in Tables 1, 2, and 3, respectively.

Table 3 Comparison results of IAIS algorithm ($n = 100$)

Instance	LB	C_{max}	IAIS Gap	Time
J4S1O1-1	1,503	1,715	14.11	60.01
J4S1O1-2	1,555	1,686	8.42	60.01
J4S1O1-3	1,441	1,722	19.50	60.01
J4S1O1-4	1,445	1,666	15.29	60.01
J4S1O1-5	1,672	1,805	7.95	60.01
Ave.			13.06	60.01
J4S2O1-1	2,536	2,674	5.44	60.01
J4S2O1-2	2,618	2,656	1.45	40.13
J4S2O1-3	2,521	2,574	2.10	46.59
J4S2O1-4	2,497	2,543	1.84	47.42
J4S2O1-5	2,587	2,616	1.12	56.68
Ave.			2.39	50.17
J4S1O2-1	2,008	2,110	5.08	60.01
J4S1O2-2	2,139	2,154	0.70	60.01
J4S1O2-3	2,075	2,100	1.20	58.44
J4S1O2-4	2,287	2,287	0.00	57.78
J4S1O2-5	1,846	1,988	7.69	60.01
Ave.			2.94	59.25
J4S2O2-1	2,528	2,797	10.64	59.39
J4S2O2-2	2,620	2,646	0.99	47.48
J4S2O2-3	2,526	2,611	3.37	51.70
J4S2O2-4	2,542	2,617	2.95	53.92
J4S2O2-5	2,623	2,639	0.61	57.68
Ave.			3.71	54.03

$$Gap = \frac{Best\ C_{max} - LB}{LB} \times 100\% \qquad (3)$$

From Table 1, the average Gap of IAIS is 9.00 with the maximum of 20.83, and the average computation time of IAIS is 2.19 s. From Tables 2 and 3, the average Gaps of IAIS are 8.41 and 5.52, respectively.

5 Conclusions and Future Research

In this paper, we have examined a two-stage hybrid flowshop problem with a single BPM in the first stage. The problem has the objective of minimizing the makespan, and is NP-hard problem. To evaluate the performance of the IAIS algorithm, it has been tested on the benchmark problems. Computational results have shown that the proposed IAIS algorithm is quite stable and efficient.

Future research may be conducted to exploit the effectiveness of IAIS to other scheduling problems, such as flowshop with multiple batch processing machines with various constraints and objectives. It is also interesting to investigate the performance of IAIS to other combinatorial optimization problems.

References

Bagheri A, Zandieh M, Mahdavi I, Yazdani M (2010) An artificial immune algorithm for the flexible job-shop scheduling problem. Future Gener Comp Sy 26(4):533–541

Bellanger A, Oulamara A (2009) Scheduling hybrid flowshop with parallel batching machines and compatibilities. Comput Oper Res 36(6):1982–1992

Chiang TC, Cheng HC, Fu LC (2010) A memetic algorithm for minimizing total weighted tardiness on parallel batch machines with incompatible job families and dynamic job arrival. Comput Oper Res 37(1):2257–2269

Chou FD, Chang PC, Wang HM (2006) A hybrid genetic algorithm to minimize makespan for the single batch machine dynamic scheduling problem. Int J Adv Manuf Technol 31(3–4):350–359

Chung TP (2011) Heuristics for single- and multi-stage identical parallel machine scheduling problems, Dissertation, National Taiwan University of Science and Technology

Engin O, Döyen A (2004) A new approach to solve hybrid flowshop scheduling problems by artificial immune system. Future Gener Comp Sy 20(6):1083–1095

Fu Q, Sivakumara AI, Lib K (2011) Optimisation of flow-shop scheduling with batch processor and limited buffer. Int J Prod Res 50(8):2267–2285

Gong H, Tang L, Duin CW (2010) A two-stage flow shop scheduling problem on a batching machine and a discrete machine with blocking and shared setup times. Comput Oper Res 37:960–969

Lee CY, Uzsoy R (1999) Minimizing makespan on a single batch processing machine with dynamic job arrivals. Int J Prod Res 37(1):219–236

Liao LM, Huang CJ (2010) Tabu search heuristic for two-machine flowshop with batch processing machines. Comput Ind Eng 60(3):426–432

Lu LF, Zhang LQ, Yuan JJ (2008) The unbounded parallel batch machine scheduling with release dates and rejection to minimize makespan. Theor Comput Sci 396(1–3):283–289

Malve S, Uzsoy R (2007) A genetic algorithm for minimizing maximum lateness on parallel identical batch processing machines with dynamic job arrivals and incompatible job families. Comput Oper Res 34(10):3016–3028

Mathirajan M, Sivakumar AI (2006) A literature review, classification and simple meta-analysis on scheduling of batch processors in semiconductor. Int J Adv Manuf Technol 29(9–10):990–1001

Oulamara A (2007) Makespan minimization in a no-wait flow shop problem with two batching machines. Comput Oper Res 34(4):1033–1050

Su LH (2003) A hybrid two-stage flowshop with limited waiting time constraints. Comput Ind Eng 44(3):409–424

Su LH, Chen JC (2010) Sequencing two-stage flowshop with nonidentical job sizes. Int J Adv Manuf Technol 47(1):259–268

Woldemariam KM, Yen GG (2010) Vaccine-enhanced artificial immune system for multimodal function optimization. IEEE Trans Syst Man Cybern Part B: Cybernetics 40(1):218–228

Xu R, Chen H, Li X (2012) Makespan minimization on single batch-processing machine via ant colony optimization. Comput Oper Res 39(3):582–593

Xuan H, Tang L (2007) Scheduling a hybrid flowshop with batch production at the last stage. Comput Oper Res 34(9):2718–2733

Modeling and Simulation on a Resilient Water Supply System Under Disruptions

X. Liu, J. Liu, S. Zhao and Loon Ching Tang

Abstract We address the resilient water supply system against disruptions in megacities. The study is aimed at developing a quantitative approach for assessing the resilience of water supply system to disruptions. In this context, we propose a decision model incorporates two determinants of system resilience both in robust level and recovery time, and discuss their relationship towards the occurrence of disruptions. Furthermore, a simulation-based model is proposed that incorporates resilience into the proper performances of water supply system, which leads to the impacts of the loss caused by those disruptions. We also present a case study in which the resource scheduling strategies are taken into account to increase the resilience level of water supply system, such as the level of inventory, pipeline-dispatched water, and dispatched water by transportation.

Keywords Disruption · Resilience · Robust · Water supply · System dynamics

---

X. Liu (✉) · S. Zhao
Department of Industry Engineering, Shanghai Jiao Tong University, 800 Dongchuan Road, Min-Hang District 200240 Shanghai, People's Republic of China
e-mail: x_liu@sjtu.edu.cn

S. Zhao
e-mail: Sixiang.zhao@hotmail.com

J. Liu
Sino-US Global Logistic Institute, Shanghai Jiao Tong University, 1954 Huashan Road, Xu-hui District 200030 Shanghai, People's Republic of China
e-mail: Liujian007@sjtu.edu.cn

L. C. Tang
Department of Industrial and Systems Engineering, National University of Singapore, 10 Kent Ridge Crescent, Singapore 119260, Singapore
e-mail: isetlc@nus.edu.sg

1 Introduction

The challenges facing today's megacities are daunting, and water management is one of the most serious concerns. The rapidly escalating demands for water, conflicts, shortages, waste, and degradation of water resources make us to rethink its emerge threatens as well as the need to make water systems more resilient (WHO 2009; UN-Habitat 2012). Although the concept of resilience is considered as multidimensional and multidisciplinary, the concept of resilience is becoming more widely recognized the act of rebounding or springing back, and able to quickly return to normal operations from events (Haimes 2009; Aven 2011).

A resilience water system can be defined in this study as the ability to recover its functions to a certain level from the impacts of disruptions, which is consisting of the robustness, redundancy, resourcefulness, and rapidity (Bruneau et al. 2003). Resilience has been applied in water resource system performance evaluation (Hashimoto et al. 1982), water resource system under impact of climatic change (Fowler et al. 2003), water distribution networks (Todini 2000) and sustainability indicators in urban water supply (Milman and Short 2008). Few papers have focus on building a quantitative resilience metric in megacities water supply system.

In this paper, we formulate a water supply system model under disruptions using system dynamics. Then, we provide a decision-making platform for simulating the system resilience of water supply system in different scenarios.

2 Model of Water System Resilience

2.1 Definition of Water System Resilience

Water system resilience under disruption offers significant advantages to promote ability of system function to prepare and response quickly, reduce losses of system. A general method of water resilience system under disruption is presented in Fig. 1.

In Fig. 1, horizontal axis represents the recovery time, and the vertical axis represents the utility function $F(t)$ (units in percentage). We assume well performance of system function is 100 %, while entire disrupted is zero. The robustness is represented by x, so the function loss of system could be represented as $1 - x$. The loss of resilience can be divided into three stages: The first stage ($t_1 \leq t \leq t_2$) is the hazard diffusion process after initial failures, which mainly reflects the absorptive capacity of the system as the degree to which it absorbs the impacts of initial damage and minimizes the consequences, such as cascading failures; The second stage ($t_2 \leq t \leq t_3$) is the delay stage, which reflects that the resource would be effective only after an inevitable delay; The third stage ($t_3 \leq t \leq t'$) is the recovery stage. t' is the most satisfied recovery time integrating

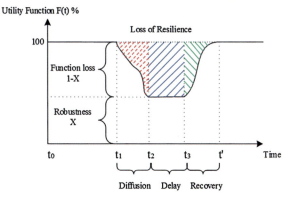

Fig. 1 Quantitative measure method for the loss of water system resilience

cost and demand factors. The loss of resilience can be calculated by the following equation:

$$LR(x,t) = \frac{\int_{t_1}^{t_2}(1-x)dt + (1-x)(t_3-t_2) + \int_{t_3}^{t'}(1-x)dt}{T} \quad (1)$$

Based on the previous definition, it's easily to get the resilience as follow:

$$R(x,t) = 1 - LR(x,t) \quad (2)$$

2.2 Water System Resilience Under Different Strategies

In a water system, the resilience is varying with different strategies during disruptions, such as resource allocation, scheduling, probability of consequence, and information uncertainty, et al. In this context, the 3-Dimensional water system resilience method can be shown as Fig. 2 by adding the strategy axis for resource

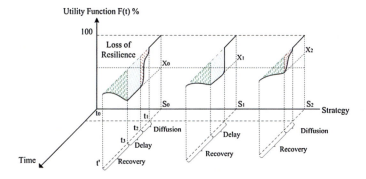

Fig. 2 Water system resilience under different strategies

axis. The key point is how to make the strategies during the different stages to minimize the total loss of system resilience.

3 System Dynamic Model of Water Supply System

The system dynamic as powerful tool could effectively modeling and simulation the strategies for water supply system. In this section, we construct system dynamic model to describe the system resilience of water supply system under disruption, also the state of system can be captured by different strategies to achieve the resilience levels.

3.1 Forming of the Causal Loop Diagram

For a water supply system, system resilience means in this study the satisfaction rate (SR) of water supply system, is often used to assess resilient of water supply. As stated above, system resilience of water supply system is to minimize the total loss area over the time horizon. In another word, the problem could be convert into maximize the satisfaction rate of water supply system. The SR depends on the total resilience level $R(x,t)$ and adjusting different strategies. The conceptual and causal diagram for water supply system under disruption is show in Fig. 3.

The causal loop diagram is described by causal and effect relationship, using the "+" "−" means that the two variables change in the same or opposite direction, the signs B represents negative feedback loop that means balance, equilibrium and stasis. System dynamics model of water supply consists of five major sub-systems: (B4) water reservoir construction; (B3) water demand management; (B1) dispatched-water by transportation water; (B2) pipeline-dispatched water; (B5) water demand and usage. Where (B4) represents the expending the water reservoir strategies caused by water supply shortage for short term; (B1), (B2), (B3) represent the different strategies under disruption for short term.

3.2 Stock Flow Diagram of Water Supply System

Based on the causal loop diagram, stock-flow diagram of water supply system is developed as Fig. 4. In normal condition, a balance between supply rate and usage rate keeps adequate water supply and a normal water level of reservoir. When loss of supply happens and exceeds its absorption capacity, water dispatch is needed to cover the supply shortfall.

Modeling and Simulation on a Resilient Water Supply System

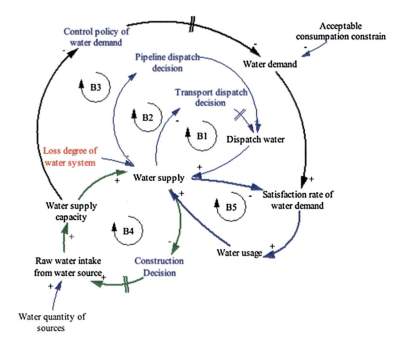

Fig. 3 Causal loop diagram for water supply system under disruption

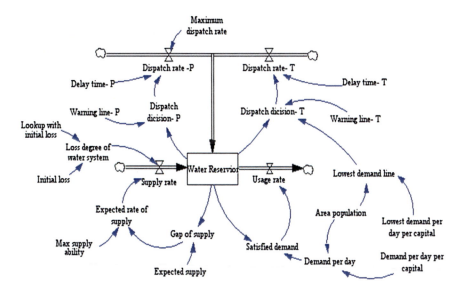

Fig. 4 Stock flow diagram for water supply system under disruption

4 Case Study

Chenhang reservoir is an important reservoir in Shanghai, which covers more than 3 million population's water demand. The reservoir capacity is designed to be 9.14 million m^3, and effective reservoir is 9 million m^3. Chenhang reservoir is facing threatens rising caused by salt tide. The objective is to improve its supply capacity and maximize the satisfaction rate of water supply under different strategies using the proposed model.

4.1 Initial Conditions and Assumptions

We assume effective reservoir as inventory that is one of decision variables of system output. Supply capacity is 1.6 million m^3/day as business as usual according to engineer design, and accounts for about 16.2 % of total supply capacity. Due to the impact of climate change, the quality of water intakes from Yangtze revers are often not up to standard from December in last year to the April in next year. The most extremely case is that continuous time of salt tide is approximately 24 days based on the history data record from Shanghai Water Authority. Assume that the water reservoir is equal to inventory, then the supply capacity is primary depend on the inventory, pipeline-dispatched water, and dispatched water by transportation during this period. So the initial loss, continuous time, supply capacity, and initial inventory level can be obtained.

4.2 Simulation Results and Discussions

Three scenarios are assumed in the sensitivity test by assigning three kinds of loss, continuous time, warning line of pipeline-dispatched. The results can be obtained and demonstrated the Table 1.

The dispatch rate of pipeline water and dispatch rate of transportation water would depend on the setting of the warning line of pipeline-dispatched and

Table 1 The result of salt tide scenario for Chenhang reservoir

Case	Loss	CT	WL (P)	Dispatch (P)	Dispatch (T)	SR
Unit	%	Day	Million m^3	Million m^3	Million m^3	%
1	50	10	450	180	0	95
2	50	24	450	750	6	76.77
			750	108	150	90.84
3	75	24	450	810	325	60.7
			750	1,050	350	65

warning line of transportation-dispatched that triggered by the lowest demand line correspondently. The dispatching strategies toward specific targets are simulated and demonstrated by satisfied demand via adjusting different parameters so that the system resilience could be calculate by the changes of satisfied demand. Assume that system resilience meet the customer demands when satisfaction rate reach 95 %, the simulation result for water reservoir, dispatch rate of pipeline water, dispatch rate of transportation water and satisfaction rate are show in the following 4 Figures.

From the above 4 Figures, one can find:

1. Assume the initial loss level is 50 %, and warning line of pipelined dispatch is 450 m^3. Figure 5 shows that water reservoir of scenario 1 is not significant influence for system resilience when continuous time is 10 days; while continuous time is 24 days, the water reservoir will continuously decline. The improve redundancy of design reservoir could enhance the robustness level that withstand disruption to water supply system.
2. Assume the initial loss level is 50 %, and continuous time is 24 days, the Fig. 6 provide the estimate level on the change in dispatch rate of pipeline water with different trigger strategies of pipeline supply. The simulation results show that the dispatch rate of pipeline water is closely related to the warning line of pipelined dispatch. Comparing Fig. 7 with Fig. 8, it is illustrated that the way of dispatch rate for pipeline water is prior to the dispatch rate for transportation water based on cost considerations. The improve redundancy of scheduling capacity could enhance the robustness level that withstand disruption to water supply system in short term.
3. Assume the continuous time is 24 days, we compare scenario 2, 3, 4, 5 under different initial loss and warning line of pipelined dispatch, the Fig. 8 shows the estimate level for satisfaction rate when the corresponding strategies are taken. Compare 75 % initial loss (scenario 4 and 5) with 50 % initial loss (scenario 2 and 3), the degree of initial loss and reduce the probability to disruption is more crucial to the system resilience.

Fig. 5 Water reservoir

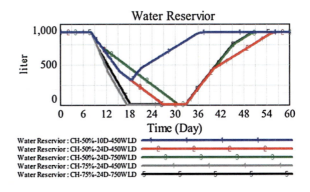

Fig. 6 Dispatch rate of pipeline water

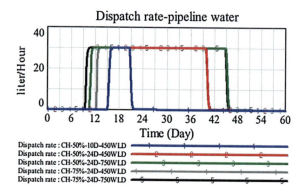

Fig. 7 Dispatch rate of transportation water

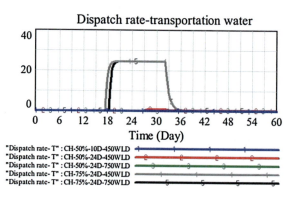

Fig. 8 Satisfaction rate

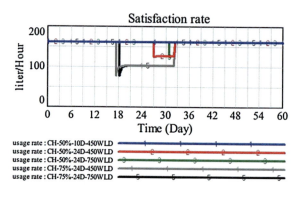

5 Conclusion and Further Research

We have developed a system dynamics model of resilience water supply system with strategies consideration under disruptions. The general properties of the system resilience are very depended on the robustness and recovery time. Based on the uncertainty of initial loss, continuous time, strategies taken consideration, we have extended both the theoretical and system dynamics models to simulate the scenarios happened in Shanghai. The resilience of water supply system has been extended by analysis of robustness, uncertainty, and adoptive supply ability.

Acknowledgments This research is supported by the National Research Foundation Singapore under its Campus for Research Excellence and Technological Enterprise (CREATE), and NSFC (91024013, 91024131), the authors also gratefully acknowledge the helpful comments and suggestions of the reviewers, which have improved the presentation.

References

Aven T (2011) On some recent definitions and analysis frameworks for risk, vulnerability, and resilience. Risk Anal 31(4):515–522
Bruneau M, Chang SE, Eguchi RT et al (2003) A framework to quantitatively assess and enhance the seismic resilience of communities. Earthquake Spectra 19(4):733–752
Fowler HJ, Kilsby CG, O'Connell PE (2003) Modeling the impacts of climatic change and variability on the reliability, resilience, and vulnerability of a water resource system. Water Resour Res 39(8):0
Haimes YY (2009) On the definition of resilience in systems. Risk Anal 29(4):498–501
Hashimoto T, Stedinger JR, Loucks DP (1982) Reliability, resiliency, and vulnerability criteria for water resource system performance evaluation. Water Resour Res 18(1):14–20
Milman A, Short A (2008) Incorporating resilience into sustainability indicators: an example for the urban water sector. Global Environ Change 18:758–767
Todini E (2000) Looped water distribution networks design using a resilience index based heuristic approach. Urban Water 2(2):115–122
UN-Habitat (2012) State of the World's Cities 2012/2013: Prosperity of Cities. http://www.unhabitat.org/pmss/listItemDetails.aspx?publicationID=3387. Accessed 10 Apr. 2013
WHO (2009) Vision 2030: The resilience of water supply and sanitation in the face of climate change. http://www.who.int/water_sanitation_health/publications/9789241598422/en/. Accessed 13 April 2013

A Hybrid ANP-DEA Approach for Vulnerability Assessment in Water Supply System

C. Zhang and X. Liu

Abstract Vulnerability reflects the potential of disrupting the whole system to some extent when the system is exposed to hazard. One of the most important issues of the indicator-based vulnerability assessment problem is to determine the weights of vulnerability indicators, especially when they are correlated with each other in multiple dimensions (i.e., physical, functional and organizational). In this paper, a framework for assessing vulnerability of critical infrastructure system is identified and applied to the evaluation in a water supply system. A complete critical infrastructure system vulnerability index is developed, which contains dimensions of "protection and defense", "quick response after disaster", "maintenance and recovery capacity" and "possible damage to system". A quantitative method, integrating analytic network process (ANP) and game cross-efficiency data envelopment analysis (DEA) model, is proposed to analyze the vulnerability of interdependent infrastructures. Finally, the assessed vulnerability level of each infrastructure in water supply system is graded into four classes.

Keywords Vulnerability · Assessment · Data envelopment analysis · Analytic network process

C. Zhang (✉) · X. Liu
Department of Industry Engineering and Logistic Management, School of Mechanical Engineering, Shanghai Jiao Tong University, Shanghai 200240, China
e-mail: xiaochongzi@sjtu.edu.cn

X. Liu
e-mail: X_liu@sjtu.edu.cn

1 Introduction

Since the interconnections (both physical and logical) between modern infrastructures become more complex, lifeline systems are more vulnerable to disasters. So it is important for the government to select critical elements of the system and protect them with highest priority. However, how to assess the vulnerability of these elements is still a huge challenge for researchers.

Vulnerability is an important attribute of critical infrastructure systems. The concept of vulnerability is still evolving and has not yet been established. Different researchers have different definitions on vulnerability. TurnerII et al. (2003) defines it as the degree to which human and environmental systems are likely to experience harm due to a perturbation or stress, Aven (2011) defines it as the manifestation of the inherent states of the system that can be subjected to a natural hazard or be exploited to adversely affect that system. Aggregating the existing definitions (TurnerII et al. 2003; Barbat and Carreño 2010), we define vulnerability in this study as follows: vulnerability reflects the potential of disrupting the whole system to some extent when the system is exposed to hazard.

There are various approaches for characterizing vulnerability. The first one is a multi-dimensional indicator framework which needs weights assignment, such as expert decision and analytic network process (ANP) (Aven 2011; Grubesic and Matisziw 2007; Piwowar et al. 2009). The second one is network modeling approaches (Qiao et al. 2007; Scaparra and Church 2008a, b), which is mainly based on the network topology of infrastructures, such as maximal flow model, shortest path model and network flow model. The third one is probabilistic modeling (Sultana and Chen 2009; Doguc and Ramirez-Marquez 2009), which is usually used when analyzing inter-dependencies and cascading failures.

In this paper, a hybrid ANP and Cross-Efficiency DEA approach is proposed to solve this multi-criteria assessment problem. The remainder of this paper is organized as follows. In Sect. 2, the indicator system for vulnerability evaluation and value functions of selected indicators are described. Section 3 presents the proposed hybrid ANP-DEA approach for vulnerability assessment. Section 4 illustrates the proposed evaluation framework via a case study. Finally Sect. 5 concludes this paper.

2 Indicators for Vulnerability Assessment

Four dimensions are considered for constructing the vulnerability index in this study: (1) Protection and Defense (reflecting the exposure of system to disasters); (2) Quick Response after disaster; (3) Maintenance and recovery capacity; (4)

A Hybrid ANP-DEA Approach for Vulnerability Assessment

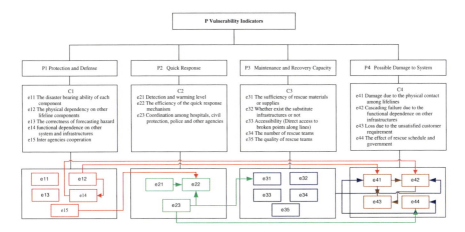

Fig. 1 The ANP decision structure for vulnerability evaluation

Possible damage to system (Ezell 2007). The vulnerability indicators are demonstrated in Fig. 1.

Values of qualitative indicators are acquired by ranking or categorizing performed by experts. For clearness, selected parameters need to be explained especially. "Functional or logical interdependency" values are obtained from the correlation matrices of infrastructures and "the sufficiency of rescue materials" are calculated using the following function: "the realistic storage of rescue materials/the expected storage of rescue materials".

3 The Proposed Methodology

The vulnerability evaluation processes are shown in Fig. 2.

3.1 Bounds of weights for each indicator by ANP

Variables:

1) CI_i ($i = 1, 2,..., n$) means the ith in infrastructure to be evaluated;
2) The total number of DMs (Decision makers) involved in the evaluation is K and $k = 1, 2, ..., K$ expresses the kth expert;
3) The weight of the ith cluster, the jth element of the kth expert is represented by $\hat{w}_{ijk}$ ($i = 1, 2,..., 4, j = 1, 2, ..., 4, k = 1, 2,..., K$).

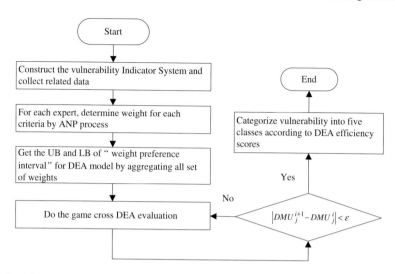

Fig. 2 Hybrid ANP-DEA vulnerability evaluation process

Analytic network process (ANP) is used for determining the upper and lower limit of weights. It is a relatively simple and systematic approach that can be used by decision makers (Khadivi and Fatemi Ghomi 2012). It allows both interaction and feedback within clusters of elements and between clusters. Such feedback well captures the complex effects of interplay in human society, especially when risk and uncertainty are involved. The Decision structure of this process is shown as Fig. 1 and Super Decision 1.6.0 can be used to complete the ANP process for each expert and get the weight of each parameter, referred to as $\hat{w}_{ijk}$ ($i = 1, 2,..., 4$, $j = 1, 2, ..., 4$, $k = 1, 2,..., K$).

Then, to eliminate the impact of subject bias of single expert, weights from all DMs should be aggregated. By calculating the mean value $\bar{w}_{ij}$ and variance E_{ij}^2, the LB and UB of weight interval can be given, as is shown in Eq. 1.

$$\hat{w}_{ij} \in [\bar{w}_{ij} - 1.96\delta_{ij}, \bar{w}_{ij} + 1.96\delta_{ij}] = [\hat{w}_{ij\min}, \hat{w}_{ij\max}] \quad (1)$$

3.2 DEA Game Cross-Efficiency Model

Charnes et al. (1978) proposed the initial DEA model (CCR model) for evaluating the relative efficiencies of a set of decision making units (DMUs). The application of DEA as an alternative multi-criteria decision making (MCDM) tool has been gaining more attentions in the literatures because it can find optimal weights for all relevant inputs and outputs of each DMU in a relatively objective and fair way (Wu et al. 2009). The DEA game cross-efficiency model, which was proposed by Wu and Liang (2012), can obtain unique game cross-evaluation scores which constitute a Nash equilibrium point. This model is used for evaluating the vulnerability index here.

In the model, the input and output indicators are represented by m_i and s_j respectively. The indicators which have positive correlations with vulnerability are defined as "output" while others are defined as "input". The ith input and rth output of DMU_j are represented as x_{ij} and y_{rj} respectively.

The steps of this DEA game cross-efficiency method are listed as follows.

Step 1: Add the bound of weights to the initial CCR model as constraints and solve the model.

The bounds of input and output indicators in DEA model are represented by $[w_{imin}, w_{imax}]$ and $[\mu_{imin}, \mu_{imax}]$ separately, and the constraint set of each indicator in this step is derived from LB and UB. The game cross DEA model is illustrated as model 2. Let $t = 1$, $\alpha_j = \alpha_j^1 = \bar{E}_j$, and solve the model.

$$max \sum_{r=1}^{s} \mu_r y_{rd} = \theta_d$$

$$s.t \sum_{i=1}^{m} w_{ij}^d x_{il} - \sum_{r=1}^{s} \mu_{rj}^d y_{rl} \geq 0 \quad l = 1, 2, \ldots, n$$

$$\sum_{i=1}^{m} \omega_i x_{id} = 1, \quad (2)$$

$$w_i - \frac{w_{imin}}{w_{1min}} w_1 \geq 0, \quad w_i - \frac{w_{imax}}{w_{1max}} w_1 \leq 0,$$

$$\mu_i - \frac{\mu_{imin}}{\mu_{1min}} \mu_1 \geq 0, \quad \mu_i - \frac{\mu_{imax}}{\mu_{1max}} \mu_1 \leq 0,$$

$$\omega_i \geq 0, i = 1, 2, \ldots, m, \ \mu_r \geq 0, r = 1, 2, \ldots, s.$$

Model 2 is solved n times (one for each d) for each alternative i, thus each DMU corresponds to a set of optimal weights: $w^*_{1d}, w^*_{2d}, \ldots, w^*_{md}, \mu^*_{1d}, \mu^*_{2d}, \ldots, \mu^*_{sd}$. Calculate the game cross-evaluation score for each DMU by Eq. 3.

$$\bar{E}_j = \frac{1}{n}\sum_{d=1}^{n} E_{dj} = \frac{1}{n}\sum_{d=1}^{n} \left(\frac{\sum_{r=1}^{s} \mu^*_{rd} y_{rj}}{\sum_{i=1}^{m} w^*_{id} x_{ij}}\right) \quad d, j = 1, 2, \ldots, n \quad (3)$$

Table 1 Organizational structure of water system

Organizational structure of water system														
Raw water system			Transmit			Treat and store				Distribution			Control
1	2	3	4	5		6	7	8	9	10	11	12	13	14
Raw water resources	Reservoir	Power supplies	Pipelines	Power supplies		Tanks	Settling pond	Facilities	Power supplies	Distribution network	Valves	Pump station	Power supplies	SCADA

A Hybrid ANP-DEA Approach for Vulnerability Assessment

Table 2 Score of vulnerability evaluation

Infrastructure no.	1	2	3	4	5	6	7
Vulnerability index	0.805	0.870	0.506	0.557	0.966	0.983	0.955
Infrastructure no.	8	9	10	11	12	13	14
Vulnerability index	0.893	0.456	0.680	0.958	0.964	1	0.442

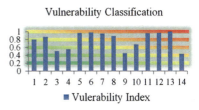

Fig. 3 The grade of infrastructure

Step 2: Add a new constraint, as is shown in Eq. 4, to the DEA model and solve this model again. Then, define $\alpha_j^{t+1} = \frac{1}{n}\sum_{d=1}^{n}\sum_{r=1}^{s}\mu_{rj}^{d*}(\alpha_d^t)y_{rj}$, where $\mu_{rj}^{d*}(\alpha_d^t)$ represents the optimal value of μ_{rj}^d.

$$\alpha_d \times \sum_{i=1}^{m}w_{ij}^d x_{id} - \sum_{r=1}^{s}\mu_{rj}^d y_{rd} \leq 0 \qquad (4)$$

Step 3: If $|\alpha_j^{t+1} - \alpha_j^t| \geq \varepsilon$ for some j, where ε is a specified small positive value, then let $\alpha_j^t = \alpha_j^{t+1}$ and return to Step 2, else if $|\alpha_j^{t+1} - \alpha_j^t| < \varepsilon$ for all j, then stop and α_j^{t+1} is the final game cross-evaluation score given to DMU_j. Finally rank the vulnerability according to the final efficiency scores.

4 Numerical Example

In this section, we examine a numerical example using the above hybrid ANP-DEA method to illustrate its validity in evaluating vulnerability for water supply system. Assume that City X locates at earthquake-prone regions, now the vulnerability of its water system against earthquake need to be assessed for prevention of disasters. As is shown in Table 1, the organizational structure of water system is composed of 14 infrastructures which need to be evaluated.

The data and specific calculation process are omitted due to the limited space, only the final DEA efficiency score of each DMU is presented in Table 2.

During the process, two virtual infrastructures, which are used as the benchmark for the most and the least vulnerable infrastructures are introduced into the analysis. In this study, the UB and LB of the vulnerability index are 1 and 0.2 and accordingly, the vulnerability is categorized into four levels. Figure 3 shows the ultimate ranking of each infrastructure.

5 Conclusions

A hybrid ANP-DEA approach for infrastructure vulnerability assessment is proposed in this paper. The difficulty of infrastructure vulnerability evaluation mainly comes from the complexity of lifeline system and the multi-dimensional nature of vulnerability. The incorporation of ANP and DEA game cross efficiency method make the evaluation more objective and fair. Since disasters are always dynamic process and evaluation indicators may keep changing with the evolution of disasters, the suggestion for future researches is that a dynamic model can be considered for treating the uncertainties and dynamics.

Acknowledgments This research is partly supported by the National Research Foundation Singapore under its Campus for Research Excellence and Technological Enterprise (CREATE) and the NSFC (91024013, 91024131).

References

Aven T (2011) On some recent definitions and analysis frameworks for risk, vulnerability, and resilience. Risk Anal 31:515–522

Barbat AH, Carreño ML (2010) Seismic vulnerability and risk evaluation methods for urban areas. A review with application to pilot area. Struct Infrastruct Eng 6:17–19

Charnes A, CooperWW Rhodes E (1978) Measuring the efficiency of decision-making units. Eur J Oper Res 2:429–444

Doguc O, Ramirez-Marquez JE (2009) A generic method for estimating system reliability using Bayesian networks. Reliab Eng Syst Saf 94:542–550

Ezell BC (2007) Infrastructure vulnerability assessment model (I-VAM). Risk Anal 27(3):571–583

Grubesic TH, Matisziw TC (2007) A typological framework for categorizing infrastructure vulnerability. GeoJournal 78:278–301

Khadivi MR, Fatemi Ghomi SMT (2012) Solid waste facilities location using of analytical network process and data envelopment analysis approaches. Waste Manage (Oxford) 32:1258–1265

Piwowar J, Châtelet E, Laclémence P (2009) An efficient process to reduce infrastructure vulnerabilities facing malevolence. Reliab Eng Syst Saf 94:1869–1877

Qiao J, Jeong D, Lawley M, Richard JPP, Abraham DM, Yih Y (2007) Allocating security resources to a water supply network. IIE Trans 39:95–109

Scaparra MP, Church RL (2008a) A bilevel mixed-integer program for critical infrastructure protection planning. Comput Oper Res 35:1905–1923

Scaparra MP, Church RL (2008b) An exact solution approach for the interdiction median problem with fortification. Eur J Oper Res 189:76–92
Sultana S, Chen Z (2009) Modeling flood induced interdependencies among hydroelectricity generating infrastructures. J Environ Manage 90:3272–3282
TurnerII BL, Kasperson RE et al (2003) A framework for vulnerability analysis in sustainability science. Proc Natl Acad Sci USA 100:8074–8079
Wu J, Liang L (2012) A multiple criteria ranking method based on game cross-evaluation approach. Ann Oper Res 197:191–200
Wu J, Liang L, Chen Y (2009) DEA game cross-efficiency approach to Olympic rankings. Omega 34:909–918

An Integrated BOM Evaluation and Supplier Selection Model for a Design for Supply Chain System

Yuan-Jye Tseng, Li-Jong Su, Yi-Shiuan Chen and Yi-Ju Liao

Abstract In a supply chain, the design of a product can affect the activities in the forward and reverse supply chains. Given a product requirement, the components of the product can be designed with different specifications. As a result, the bill of material and the manufacturing activities will be different. Therefore, in different design alternative cases, there can be different decisions of supplier selection for producing the product. In this research, a new model for supplier selection and order assignment in a closed-loop supply chain system is presented. First, the design information of the design alternative cases are analyzed and represented in the form of a bill of material model. Next, a mathematical model is developed for supplier selection and order assignment by evaluating the design and closed-loop supply chain costs. Finally, a solution model using the particle swarm optimization method with a new encoding scheme is presented. The new model is developed to determine the decisions of design evaluation and supplier selection under the constraints of capacity and capability to achieve a minimized total cost objective. In this paper presentation, an example product is illustrated. The test results show that the model and solution method are feasible and practical.

Keywords Supply chain management · Closed-loop supply chain · Product design · Bill of material · PSO

1 Introduction

To design a product, there may be different ways to design the detailed specifications of the components and product. In order to satisfy the product requirement, different design alternative cases can be utilized to design the product. In the

Y.-J. Tseng (✉) · L.-J. Su · Y.-S. Chen · Y.-J. Liao
Department of Industrial Engineering and Management, Yuan Ze University, Chung-Li, Taoyuan, Taiwan
e-mail: ieyjt@saturn.yzu.edu.tw

different design alternative cases, the components of the product can be designed with different shapes, dimensions, different materials, and other specifications. If the components and product are designed differently, the bill of material (BOM) of the product will be different. As a result, the downstream manufacturing activities will be affected. Therefore, the different design alternative cases can affect the decisions in the supplier selection and order assignment. It is necessary to evaluate how the different design alternative cases affect the supply chain.

In a product life cycle, the supply chain that performs the common activities such as manufacturing, assembly, transportation, and distribution can be described as a forward supply chain. In the reverse direction, the reverse supply chain performs the activities to process a product at the end of the product life cycle. The activities in a reverse supply chain can include recycle, disassembly, reuse, remanufacturing, and disposal.

If several design cases are available, the different design alternative cases can affect the decisions in supplier selection in the closed-loop supply chain. It is necessary to consider the design alternative cases prior to the actual production of the product. The different design alternatives cases can be modeled and analyzed based on different objectives to determine the best design. In addition, the order assignment and supplier selection can be analyzed and evaluated for the design.

In this research, the information of design of the components and the product is represented as BOM information. In the forward supply chain, the information of the components and product are represented as a BOM model. In the reverse supply chain, the components and product are represented as a reverse bill of material (RBOM) model. A design for closed-loop supply chain model is developed to find the suitable design case and the suitable suppliers in an integrated way (Fig. 1).

In the previous research, the problem and models of supplier selection have been discussed in the papers of Kasilingam and Lee (1996) and Humphreys et al. (2003). A literature review of supply chain performance measurement was presented in Akyuz and Erman (2010). The topics of reverse logics and closed-loop supply chains have been presented in French and LaForge (2006) and Kim et al. (2006). In Schultmann et al. (2006) and Alshamrani et al. (2007), the problems of close-loop supply chain were modeled and discussed. In Sheu et al. (2005) and Lu et al. (2007), the models for green supply chain management were presented. In Ko and Evans (2007), a genetic algorithm-based heuristic for the dynamic integrated forward and reverse logistics network was developed. In Yang et al. (2009), an optimization model for a closed-loop supply chain network was presented. In Kannan et al. (2010), the closed loop supply chain model was modeled and solved using a genetic algorithm approach. Based on the review, the product design and the supplier selection in the closed-loop supplier chain have not been evaluated in an integrated way. Therefore, in this research, a model for integrated evaluation of product design and the closed-loop supply chains is developed.

The cost items for the design activities include function, dimension, material, assembly operation, and manufacturing operation. The cost items for the forward supply chain activities include manufacturing cost, purchase cost, and transportation cost. The cost items for the reverse supply chain activities include recycle

An Integrated BOM Evaluation and Supplier Selection Model

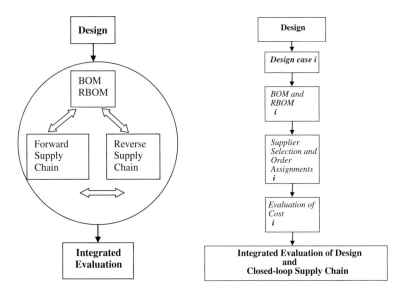

Fig. 1 The concept and model of design for closed-loop supply chain

cost, disassembly cost, reuse cost, remanufacturing cost, and disposal cost. The total cost is the sum of the above cost items. The constraints include the capacity and capability constraints. The mathematical model is developed to select the supplier and determine the order assignment for each supplier under the constraints of capacity and capability to achieve a minimized total cost objective.

In the previous research, the PSO algorithm has been successfully applied to many continuous and discrete optimizations (Kennedy and Eberhart 1997). The research in Banks et al. (2008) reviewed and summarized the related PSO research in the areas of combinatorial problems, multiple objectives, and constrained optimization problems. In this research, the solution method using the PSO approach is utilized. In the presentation, the test results are presented and discussed.

In this paper, Sect. 1 presents an introduction. Section 2 describes the mathematical model and PSO model of design for closed-loop supply chain. In Sect. 3, the implementation and application are illustrated and discussed. Finally, a conclusion is presented in Sect. 4.

2 The Design for Closed-Loop Supply Chain Model

2.1 Bill of Material and Reverse Bill of Material Models for Representation Design of a Product

In this research, the information of design of the components and product is represented in the form of a bill of material (BOM). In the forward supply chain,

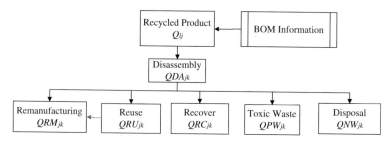

Fig. 2 The format of a reverse bill of material (RBOM)

the information of the components and product are represented as a BOM model. In the reverse supply chain, the components and product are represented as a reverse bill of material (RBOM) model (Fig. 2).

2.2 The Design for Closed-Loop Supply Chain Model

In this research a design for closed-loop supply chain model is presented. A brief description of the notations is as follows.

l	An order $l \in L$, $L = L_F \cup L_R$
L	All the orders
LF	Orders in the forward supply chain
LR	Orders in the reverse supply chain
i	A plant $i \in I, I = I_p \cup I_o$
I	All the plants
j	A product $j \in J_l$, J_l represents the products in order l
k	A component
x_{li}	A 0–1 decision variable representing whether order l is assigned to plant i
TC	Total cost
TPC	Cost of forward supply chain
TRC	Cost of reverse supply chain
TRB	Value of reverse supply chain
TUB	Value of reuse supply chain
TMB	Value of remanufacturing
TCB	Value of recovery
TUC	Cost of reuse operation
TNC	Cost of remanufacturing operation
TCC	Cost of recovery operation
TAC	Cost of disassembly operation
TWC	Cost of disposal operation
TSC	Transportation cost
TMC	Material cost

TFC Manufacturing cost

The model is briefly described as follows.

$$\text{Min } TC = TPC + TRC - TRB \tag{1}$$

$$TPC = TOC + TSC + TMC + TFC \tag{2}$$

$$TRC = TUC + TNC + TCC + TAC + TWC \tag{3}$$

$$TRB = TUB + TMB + TCB \tag{4}$$

$$TUC = \sum_{l \in L_R} \sum_{i \in I_p} \sum_{j \in J_l} \sum_{k \in K_{lj}} Q_{lj} \times QRU_{jk} \times CRU_{ijk} \times x_{li} \tag{5}$$

$$TNC = \sum_{l \in L_R} \sum_{i \in I_p} \sum_{j \in J_l} \sum_{k \in K_{lj}} Q_{lj} \times QRM_{jk} \times CRM_{ijk} \times x_{li} \tag{6}$$

$$TCC = \sum_{l \in L_R} \sum_{i \in I_p} \sum_{j \in J_l} \sum_{k \in K_{lj}} Q_{lj} \times QRC_{jk} \times CRC_{ijk} \times x_{li} \tag{7}$$

$$TAC = \sum_{l \in L_R} \sum_{i \in I_p} \sum_{j \in J_l} \sum_{k \in K_{lj}} Q_{lj} \times QDA_{jk} \times CDA_{jk} \times x_{li} \tag{8}$$

2.3 The Solution Model Using the Particle Swarm Optimization Method

The PSO algorithm is utilized for finding the solutions in the model. The PSO algorithm is an evolutionary computation method introduced by Kennedy and Eberhard (1997). In the PSO method, a particle is defined by its position and velocity. To search for the optimal solution, each particle adjusts its velocity according to the velocity updating equation and position updating equation.

$$v_{id}^{new} = w_i \cdot v_{id}^{old} + c_1 \cdot r_1 \cdot (p_{id} - x_{id}) + c_2 \cdot r_2 \cdot (p_{gd} - x_{id}) \tag{9}$$

$$x_{id}^{new} = x_{id}^{old} + v_{id}^{new} \tag{10}$$

This research applies the PSO method to the problem by developing a new encoding and decoding scheme. In the developed encoding scheme, a particle is represented by a position matrix. The elements in the position matrix are denoted as h_{pj}, where p represents a particle and l represents an order. The encoding of the position matrix is as follows

$$PL = \begin{array}{c} 1 \\ 2 \\ \vdots \\ p \end{array} \begin{bmatrix} \overset{1}{h_{11}} & \overset{2}{h_{12}} & \cdots & \overset{l}{h_{1l}} \\ h_{21} & h_{22} & \cdots & h_{2l} \\ \vdots & \vdots & \ddots & \vdots \\ h_{p1} & h_{p2} & \cdots & h_{pl} \end{bmatrix}_{P \times L} \quad (11)$$

where h_{pj} represents the position matrix of particle p, where $p = 1, \ldots, E$, and $l = 1, 2, \ldots, L$.

The fitness function represents the objective of the PSO enumeration. The objective as shown in (12) is used as the fitness function.

$$\text{Min } Fitness = \text{Min } TC = TPC + TRC - TRB \quad (12)$$

The position matrix of a particle can be decoded into an order assignment set where each order l is assigned to a suitable plant i to achieve the objective of the fitness function. The assigned plant i can be determined by decoding h_{pj}. The total number of plants I is divided into several numerical zones. The assignment of order l to plant i can be determined by searching the numerical zone where h_{pj} falls in. After the PSO enumeration, the final particle represents an order assignment arrangement set in the closed-loop supply chain.

3 Implementation and Test Results

The mathematical model and the PSO solution model was implemented and tested. An example is modeled and illustrated. Product A is a simplified mobile phone as shown in Fig. 3. The BOM is shown in Fig. 4. The related data are modeled for implementation and testing. The cost items in the forward and supply chain are also modeled. Using the input data of the example product, the mathematical model are modeled and solved using the PSO model. The test results show that the models are feasible and efficient for solving the problem.

4 Conclusions

In this research presentation, the concept of design for closed-loop supply chain is developed. A mathematical model and PSO solution model is presented for integrated evaluation of the design and its closed-loop supply chain. The related BOM and RBOM are developed to represent the design information. The costs in design, forward supply chain, and reverse supply chain are modeled. A mathematical

An Integrated BOM Evaluation and Supplier Selection Model

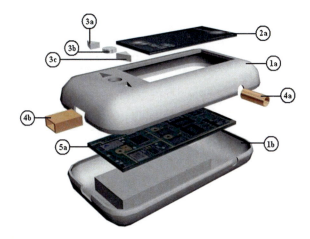

Fig. 3 Product A is a simplified mobile phone used for testing and illustration

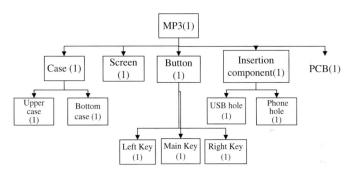

Fig. 4 The BOM of product A

model is presented for use in supplier selection and order assignment. The PSO solution model is presented to find solutions. An example is demonstrated in the presentation. The test results show that the models are feasible and useful for solving the design for closed-loop supply chain problem. Given a design, the order assignment and suppler selection can be analyzed and evaluated to achieve the objective of the fitness function of minimizing total costs of design, forward supply chain, and reverse supply chain. Future research can be directed to explore more detailed value and cost functions of design, and forward and reverse supply chain activities.

Acknowledgments This research is funded by the National Science Council of Taiwan ROC with project number NSC 100-2221-E-155-021-MY3.

References

Akyuz G, Erman ET (2010) Supply chain performance measurement: a literature review. Int J Prod Res 48:5137–5155

Alshamrani A, Athur K, Ballou RH (2007) Reverse logistics: simultaneous design of delivery routes and returns strategies. Comput Oper Res 34:595–619

Banks A, Vincent J, Anyakoha C (2008) A review of particle swarm optimization. Part II: hybridization, combinatorial, multicriteria and constrained optimization, indicative applications. Nat Comput 7:109–124

French ML, LaForge RL (2006) Closed-loop supply chains in process industries: an empirical study of producer re-use issues. J Oper Manage 24:271–286

Humphreys PK, Wong Y, Chan FTS (2003) Integrating environmental criteria into the supplier selection process. J Mater Process Technol 138:349–356

Kannan G, Sasikumar P, Devika K (2010) A genetic algorithm approach for solving a closed loop supply chain model: a case of battery recycling. Appl Math Model 34:655–670

Kasilingam RG, Lee CP (1996) Selection of vendors: a mixed-integer programming approach. Comput Ind Eng 31(1–2):347–350

Kennedy J, Eberhart RC (1997) A discrete binary version of the particle swarm algorithm. In: 1997 Proceedings of international conference systems, man and cybernetics, Piscataway, NJ, 1997

Kim K, Song I, Kim J, Jeong B (2006) Supply planning model for remanufacturing system in reverse logistics environment. Comput Ind Eng 51:279–287

Ko HJ, Evans GW (2007) A genetic algorithm-based heuristic for the dynamic integrated forward/reverse logistics network for 3PLs. Comput Oper Res 34:346–366

Lu YY, Wu CH, Kuo TC (2007) Environmental principles applicable to green supplier evaluation by using multi-objective decision analysis. Int J Prod Res 45:4317–4331

Schultmann F, Zumkeller M, Rentz O (2006) Modeling reverse logistic tasks within closed-loop supply chains: an example from the automotive industry. Eur J Oper Res 171:1033–1050

Sheu JB, Chou YH, Hu CC (2005) An integrated logistics operational model for green-supply chain management. Transp Res Part E: Logistics Transp Rev 41:287–313

Yang GF, Wang ZP, Li XQ (2009) The optimization of the closed-loop supply chain network. Transp Res Part E: Logistics Transp Rev 45:16–28

Estimation Biases in Construction Projects: Further Evidence

Budi Hartono, Sinta R. Sulistyo and Nezar Alfian

Abstract Construction projects are characterized by their unique and temporary features. Very limited data, if any, would be available and readily transferable for the analysis of subsequent projects. Hence most project analysts would depend on intuitions and gut feelings to make judgment and estimations. This experimental study is a follow-up study on investigating the possible existence of systematic estimation errors (biases in estimations) in an Indonesian context. It is focused on the two suspected types of biases; the anchoring accuracy and overconfidence biases in project time duration estimates. Two groups of estimators (experienced, n = 20 vs. non-experienced, n = 20) were involved in the study. A hypothetical project case based on an actual construction project was developed. The estimators were then individually requested to provide duration estimates (for each project activity and overall) for the project case. The estimates were then compared against the actual duration of the project. The result of suggests that anchoring bias is not statistically observable for both non-experienced and experienced estimators. This study finds that overconfidence bias is identifiable when making the range estimation of the project duration.

Keywords Construction project · Judgmental bias · Duration estimate · Experiment

---

B. Hartono (✉) · S. R. Sulistyo · N. Alfian
Industrial Engineering Program, Mechanical and Industrial Engineering Department,
Universitas Gadjah Mada, Jl. Grafika 2, Yogyakarta, Indonesia
e-mail: boed@gadjahmada.edu

S. R. Sulistyo
e-mail: sintarahmawidya@gmail.com

1 Introduction

Project is a temporary group of activities designed to create a unique product, service, or result (PMI 2000). Due to the unique and temporary characteristics, the availability of the historical data becomes limited or not easily accessible. This brings various problems to the analysts in project planning process, including time and cost estimations.

Subjective expert judgment becomes a viable option for the project planners and analysts to estimate time and cost. The subjective expert judgment uses the cognitive aspect of decision making process which according to studies (Bazerman 1998; Hastie and Dawes 2001) prone toward systematic errors or biases.

According to Tversky and Kahneman (1974), bias which affects judgment process is specific and systematics. Specific means the natural error; whereas systematic means the errors made by the research subjects is statistically consistent hence they are predictable (Mak and Raftery 1992). In many occasions, decision makers use some simplifying strategies in estimation or decision-making, called heuristics to make the process faster and require lower cognitive loads (Tversky and Kahneman 1974).

Duration estimation for project activities is an essential stage in project planning. An estimation biases in the process, if any, could result in inaccuracy of the overall project planning which in turn could affect the project overall performance.

Various past studies had been administered with the aims of identifying systematic errors or biases in estimation within general management as well as in project management context [e.g. (Cleaves 1987; Flyvbjerg 2003, 2006; Flyvbjerg et al. 2002, 2004a, b, 2005)]. For the specific context of Indonesian projects, Nugroho (2011) identified the possible existence of accuracy, anchoring, and overconfidence biases of project duration estimates. The study was administered by means of experiments using mini-projects of building Lego® blocks. The study found that anchoring biases were not observable while the accuracy biases and over confidence biases are both statistically observable from the subjects. Saputra (2012) and Aji (2012) respectively conducted follow-up experiments with Lego® software to observe possible biases for estimators with a different level of experiences. Result shows that accuracy bias and overconfidence biases are observable, while anchoring bias is not identifiable among the experienced respondents. Handayani (2012) attempted to identify the same estimation biases by means of a survey. In this study, data was collected from real IT projects from various companies. Respondents were required to estimate the project duration and activities' times prior to the project execution. Actual project durations were collected after the project completion. The survey method provides a more realistic setting for the study and could indicate a better external validity if compared to the experiments. Results identified two biases: overconfidence and anchoring. The main drawback of the survey study is that the project size and complexity varies across companies and estimators. Accordingly, project size and complexity which may affect the estimation accuracy cannot be controlled.

This reported, follow-up study utilizes an experimental approach by using a real project case on which the estimators are required to provide their estimations. This study attempts to address the two drawbacks on the previous pertinent studies. By using a case adopted from a real construction project, the study could provide a more realistic setting than those experimental studies using mini projects of building Lego® blocks. Furthermore, since all estimators were exposed to the same case, size and complexity of the project are controllable and hence provides more conclusive results compared to the survey study.

2 Research Method

The subjects of the experimental study are classified into: experienced and non-experienced respondents (n = 20 for each). The experience group comprises contractor practitioners with a minimum experience of five (5) years in managing similar projects. They were mid-level manager who were actively involved in the decision making process in project bidding. On the other hand, the un-experienced group comprises vocational school teachers and students with some knowledge exposures but no experience in project management.

A hypothetical construction project is developed on the basis of a real construction (housing) project. Some pertinent documents such as: the engineering drawing, the plot plan, the project activity list and the bills of materials are developed. By referring to the documents, individual respondents were required to provide time estimates for each project activities as well as the estimates for the total project duration. The order of tasks for each respondent was randomly sequenced to eliminate noises.

3 Results

3.1 Anchoring Biases

Anchoringg refers to a cognitive bias in which an individual's judgment or estimation is strongly tied to some arbitrary (or irrelevant) value or reference point or anchoring stimuli (Tversky and Kahneman 1974). As a consequence, the estimation value would be statistically correlated to the irrelevant value.

In the study, anchoring stimuli was given by asking the respondents to response the following questions prior to their assignment to provide project estimation: "how much time needed to work on a small grocery store?". Figure 1 shows the value of the responses on anchoring stimuli against the value of project duration estimation across the experienced respondents.

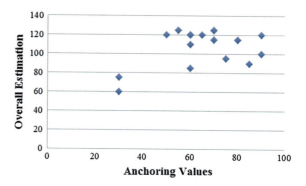

Fig. 1 Anchoring versus overall estimation for experienced respondents (n = 15)

Table 1 indicates the correlation test for both groups of experienced and non-experienced estimators. It was suggested from the statistical analysis that no significant correlation between anchoring responses and project estimation is observable. Hence in this study anchoring effect is not observable for both groups.

3.2 Accuracy Biases

The second point to observe is possible accuracy biases. There are two sets of estimation data used in this analysis, overall project estimation and activity based project estimation, for both respondents. Overall project estimation is the estimation of the total project duration which is derived from the estimators at once. Activity-based project estimation is the estimation of the total project duration which is derived from the sum of the estimation of individual activities in the lowest level of the work breakdown structure.

Error in accuracy is computed by using Eq. (1).

$$\text{Errors in accuracy}\,(\Delta) = \text{the project estimation} - \text{the actual duration} \quad (1)$$

The one sample t test is used to analyze the possible existence of systematic errors in accuracy of the estimation The hypotheses are indicated by Eqs. (2) and (3).

$$H_0 : \mu_\Delta = 0; \text{ there is no systematic error} \quad (2)$$

$$H_1 : \mu_\Delta \neq 0; \text{ there is a systematic error} \quad (3)$$

The summary of the one sample t-test result is depicted in Table 2.

Table 1 Anchoring effect results

	Experienced	Inexperienced
Spearman correlation coefficient	0.041 (p-value = 0.864)	0.221 (p-value = 0.350)
Interpretation	Anchoring effect is **not** observable	

3.3 Overconfidence Biases (Range Estimates)

The third point to be observed in this study is to identify the possible existence of overconfidence biases for range estimates. Observation was administered by giving task to the respondents to make the range of estimation (the min and the max) for the project duration with a 90 % confidence level. The actual overall project time is then used to calibrate the data range estimation which have been made. A range estimate by an estimator is considered a hit if the actual project duration is located within the minimum and maximum values of project duration estimations.

Figure 2 shows the results for experienced respondents. Analysis indicates that the hit ratio is 65 % for experienced and 20 % for inexperienced respondents respectively which are much lower than the designated 90 % confidence level. The low hit ratio indicates that most of the estimated ranges were too narrow indicating overly-confident estimators.

4 Comparisons with Past Studies and Concluding Remarks

Tables 3 and 4 provide the profile and result for four pertinent studies as well as the current study in estimation bias identifications. Across the studies, the results for anchoring biases tend to be negative. All studies could not identify the anchoring biases except for Study A. Study C which utilized two methods of anchoring shows different results. It suggests that anchoring bias is highly affected by the way the anchoring stimuli are administered. This is consistent to the finding by past studies indicating that not all types of anchoring stimuli would yield the same level of anchoring biases (Brewer and Chapman 2002; Chapman and Johnson 1994).

Accuracy biases are observable in most of the cited studies except for the current study of experienced estimators. Experienced estimators in construction industry seem to exhibit less accuracy biases. The fact that the current study is administered within the construction context while the other studies are carried out in IT settings may affect the accuracy biases. Complexity and culture of the specific project domains may matter. However, further studies are required to

Table 2 One sample t-test summary for systematic error in accuracy

One sample t-test	Experienced		Non-experienced	
	t	Sig. (2-tailed)	T	Sig. (2-tailed)
Overall estimates	−0.721	0.48	−4.751	0
	Not observable		**Observable, aggressive**	
Activity-based estimates	5.835	0	3.188	0.005
	Observable, conservative		**Observable, conservative**	

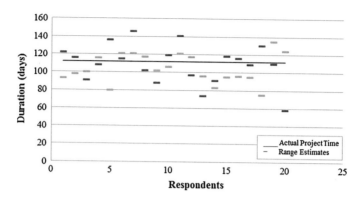

Fig. 2 Estimated ranges by experienced respondents

Table 3 Profile of pertinent studies

	Study A	Study B	Study C	Study D	Current study
Author	Aji (2012)	Saputra (2012)	Handayani (2012)		
Method	Experiment	Experiment	Survey	Survey	Experiment using case from a real project
Domain	IT	IT	IT	IT	Construction
Type of project	Simple, simulated, LEGO software	Simple,	simulated, LEGO software	Complex, real, variety	Complex, real, variety
Respondents	Non-experience	Experience	Non-experience	Experience	Real, single case Experienced and Non experienced
Respondent criteria	New for the task	Learning curve	Zero experience	>5 yrs. experience	Zero experience and >5 yrs. experience

examine such determinants affecting estimation biases. Observations for overconfidence biases across all studies provide consistent results. It is found that estimators exhibit similarity on overconfidence biases, regardless the domain (IT, Construction) and the level of experiences.

Table 4 Results of pertinent studies

Study A	Study B	Study C	Study D		Current study
Non-experienced	Experienced	Non-experienced	Experienced	Non-experienced	Experienced
Anchoring biases					
Observable	Not observable	Mixed	Not observable	Not observable	Not observable
Accuracy biases					
Observable, conservative (for pessimistic estimates)	Observable, conservative	Observable, conservative	Observable, conservative	Observable, aggressive	Not observable
Over confidence biases (range estimates)					
Observable, over confidence	Observable, over confidence	Observable, over confidence	Observable, over confidence	Observable, over confidence	Observable, over confidence

References

Aji RK (2012) judgmental biases on project duration estimations for inexperienced estimators. Mechanical & Industrial Engineering. Universitas Gadjah Mada, Yogyakarta

Bazerman MH (1998) Judgment in managerial decision making. Wiley, New York

Brewer NT, Chapman GB (2002) The fragile basic anchoring effect. J Behav Decis Making 15:65–77

Chapman GB, Johnson EJ (1994) The limits of anchoring. J Behav Decis Making 7:223–242

Cleaves DA (1987) Cognitive biases and corrective techniques: proposals for improving elicitation procedures for knowledge-based systems. Int J Man Mach Stud 27:155–166

Flyvbjerg B (2003) How common and how large are cost overruns in transport infrastructure projects? Transp Rev 23:71–88

Flyvbjerg B (2006) From nobel prize to project management: getting risks right. Proj Manage J 37:5–15

Flyvbjerg B, Holm MS, Buhl S (2002) Underestimating costs in public works projects: error or lie? J Am Plann Assoc 68:279–295

Flyvbjerg B, Glenting C, Rønnest AK (2004a) Procedures for dealing with optimism bias in transport planning: guidance document. The British Department for Transport, London

Flyvbjerg B, Holm MKS, Buhl SL (2004b) What causes cost overrun in transport infrastructure projects? Transp Rev 24:3–18

Flyvbjerg B, Holm MKS, Buhl SL (2005) How (in) accurate are demand forecasts in public works projects?: the case of transportation. J Am Plann Assoc 71:131–146

Handayani D (2012) A survey of judgmental biases in estimating the duration of information technology. Projects mechanical and industrial engineering department. Universitas Gadjah Mada, Yogyakarta

Hastie R, Dawes RM (2001) Rational choice in an uncertain world: the psychology of judgment and decision making. Sage Publications, Inc., Thousand Oaks, CA

Mak S, Raftery J (1992) Risk attitude and systematic bias in estimating and forecasting. Constr Manage Econ 10:303–320

Nugroho FI (2011) Initial study on judgmental biases of project time estimation using experts judgment. Mechanical and Industrial Engineering. Universitas Gadjah Mada, Yogyakarta

PMI (2000) Guide to the project management body of knowledge, 2000 edn. Project Management Institute, Pennsylvania, USA

Saputra BA (2012) Judgemental biases analysis in project time estimation based on individual and group of estimators. Mechanical and Industrial Engineering. Universitas Gadjah Mada, Yogyakarta

Tversky A, Kahneman D (1974) Judgment under uncertainty: heuristics and biases. Science 185:1124–1131

Exploring Management Issues in Spare Parts Forecast

Kuo-Hsing Wu, Hsin Rau and Ying-Che Chien

Abstract According to literature, we know that research in spare parts forecast is a very popular topic in recent years due to the need of higher service level of customer demand. Most approaches in these researches focus on how to forecast spare parts more accurately under various conditions with some mathematical techniques. However, few attentions are paid to the management issues to control or maintain some factors to keep the spare parts forecast useful. For example, Engineering Change (EC) often introduce lots of planed items no longer valid and consume mass efforts trying to continue the spare parts support. This motivates our study. This study starts from service process, spare parts requirement, and spare parts forecast flow. When we come to investigate management issues in spare parts forecast, besides EC, such as issues like different service models for OEM, ODM or Brander, spare parts information correctness and service organization structures are worth exploring.

Keywords Spare parts forecast · Management issues · Spare parts lifecycle

1 Introduction

Spare parts forecast is popular research topic in recent years following the increasing customers' service level demand, especially for those manufacturer which is seeking to improve their service level in coping with customer increasing

K.-H. Wu · H. Rau (✉) · Y.-C. Chien
Department of Industrial and Systems Engineering, Chung Yuan Christian University,
Chungli 320, Taiwan, Republic of China
e-mail: hsinrau@cycu.edu.tw

K.-H. Wu
e-mail: 10102405@cycu.edu.tw

Y.-C. Chien
e-mail: cyc@cycu.edu.tw

demand (Tan et al. 2010), and at the same time, an internal pressure in lower the service cost is pressing in Dutta (2009).

The results of those researches are forecast methodologies for solving different kinds of service environments and conditions. But most of them stand in forecast technical point of view, in general spare parts forecast implementation practice, there are issues other than forecast technical aspect needs to be considered in order the forecast implementation can function as they are expected. Giving an example, in case forecast methodologies use not updated Engineering Change (EC) spare parts price list as the base in forecast, the result of the forecast will not be correct and may with those useless substituted parts, In the worst case with prohibited parts which may introduce not just inventory issue, but also field service issue.

In this study, we aim to explore those issues which may impact the forecast implementation result other than the forecasting technical mechanism itself, those issues are treated as the management issues, because by those methodologies found, mostly they focus in forecast aspect with fixed conditions, and not looking into the surrounding management issues.

2 The Service Process and Spare Parts Forecast

For exploring management issue, it is necessary to look back to the overall service process from the beginning of the service. The service process lifecycle, spare parts support cycle and information requirement for creating forecast methodology will be introduced and reviewed. By the overall review, it starts to explore the possible management issue which leads the well-developed spare parts forecast performance not as expected or simply in vain.

2.1 Service Process Lifecycle

Service process lifecycle describes how a service is formed from offer, setup, maintenance, and reaches the end-point; the service process lifecycle is shown in Fig. 1. Referring to the models of Aurich et al. (2010) and Maxwell and Vorst (2003), we propose the service process lifecycle in five different stages, each stage has its activities that need to be fulfilled; i.e. the requirement for service, a solid

Fig. 1 Service process lifecycle

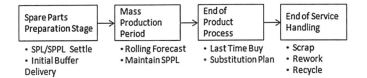

Fig. 2 Spare parts support cycle

customer service requirement has to be provided, and for the rest of the stages which have their own activities to be fulfilled as shown in the Fig. 1.

2.2 Spare Parts Support Cycle

As shown in service process lifecycle, spare parts support plays a vital position; it backs up the service activities, so, examining spare parts support cycle becoming a necessary. Sanborn et al. (2011) proposed a spare parts life cycle model, Fig. 2 we revised Sanborn's and proposed a model based on parts acquiring characteristics, this model describes the cycle into four stages.

In each stage, similar to service lifecycle there are activities need to be fulfilled, giving example, End of Life (EOL) process starts to secure the continued support for spare parts been described by Pourakbar and Dekker (2012) service agreements oblige it means EOL is a job not avoidable and for resulting this issue two actions will follow; Last Time Buy (LTB) or substitution plan. These two actions all invent different departments, such as sales, purchasing, component engineering, quality assurance, etc. especially LTB is truly a burden either to the vendor or customer, so it always happens that LTB liability becoming a tough issue between vendor and customer negotiation, simply because none of them is willing to bear the responsibility in carrying spare parts inventory. The other activity, spare parts substitution will consume company large resources and often bring internal conflict in resource allocation.

2.3 Spare Parts Forecast

By reviewing current paper there is no doubt, there lots of different forecast methodologies had been developed. No matter which spare parts forecast methodology and model, their creation must be based on existing information and assumptions. The information needed for creating forecast as Huiskonen (2001) proposed in four different catalogs, in this study, the information required is separated into catalogs, it can be temporal such as past usage history, and sales forecast, it may also with some condition such as one layer distribution, and one

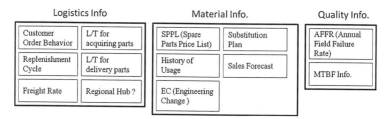

Fig. 3 Information needed for spare parts forecast chart

central hub for the source of distribution, beside those information, Sanders and Larry (1995) also point out human judgment/factor also play its role.

In this study, forecast methodology is not the focus, but information needed and used is the aim. Figure 3 shows the summary of information used in developing forecast methodology. By viewing this chart, some of the information actually came from a complex combination, for example Spare Parts Price List (SPPL) is mixed by balancing between usability (kitting parts), flexibility (EOQ, MOQ), compatibility (substitution), and cost (Huiskonen 2001).

3 The Management Issues

There are many management issues that can impact well developing forecast methodologies; they impact the forecast in two major catalogs: false information and forecast fulfillment. In Fig. 4, it illustrates the relationships between those items, and they all come to the end with forecast malfunction. When we look at the causes, totally 9 different causes are listed, none of them is related to the forecast methodology itself, but any one of them can distort or even cause failure to the forecast implementation objective. Even though there are many factors that may give impacts to the forecast, we focus on the Sales forecast in this section, Engineering Change and service provider will be discussed.

3.1 Sales Forecast

All the forecast methodology must consider the sales forecast in handling the spare parts forecast; a false information input to the forecast methodology will not produce the correct result, because the demand for spare parts in the field will follow the increasing sales and increases. In the study of Michael and O'Connor (2000), they concluded that forecast accuracy is hard to reach even amendments apply. An unreliable sales forecast induce fluctuations in forecast result and such situation makes spare parts hard to catch the real demand, and will cause spare parts to be in either shortage or surplus to the demand.

Exploring Management Issues in Spare Parts Forecast

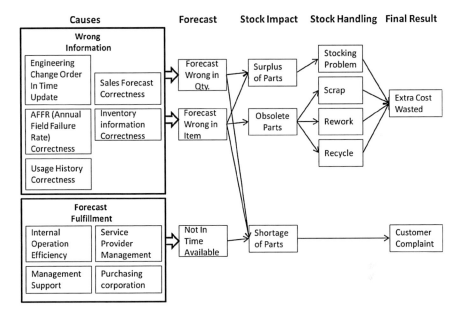

Fig. 4 The causes of forecast malfunction chart

Figure 5 shows the sales forecast formation, a similar approach has been proposed by Danese and Kalchschmidt (2011), they named it forecasting process management, it describes the sales forecast generally comes from the information collected from the second tier or customer's forecast, usage history or demand for existed product (Armstrong et al. 2000), or marketing survey for new launching product, so we can easily imagine that it may contain uncertainties. To amending such unreliable factor, in general practice a confidence level/factor is added to make such forecast into a range. But beside the major information, other aux. information shown in Fig. 5 are also needed to be considered to amend the forecasting such as manufacturing capacity.

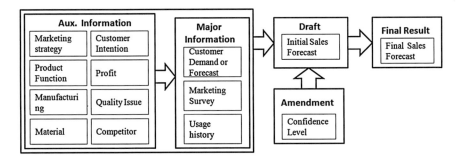

Fig. 5 The formation of sales forecast chart

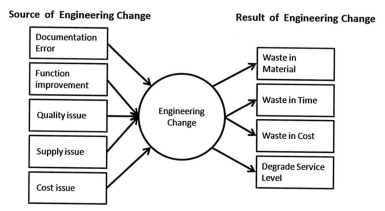

Fig. 6 EC demand and the consequence chart

3.2 Engineering Change Management

Wright (1997) in his paper directly describes the engineering change (EC) resulting EC demands are an evil. EC is an issue which all the companies hate to issue, but it does exist. Since EC implementation means lots of actions will be followed up and to be completed. In reality it just cannot be avoided, especially it is highly unpredictable as it is indicated by Tavcar and Duhobnik (2005) since in real world nothing is perfect, and EC is the action trying to make product perfect no matter in function or in cost. Figure 6 illustrates the EC sources and the consequences.

The reasons demanding for issuing EC may be for resulting quality issue, improving function, correcting documentation error, material supply issue, or saving cost. Huang et al. (2003) provide their view in the source of the EC, and Prabhu et al. conduct a survey in automotive OEM about the EC initiation of the change showing that cost reduction occupies the highest percentage of all cause. In general industrial practice, different departments may request the EC to solve the problem they are facing, example, purchasing may ask for EC to replace current used component with same function with lower cost, Engineering department may ask EC to change some component to enhancing the product function according to the sales request (Fig. 7).

3.3 Service Provider

Service provider may directly or indirectly seriously influence the spare parts forecast result by either lack of capability or wiliness in spare parts forecast fulfillment, and especially in a international company, service provider (self own

Exploring Management Issues in Spare Parts Forecast

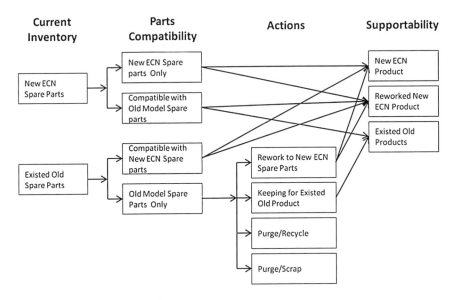

Fig. 7 On hand inventory handling flow chart

or out-sourcing) is part of the service net which spare parts may be stocked (Dekker et al. 2013). It is often seen that company outsources the service delivery to external service provider.

Service delivery can be separated into three categories. Their content can be separated into service operation, spare parts support, and service logistics. The service provider spare parts support model can be arranged either vendor consign or service provider purchases first and charged back after the consumption. This shows the difference of the spare parts ownership, because none will be pleased to hold the spare parts inventory, since spare parts inventory is always be treated as the most important factor impacts the operation and financial performance of the service organization as the survey done by Dutta and Pinder (2012).

How the vendors perform service outsourcing, it is necessary to understand the possible service model in practice, if standing in supplier/vendor point of view, the possible service model is shown in Fig. 8. Since HQ holds communication between spare parts supplier, and supplier will ask for a forecast in order for suppliers to plan and manage the parts availability and delivery. In such cases, even when HQ is not responsible for making forecast, but region office does, a forecast is still a must in planning spares. In the mean time, for a small service organization, own office or outsourcing service provider, it is hard to ask for making good quality forecast, and this even messes the total spare parts supply. We saw such situation often happens in practice.

So without optimal service provider management, this will cause spare parts forecast deterioration, it may be that spare parts surplus and abuse in the first model or shortage parts caused lower service level in second model.

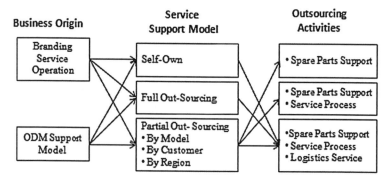

Fig. 8 Service operation model and service activity chart

4 Conclusion

This study discussed the service management issues which impact the spare parts forecast performance from beginning from the review of service process, the spare parts support flow, the spare parts forecast methodology creation requirement, a listed of management issues, and do further 3 issue discussions; sales forecast, EC and service provider.

This study concludes that having well design spare parts forecast methodology does not guarantee expected forecast result, as well as good fulfillment and implementation. Lots of information has to be supported and fed into the forecast mechanism, but without proper management control in securing correctness of the information, the first step of the forecast fails simply because of the output of the forecast is incorrect. There are even more management issues involved in forecast implementation. A continue well management support for the fulfillment in implementation plays vital key to ensure the set forecast result is achieved.

Acknowledgments This work is supported in part by National Science Council of Republic of China under the grants NSC 101-2221-E-033-037-MY3. The authors also gratefully acknowledge the helpful comments and suggestions of the reviewers.

References

Armstrong JS, Morwitzb VG, Kumar V (2000) Sales forecasts for existing consumer products and services: do purchase intentions contribute to accuracy. Int J Forecast 16:383–397

Aurich JC, Mannweiler C, Schweitzer E (2010) How to design and offer services successfully. CIRP J Manuf Sci Technol 2:136–143

Danese P, Kalchschmidt M (2011) The role of the forecasting process in improving forecast accuracy and operational performance. Int J Prod Econ 131:204–214

Dekker R, Pinçe Ç, Zuidwijk R, Jalid MH (2013) On the use of Installed base information for spare parts logistics: a review of idea and industry practice. Int J Prod Econ 143:536–545

Dutta S (2009) Delivering customer service via contact center and the web. Aberdeen Group, Boston

Dutta S, Pinder A (2012) Optimizing the service supply chain. Aberdeen Group, Boston

Huang GQ, Yee WY, Mak KL (2003) Current practice of engineering change management in Hong Kong manufacturing industries. J Mater Process Technol 139:481–487

Huiskonen J (2001) Maintenance spare parts logistics: special characteristics and strategic choices. Int J Prod Econ 71:125–133

Lawerance M, O'Connor M (2000) Sales forecasting updates: how good are they in practice? Int J Forecast 16:369–382

Maxwell D, Vorst R (2003) Developing sustainable product and services. J Cleaner Prod 11:883–895

Pourakbar M, Dekker R (2012) Customer differentiated end-of-life inventory problem. Eur J Oper Res 222:44–53

Sandborn P, Prabhakar V, Ahmad O (2011) Electronic part life cycle concepts and obsolescence forecast. IEEE Trans Compon Packag Technol 51:707–717

Sanders NR, Ritzman LP (1995) Bringing judgment into combination forecasts. J Oper Manage 13:311–321

Tan AR, Matzen D, McAloone TC, Evans S (2010) Strategies for designing and developing services for manufacturing firms. CIRP J Manuf Sci Technol 3:90–97

Tavcar J, Duhobnik J (2005) Engineering change management in individual and mass production. Rob Comput Integr Manuf 21:205–215

Wright IC (1997) A review of research into engineering change management: implications for product design. Des Stud 18:33–42

Artificial Particle Swarm Optimization with Heuristic Procedure to Solve Multi-Line Facility Layout Problem

Chao Ou-Yang, Budi Santosa and Achmad Mustakim

Abstract The facility layout problem (FLP) has an important effect on the efficiency and the profitability of the manufacturing system from the standpoint of the cost and time. This research has objective to minimize total material handling cost. Multi-line facility layout problems (MLFLP) is FLP that assigns a few facilities in the two or more lines into industrial plant, where the number of the facilities is less than the number of the locations with no constraint for placing the facilities. This study present Heuristic Artificial Particle Swarm Optimization (HAPSO) a hybrid meta-heuristic algorithm to solve MLFLP and it consider the multi-products. The proposed algorithm applied to the case study from other paper. The computational results indicate that the proposed algorithm more effective and efficient to solve the case.

Keywords Multi-line facility layout problem · Estimation distribution algorithm · Particle swarm optimization · Tabu search · Heuristic procedure

1 Introduction

The facility layout problem (FLP) has an important effect on the efficiency and the profitability of the manufacturing system from the standpoint of the cost and time. Multi-lines facility layout problem (MLFLP) assign a few facilities in the two or

---

C. Ou-Yang (✉) · A. Mustakim
Department of Industrial Management, School of Management, National Taiwan University of Science and Technology, Taipei, Taiwan, R.O.C
e-mail: ouyang@mail.ntust.edu.tw

A. Mustakim
e-mail: mustakimachmad@gmail.com

B. Santosa
Department of Industrial Engineering, Sepuluh Nopember Institute of Technology, Surabaya, Indonesia
e-mail: budi_s@ie.its.ac.id

more rows into industrial plant. Where the number of the facilities is less than the number of the locations and the total facilities area is less than total area of industrial plant (Sadrzadeh 2012). This research has objective to mini-mize total material handling cost for MLFLP.

Meta-heuristic and its hybrid algorithm have been proposed by many recent researchers to solve FLP. El-Baz (2004) use Genetic Algorithm (GA) to solve FLP. Miao and Xu (2009) use Hybrid GA and Tabu Search (TS) to solve Multi-row FLP. Zhou et al. (2010) use GA to solve Multi-row FLP. Sadrzadeh (2012) use GA with heuristic procedure to solve MLFLP. Utamima (2012) use Hybrid meta-heuristic to solve single-row FLP. This paper proposes hybrid algorithm Estimation Distribution Algorithm (EDA), Particle Swarm Optimization (PSO), TS and it enchanted by heuristic procedure.

2 Mathematical Model

To determine the material handling cost for one of the possible layout plans, the production volumes, production routings, and the cost table that qualifies the distance between a pair of machines/locations should be known. The following notations are used in the development of the objective function: Fij amount of material flow among machines i and j (i, j, ..., M). M is the number of machines. C_{ij} unit material handling cost between locations of machines i and j. D_{ij} rectilinear distances between locations of machines i and j. C total cost of material handling system. The total cost function is defined as:

$$C = \sum_{i=1}^{M} \sum_{j=1}^{M} F_{ij} C_{ij} D_{ij} \qquad (1)$$

This research has objective function to minimize C.

3 The Proposed Algorithm

This research want to develop the proposed algorithm to solve multi-rows Facility Layout Problems. The proposed algorithm is combining Particle Swarm Optimization (PSO), Estimation Distribution Algorithm (EDA) and Tabu Search(TS). Figure 1 describes flowchart of the proposed algorithm. The proposed algorithm divided by four main processes. These are Heuristic Procedure, EDA, PSO, and TS. It is enchanted by elitism to keep 10 % of the best particles for next iteration. The proposed algorithm is named as Heuristic Artificial Particle Optimization (HAPSO).

Artificial Particle Swarm Optimization

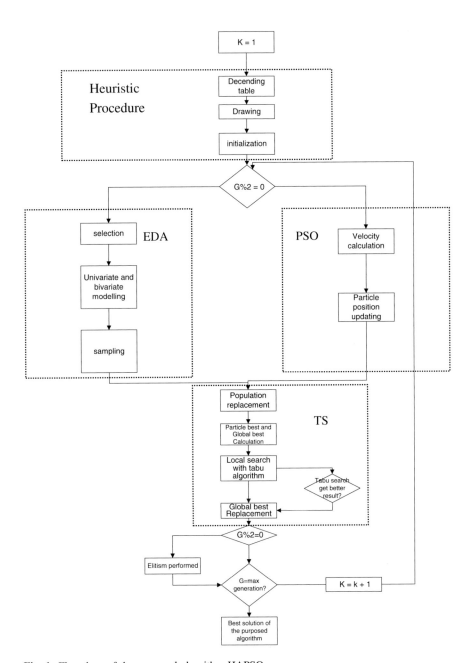

Fig. 1 Flowchart of the proposed algorithm HAPSO

3.1 Heuristic Procedure

This paper adopt (Sadrzadeh 2012) heuristic procedure for generate initial particles for EDA procedure and PSO procedure. For generating the initial particle for this purpose algorithm, this paper uses three steps. The steps are : production descending table, drawing the facilities, generate heuristic feasible initial particles.

3.1.1 Production of Descending Table

Almost all of the FLP researches have objective function minimizing the total material handling. Table 1 is flow data of material between facilities. The cost da-ta is 1 each flow of material. Then firstly Table 1 multiply with cost data. Then the relationship between facilities is obtained. The relationship numbers are arranged from the largest to smallest. Then it called descending table in Table 2

3.1.2 Drawing the Facilities

By using the column of "relationship" in Table 2, a schematic representation created, which is called drawing. The following procedure of drawing are : first, If there are two facilities in the drawing, a link line drawing between them. Second, if there is only one facility in the drawing, another facility with a link line between

Table 1 Flow data of material between facilities

From/to	1	2	3	4	5	6	7	8
1	–	100	0	0	0	0	0	0
2	–	–	0	0	0	45	0	0
3	–	–	–	0	0	0	87	0
4	–	–	–	–	0	33	0	0
5	–	–	–	–	–	0	63	0
6	–	–	–	–	–	–	0	0
7	–	–	–	–	–	–	–	150
8	–	–	–	–	–	–	–	–

Table 2 Descending table

No	Facilities	Arranged number
1	7–8	150
2	1–2	100
3	3–7	87
4	5–7	63
5	2–6	45
6	4–6	33

Fig. 2 Drawing the facilities

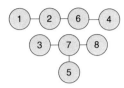

them is added to the drawing. Third, if there is no facility in the drawing, the two facilities with a link line between them are added to the drawing. The procedure is continued until all facilities assigned to the drawing. From Table 2, the drawing obtained in Fig. 2

3.1.3 Generate Heuristic Initial Particle

From Fig. 2, heuristic initial particles can be obtained by changing the configuration of the drawing and place them in the locations. For example Fig. 3 shows the two created initial particles corresponding in the drawing considered Fig. 2.

3.2 Estimation Distribution Algorithm

Initial particles is obtained from Heuristic procedure. The particles from heuristic procedure are labeled as $X^1, X^2, X^3, \ldots$ and X^N which N is half number of particles that is generated from heuristic procedure. Then univariate and bivariate probabilistic model (Chen et al. 2012) are developed from the initial particles of Heuristic procedure. The solutions generated by two probabilistic models are called as artificial particles.

This research adopted univariate and bivariate probabilistic model from (Utamima 2012). The univariate probabilistic model $\phi_{i\{i\}}i[i]$ in Eq. 2 shows the importance of facilities sequence. It represents how many times facility i is placed at position [i] at current generation. $A^k_{i[j]}$ is set to 1 if facility I is placed at position [i], otherwise it is set to 0.

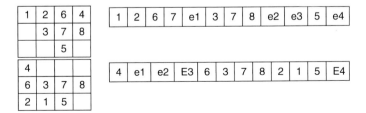

Fig. 3 Heuristic initial particles

$$\phi_{i[i]} = \sum_{k=1}^{N} A_{i[i]}^{k}, \quad i = 1, 2, \ldots, n; \; k = 1, 2, \ldots, N \qquad (2)$$

The bivariate probabilistic model ψ_{ij} in Eq. (3) represents how many times facility j is placed after facility j. B_{ij}^{k} has value 1 if facility j is placed next to facility I, otherwise it is set to 0. This proposed algorithm replace the 0 value in $\phi_{i\{i\}}$ and ψ_{ij} with 1/N for maintaining the diversity of the purposed algorithm.

$$\psi_{ij} = \sum_{k=1}^{N} B_{ij}^{k}, \quad i = 1, 2, \ldots, n; \; k = 1, 2, \ldots, N \qquad (3)$$

Let Pi[i](g) in Eq. (4) be the probability value of facility i at position [i]. Selecting facility i which has better probability value than other facility when both probabilistic models statistic information used. For every offspring $O^1, O^2, \ldots$ and O^{2N}, this method according to Utamima to assign the first facility and the rest facilities according to Chen et al. (2012) using Roulette wheel with the probability Pi[i].

$$P_{i[j]}(g) = \frac{\phi_{i[j]}(g) x \psi_{ij}(g)}{\sum_{f \in \Omega} \phi_{i[j]}(g) x \psi_{ij}(g)} \qquad (4)$$

where,
[i] 2,3, …, n; i = 1, 2,3, …, n
Ω set of unassigned facilities

3.3 Particle Swarm Optimization

Initial particles from heuristic procedure will be used for Particle Swarm Optimization (PSO) initial particles. Then the velocities are calculated for every particles. In standard PSO, the velocity is added to the particles on each dimension to update the particles. If the velocity is larger, the particle may explore more distant area. Sometimes the particle pass the best solution then cannot find the best solution. So it needs the inertia weight which is reference from is listed below :

$$w = \frac{\text{maxgen} - \text{currentgen}}{\text{maxgen}} \qquad (4)$$

Velocity formulation below:

$$V_i(t+1) = wxv_i(t) + r_1 c_1 (x_{pbest} - x_i(t)) + r_2 c_2 (x_{gbest} - x_i(t)) \qquad (5)$$

Artificial Particle Swarm Optimization

velocity	-3.8	5.6	0.5	6.0	4.1	7.3	3.6	-4.7	7.9	9.5	8.2	-4.2
Absvel	3.8	5.6	0.5	6.0	4.1	7.3	3.6	4.7	7.9	9.5	8.2	4.2
gBest	5	7	E1	2	E2	8	1	3	E3	4	E4	6
Particle	2	3	4	8	E1	5	1	6	E2	7	E3	E4
Updated particle	2	3	7	8	E1	5	1	6	E2	4	E3	E4

Fig. 4 Steps of update the particles

where V is velocity, w is weight inertia, t is iteration, r_1 and r_2 are random number, c_1 and c_2 are parameter for update the velocity, x_{pbest} is best particle each iteration, x_i current particles, x_{gbest} is the best particle among other particles.

This part use permutation-based particle updating based on concept from (Hu et al. 2003). The steps explain at Fig. 4. Get highest velocity value within particle. Get the corresponding gBest position which has the highest velocity. Find the position of value in current particle. Update the particle position that is to set the value of same position gbest by swapping the value.

3.4 Tabu Search

The best solution from EDA or PSO part each generation is the input for tabu search (TS) part. The procedure for TS is according to Utamima. If the output from the TS is better, so update the best solution.

Then performing the elitism to keep 10 % best particles of total number particles. Elitism try to keep the best solution in next iteration so that better solution can be obtained.

Table 3 Comparison between HAPSO and other approaches

Method	Best (30 runs)	Average	Worst	Succesful hits
Proposed HAPSO	11,632	11,764	11,981	12
Sadrzadeh (2012)	11,662	11,676.1	11,951	28
El-Baz (2004)	11,862	11,871.8	13,373	23
Mak et al. (1998)	12,892	15,087.7	18,657	11
PMX Chan and Tansri (1994)	14,947	18,355.9	20,654	0
OM Chan and Tansri (1994)	22,406	24,301.7	26,926	0
CX Chan and Tansri (1994)	14,717	17,216.5	20,654	0

Fig. 5 Best solution

	9	17	10		
3	20	12	8	18	6
23	24	16	4		15
7	14	2	11	19	5
		21	13	1	22

4 Case Study

This case study which has been introduced by (Kazerooni et al. 1996), 24 fa-cilities are located in a 5 × 6 grid. The number of product is 38 products. The cost needed for handling one unit of the material between the facilities is assumed to be the same. By using a comprehensive search, the determination of the optimal solution amongst the 3.68 × 1,029 feasible solutions is impossible.

HAPSO applied in this case. It start from heuristic procedure then EDA, PSO and the last is TS. We obtain the best result is **11,632**. This result is better than the other papers. Table 3 represents the comparison between HAPSO and other ap-proach from other researches. The results show that HAPSO obtains a more efficient solution as compared to the other approach. Figure 5 shows best solution.

5 Conclusion and Future Research

The proposed algorithm HAPSO can solve MLFLP more efficient than other method. The hybridization of PSO and EDA can avoid the premature convergence. It is enchanted by Tabu Search to get better solution from the EDA and PSO. The heuristic initialization make the searching area become near to the optimal solution. So the PSO, EDA and Tabu can focus to certain area. The result of HAPSO indicate that HAPSO reliable enough for solving the MLFLP with total cost 11,632. It gives better result than other method. Better solution from HAPSO is obtained from the heuristic initialization. Heuristic initialization put the higher "relationship" facilities always together.

Future research of this research can apply this method to different environment of FLP. Develop interface to help decision maker of FLP within HAPSO algorithm. HAPSO can be improved again by other approach to get more efficient results because the number of hits still below the expectation.

Acknowledgments This work has been supported by National Science Council under contract NSC 101-2221-E-011-078.

References

Chan KC, Tansri H (1994) A study of genetic crossover operations on the facility layout problem. Comput Ind Eng 26:537–550

Chen YM, Chen MC, Chang PC, Chen SH (2012) Extended artificial chromosomes genetic algorithm for permutation flowshop scheduling problems. Comput Ind Eng 62:536–545

Kazerooni M, Luonge L, Abhary K, Chan F, Pun F (1996) An integrated method for cell layout problem using geneticalgorithms. In: Proceedings of the 12th international conference on CAD/CAM robotics and factories of the future 1996

Mak KL, Wong YS, Chan TS (1998) A genetic algorithm for facility layout problems. J Comput Integr Manuf Syst 1:113–123

El-Baz MA (2004) A genetic algorithm for facility layout problems of different manu-facturing environments. Comput Ind Eng 47:233–246

Hu X, Eberhart R, Shi Y (2003) Swarm intelligence for permutation optimization, a case study on N-Queens problem. In IEEE swarm intelligence symposium. Indianapolis, USA 2003

Miao Z, Xu K-L (2009) Research of multi-rows facility layout based on hybrid algorithm. In: International conference on information management, innovation management and industrial engineering 2009

Zhou N et al (2010) Research on Multi-rows layout based on genetic algorithm. In: 3rd international conference on information management, innovation management and industrial engineering 2010

Sadrzadeh A (2012) A genetic algorithm with the heuristic procedure to solve the multi-line layout problem. Comput Ind Eng 62:1055–1064

Applying a Hybrid Data Preprocessing Methods in Stroke Prediction

Chao Ou-Yang, Muhammad Rieza, Han-Cheng Wang, Yeh-Chun Juan and Cheng-Tao Huang

Abstract Stroke has always been highlighted as big threat of health in the worldwide. Brain image examination and ultrasound are some alternatives to discover stroke disease. Data mining has been used widely in many areas, include medical industry. The uses of data mining methods can help doctors to make prediction of certain diseases. Therefore, in this research, a hybrid model integrating imbalance data preprocessing, feature selection, and back propagation network, decision tree for stroke prediction. The dataset used is brain examination data which collected from 2004 to 2011. However, highly imbalance dataset available can impact the performance of prediction as well as feature selected. The study firstly "rebalance" the dataset by comparing sampling methods; RUSboost and MSmoteBoost. In addition, important features of balance training dataset would be selected by information gain, stepwise regression based feature selection. Towards the end, selected features would be processed using Back Propagation

---

C. Ou-Yang (✉) · M. Rieza · C.-T. Huang
Department of Industrial Management, National Taiwan University of Science and Technology, Taipei, Taiwan
e-mail: ouyang@mail.ntust.edu.tw

M. Rieza
e-mail: muhd_rieza@hotmail.com

C.-T. Huang
e-mail: M10001015@mail.ntust.edu.tw

H.-C. Wang
Department of Neurology, Shin Kong Wu Ho-Su Memorial Hospital
College of Medicine, National Taiwan University, Taipei, Taiwan
e-mail: drhan@ms1.hinet.net

Y.-C. Juan
Department of Industrial Engineering and Management, Ming Chi University of Technology, New Taipei City, Taiwan
e-mail: ycjuan@mail.ncut.edu.tw

Network and Decision Tree to predict the stroke. These hybrid methods can assist doctor to provide some possibilities information to the patient.

Keywords Stroke · Preprocessing methods · Imbalance data · Feature selection · BPN

1 Introduction

Data preprocessing or preparation is an important and critical step in data mining process and it has a huge impact on the success of a data mining project (Hu 2003). Data mining method has been used widely in medical data diagnosis, include stroke prediction. Brain image examination and ultrasound are some alternatives to discover stroke disease. This paper has been used brain medical data collected from 2004 to 2011.

Because the data distribution is not balance, the existing classification techniques face some difficulties for correctly predicting the minority class. Therefore, imbalance classification becomes major problem. Sampling is a most common method to process the imbalance data sets. Under-sampling and over-sampling are two kinds of modes of sampling. Under-sampling method is used by reducing number of majority class, on the other hand over-sampling used by duplication minority class samples (Liang et al. 2009).

Before any classification method, a feature selection algorithm would be applied in order to reduce the size of search space. The objective of variable selection is threefold: improving the prediction performance of the predictors, providing faster and more cost-effective predictors, and providing a better understanding of the underlying process that generated the data (Guyon and Elisseeff 2003). Feature Selection analysis is classified into three main categories, there are Filter, Wrapper, and Embedded.

Various diagnosis model based on statistic and classification have been proposed as a decision making process. This study proposed hybrid methods by "rebalance" imbalance dataset and perform feature selection in data preprocessing. In the end, back propagation network and decision tree used to diagnose and predict the stroke.

2 Algorithm

2.1 "Rebalance" Algorithm

A. RusBoost

Random Undersampling method balances a data set by removing examples from the majority class. In addition, a boosting method can be performed either by "reweighting" or "resampling" (Seiffert et al. 2010). This paper adopted a proposed method by Seiffert et al. (2010) by combining RUS and Boost in order to resample the training data according to examples assigned weight. The performance of this method presents a simpler, faster and less complex alternative to other method from imbalanced data.

B. MSmoteBoost

Compare to under sampling method, MSmote generate data and increase the sampling weight for the minority class. MSmote is a variant of the smote algorithm, so the basic flow is consistent with Smote (Chawla et al. 2002). This method hybrid with boosting algorithm in order to enable each learner to be able to sample more of the minority class cases, and also learn better and broader decision regions for the minority class (Chawla et al. 2003). Liang et al. (2009) proposed MSmoteBoost to improve performance when training data is imbalance, this techniques is implemented in this study.

2.2 Feature Selection Algorithm

A. Information Gain (IG)

Information gain calculated by using entropy measurement. High entropy means the distribution is uniform, low entropy means the distribution is gathered around a point (Dağ et al. 2012).The range of entropy from 0 to 1. The formula of information gain:

$$Info\ Gain\ (Class,\ Attribute) = Entropy(Class) - Entropy(Class|Atrribute) \quad (1)$$

$$Entropy(S) = -\sum_{j=1}^{j=m} p_j \log_2 p_j \quad (2)$$

In this paper, the information gain uses ranker method. This algorithm ranks attributes according to average merit. It has an option to assign a cutoff point (by the number of attributes or by a threshold value) to select the attributes (Witten and Frank 2005).

B. Stepwise Regression Analysis (SRA)

Stepwise regression analysis is applied to determine the set of independent variable that most closely affect the dependant variable (Fan et al. 2011). The selected features are generated by running statistical software SPSS 17. In this method, we calculate F value, according to reference (Chang and Fan 2008; Chang and Liu 2008) if F value of a specific variable is greater than the user defined threshold, it is added as significant factor, and otherwise it is removed from the model.

$$SSR = \sum (\hat{Y}_i - \bar{Y})^2 \qquad (3)$$

$$SSE = \sum (\hat{Y}_i - Y_i)^2 \qquad (4)$$

$$F_j = \frac{MSR(X_j|X_i)}{MSE(X_j|X_i)} = \frac{SSR/1(X_j|X_i)}{SSE/(n-2)(X_j|X_i)}, i \in I \qquad (5)$$

2.3 Prediction Algorithm

A. Back Propagation Network (BPN)

Neural network has attracted many researchers to perform any classification. Classification rules are useful for solving medical problems and have been applied particularly in the area of medical diagnosis (Freitus 2002). This paper adopted Back Propagation Network as prediction techniques. A simple BPN consist of 3 layers: input node, hidden node and output node. One of the concern in determining the number of hidden nodes, the most commonly way to set by $(x+y)/2$, $(x+y)/2 \pm 1$, $(x+y)/2 \pm 2$ (Ripley 1993) or $\sqrt{xy}$, $\sqrt{xy} \pm 1$, $\sqrt{xy} \pm 2$ (Kaastra and Boyd 1996) where x is input node and y is output node.

B. C4.5 Decision Tree

This method is one of the most widely used in data mining. C4.5 method for approximating discrete-value function that is robust to noisy data and capable of learning disjunctive expression (Mitchell 1997; Quinlan 1986). There are many algorithm developed in decision tree. However, the main difference between C4.5 and other similar decision tree building algorithms is in the test selection and evaluation process (Ture et al. 2009). C4.5 decision tree algorithm uses modified splitting criteria, called gain ratio. C4.5 selects the test that maximizes gain ratio value (Benjamin et al. 2000).

3 The Proposed Hybrid

In this paper, we propose hybrid methods including data preprocessing and prediction. Data preprocessing start with removing outlier of brain medical data. In addition, the imbalance brain medical data was analyzed by applying sampling techniques (RUSBoost and MSmoteBoost). The sampling method would generate two types of *balance* data sets, and then important feature is selected by Information Gain and Stepwise Regression Analysis. These preprocessing would execute 4 sample dataset (combination of RUSBoost & SRA, RUSBoost & IG, MSmoteBoost & SRA, and MsmoteBoost & IG) as shown in Fig. 1.

Applying a Hybrid Data Preprocessing Methods in Stroke Prediction 1445

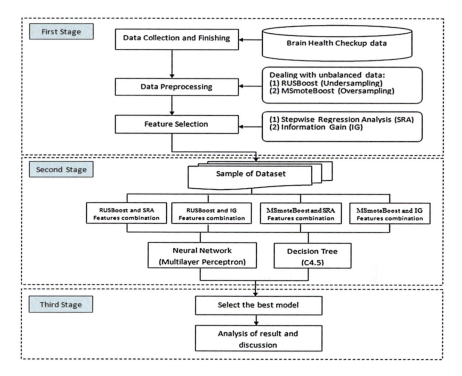

Fig. 1 Hybrid process stage

The combination of data preprocessing would be predicted by neural network and decision tree. All of these combinations have been normalized between range 1 and 0. In the later stage, we could compare 8 outputs from this hybrid method.

4 Experiments

4.1 Data Sets

Our experiments were performed on the brain examination data which collected from 2004 to 2011. The data sets contain 10,037 examples with 24 features and 1 class. The aim of data sets is to predict MRI brain examination for normal (majority class) and stroke (minority class), where 9,495 examples are from majority "normal" class and 542 examples are from the minority "stroke" class (Table 1).

Table 1 Original features of data set

Feature	Range	Feature	Range
Gender	Male/female	Fasting glucose	48–410
Age	12–93	HbA1C	3.3–16.5
Waistline	52.5–139.5	GOT	9–392
Diastolic	37–149	GPT	4–634
Systolic blood pressure	66–237	BUN	5.0–60.0
BMI	12.9–47.3	Creatinine	0.3–5.8
Hypertension	Yes/no	Total cholesterol	88–580
High cholesterol	Yes/no	HDLc	5–135
Diabetes	Yes/no	LDLc	20–442
Heart disease	Yes/no	TC/HDL	1.3–39.2
Anemia	Yes/no	Triglyceride	20–1938
Prothrombin time	8.9–37.0	MRI brain (class)	Stroke/normal
APTT	10.8–72.5		

4.2 Experimental Result and Performance Evaluation

In performance evaluation, we specify the classification accuracy, sensitivity and specificity analysis, and 10-fold cross validation to evaluate proposed method.

Classification Accuracy calculated by using equation:

$$accuracy(t) = (correctCount/transCount) * 100 \qquad (6)$$

where *correctCount* is number of correct prediction, *transCount* is total of class label.

$$sensitivity = \frac{TP}{TP + FN} \ (\%) \qquad (7)$$

$$specificity = \frac{TN}{FP + TN} \ (\%) \qquad (8)$$

where TP = True Positive, TN = True Negative, FP = False Positive and FN = False Negative (Table 2).

As can be seen from above result, MSmoteBoost_SRA_C4.5 combination achieved 95.50, 96.15 and 94.48 for classification accuracy, sensitivity and specific. However, C4.5 predicted lowest accuracy in minority data (RUSBoost_IG features combination). In feature selection techniques, indicate that there is least difference result of predicted accuracy between Information Gain (IG) and Stepwise Regression Analysis (SRA) in each combination for different sampling methods.

For prediction accuracy, comparative between BPN and C4.5 obtained different result while computing under sampling and oversampling dataset. Gap result obtained from BPN Prediction in minority and majority data is less than prediction of C4.5.

Table 2 Result comparison based on classification accuracy, sensitivity and specificity

	Prediction method Preprocessing	BPN Classification accuracy (%)	Sensitivity (%)	Specificity (%)	C4.5 Classification accuracy (%)	Sensitivity (%)	Specificity (%)
Undersampling	RUSBoost and SRA	72.80	72.52	73.08	67.13	71.10	63.17
	RUSBoost and IG	71.39	68.83	73.93	66.28	69.12	63.45
Oversampling	MSmoteBoost and SRA	83.39	78.69	88.08	95.50	96.15	94.48
	MSmoteBoost and IG	85.25	81.36	89.11	95.27	95.79	94.75

5 Conclusions and Discussion

In this paper, we reported data preprocessing techniques able to improve prediction performance of imbalance data set. A hybrid approach of sampling methods and feature selection provide four type combinations include RUSBoost & IG, RUSBoost & SRA, MSmoteBoost & IG, and MSmoteBoost & SRA. The sample data generated of these combinations goes to prediction method (Neural Network and Decision Tree) to classify the correct result.

In the process of running hybrid method, sampling with majority examples provides a better result compare to minority. However, majority examples consumes more time. This indicates that future research would be able to increase the performance of under sampling method and consuming less time while processing the experiment.

Acknowledgments This study has been supported by Shin Kong Wu Ho-Su Memorial Hospital and National Science Council, Taiwan, under contract NSC 101-2221-E-011-078.

References

Benjamin KT, Tom BYL, Samuel WKC, Weijun G, uegang Z (2000). Enhancement of a Chinese discourse marker tagger with C4.5. In Annual Meeting of the ACL Proceedings of the 2nd workshop on Chinese language processing: Held in conjunction with the 38th Annual Meeting of the Association for Computational Linguistics, vol 12. Association for Computational Linguistics, Morristown, NJ, USA, 38–45

Chang PC, Fan C-Y (2008) A hybrid system integrating awavelet and TSK fuzzy rules for stock price forecasting. IEEE Trans Syst Man Cybern Part C Appl Rev 38(6):802–815

Chang PC, Liu C-H (2008) A TSK type fuzzy rule based system for stock price prediction. Expert Syst Appl 34(1):135–144

Chawla NY, Bowyer KKW, Hall LO, Kegelmeyer WP (2002) SMOTE: synthetic minority over-sampling technique. J Artif Intel Res 16:321–357

Chawlalal NV, Lazarevic A, Hall O (2003) SMOTEBoost: improving prediction of the minority class in bosting. In: The 7th European Conference on Principles and Practice of Knowledge Discovery in Databases. Springer, Berlin, pp 107–119

Dağ H, Sayın KE, Yenidoğan I, Albayrak S, Acar C (2012) Comparison of feature selection algorithms for medical Data. In: Innovations in intelligent systems and applications (INISTA), 2012 international symposium on digital object identifier: 10.1109/INISTA.2012.6247011. IEEE, 2012:1-5

Fan C-Y, Chang P-C, Lin J-J, Hsieh JC (2011) A hybrid model combining case-based reasoning and fuzzy decision tree for medical data classification. Appl Soft Comput 11(1):632–644

Freitus AA (2002) A survey of evolutionary algorithms for data mining and knowledge discovery. In: Ghosh A, Tsutsui S (eds) Advances in Evolutionary Computation. Springer, Berlin

Guyon I, Elisseeff A (2003) An introduction to variable and feature selection. J Mach Learn Res 3:1157–1182

Hu X (2003) DB-reduction: a data preprocessing algorithm for data mining applications. Appl Math Let 16:889–895

Hu S, Liang Y, Ma L, He Y (2009) MSMOTE: improving classification performance when training data is imbalanced. In: 2nd international workshop on computer science and engineering (WCSE 2009), Qingdao, China, pp 13–17

Kaastra I, Boyd M (1996) Designing a neural network for forecasting financial and economic timeseries. Neurocomputing 10:215–236

Mitchell MT (1997) Machine learning. McGraw-Hill, Singapore

Quinlan JR (1986) Induction of decision trees. Mach Learn 1:81–106

Ripley BD (1993) Statistical aspects of neural networks. In: Barndoff-Neilsen OE, Jensen JL, Kendall WS (ed.) Networks and Chaos—statistical and probabilistic aspects, Chapman & Hall, London, pp 40–123

Seiffert C, Khoshgoftaar T, Van Hulse J, Napolitano A (2010) Rusboost: a hybrid approach to alleviating class imbalance. IEEE Trans Syst Man Cybern Part A 40(1):185–197

Ture M, Tokatli F, Kurt I (2009) Using Kaplan–Meier analysis together with decision tree methods (C&RT, CHAID, QUEST, C4.5 and ID3) in determining recurrence-free survival of breast cancer patients. Expert Syst Appl 36(2):2017–2026

Witten IH, Frank E (2005) Data mining: practical machine learning tools and techniques, Morgan Kaufmann series in data management systems, 2005

Applying a Hybrid Data Mining Approach to Develop Carotid Artery Prediction Models

Chao Ou-Yang, Inggi Rengganing Herani, Han-Cheng Wang, Yeh-Chun Juan, Erma Suryani and Cheng-Tao Huang

Abstract This paper performs a hybrid method for imbalanced medical data set with many features on it. A synthetic minority over-sampling technique (SMOTE) is used to solve two-class imbalanced problems. This method enhanced the significance of the small and specific region belonging to the positive class in the decision region. The SMOTE is applied to generate synthetic instances for the positive class to balance the training data set. Another method that used is Genetic Algorithm for feature selection. The proposed of this method is to receive the reduced redundancy of information among the selected features. On the other hand, this method emphasizes on selecting a subset of salient features with reduced

---

C. Ou-Yang (✉) · I. R. Herani · C.-T. Huang
Department of Industrial Management, National Taiwan University
of Science and Technology, Taipei 106, Taiwan
e-mail: ouyang@mail.ntust.edu.tw

I. R. Herani
e-mail: inggi.herani@gmail.com

C.-T. Huang
e-mail: M10001015@mail.ntust.edu.tw

H.-C. Wang
Department of Neurology, Shin Kong Wu Ho-Su Memorial Hospital,
Taipei, Taiwan, R.O.C
e-mail: drhan@ms1.hinet.net

H.-C. Wang
College of Medicine, National Taiwan University, Taipei, Taiwan, R.O.C

Y.-C. Juan
Department of Industrial Engineering and Management, Ming Chi University
of Technology, New Taipei City, Taiwan, R.O.C
e-mail: ycjuan@mail.ncut.edu.tw

E. Suryani
Department of Information System, Sepuluh Nopember Institute of Technology,
Surabaya, Indonesia
e-mail: erma@is.its.ac.id

number using a subset size determination scheme. Towards the end, selected features would be processed using back Propagation Network (NN) and Decision Tree to predict the accuracy of Carotid Artery Disease. Experimental results show that these methods achieved a high accuracy, so it can assist the doctors to provide some possibilities information to the patient.

Keywords Carotid artery disease · Imbalanced data · SMOTE · Feature selection · GA · Over-sampling · BPN · Decision tree

1 Introduction

Data pre-processing is a data mining technique that involves transforming raw data into an understandable format. Real-world data is often incomplete, inconsistent, imbalanced, and is likely to contain many errors. Data pre-processing is a proven method of resolving such issues. Data pre-processing prepares raw data for further pre-processing.

The class imbalanced problem has been recognized in many real world application. There are many methods to deal with imbalanced problems, includes oversampling minority class and downsizing majority class (Farquad and Bose 2012). Typically, the approaches for solving the imbalanced problem can divided into two categories: re-sampling methods and imbalanced learning algorithms. Re-sampling techniques are attractive under most imbalanced data. This is because re-sampling adjusts only the original data set, instead of modifying the learning algorithm. In the SMOTE (Synthetic Minority Over-Sampling Technique), the positive class is oversampled by creating synthetic instances in the feature space formed by the positive instances and their K-nearest neighbors (Gao et al. 2011).

After balancing the data, feature selection is used to select a subset of terms occuring in the training set. A large number of irrelevant and/or redundant features generally exist in the real world datasets that may significantly degrade the accuracy of learned models and reduce the learning speed of the models (Kabir et al. 2011). The traditional approaches in feature selection can be broadly categorized into three approaches: filter, wrapper, and embedded methods. GA (Genetic Algorithm) can reduce redundancy of information among the selected features. Oh the other hand, this method emphasizes on selecting a subset of salient features with reduced number using a subset size determination scheme. Towards the end, to predict the accuracy of data, Back Propagation Network (NN) and Decision Tree (J48) are used.

2 Methodology

2.1 SMOTE

The SMOTE, over-samples the positive class by creating synthetic instances by a specified over-sampling ratio of the minority data size (Gao et al. 2011). The over-sampling denoted as β %, each minority data sample denoted as x_0, and randomly selecting, data points linking x_0 with K-nearest network (K is predetermined), and synthetic instance is denoted as x_s.

$$x_s = x_0 + \delta \cdot \left(x_0^{[t]} - x_0 \right) \tag{1}$$

The detailed of SMOTE can be seen as below:

1. SMOTE initialization: Specify the balanced degree β % and the value of K.
2. Create the new training data set $\tilde{D}_N$ by appending the generated positive training data points to the original training data set via the SMOTE.

2.2 Genetic Algorithm

Genetic Algorithms have received significant interest in recent years and are being increasingly used to solve real-world problems. A GA is able to incorporate other techniques within its framework to produce a hybrid that reaps the best from the combination (El-mihoub et al. 2006).

The steps of Hybrid GA can be explained further as follows:

Step 1: Initialize a feature set F of f features, a subset K of k salient features, and a population set P of p strings.
Step 2: Calculate the fitness value of each string p in P using the feed-forward NN training model.
Step 3: Perform the standard crossover operation upon the pair of fitter strings sequentially. The selection of possible fitter strings is made followed by standard rank-based selection procedure with utilizing the predefined crossover probability.
Step 4: Perform the standard mutation operation over the newly generated off-springs. Each bit of every string follows the predefined mutation probability whether it likes to be mutated or not.
Step 5: Perform the local search operation upon an offspring among the all newly generated ones sequentially in order to readjust the number of 1-bits.
Step 6: Replace the strings of lowest rank order in P by the whole generated offsprings.

Step 7: Check the stopping criterion to stop the genetic process. If the criterion is satisfied, then continue; otherwise, go to Step 2.
Step 8: Find the best string from the final generation according to its fitness value that signifies the desired subset of salient features for the given dataset.

2.3 Back Propagation Network

Back Propagation is a common method of training Artificial Neural Networks (ANN). From a desired output, the network learns many inputs. This method is not only more general that the usual analytical derivations, which handle only the case of special network topologies, but also much easier to follow. It also shows how the algorithm can be efficiently implemented in computing systems in which only local information can be transported through the network.

The algorithm of BPN applies the fundamental principle of the gradient steepest descent method to minimize the error function. It compares the outputs of the processing units in the output layer with desired outputs to adjust the connecting weights (Chen et al. 2010). There are three layers inside BPN:

1. Input layer: To demonstrate the input nodes of variables. The input layer receives features of input data and distributes them to hidden layer. The number of input nodes depends on different problems, $f(x) = x$.
2. Hidden layer: To show the interactions among input layer and output layer. There is no standard rule to define the number of hidden nodes. The usual way to get the best number of nodes is by experiments.
3. Output layer: To indicate the output nodes of variables. The number of output nodes depends on different problems to be solved. where X_i is the input vectors, $i = 1, 2, ..., m$, Y_j is the output vectors, $j = 1, 2, ..., n$ and H_t is the hidden nodes, $t = 1, 2, ..., k$.

2.4 Decision Tree

Decision Tree is a decision support tool that uses a model of decision and their possible consequences. These methods are commonly used in operations research, specifically in decision analysis and to help identify a strategy. C4.5 Decision Tree classifier and one-against-all method were combined to improve the classification accuracy for multi-class classification problems including dermatology, image segmentation, and lymphog- raphy datasets (Polat and Güneş 2009).

3 The Proposed Method

In this paper, we proposed methods that include pre-processing and prediction data. The study firstly rebalances the imbalanced dataset by using SMOTE (Synthetic Minority Over-Sampling Technique). After generated the balanced data set, then important features of balance training dataset would be selected by Hybrid Genetic Algorithm Feature Selection. These pre-processing and feature selection can be shown in first stage Fig. 1.

Towards the end, the combination of pre-processing and selected features would be processed using Back Propagation Network (NN) and Decision Tree to predict the best model.

4 Experiments

This study used North Medical Center Brain Health Checkup between 2004 and 2011, seven years of cerebral vascular health and diagnosis data set. It contains general inspection, blood and urine checks, and also professional brain check data (Table 1).

The datasets contain 9.892 unbalanced data, which is 9.578 data is Stroke (majority class) and 314 data is Normal (minority class). This dataset has 21 features that taken from General Inspection and Blood and Urine Checkup (Table 2).

For the evaluation, we specify the classification accuracy, sensitivity, and specificity for proposed methods. The equation of classification accuracy, sensitivity, and specificity (Table 3):

$$fp_{rate} = \frac{FP}{N} \qquad (2)$$

$$precision = \frac{TP}{TP + FP} \qquad (3)$$

$$accuracy = \frac{FP}{N} \qquad (4)$$

$$tp_{rate} = \frac{TP}{N} \qquad (5)$$

$$recall = \frac{TP}{P} \qquad (6)$$

$$F_{measure} = \frac{2}{1/precision + 1/recall} \qquad (7)$$

Additional terms associated with ROC are:

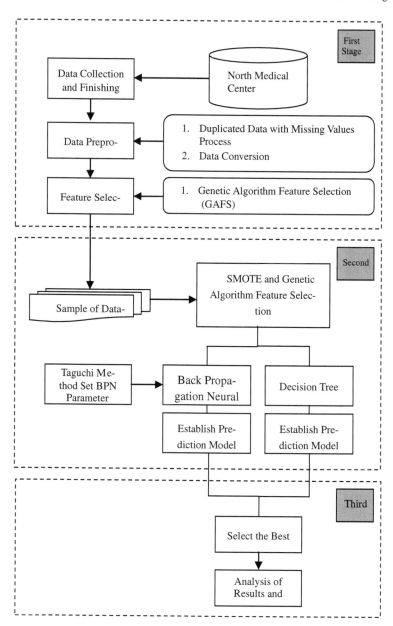

Fig. 1 Framework of the methodology

Table 1 Brain health checkup

Testing category	Test items
General Inspection	Waist circumference, blood pressure (left/right), respiration, pulse, temperature, height, weight, BMI, and past history, habits, and others
Blood and urine checkup	Thrombin original time (prothrombin time), and part coagulation blood live tenderloin time (APTT), and before meals blood glucose, and mashing hemoglobin (HbA1c), and bran amino acid grass acetate go amino enzymes (GOT), and bran amino acid Coke grape go amino enzymes (GPT), and urea nitrogen (BUN), and muscle anhydride (Creatinine), and total cholesterol (Total Cholesterol), and high-density fat protein (HDLc), and low density fat protein (LDLc), and triglyceride (Triglyceride), cysteamine acid (Homocysteine), and others
Professional brain checkup	Neck vascular ultrasound, vascular ultrasound, color wearing a cranial ultrasound, MRI, nasal MRI, MRI brain blood vessels in the brain, EEG brain wave, and others.

Table 2 Important features dataset

General Inspection		Blood Tests	
Name	Scope	Name	Scope
Sex	Male/female	Prothrombin time	8.9–37
Age of health inspection	12–93	The active part thrombosis time	10.8–115.8
Waist circumference	16–139.5	Fasting glucose	48–410
Diastolic blood pressure	40–146	Glycosylated hemoglobin	3.3–17.1
Systolic blood pressure	73–237	Total cholesterol	76–580
BMI	12.9–47.3	High-density lipoprotein cholesterol	5–125
Hypertension (history)	With/without	Low-density lipoprotein cholesterol	20–442
Anemia (medical history)	With/without	Total cholesterol/high-density lipoprotein cholesterol	1.3–39.2
High blood cholesterol (history)	With/without	Heart disease (medical history)	With/without
Diabetes (history)	With/without	Triglycerides	20–1,938

$$\text{Sensitivity} = \text{recall} \quad (8)$$

$$specificity = \frac{TN}{FP + TN} = 1 - fp_{rate} \quad (9)$$

$$\text{Positive predictive value} = \text{precision} \quad (10)$$

where TP = True Positive, TN = True Negative, FP = False Positive and FN = False Negative.

Table 3 Result comparison based on classification accuracy, sensitivity, and specificity

Methodology	Correctly classified					
	Back propagation neural network			Decision tree (C4.5)		
	Classification (%)	Sensitivity	Specificity	Classification (%)	Sensitivity	Specificity
SMOTE and genetic algorithm feature selection attributes combination	80.08	0.743	0.859	93.435	0.934	0.934

As we can see from the Table 3, SMOTE and Genetic Algorithm Feature Selection achieved **80.08 %, 0.743, 0.859** for classification accuracy, sensitivity and specificity using BPN, and **93.435 %, 0.934, 0.934** for classification accuracy, sensitivity and specificity using Decision Tree.

5 Conclusions & Discussion

Our study in this paper show that activity modeling that is developed using hybrid methods can provide a higher accuracy for predicting Carotid Artery Disease based on patient instances data and features. These hybrid methods include SMOTE that used to solve two-class imbalanced problems, Genetic Algorithm used for receive the reduced redundancy of information among the selected features, and for selected features would be processed using back Propagation Network (NN) and Decision Tree to predict the accuracy.

Genetic Algorithm that is used for independent feature selection provides a good result for the prediction. However, the result of independent feature selection may not so satisfactory than the dependent or interdependent ones. This indicates that future research might be able to increase the accuracy of feature selection method by using other feature selection method.

Acknowledgments This study has been supported by Shin Kong Wu Ho-Su Memorial Hospital and National Science Council, Taiwan, under contract NSC 100-2221-E-011-036.

References

Chen F-L, Chen Y-C, Kuo J-Y (2010) Applying moving back-propagation neural network and moving fuzzy neuron network to predict the requirement of critical spare parts. Expert Syst Appl 37(6):4358–4367

El-mihoub TA, Hopgood AA, Nolle L, Battersby A (2006) Hybrid genetic algorithms: a review. Eng Lett 13:124

Farquad MAH, Bose I (2012) Preprocessing unbalanced data using support vector machine. Decision Supp Syst 53(1):226–233

Gao M, Hong X, Chen S, Harris CJ (2011) A combined SMOTE and PSO based RBF classifier for two-class imbalanced problems. Neurocomputing 74(17):3456–3466

Kabir MM, Shahjahan M, Murase K (2011) A new local search based hybrid genetic algorithm for feature selection. Neurocomputing 74(17):2914–2928

Oh I-S, Lee J-S, Moon B-R (2004) Hybrid genetic algorithms for feature selection. IEEE Trans Pattern Anal Mach Intell 26(11):1424–1437

Polat K, Güneş S (2009) A novel hybrid intelligent method based on C4.5 decision tree classifier and one-against-all approach for multi-class classification problems. Expert Syst Appl 36(2):1587–1592

Comparing Two Methods of Analysis and Design Modelling Techniques: Unified Modelling Language and Agent Modelling Language. Study Case: A Virtual Bubble Tea Vending Machine System Development

Immah Inayati, Shu-Chiang Lin and Widya Dwi Aryani

Abstract The developing of internet worldwide encourages the research in Software Engineering fields in the development of numerous analysis and design methods used in Software Development. Among these methods, Unified Modelling Language (UML) has been known to Software developers as a popular object-oriented tool to analyze and design a system. On the other hand, a less known tool, Agent Modeling Language (AML), is a semi-formal visual modeling language for specifying, modelling and documenting system that incorporate features drawn from multi-agent system theory. This paper presents an overview of UML and AML using the case of Virtual Bubble Tea Vending Machine Software development. This paper also compares UML and AML methods in analyzing phase and designing phase in system development.

Keywords Software engineering · Unified modelling language · Agent modeling language · Virtual bubble tea vending machines

---

I. Inayati (✉) · S.-C. Lin · W. D. Aryani
Industrial Management, National Taiwan University of Science and Technology,
Keelung Road Sec. 4 no. 43 Daan District, Taipei 106, Taiwan, Republic of China
e-mail: immah.inayati@yahoo.com

S.-C. Lin
e-mail: slin@mail.ntust.edu.tw

W. D. Aryani
e-mail: widyadwiaryani@yahoo.co.id

1 Introduction

Modelling is used in many walks of life and is also widely used in science and engineering to provide abstractions of a system at some level or precision and detail. After a model is built, it is then analyzed in order to obtain a better understanding of the system being developed. According to Object Modelling Group (OMG) "Modelling is the designing of software applications before coding" (Gomaa 2011).

Many modelling techniques can be used to develop a system or a software. Those models have their own notations and diagrams to make a better understanding of a system. Object based modelling and agent based modelling are two such techniques that have been used widely. This paper will present the analyzing and design phase of the development of Virtual Bubble Tea Vending Machine (VBTVM) system using two modellings: the object based modelling (using the UML diagram) and the agent based modelling (using Gaia method).

1.1 Object Oriented and Agent Oriented Modelling

Object oriented modelling methods apply object oriented concepts to the analysis and design phase of the software lifecycle. The emphasis is on identifying real-world objects in the problem domain and mapping them to software objects.

Agent oriented modeling applies agent based concept to the analysis and design phase of software lifecycle. Agent is a computer system that is capable of independent action on behalf of its user or owner (Wooldridge 2004). This model is normally used to specify, model and document systems by applying the concept of *Multi-Agent Systems* (*MAS*) theory.

1.2 Unified Modelling Language and Agent Modelling Language

Unified Modelling Language (UML) is a standard language for writing software blueprints. Booch et al. (2005). It is a visual language that provides a way for people to visualize, construct, and document the artifacts of software system (Bennt et al. 2005).

There are three important characteristics inherent in UML, i.e., sketches, blueprints and programming languages. As a sketch, UML can serve as a bridge to communicate some aspects of the system. Thus all members of the team will have the same picture of a system. UML can also serve as a blueprint for a very complete and detailed. With this blueprint will be known to compile detailed information program code (forward engineering) or even to read and interpret the program back to the diagram (reverse engineering).

The Agent Modeling Language (AML) is specified as an extension to UML 2.0 in accordance with major OMG modeling frameworks (MDA, MOF, UML, and OCL). The ultimate objective of AML is to provide software engineers with a ready-to-use, complete and highly expressive modeling language suitable for the development of commercial software solutions based on multi-agent technologies (Trencansky and Cervenka 2005). There were many Methodology using this way of Modelling and Gaia is one of them.

Gaia Methodology is intended to allow an analyst to go systematically from a statement of requirements to a design that is sufficiently detailed so it can be implemented directly (Wooldridge 2004). The objective of the Gaia analysis process is the identification of the roles and modeling the interactions between them (Dennis et al. 2012).

The Analysis phase starts with the definitions of the *global organization goal* which includes the decomposition of the global organization into sub-organizations. The next step is to build the *environmental model* that lists all resources in which one agent can access. The *preliminary role model* is built to capture the basic skills. In this model, a role is represented with an abstract and semiformal description (Blanes et al. 2009).

2 Proposed Model of UML and AML for Virtual Bubble Tea Vending Machine

2.1 Method Used in Modelling Analysis and Design Phase for Bubble Tea Vending Machine Using UML

Figure 1 is the Unified Modelling Language (UML) diagram used to analyze and Design Virtual Bubble Tea Vending Machines. Not all the UML diagram used in this development, as shown below:

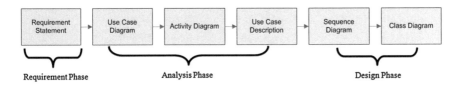

Fig. 1 UML model virtual bubble tea vending machines

2.2 Method Used in Modelling Analysis and Design Phase for Bubble Tea Vending Machine Using AML

Figure 2 illustrates the Agent Modelling Language (AML) using Gaia model. This model will be used in analyzing and designing Virtual Bubble Tea Vending Machine.

3 Preliminary Results and Discussion

Among the three stages of developing proposed model of UML and AML for Virtual Bubble Tea Vending Machine, the authors have completed the first 2 stages, which include the requirement phase and the analysis phase. The analysis phase that has been done are Use case diagram and Activity diagram for UML Diagram and Role model and Interaction model for AML (Gaia) Method.

3.1 Requirement Statement for Virtual Bubble Tea Vending Machine

Table 1 shows our proposed Requirement statement for virtual Bubble Tea Vending Machine. This Requirement will be continued with the Analyzing and Designing steps with two methods, Agent Based and Object Based.

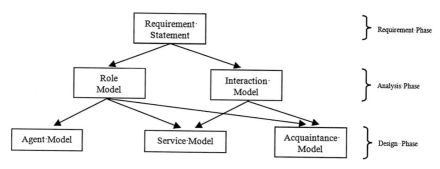

Fig. 2 Gaia model for virtual bubble tea vending machines

Table 1 Requirement statement of bubble tea vending machine

No	Requirement
1	System can identify different user
2	System can save the data of user (including name and email)
3	System facilitate a log management system
4	System can give service to a registered user on a non registered user
5	User can see Bubble tea catalog
6	System can receive 2 kinds of payment: cash and easy card
7	System can give change to cash payment
8	System have to ask user the Bubble tea he/she wants to buy before payment

3.2 Method Used in Modelling Analysis and Design Phase for Bubble Tea Vending Machine Using UML

Figure 3 (revised from Bennet et al. 2005) shows the UML diagrams in Analysis phase based on the requirement statements in the previous phase. The Diagram completed currently are Use case diagram and Activity diagram.

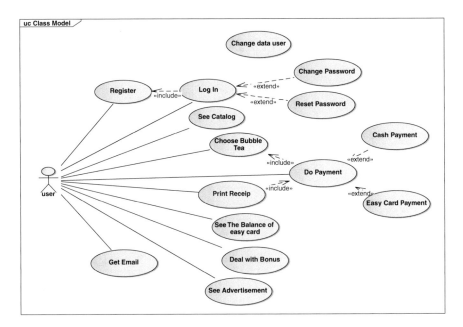

Fig. 3 Use case diagram for virtual bubble tea vending machine system

3.2.1 Use Case Diagram

As can be shown in Fig. 3, Use case diagram shows actors, use cases, and the association among them. Use cases attempt to capture the functional requirements of the system by describing the different ways in which an actor (essentially a type of user) can interact with the system (Hunt 2000). This diagram shows the "user" actor that can access 10 use cases. *Extends* association in the diagram means that actor can choose between some use case and *include* association means that before doing one use case other use cases needed to be accomplished first.

3.2.2 Activity Diagram

Activity diagram is one of the five diagrams used in UML for modeling the dynamic aspects of the system. An activity diagram is essentially a flow chart, showing flow of control from activity to activity. In this development, the activity diagram describes the detail activity of each use case. The picture above shown the activity diagram that show the detail activities of "Choose Bubble Tea" use case (Fig. 4).

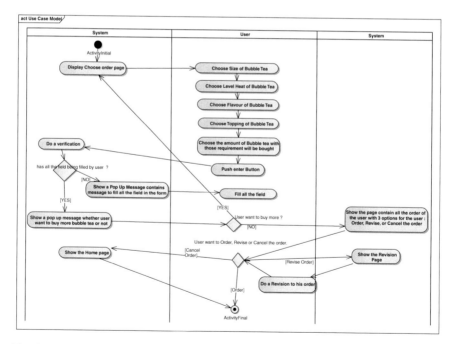

Fig. 4 Activity diagram for virtual bubble tea vending machine system

3.3 Method Used in Modelling Analysis and Design Phase for Bubble Tea Vending Machine Using AML

The Organization

The first phase in Gaia analysis is to determine whether multiple organizations have to coexist in the system and become autonomous interacting MASs (Jain and Dahiya 2011). In our system, the organizations can be easily identified as:

- The one that takes care Transaction
- The one that Make the Bubble Tea

3.3.1 Environmental Model

The environment is treated in terms of abstract computation resources, such as variables or tuples, made available to the agents for sensing. The environment model for the Bubble Tea Vending Machine can be depicted as (Table 2):
  Role Model
  In the Bubble Tea Vending Machine System, the roles can be identified as:

1. Bubble_manager
2. Register
3. Authorization
4. Solve_query
5. Order_manager
6. Display
7. Bubble_producer
8. Bonus_manager
9. Payment _manager
10. Send_email

Bubble_manager role keeps all the information about bubble_catalogue. The *register* role is responsible to register new buyer. The buyer will give his/her name, password and email address. If the buyer is new, then the registration process is handled by this agent, otherwise the request is sent to *authorization* agent. The Authorization role is used to get the username and password from the user. The details are matched with the database. Once the *Authorization* is done, a message is passed to the *solve_query* agent. The *solve_query* agent then takes the query

Table 2 Environmental model of bubble tea vending machine

bubble_catalog	The list of Bubble tea provided
bubble_recommended	The list Bubble tea recommended for the user
buyer_order	The item Bubble tea that is ordered by the buyer
buyer_detail	The data of buyer

Table 3 Role model of bubble tea vending machine

Role schema: order_manager	
Description:	
This preliminary role involves keeping the order of each buyer before it is being paid. It uses the data structures called bubble_catalogue that contains information of the detail of bubble tea needed to be produced. The details are: the size, the heat level, the taste, and the topping of bubble tea. This role also change the data of buyer	
Protocols and activities	
add_bubble_order, change_bubble_order, change_buyer_bonus	
Liveness:	
Bubble_order = (add_bubble_order, change_bubble_order)	

from the user and solves it. After it is solved, the agent will send a message to the *display* role, which will display the Order page. The Buyer will then choose his/her Bubble tea order.

The *order_manager* role will enable user to make a choice to the bubble tea with detailes of the size, heat level, flavor, topping, and the number of bubble tea. After the *order_manager* role is done, *bubble_producer* produces the bubble tea. *Bonus_manager* role enables buyer to choose the bonus options. *Payment_manager* role will help buyer to choose pay options. *Send_email* role enables system to send email to buyer advertised information such as product promotion, new product launch, and lottery winner of promoted products.

4 Result and Future Work

This paper discusses the differences between Unified Modelling Language and Agent Modelling Language in Analyzing phase. Since UML is based on object, it begins with Use Case which explains what the system does. On the other hand, the Gaia model, an Agent Modelling Language, is based on agents and it begins with the role of organization.

Our future work will encompass the design phase and the final implementation phase as well so we can make a more thorough comparison based on rules and guidelines, notation, techniques used, system developed, and the complete documentation (Table 3).

References

Bennet S, Skeleton J, Lunn K (2005) Schaum's outlines UML, 2nd edn. Mc Graw Hill, Singapore pp 5–10

Blanes D, Insfran E, Abrahão S (2009) RE4Gaia: a requirements modeling approach for the development of multi-agent systems. Springer, Berlin Heidelberg ASEA 2009, CCIS 59, pp 245–252

Booch G, Rumbough J, Jacobson I (2005) The unified modelling language user guide, 2nd edn. Addison Wesley, United States, p 267

Dennis LA, Boissier O, Bordini RH (2012) A Gaia-driven approach for competitive multi-agent systems. Springer, Berlin Heidelberg, pp 208–216

Gomma H (2011) Software modelling and design. Cambridge University Press, New York, p 3

Hunt J (2000) The unified process for practitioners. Springer, London, p 63

Jain P, Dahiya D (2011) Architecture of a library management system using Gaia extended for multi agent systems. ICISTM 2011, CCIS 141, pp 340–349

Trencansky I, Cervenka R (2005) Agent modeling language (AML): a comprehensive approach to modeling. Informatica 29:391–400

Wooldridge M (2004) An introduction to multiagent system. Wiley, LTD, England, pp 228–230

Persuasive Technology on User Interface Energy Display: Case Study on Intelligent Bathroom

Widya Dwi Aryani, Shu-Chiang Lin and Immah Inayati

Abstract Computing products for creating persuasive technology are getting easier to use with innovations in online video, social networks, and metrics, among others. As a result, more individuals and organizations can utilize different media to influence people's behaviour via technology channels. In-home displays (IHDs) are one of these trendy and powerful media that have the potential to communicate energy usage feedback and to persuade energy saving action to householders. By providing real-time information on energy consumption, IHDs can persuade householders to change into target behaviour. IHDs user interface design plays a significant role in the success of persuading user behaviour change. This paper presents a laboratory study to investigate how the user interface design of IHDs might persuade people to save energy. A model was proposed, questionnaires, including open-ended questionnaire and closed-ended questionnaire, were created at a first phase of the lab study to gather information regarding user preference among 35 information displays and icon displays. In the second phase study, two user interface prototypes, one with target behaviour feedback and one without, will be developed to investigate user's behaviour change in intelligent bathroom with regard to energy conservation.

Keywords Persuasive technology · Intelligent bathroom · Energy consumption · In-home displays

---

W. D. Aryani (✉) · S.-C. Lin · I. Inayati
Industrial Management, National Taiwan University of Science and Technology, Keelung Road Sec. 4 no. 43 Daan District, Taipei 106, Taiwan, Republic of China
e-mail: widyadwiaryani@yahoo.co.id

S.-C. Lin
e-mail: slin@mail.ntust.edu.tw

I. Inayati
e-mail: immah.inayati@yahoo.com

1 Introduction

Energy conservation of household has being an interesting topic within applied social and environmental psychological research for number of decades (Abrahamse et al. 2005). People's interaction with indoor environment plays a significant role in energy consumption (Yao and Zheng 2010). Studies such as (Doukas et al. 2007) further pointed out that the requirements for necessary thermal comfort, visual comfort and indoor air quality assurance are increasing. There is a need to efficiently manage energy inside building and to support building energy conservation. This effort aims to assure the operational needs with the minimum possible energy cost.

The major barriers to efficient energy management are mismatching and delaying feedback information to the building energy management system (Yao and Zheng 2010). Energy consumption feedback is typically provided in form of monthly bills report. Studies have shown that implementing real-time feedback information inside an ordinary house has reduced individual household electricity consumption by 4 to 12 % (Ehrhardt-Martinez et al. 2010). By providing real-time feedback information, user will be aware of the correlation between their everyday behaviour and energy consumption (Abrahamse et al. 2005). This awareness will persuade user to change behaviour with regards to energy consumption. Presenting real-time feedback information by monitoring and display technologies has more potential for achieving energy savings than giving information alone.

Display technologies can be supported with an interactive computing design to change people's attitude and behaviour. This kind of interactive computing design has been known as persuasive technology. Persuasive technology can take on many forms from mobile phones to smart devices. In this research, we use intelligent bathroom user interface as our case study. Our persuasive system was developed based on the "Functional Triad", a framework for thinking about the roles that computing products play from the perspective of the user (Fogg 2003). The functional triad proposes that interactive technologies can operate in three basic ways: as a tool, as a media, and as a social actor.

This research focused on social actor role because, as Fogg (2003) acknowledged, humans apply the same persuasive principle to influence others. The fact that people responds socially to computer products has significant implications for persuasion. As a social actor, computing product can persuade people to change their attitudes or behaviours by rewarding them with positive feedback, modelling a target behaviour or attitude, or providing social support (Fogg 2003).

The objective of this research is to investigate whether computing product as a social actor has a significant implication in persuading people and hence yields to people's attitude and behaviour change. Figure 1 illustrates our study's research framework.

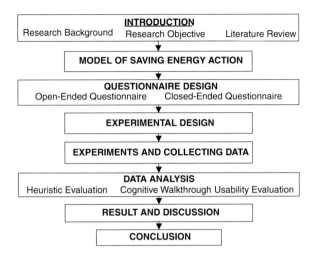

Fig. 1 Research framework

2 Model of Saving Energy Action

Figure 2, a revised model based on EBM Consumer Decision Process-Model (Engel et al. 1995), shows how users are being triggered to choose saving energy actions, a decision making process that user will go through. This model is adapted because it describes in detail the system of internal and external forces that interact and affect how the user thinks, evaluates, and acts.

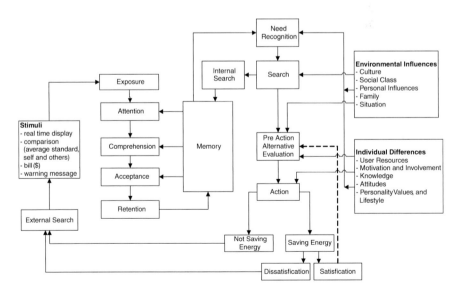

Fig. 2 Model of saving energy action

EBM model proposed five stages that consist of need recognition, information search, pre-action alternative evaluation, action, and post action evaluation. We consider need recognition is a stage when user realized to minimize the energy used which occur as effect from information searching. In action stage, user starts to evaluate the alternative action and decide whether to choose saving energy or not, in terms of expected benefits and narrowing choice to the preferred one. If user chooses not to save energy, the decision process will go back to information searching process. If user chooses to save energy, there will be a post action evaluation whether user is satisfied or dissatisfied. Satisfied user will do action repetitively while dissatisfied user's decision making process will go back to an external search. Satisfaction in this research is shown by how user responds when they had minimised or efficiently used the household appliances.

Model of Saving Energy Action is reciprocal with the three principal factors in Fogg Behaviour Model (FBM) (Fogg 2009) which become a way to understand the drivers of human behaviour. The target behaviour in this research refers to save energy action. Motivation and ability in FBM are represented by Individual Differences while ability becomes one of user resources in our proposed model. Meanwhile, triggers are represented by stimulation in form of real time display, comparison information, bill, and warning message.

3 Implementation of Persuasive Technology in Energy Display

In first phase of our study, questionnaires were developed to gather user preferences for user interface design. The first questionnaire, an open-ended questionnaire, was distributed to five participants, 3 females and 2 males, with ages ranging from 23 to 40 years old. Participants are graduate students specializing in human factors and all participated voluntarily. Participants were interviewed individually to answer his/her preferences regarding energy display information and icon displays. It took an average of 30 min to complete the questionnaire.

A second questionnaire, closed-ended questionnaire, was employed based on the results of the first questionnaire to gather more precise assessments of the information display (35 contents) and icon display (35 icons). Likert-type seven point scales ranging from very important, important, slightly important, neutral, slightly unimportant, unimportant, and very unimportant were applied. Questionnaire was distributed via online questionnaire system and a total of 123 responses were collected with age ranging from 19 to 42 years old and with various occupations.

In the second phase, we will design a persuasive user interface based on functional triad roles. Two user interface prototypes will be developed, including one with social actor role and the other one without social actor role. Participants will be asked to finish 15 tasks based on three different task difficulty levels (easy, medium, and difficult). A post-experiment general satisfaction questionnaire will be distributed as a subjective measure compared to objective measure results gathered during the user interface experiment to examine the effect of persuasive technology.

4 Preliminary Result and Discussion

Summarized results from the open-ended questionnaire reveal that an intelligent bathroom user interface display should provide not only common intelligent bathroom functions such as water temperature or water filling automation, but also provide broader functions such as In-Home Display (IHD), and it would be best to provide smart phones functions. These functions are further breakdown into 35 different functions. The questionnaire also found five alternative icons preferred by the users in information display.

Closed-ended questionnaire further reveals that among 35 information to be displayed in intelligent bathroom, "bathroom indoor air quality" and "home security" are ranked as very important information while "bed temperature", "location recognition", "home appliances automation", "education/learning", "picture", "e-books", and "GPS" are ranked as "neutral". Participants also have chosen the most represented icon for each intelligent bathroom functions. These results were applied to our next phase of implementing our user interface display design.

Since the responses were at least as good as neutral, we plot all of these 35 information displays into user interface design. The first functional triad, *as a tool*, has a goal to make activities in bathroom more user friendly. One example of this is shown in Fig. 3's Lamp setting, users can easily switch lights on and off and adjust the light level by finger touching the icons.

The second functional triad, *as a media*, has two categories, symbolic and sensory. Symbolic media category uses symbols (text, graphics, charts, and icon) to convey information. Sensory media provides sensory information such as audio, video, and even smell and touch sensations. The third functional triad, *as a social actor*, means user will respond to the user interface display as though it were a living being and people will get emotionally involved with user interface display through social actor role. In our user interface design, we adapted this role in giving advice to user to lead to energy saving actions. Figure 4 shows the still under-developing user interface display when the display is placed in standby position for 5 min.

Once the user interface prototypes are completed developed, the respondents will be asked to finish designated tasks in our lab and to what extent the persuasive technology plays a role in our intelligent bathroom user interface will be investigated.

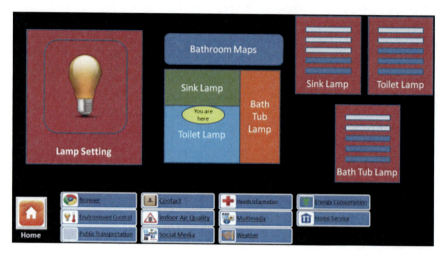

Fig. 3 Persuasive user interface design—as a tool

Fig. 4 Persuasive user interface design—as a social actor

5 Conclusions

This paper presents a case study in which the user interface display could persuade people in saving energy action. Questionnaire design and analysis were conducted to gather which information display should be provided in user interface display. Preliminary study results 35 different information displays and icon displays which were applied in designing user interface design. This research will then test user interface design by evaluating the usability and user preference.

References

Abrahamse W, Steg L, Vlek C, Rothengatter T (2005) A review of intervention studies aimed at household energy conservation. J Environ Psychol 25:273–291

Doukas H, Patlitzianas KD, Iatropoulos K, Psarras J (2007) Intelligent building energy management system using rule sets. Build Environ 42:3562–3569

Ehrhardt-Martinez K, Donnelly KA, John A (2010) Advanced metering initiatives and residential feedback programs: a meta-review for household electricity-saving opportunities. Am Counc Energ Efficient Econ

Engel JF, Blackwell RD, Miniard PW (1995) Consumer behavior. International edition, 8th edn. The Dryden Press, Orlando

Fogg BJ (2003) Persuasive technology: using computers to change what we think and do. Morgan Kaufmann Publishers, United States of America

Fogg BJ (2009) A behavior model for persuasive design. Persuasive Technology Lab, Stanford University

Yao R, Zheng J (2010) A model of intelligent building energy management for the indoor environment. Intell Buildings Int 2:72–80

Investigating the Relationship Between Electronic Image of Online Business on Smartphone and Users' Purchase Intention

Chorng-Guang Wu and Yu-Han Kao

Abstract The extensive popularity of smartphones has been recently providing online business practitioners with an alternative useful channel to communicate product information with potential and existing customers. Previous research on Internet shopping suggests that electronic image, the quality of product information within the shopping website, plays a critical role in influencing online shoppers' purchase decisions substantially. Therefore, by leveraging prior studies on website quality and web customer behavior, this study aims to identify the influential factors that contribute to the electronic image of shopping websites on smartphones, and investigates the effect of smartphone users' perception of the electronic image upon their intended purchases. The arguments proposed for this study were empirically validated by using the data from a web survey of 321 smartphone users in the context of two online bookstores. The findings suggest that users' perception of the electronic image on a smartphone related to the usefulness and scope of product information influences their purchase intention.

Keywords Smartphone · Electronic image · Website quality · Purchase intention

1 Introduction

Over the past few years, smartphones, more than just a cellphone, have been developed as mobile gadgets for connecting to the Internet. Not only can they work as media players, GPS, and cameras but also they are tools for surfing the

---

C.-G. Wu (✉) · Y.-H. Kao
College of Management, Yuan Ze University, 135 Yuan-Tung Road, Chung-Li,
32003 Taoyuan, Taiwan, Republic of China
e-mail: chuckwu@saturn.yzu.edu.tw

Y.-H. Kao
e-mail: S1007117@mail.yzu.edu.tw

Web, managing emails and even checking social media. Although desktop and laptop computers are so far the major devices used to get online, it is worth noting that more and more people are accessing the Web from their smartphones. Therefore, various online businesses gradually deploy their smartphone-based websites for phone holders to easily access their online services in terms of obtaining potential business opportunities. For this, providing appropriate information necessary to satisfy smartphone users is one of the critical issues facing online business practitioners.

According to the e-commerce literature, electronic image (e-image), consisting of website appearance and information content, is considered as one of the most important influential factors of purchase intention related to online business websites (Gregg and Walczak 2008). However, previous research largely focused on examining online purchasers' perception of e-image in the context of computers rather than smartphones. And, e-image in this study that refers to the quality of product information within a shopping website has not been fully defined for the smartphone setting. Thus, based on prior studies on website quality and online shopper attitudes, we attempted to identify several factors that formed the e-image of shopping websites within a smartphone and study the relationship between the factors and smartphone users' online purchase intention.

2 Literature

2.1 Electronic Image and Online Shopping

As an information signal delivered by online businesses to their customers, electronic image (e-image) within the web settings affects online shoppers' overall perception of the value offered by an online store (Lohse and Spiller 1998), and has been considered to be a critical factor that influences the success of e-business (Gregg and Walczak 2008). Unlike online image which presents the photographs or graphics of products online, e-image is viewed as a personality of website that refers to the presentation and quality of product information and an online firm's policies information (Gregg and Walczak 2008). It helps an online business shape its unique online identity as the business can develop a suitable e-image (Hesketh and Selwyn 1999; Sand 2007).

Literature suggests that an online business's e-image is significantly influenced by the quality of information content related to its website which in turn will have a direct and considerable impact on online shoppers' purchase intention (Gregg and Walczak 2008). Online shoppers usually cannot have physical contact with the products demonstrated on the website while purchasing online; under the circumstances, they largely rely on the website information to establish their shopping experience (McKinney et al. 2002). As a result, the enhancement of information quality can foster their intention to purchase online (Gregg and Walczak 2008).

Similarly, prior studies on e-commerce indicate that website quality including design, scope of information, security and privacy, serves as a very useful metric to evaluate online shoppers' purchase intention (Ranganathan and Ganapathy 2002; Park et al. 2004).

2.2 Attributes of Electronic Image on Smartphone

Online purchase refers to the process of searching and transferring product information and making purchases in the online shopping environment (Pavlou 2003). Previous research has been identifying a variety of factors that influence consumers' intention to purchase online. In the context of smartphone, we deliberately chose three of them—usefulness, navigation, scope—to form the e-image of website on smartphone and applies these variables to analyze the user's intention to purchase online via the website displayed in smartphone.

Usefulness refers to the degree to which a person believes that a particular information system would enhance his/her job performance (Davis 1989). In the setting of online shopping, it represents what a consumer can benefit from getting access to shopping website to make a purchase (Abadi et al. 2011), and serves as an important determinant of a consumer's attitude toward purchasing online (Chen et al. 2002). Numerous empirical studies have been investigating the relationship of usefulness and consumers' intention to online purchase; their findings demonstrate that usefulness is positively related to online purchase intention (Kourfaris 2002; van der Heijden and Verhagen 2004; Wen et al. 2011).

Navigation has been used to evaluate the links to required information stored in a website (Alba et al. 1997; Wilkinson et al. 1997). It is concerned with whether a user can easily go back and forth to find a target website in a very short period with no disorientation (Ghalib Al-Masoudi and Chandrashekara 2010). The navigation function of a website has been drawn lots of attentions in the e-commerce sector since making users convenient and time-saving is the key to the success of online shopping (Deck 1997; Huizingh 2000; Ranganathan and Ganapathy 2002).

Scope in this study is related to the level of detail and range of the information provided by the website. The scope of information within an e-commerce website may include firm, brand, product variety, price, quality, service, and so on. Aaker (1991) suggested that the quantities of product or brand information associated with a store would affect consumers' confidence in their purchase decisions. Hymers (1996) also pointed out that consumers prefer surfing the website that is informative for making purchase decisions.

3 Methods

A web-based and cross-sectional survey was conducted to evaluate the research arguments developed for investigating the relationship between the e-image's attributes of a smartphone's website and users' purchase intention. In this section, instrument construction, sampling method, construct measurement, analytical methods and test results are described as follows.

3.1 Instrument Development and Data Collection

Our survey items was developed from previous validated measures in online store image and website quality studies, and reworded to refer particularly to the context of smartphone use. In this study, there are three e-image constructs measured using scales adapted from McKinney et al. (2002), van der Heijden and Verhagen (2004). Purchase intention was assessed using items extended from van der Heijden and Verhagen's (2004) intention to purchase online. All scale items used five-point Likert scales anchored between "strongly agree" and "strongly disagree." The instrument was pre-tested with 30 graduate students to examine the psychometric properties of the scale items. A test of factor analysis and standard reliability indicated that the piloted instrument was reliable with no major bias.

The online questionnaire for this study was composed of two parts in which two online bookstores showing different e-images were illustrated. The first part solicited survey participants to provide their responses for an online bookstore with limited information about its products while the other part asked respondents the same questions as those in the first part for their ideas concerning the online bookstore with detailed information related to its products. The sample consisted of students at several universities in northern Taiwan. Following a single round of data collection with a response rate of 65 %, 321 usable responses were collected.

3.2 Construct Reliability

In this study, the Cronbach's alpha was used to measure the internal consistency of responses. The Cronbach' alpha greater than 0.70 is considered acceptable in confirmatory research and the value should exceed 0.60 for exploratory research (Gefen et al. 2000). Table 1 demonstrates the descriptive statistics and the Cronbach's alpha for each construct of the two different types of e-image. All constructs revealed a reasonable distribution across the ranges, and the alpha value of each construct greatly exceeded 0.70, suggesting adequate reliability.

Table 1 Descriptive statistics and reliability measurements for constructs

E-image type	Construct	Mean	Standard deviation	Cronbach's alpha
Limited	Usefulness	3.40	0.93	0.84
	Navigation	3.76	0.88	0.86
	Scope	3.32	0.92	0.89
	Purchase intention	2.96	0.77	0.86
Detailed	Usefulness	3.86	0.88	0.92
	Navigation	3.85	0.86	0.91
	Scope	3.96	0.83	0.94
	Purchase intention	3.38	0.91	0.91

3.3 Data Analysis and Results

In assessing the proposed research model, construct validity and hypothesis testing were measured using LISREL, a structural equation modeling (SEM) technique. Based upon recommendations by Gerbing and Anderson (1988), data analysis for this study was performed to first test the convergent and discriminant validity of each individual measurement model for the two different types of e-image before making any attempt to evaluate their structural models. Basically, if a measurement model is operating adequately, one can then have more confidence in findings related to the assessment of the hypothesized structural model (Byrne 2001).

To evaluate convergent validity for the measurement models, selected goodness-of-fit statistics related to confirmatory factor analysis (CFA) and standardized indicator factor loadings were examined. The analysis gave a Chi square of 259.61 with 113° of freedom ($\chi^2/d.f. = 2.297$; $p < 0.001$) for the limited type of e-image and a Chi square of 287.81 with 113° of freedom ($\chi^2/d.f. = 2.547$; $p < 0.001$) for the detailed type, both of which were well within the recommended ratio of 3:1. For model fit of the limited type, the goodness of fit index (GFI) was 0.91, adjusted goodness of fit index (AGFI) was 0.88, comparative fit index (CFI) was 0.99, incremental index of fit (IFI) was 0.99, normed fit index (NFI) was 0.98, root mean square residual (RMR) was 0.039, and root mean square error of approximation (RMSEA) was 0.064; those values were within the commonly acceptable benchmarks. As for the detailed type, GFI at 0.90, AGFI at 0.87, CFI at 0.99, IFI at 0.99, NFI at 0.98, RMR at 0.028, and RMSEA at 0.070, suggested adequate model fit as well. Further, the factor loadings of each individual construct's indicator load higher on the construct of interest than on any other variable for both types of e-image. Therefore, convergent validity for the measurement models is met by adequate model fit and high factor loadings.

The assessment of discriminant validity was performed by comparing the average variance extracted (AVE) against the correlation of constructs. To meet discriminant validity, AVE of each construct exceeding the variance due to measurement error is required, i.e., greater than the squared correlation between that and any other construct (Fornell and Larcker 1981). Table 2 shows the AVE and the correlation of constructs for the two types of e-image. The square root of

Table 2 AVE and correlation of constructs

E-Image Type	Construct	Correlation Matrix			
		1	2	3	4
Limited	Usefulness	**0.73**			
	Navigation	0.48	**0.77**		
	Scope	0.68	0.64	**0.86**	
	Purchase intention	0.59	0.63	0.55	**0.80**
Detailed	Usefulness	**0.84**			
	Navigation	0.71	**0.85**		
	Scope	0.81	0.66	**0.91**	
	Purchase intention	0.66	0.53	0.67	**0.86**

[1] Diagonal elements (in bold) are the square root of the average variance extracted
[2] All correlations are significant at 0.01 level

the AVE for each construct is greater than the correlation of the construct to others; thus, discriminant validity is met.

To test the hypotheses, the structural models that specified the hypothesized relationships were evaluated based upon the measurement models. Structural path estimates were measured by examining the statistical significance of all structural parameter estimates. Statistically significant values at $p < 0.05$ were considered to be acceptable in this study. Figures 1 and 2 show the standardized path coefficients, the overall fit indexes and levels of significance for both types of e-image respectively. These metrics provided evidence of adequate fit between the hypothesized model and the observed data.

Based on the analyses, the R^2 value for purchase intention in the model of the limited type of e-image is 0.68, indicating strong predictive power. Likewise, the R^2 value for the model of the detailed type of e-image, 0.56, demonstrates equivalent predictive power. This suggests that the research model is adequate to estimating

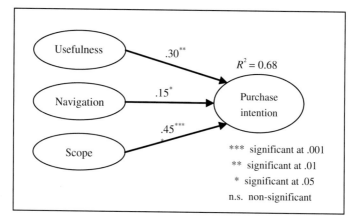

Fig. 1 LISREL estimations of the structural model for the limited type of e-image

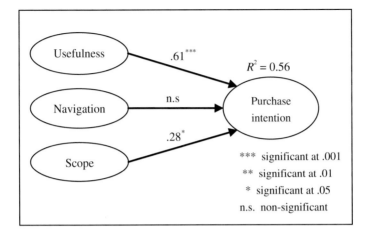

Fig. 2 LISREL estimations of the structural model for the detailed type of e-image

users' intention to make purchases via shopping websites on smartphone. For the limited version, users' purchase intention is well predicted by scope ($\beta = 0.45$, $p < 0.001$), then followed by usefulness ($\beta = 0.30$, $p < 0.01$) and navigation ($\beta = 0.15$, $p < 0.05$). With respect to the detailed version, usefulness ($\beta = 0.61$, $p < 0.001$) and scope ($\beta = 0.28$, $p < 0.05$) show direct and statistically significant effects on purchase intention while navigation has no relationship with intention.

4 Conclusions

The purpose of our study was to identify e-image attributes of a smartphone's shopping website and test the relationship between those attributes and smartphone users' online purchase intention. The findings show that smartphone users are affected by the usefulness and scope attributes regardless of a website's e-image, which provides online business practitioners with the idea regarding how to demonstrate their product information appropriately in the context of smartphones.

References

Aaker DA (1991) Managing brand equity. Free Press, New York
Abadi D, Rezaee H, Hafshejani A, Nasim S, Kermani F (2011) Considering factors that affect users online purchase intentions with using structural equation modeling. Interdisc J Contemp Res Bus 3:463–471
Alba J, Lynch J, Weitz B, Janiszewski C, Lutz R, Sawyer A, Wood S (1997) Interactive home shopping: consumer, retailer, and manufacturer incentives to participate in electronic marketplaces. J Market 39:38–53

Byrne BM (2001) Structural equation modeling with AMOS: basic concepts, applications, and programming. Lawrence Erlbaum Associates, Mahwah, New Jersey

Chen LD, Gillenson ML, Sherrell DL (2002) Enticing online consumers: an expected technology acceptance perspective. Inform Manage 39:705–719

Davis FD (1989) Perceived usefulness, perceived ease of use, and user acceptance of information technology. MIS Q 13:319–339

Deck S (1997) Ease of navigation key to successful emails. Computerworld 31:4

Fornell C, Larcker DF (1981) Evaluating structural equations with unobservable variables and measurement error. J Market Res. doi:10.2307/3151312

Gefen D, Straub DW, Boudreau MC (2000) Structural equation modeling and regression: guidelines for research practice. Commun Assoc 4:2–77

Gerbing DW, Anderson JC (1988) An updated paradigm for scale development incorporating unidimensionality and its assessment. J Market Res. doi:10.2307/3172650

Ghalib Al-Masoudi MA, Chandrashekara M (2010) Usability satisfaction of open source eLearning courseware websites. J Inf Sci Technol 7:21–35

Gregg DG, Walczak S (2008) Dressing your online auction business for success: an experiment comparing two EBay business. MIS Q 32:653–670

Hesketh AJ, Selwyn N (1999) Surfing to school: the electronic reconstruction of institutional identities. Oxf Rev Educ 25:501–520

Huizingh EKRE (2000) The content and design of web sites: an empirical study. Inf Manage 37:123–134

Hymers J (1996) Integrating the internet into marketing strategy. J Target Measure Anal Market 4:363–371

Koufaris M (2002) Applying the technology acceptance model and flow theory to online consumer behavior. Inf Syst Res 13:205–223

Lohse GL, Spiller P (1998) Electronic shopping. Commun ACM 41:81–87

McKinney V, Yoon K, Zahedi F (2002) The measurement of web-customer satisfaction: an expectation and disconfirmation approach. Inf Syst Res 13:296–315

Park J, Lee Y, Widdows R (2004) Empirical investigation on reputation and product information for trust formation in consumer to consumer market. J Acad Bus Econ 3:231–239

Pavlou PA (2003) Consumer acceptance of electronic commerce: integrating trust and risk with the technology acceptance model. Int J Eletron Commer 7:197–226

Ranganathan C, Ganapathy S (2002) Key dimensions of business-to-consumer web sites. Inf Manag 39:457–465

Sand S (2007) Future considerations: interactive identities and the interactive self. Psychoanal Rev 94:83–97

Van der Heijden H, Verhagen T (2004) Online store image: conceptual foundations and empirical measurement. Inf Manag 41:609–617

Wen C, Prybutok VR, Xu C (2011) An integrated model for customer online repurchase intention. J Comput Inf Syst 52:14–23

Wilkinson GL, Bennett LT, Oliver KM (1997) Evaluation criteria and indicators of quality for internet resources. Educ Technol 37:52–59

Forecast of Development Trends in Big Data Industry

Wei-Hsiu Weng and Wei-Tai Weng

Abstract Big Data technology is used to store, convert, transmit and analyze large quantities of dynamic, diversified data, which may be structured or unstructured data, for the purpose of commercial or social benefit. Big data technology applications need to be able to undertake real-time, high-complexity analysis of vast amounts of data, to help business enterprises perform decision-making within the shortest possible timeframe. With the rapid pace of development in cloud computing applications, both public cloud and private cloud data centers are continuing to accumulate enormous volumes of data. As a result, big data technology applications are becoming ever more important. This paper will analyze recent development trends in the field of big data industry for the reference of interested parties.

Keywords Big data · Industry development · Forecasting · Innovation · SWOT

1 Introduction

Big Data is an emerging terminology to represent the fast growing data size encountered in organizations and societies (Bollier 2010; Brown et al. 2011). Big Data Analytics (BDA) refers to a technology and framework for quickly storing, converting, transferring and analyzing massive amounts of constantly updated,

W.-H. Weng (✉)
Department of Management Information Systems, National Chengchi University, Taipei, Taiwan
e-mail: wh.weng@msa.hinet.net

W.-T. Weng
Department of Industrial Engineering and Management, Ming Chi University of Technology, Taishan, Taiwan
e-mail: wtweng@mail.mcut.edu.tw

huge, varied, structured and unstructured data for commercial gain (Russom 2011). BDA has now evolved from large database storage systems to cloud technology in order to analyze and process data in a way that is more economical, more effective and easier for the customer to manipulate (Baer 2011). The key global vendors today include IBM, SPS, Oracle, Teradata and EMC. Solutions currently offered by these BDA vendors include Data Warehouse, Data Mining, Analytics, Data Organization, Data Management, Decision Support and Automation Interface, and so on. Innovative technologies and solutions in this field are currently under rapid development (Mukherjee 2012).

Currently, major IT firms worldwide are exploring possible business opportunity in the Big Data generated market. The Taiwanese IT vendors are strong players worldwide in the manufacturing and integration of IT devices and services. To assist the Taiwanese IT vendors moving forward towards the emerging Big Data market, this research aims to address the question of deriving future trends for the Taiwanese IT industry.

2 Research Method

Qualitative analysis is employed instead of quantitative analysis, by way of expert panel, vendor interviews and focus groups.

2.1 Expert Panel

The expert panel from industry experts is to assist the convergence process of data analysis. To this objective, industrial experts panel of eleven people were selected. The panel consists of CEO, CIO and line of business managers from various domains of Taiwanese IT industry. All of them are from publicly listed firms. Their business domains include System Integration (SI), Independent Software Vendor (ISV), Internet Service Provider (ISV), device manufacturer, and data center operator. These are the major participants in the Big Data industry. The main function of the Expert Panel is to help determining the research framework and deriving strategy. In particular, the following questions are discussed.

1. What are the innovative Big Data technologies potentially important to the Taiwanese IT industry?
2. Who are the major vendors worldwide that could be the benchmark for the Taiwanese Big Data industry?
3. What are the business environments in terms of internal strengths and weaknesses, as well as external opportunities and threats of the IT firms?

4. Recognizing the internal and external business environments, what are the possible production value that could be estimated for the Taiwanese Big Data industry?

2.2 Vendor Interviews and Focus Groups

Representative IT firms from Taiwan are selected as the objects of this study. The selection process is based on the rank of the revenue of the firms as well as their reputation in terms of technology innovation and market visibility. IT vendors of Taiwan enjoy high market share worldwide in the sectors such as computer, communication and consumer electronics manufacturing. Currently the Taiwanese vendors participate in Big Data market include IT device manufacturers, IT service providers, and Internet datacenter operators. We collect and analyze business proposal data of 62 Taiwanese IT vendors. The selection criteria are as follows.

1. Revenue of the firm is among the top five in its industry domain.
2. The firm has announced in public its vision, strategy, products or service toward Big Data market.

Based on these criteria, representatives from 33 IT firms are selected for vendor interviews and focus groups. These firms are summarized in the following Table 1.

3 Development of Big Data Industry

3.1 Development of Innovative Big Data Technology

Big Data technology can be divided into two broad categories: Advanced SQL technology, which is oriented towards the use of relational databases, and NoSQL technology, the emphasis in which is on non-relational databases (Baer et al. 2011). Advanced SQL is specifically designed to provide real-time analysis results with large quantities of structured data. However, as the scale of data collection

Table 1 Selected cases for vendor interviews and focus groups

Business domain	Number of firms
Independent software vendor (ISV)	7
System integration provider (SI)	9
Telecom operator	3
Server and storage device manufacture	6
Networking device manufacture	3
Mobile device manufacture	5
Total	33

grows ever larger, and as the different categories of data that need to be processed become ever more complex, non-structured data is presenting business enterprises with new challenges in terms of data storage and analysis (Borkar et al. 2012). NoSQL non-relational database systems (Adrian 2012) offer enhanced performance and extensibility, making them ideally suited to processing large amounts of non-structured, highly variable data. There are four main types of NoSQL database: key-value databases, in-memory databases, graphics databases, and document databases. The Advanced SQL database platform segment is currently going through a period of market consolidation, indicating that this segment is entering the mature phase in its evolution, characterized by slow but steady growth. By contrast, NoSQL market is still very much in the growth stage, and be expected to play an increasingly important role in big data technology in the future.

Hadoop is a big data technology that is attracting growing interest from enterprises; more specifically, it is a form of key-value database technology (Baer 2011). Leading U.S. retailer Wal-Mart is using Hadoop to analyze sales data and identify new business opportunities; online auction site eBay has been using Hadoop to process unstructured data and reduce the burden of database storage requirements. Hadoop is a software platform that enables users to rapidly write and process large quantities of data. Designed by Doug Cutting, Hadoop was originally an Apache research project, but has since been adopted for commercial applications involving Petabyte-scale big data.

The Hadoop platform architecture comprises three main elements: the Hadoop Distributed File System (HDFS), the Hadoop MapReduce model, and the HBase database system (Kobielus et al. 2011). The operational model of the Hadoop computing framework basically involves coordinating the use of MapReduce distributed computing algorithms with the HDFS distributed file system, which makes it possible to transform ordinary commercial servers into distributed computing and storage groups, creating the capability to store and process huge volumes of data. Supporting this capability is HBase, which is a distributed database system capable of coping with Petabyte-scale data volumes.

Big Data technology are classified and summarized in the following Table 2.

3.2 Development of Global Big Data Industry Leaders

Recognizing the potential business opportunities presented by big data, leading international corporations such as IBM, Oracle, Microsoft, SAP/Sybase, Teradata and EMC have all been moving aggressively into this new field, working to provide enterprises with practical commercial applications for the effective storage and analysis of large volumes of data (IDC 2012). IBM, which has for some years now been focusing heavily on the business intelligence market, recently acquired data warehousing firm Netezza, and will be incorporating Netezza's technology into its Business Analytics Optimization (BAO) solutions as a means of enhancing

Table 2 A taxonomy of big data technology

Category	Technology
Data warehouse	Central enterprise data warehouse
	Data warehouse appliance
	Data marts for analytics
	Analytics processed within the EDW
	Extract, transform, load (ETL)
NoSQL BDA	MapReduce
	Hadoop
	NoSQL or non-indexed DBM
	Column oriented storage engine
	Text mining
Advanced SQL BDA	Complex SQL
	Distributed SQL
	OLAP
	Advanced SQL appliance
	SQL accelerator
Cloud analytics	Public cloud analytics
	Private cloud analytics
	Social analytics
	Software as a service (SaaS)
	Internet of things (IoT)
Embedded analytics	Predictive analytics
	Complex event processing (CEP)
	In-memory database
	In-database analytics
	In-line analytics
Big data visualization	Advanced data visualization
	Real-time reports
	Dashboards
	Visual discovery
	Infographics

their competitiveness. Similarly, Teradata has acquired Aster Data, and EMC has purchased Greenplum (Cohen et al. 2009). This wave of M&A activity reflects leading IT companies' determination to develop the big data market.

BDA vendors have accelerated their expansion and acquisitions. HP for example acquired 3PAR in September, 2010. In November 2010, EMC acquired Isilon. In December 2010, Dell acquired Compellent. In March 2011, Teradata acquired Asterdata. In June 2011, Oracle acquired PillarData. In September 2011, HDS acquired Blue Arc. In April 2012, EMC acquired Pivotal as well.

The wave of acquisitions can be divided into two types. One is system manufacturer's acquisition of a storage system manufacture to provide customers with a one-stop solution. Oracle's acquisition of PillarData and IBM's acquisition of XIV were both of this type. The consolidation enabled large vendors to acquire complementary technologies that make their own system more complete. The other type is buy-outs between storage system manufacturers. If a storage system manufacturer was big enough and was not acquired by a system manufacturer, it began acquiring other smaller storage manufacturers with innovative technologies in order to flesh out their own product lines and technology. EMC's acquisition of Isilon and HDS' acquisition of Blue Arc were both of this type. Whether it was a system manufacturer acquiring a storage system manufacturer or storage manufacturers acquiring each other, in both cases, vendors used their acquisition strategy to acquire key technologies. Continued innovations in information technology means that, large vendors will continue to make new acquisitions in order to make their products more complete and competitive.

4 Analysis on Forecasting Results

4.1 SWOT Analysis of Taiwan's Big Data Industry

The SWOT analysis is an established method for assisting the formulation of strategy. SWOT analysis aims to identify the strengths and weaknesses of an organization and the opportunities and threats in the environment (Pickton and Wright 1998). The strengths and weaknesses are identified by an internal appraisal of the organization and the opportunities and threats by an external appraisal (Dyson 2004). Having identified these factors strategies are developed which may build on the strengths, eliminate the weaknesses, exploit the opportunities or counter the threats (Weihrich 1982).

By applying the analysis method of Weihrich (1982), the SWOT matrix of Taiwan's Big Data industry is derived as follows Table 3.

4.2 Forecast of Taiwan's Big Data Production Value

Big data mainly comes from enterprise operational data warehouses, cloud computing data centers, shared information on social network websites and intelligent sensor networks. Taiwan is currently in a cloud computing data center expansion phase. Big data access and analysis is the critical technology application for cloud computing data centers and drives the development of Taiwan's server and storage equipment market. One of the driving forces behind the current wave of growth in big data applications is the Taiwanese government's strong support for the cloud computing industry. Cloud computing can be considered one of the emerging smart industries that the Taiwanese government is actively promoting. Many government units have begun deploying cloud services and related applications as a result through the rebuilding of basic infrastructure such as the disaster prevention system, the Ministry of Finance's electronic receipt system and household

Table 3 Competitiveness analysis of Taiwan's big data industry

Strengths	Weaknesses
A. Possesses both hardware and software solutions and provides professional consultancy services experience	A. Limited in the big data related business experience
B. Cooperated with the global leading companies for many years, and prices are flexible	B. Weak research and development of key software technologies such as NoSQL and big data analytics
C. Possesses in-depth vertical domain knowledge, and localized Know-How as well	
D. As an important OEM center of server and storage equipment, Taiwan has the capability of autonomous manufacturing and cheap supply of hardware equipment	
Opportunities	Threats
A. Large-scale cloud data center continues to expand in scale, placing orders directly to the server and storage equipment OEM industry	A. Global leading companies lead technologies and standards
B. Develop the SaaS model of software to attract new customers	B. Part of the business is replaced by emerging cloud computing solutions
C. Enterprise data continue to grow and support the needs of storage devices	C. Industries rise in the emerging markets such as Mainland China, India, and so on
D. Enterprise mobile application software and services with big data needs are growing	

registration system. The adoption of cloud technology by government IT systems will facilitate the introduction of big data development tools in the future.

With these effects, the production value of Taiwanese Big Data industry is expected to achieve steady growth through 2018, as shown in Table 4.

Table 4 Production value of Taiwan's big data industry (USD Million)

Big Data product/service	2013	2014	2015	2016	2017	2018
Server	512	654	773	912	1,028	1,137
Storage equipment	421	511	613	705	803	902
IT service	131	137	142	148	154	161
Software	141	152	160	168	175	182
Total production value	1,205	1,454	1,688	1,933	2,160	2,382
YoY growth (%)	–	20.7 %	16.1 %	14.5 %	11.7 %	10.3 %

5 Conclusions

The development trends in Big Data industry is investigated with innovated technology and global leading vendors. The internal strength and weakness, the external opportunity and threat, as well as the production value forecast for Taiwanese Big Data industry are analyzed and presented.

Vendors interested in exploring the market opportunities of Big Data can use this analysis process and outcome of this research as a reference for their strategic planning, and avoid many unnecessary trial and error efforts. In particular, with a clear picture of the Big Data opportunities and threats, vendors can position themselves more precisely for a market sector of their competitive advantage.

Acknowledgments The authors gratefully acknowledge the helpful comments and suggestions of the reviewers, which have improved the presentation.

References

Adrian M (2012) Who's who in NoSQL DBMSs. Gartner report G00228114
Baer T (2011) 2012 Trends to watch: big data. Ovum report, OI00140-041
Baer T, Sheina M, Mukherjee S (2011) What is big data? The big architecture. Ovum report, OI00140-033
Bollier D (2010) The promise and peril of big data. The Aspen Institute
Borkar V, Carey M, Li C (2012) Inside "big data management": ogres, onions, or parfaits? In: Proceedings of ACM EDBT/ICDT joint conference, Berlin, Germany
Brown B, Chui M, Manyika J (2011) Are you ready for the era of "big data"? McKinsey Q 4:24–35 (McKinsey & Company)
Cohen J, Dolan B, Dunlap M, Hellerstein J, Welton C (2009) MAD skills: new analysis practices for big data. In: Proceedings of ACM VLDB conference, Lyon, France, 24–28 Aug 2009
Dyson RG (2004) Strategic development and SWOT analysis at the University of Warwick. Eur J Oper Res 152:631–640
IDC (2012) Worldwide big data technology and services 2012–2015 forecast. IDC #233485, Volume: 1, Tab: Markets
Kobielus K, Moore C, Hopkins B, Coyne S (2011) Enterprise hadoop: the emerging core of big data. Forrester Res
Mukherjee S (2012) Deploying big data systems. Ovum report, OI00140-035
Pickton DW, Wright S (1998) What's SWOT in strategic analysis? Strateg Change 7(2):101–109
Russom P (2011) Big data analytics. TDWI research, 4th Quarter, 2011
Weihrich H (1982) The TOWS matrix: a tool for situational analysis. Long Range Plan 15(2):54–66

Reliability Analysis of Smartphones Based on the Field Return Data

Fu-Kwun Wang, Chen-I Huang and Tao-Peng Chu

Abstract In recent years, smartphones have become indispensable tools and media for obtaining information. Consequently, investigation and data analysis studies exploring smartphone reliability and failure rates have become increasingly important for mobile phone retailers and manufacturers. The design capabilities and manufacturing technologies for smartphones are continuously upgraded, and customer requirement for reliability increase correspondingly. Therefore, this study investigated after-sales repair and maintenance data obtained from a Taiwanese mobile phone maintenance provider for three brands of smartphones (A, B, and C). We assumed that the quality of the different models of the three brands is consistent. Statistical analysis techniques and software were used to calculate the parameter values of four failure probability distribution functions (i.e., exponential, log-normal, log-logistic, and Weibull). The maximum likelihood estimation (MLE) method was also employed to assess the log-likelihood value and determine the most appropriate failure probability distribution for each smartphone. Finally, we calculated the predicted market failure rate for the three smartphone brands. The results and conclusions obtained in this study can benefit important market players, such as consumers, mobile phone retailers, and manufacturers, regarding smartphone quality, sales strategies, after-sales warranty service packages, and manufacturing process improvements.

F.-K. Wang (✉)
Department of Industrial Management, National Taiwan University of Science and Technology, No.43 Keelung Rd., Sec.4, Taipei 106, Taiwan Republic of China
e-mail: fukwun@mail.ntust.edu.tw

C.-I. Huang · T.-P. Chu
Graduate Institute of Management, National Taiwan University of Science and Technology, No.43 Keelung Rd., Sec.4, Taipei 106, Taiwan Republic of China
e-mail: D9816909@mail.ntust.edu.tw

T.-P. Chu
e-mail: D9816907@mail.ntust.edu.tw

Keywords Smartphones · Field return data · Interval censored data · Maximum likelihood estimation

1 Introduction

Wireless communication technology has advanced continuously in recent years. The widespread use of smart mobile devices in the communications market has ushered in an era of ubiquitous mobile networks. Cisco Visual Networking Index (Cisco 2012) predicted that more than 10 billion mobile network devices would exist worldwide by 2016, exceeding the current world population of 7.3 billion. With their rapid popularization and the maturity of mobile Internet and networking, smartphones have become an important medium closely connected to the everyday lives of users. In addition to providing sales growth for brand companies and manufacturers, consumers have also become increasingly reliant on smartphones. This indicates that the market opportunities for mobile application services are unlimited. However, hardware brand companies and mobile application service developers must continuously address new changes and challenges in usage behavior.

Smartphones have become indispensable tool for obtaining information, the standard usage and quality stability factors of smartphones significantly influence consumers, mobile phone retailers, and manufacturers on varying levels. Therefore, calculating smartphone life is essential. The purpose of this paper is to calculate the failure rate and MTBF of a smartphone based on the filed return date. The study investigated the after-sales repair data of a Taiwanese mobile phone maintenance provider for three brands of smartphones. Using statistical analysis techniques, the failure probability distribution function for the three smartphone brands was calculated and the predicted failure rate was also investigated.

2 Literature Review

Generally, a traditional reliability prediction method is adopted for electricity production. For example, British Telecom (BT) employed HRD5 (1994) for a number of years, but with limited value because that prediction method addresses only electrical and electronic components, which provided the lowest contribution to field returns. This increased the need to develop a method for estimating the field failure rate of a product prior to its launch in the field. An estimate of a product's field failure rate is a key requirement for product development.

Although laboratory reliability testing is commonly employed for product design decisions, real reliability data is obtained from the field, typically in the form of warranty returns and field tracking studies. Field data generally shows greater variability in component and product failure times compared to data from laboratory testing. Hong and Meeker (2010) asserted that the differences between carefully controlled laboratory ALT experiments and field reliability test results are caused by the uncontrolled field variation (unit-to-unit and temporal) of variables, such as the use rate, load, vibration, temperature, and humidity. Oh and Bai (2001) and Suzuki (1995) noted that field data provide more reliable information regarding life distribution compared to laboratory.

Wu and Meeker (2002) examined the use of field returns to detect potentially serious in-warranty problems. Appropriate model assumptions and robust model estimations are crucial for accurately predicting product field returns. The parametric distribution models traditionally employed for product reliability analyses do not effectively describe product field survival. Zhang et al. (2010) used the Bertalanffy-Richards biological growth model for modeling hard disk drive (HDD) field survival within the warranty period to avoid insufficient field returns observations. The discrepancy in time-to-event measurements further increased the difference between the HDD field survival process and the reliability process. Therefore, Pan (2009) developed models that appropriately describe laboratory reliability data; however, they may be unsuitable for field survival data. Marcos et al. (2005) stated that a product that has been in the field for only a short time compared to the entire warranty period cannot provide sufficient direct information to guarantee good model estimation.

3 Methodology

The methodology regarding the life or lifespan analysis of smartphone market failure statistics conducted is explained. The section explains the analysis process for the smartphone field failure return.

Step 1. Data Collection

Before the data analysis, the smartphone field failure return data was first processed and filtered. Because the information recorded in the maintenance database was complex, only the defect units categorized as warranty data was extracted.

Step 2. Interval-Censored Data Processing

After collecting and processing the data, the number of monthly shipments for each of the three smartphone brands and the number of failures and repairs within the same month were obtained. The data we collected for this study was interval-censored data shown in Fig. 1.

Step 3. Distribution fitting—MLE

The MLE statistical values from each distribution were subsequently used to identify which distribution possessed the greatest observed log-likelihood value for the brands and to determine the optimal failure probability distributions.

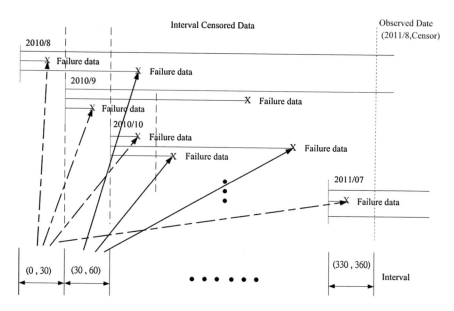

Fig. 1 The interval censoring data of smartphone

Step 4. Identifying the optimal failure probability distribution

MLE was also used to obtain the estimated parameter values employed in the optimal failure probability distributions of the three brands.

Step 5. Predicting and evaluating the field return rate

The estimated parameter values obtained in Step 4 were used to illustrate the actual and predicted failure rates for the three smartphones. MAPE was then applied to calculate the prediction or forecasting efficacy for the three smartphone brands.

4 Data Analysis

This study analyzed cumulative data for the three brands of smartphones from the beginning of shipping in August 2010 to July 2011. A total of 567,860 smartphones were shipped, and the maintenance provider received 6,958 of these smartphones for repair during this period. Table 1 shows the number of shipments and failures each month. The number of failures represents the number of smartphones shipped that month that required repair.

August 2011 was set as the time-censored cutoff point for failure data, with a unit interval comprising 30 days. Table 2 shows the censored failure statistics.

Reliability Analysis of Smartphones

Table 1 The number of shipments and failures each month

Time	Brand A		Brand B		Brand C	
	Shipped units	Failure units	Shipped units	Failure units	Shipped units	Failure units
Aug-10	4,138	236	12,802	503	18,046	504
Sep-10	15,363	245	14,111	348	17,657	394
Oct-10	13,365	238	14,350	333	11,643	416
Nov-10	15,813	167	14,963	231	9,394	388
Dec-10	16,737	177	16,701	231	10,006	250
Jan-11	22,770	125	14,954	202	16,494	241
Feb-11	21,023	150	15,551	160	16,534	214
Mar-11	16,591	86	14,295	103	21,935	142
Apr-11	14,751	82	14,136	117	17,006	120
May-11	21,733	100	12,069	91	15,988	97
Jun-11	19,265	78	11,970	46	23,074	26
Jul-11	15,506	36	13,047	14	24,079	67
Total	197,055	1,720	168,949	2,379	201,856	2,859

This study employed MLE for smartphone failure data from three brands of smartphones to determine the cumulative failure probabilities for four types of distributions, including the exponential, log-normal, log-logistic, and Weibull.

Table 3 shows that when Brand A, B, and C are all log-normal, their log-likelihood values are highest. Therefore, the three observed smartphone brands have identical optimal failure probability distribution.

The statistical analysis results indicate that log-normal is most appropriate to the failure probability distribution function of the three brands.

This study determined the most appropriate failure distributions for the three brands of smart phones using the mentioned experimental steps, and produced Fig. 2 to show the actual and predicted failure rates of the three smartphones.

MAPE was further employed for prediction or forecasting efficacy assessments. The formula for this method is as follows:

$$MAPE = \frac{\sum_{t=1}^{n} |Y_t - Y_t'|/Y_t}{t} \times 100\ \% \tag{1}$$

In this formula, t is the number of periods for prediction, Y_t is the actual value, and Y_t' is the predicted value.

The calculation results indicate that the MAPE values for Brand A, B and C are 15.89, 8.7, and 16.23 % separately. The typical MAPE values for model evaluation developed by (Lewis 1982). Evidently, Brand B is a high accuracy forecasting model, and Brands A and C are good forecasting models. This indicates that the forecasting efficacy assessment conducted for the developed models in this study provided good forecasting capabilities or better.

Table 2 The censored failure statistics

Time (day)	Brand A Censored units	Brand A Failure units	Brand B Censored units	Brand B Failure units	Brand C Censored units	Brand C Failure units
(0,30)	15,470	235	13,033	247	24,012	362
(30,60)	19,187	294	11,924	335	23,048	418
(60,90)	21,633	202	11,978	221	15,891	313
(90,120)	14,669	190	14,019	248	16,886	335
(120,150)	16,505	129	14,192	112	21,793	308
(150,180)	20,873	184	15,391	226	16,320	307
(180,210)	22,645	151	14,752	146	16,253	231
(210,240)	16,560	126	16,470	205	9,756	198
(240,270)	15,646	74	14,732	243	9,006	107
(270,300)	13,127	64	14,017	216	11,227	177
(300,330)	15,118	51	13,763	123	17,263	71
(330,360)	3,902	20	12,299	57	17,542	32

Table 3 The log-likelihood values for each distribution of the three brands

	Brand A	Brand B	Brand C
Exponential	−19,091.17	−25,039.73	−30,360.24
Log-logistic	−19,016.29	−25,005.83	−30,335.86
Log-normal	**−19,001.19**	**−24,954.31**	**−30,260.19**
Weibull	−19,016.29	−25,007.17	−30,338.12

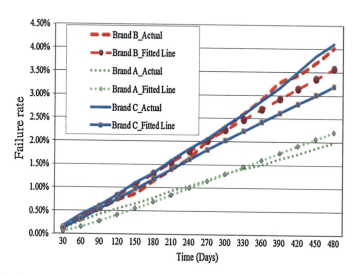

Fig. 2 The interval censoring data of smartphone

Table 4 Predicted failure of Brand A, B, C with log-normal distribution

Years	days	Predicted failure rates (%)		
		Brand A	Brand B	Brand C
1	360	1.61	2.70	2.43
2	720	3.38	5.18	4.61
3	1,080	5.02	7.32	6.49

Figure 2 demonstrates the relationship between the cumulative failure rates and time for the three smartphone brands based on analysis of the field return data. The figure indicates that the failure rate of Brand B for the first year was 2.7 %, the highest among the brands. Brand C was next with a first-year failure rate of 2.43 %. Brand A had the optimum first-year failure rate at 1.61 %. Table 4, shows the three-year predicted failure rates for the three brands.

5 Conclusions

This study used smartphone field return data to investigate the reliability of three brands of smartphones. Relevant analysis results indicated the following:

The goodness-of-fit test performed on the field return data showed that log-normal distribution was the most optimal for the failure probability distributions of all three brands. This differs from the common assumption that the majority of electronic products demonstrate a Weibull distribution. The result also indicates that the structural designs of the three smartphones may be similar. Thus, their optimal failure probability distributions are identical. Consequently, if the smartphone failure rate estimation in this study was performed directly using a Weibull distribution, the results would have demonstrated error compared to the failure rate estimations obtained using log-normal distribution. This would have led to erroneous cost estimations for future smartphone sales, repairs, and warranties for spare parts.

MLE was also employed to identify and demonstrate the relationships between the cumulative failure probabilities and time based on the parameters of the cumulative failure probability density function for the log-normal distributions of the three brands, as shown in Fig. 2. The results indicate that Brand A possessed the optimal product reliability, followed by Brand C and finally Brand B. Therefore, the three brands of smartphones possessed differing levels of reliability. This result is identical to that reported by the maintenance and repair sector from which the data for this study were obtained. And mobile phone maintenance providers can use the product failure analysis data from this study to forecast or predict the material procurement and preparation requirements for spare parts.

References

Cisco Visual Networking Index (2012) Global mobile data traffic forecast update, 2011–2016. http://www.cisco.com/en/US/solutions/collateral/ns341/ns525/ns537/ns705/ns827/white_paper_c11-520862.html. Accessed 10 Dec 2012

HRD 5 (1994) Handbook of reliability data for components used in telecommunications systems. http://infostore.saiglobal.com. Accessed 23 Oct 2012

Hong Y, Meeker WQ (2010) Field-failure and warranty prediction based on auxiliary use-rate information. Technometrics 52:148–159

Lewis C (1982) International and business forecasting methods. Butterworth-Heinemann, London

Marcos E, Philip G, Kosuke I (2005) A framework for warranty prediction during product development. In: Proceedings of 2005 ASEA international mechanical engineering congress and exposition, Orlando Florida USA, 5–11 Nov 2005

Oh YS, Bai DS (2001) Field data analysis with additional after-warranty failure data. Reliab Eng Syst Saf 72:12–25

Pan R (2009) A Bayes approach to reliability prediction using data from accelerated life tests and field failure observations. Qual Reliab Eng Int 25:229–240

Suzuki K (1995) Role of field performance data and its analysis. CRC Press, New York

Wu H, Meeker W (2002) Early detection of reliability problems using information from warranty databases. Technometrics 44:120–133

Zhang S, Sun F, Gough R (2010) Application of an empirical growth model and multiple imputation in hard disk drive field return prediction. Int J Reliab Qual Saf Eng 17:565–577

The Impact of Commercial Banking Performance on Economic Growth

Xiaofeng Hui and Suvita Jha

Abstract The financial sector and economic growth spark hot debates. This paper addresses the question of whether commercial banking performance in Nepal reasons to economic growth. In order to answer this question, the present study focuses on analyzing the relationship between deposit, loan and advances and assets as proxy for performance of commercial banks while gross domestic product proxies economic growth over the period of 1975–2010. Using Augmented Dickey Fuller and Ordinary Least Square, the regression results indicated that deposits and assets have significant impact on the economic growth of Nepal while loan and advances has insignificant impact on the economic development. Furthermore, the Granger-Causality test suggests that there was no causality with deposit, loan and advances and assets with the economic acceleration. It can be concluded that not only commercial banking performance but also other variables political stability and technology play the important role in the economic advances in Nepal.

Keywords Banking performance · Nepalese commercial banks · Economic growth · Gross domestic product

---

X. Hui (✉) · S. Jha
School of Management, Harbin Institute of Technology, Harbin 150001, People's Republic of China
e-mail: xfhui@hit.edu.cn

S. Jha
e-mail: sovi1977@yahoo.com

1 Introduction

Linking financial sector with economic growth in an economy is a major concern among economists, financial analysts, researchers and policymakers. An extensive number of literatures suggest that financial institutions are to be the best indicator of a country's real development potential and significantly influence the economic development (Talliman et al. 1998; Goldsmith 1969). Banking industry has much benefit over non-banking markets sector in the developing economies as like of Nepal with weak legal and accounting structures. In this environment, banks can be able to formulate firms disclose information and pay back their debts thereby facilitating spreading out and long-run development (Rajan and Luigi 1999).

Those countries with a better financial system have a trend to increase its economic growth faster. The bank-based financial system emphasizes that assets has the positive role in the economic progress as banks can mobilize resources and reduce risk (Beck and Levine 2004). The study of Hsu and Lin (2000) for Taiwan, and Anwar and Nguyen (2011) for Vietnam reported that both banking and stock market development are positively related to short-run and long-term economic growth. Nepal has established up comprehensive financial infrastructures such as commercial banks, development banks, finance companies, cooperatives, non-governmental organizations. Among the financial institutions, the common resource of supplies funds and the main source of financing to support the national economic performance are commercial banks (Poudel 2005). Commercial banks play an important role in financial sector and occupy 76.7 % of total assets (Nepal Rastra Bank 2011). However, some studies have investigated the impact of financial sector with economic growth of Nepal (Poudel 2005; Kharel and Pokhrel 2012), there are no any reports that focused on the subject of commercial banks performance and their impact on Nepalese economic development. The objective of this paper is to investigate the relationship of commercial banking performances with the economic growth, considering the impact of deposit, loan and advances and assets on the Nepalese economic growth.

2 Data and Methodology

2.1 Data and Variables

Yearly time-series data for the period of 1975–2010 were used in this study. Deposit, loan and advances, and assets were used as commercial banking performance while real GDP per capita at constant prices (RGDP) was used as economic growth. Data of the commercial banks performance were collected from the Nepal Rastra Bank Quarterly Economic Bulletin (published by the Central Bank of Nepal) while RGDP was noted from various issues of Economic Survey of Ministry of Finance whereas population was taken from the Statistical Pocket

Book of CBS. Monetary management turned out to be a key concern for the banking sector in Nepal only since 1975 (Acharya 2003) and that is the reason of to consider data from 1975.

2.2 Details of the Model

2.2.1 Testing for Stationary

In order to check the variables are stationary or non stationary, this study used the Augmented Dickey Fuller (ADF) tests of stationary. In this test null hypothesis means non-stationary in the data and alternative hypothesis means stationary.

2.2.2 Ordinary Least Square Method

In formulating model, it assumes that the RGDP is a function of deposit (DEP), loan and advances (ADV) and commercial banks assets. All the variables were measured in log (LN) value. Given the above theoretical considerations, the behavioral equation of the model was formulated as RGDP = f(DEP, ADV, ASSET) and consequently equation could be written in the following form:

$$\ln(\text{RGDP}) = \beta_0 + \beta_1 \ln(\text{DEP}) + \beta_2 \ln(\text{ADV}) + \beta_3 \ln(\text{ASSET}) \quad (1)$$

All the independent variables were expected to have a positive impact on economic growth. This research was based on the following hypotheses that clearly define the research criteria.

H1. There is a significant relationship between commercial banks deposit and economic growth.
H2. There is a significant relationship between commercial banks loan and advances and economic growth.
H3. There is a significant relationship between commercial banks assets and economic growth.

2.2.3 Testing for Causality

This study used Granger-Causality test proposed by Granger (1969) for testing whether a causal relationship existed among the Real GDP per capita and the deposit, loan and advances, and assets. The test involved estimating the regressions for each variable on the other variable past observations. An F- test was then applied on the residuals of the regressions, and the value was compared to tabulated F- values. If the computed F- value exceeded the critical F- value of the chosen level of significance, the null hypothesis was rejected and concluded that a causality relation existed.

All the estimations for Ordinary Least Square, Unit Root test and Granger Causality test have been performed in the E-Views7 program whereas the ordinary calculations were done in Excel.

3 Results and Discussion

3.1 Descriptive Statistics of the GDP and Commercial Banking Variables

A growth in the real GDP per capita from 1975 to 2010 (Table 1) indicates an improvement in the financial sector due to increasing inflow of remittance as well as the country's macroeconomic policy initiatives. The overall mean value of the RGDP was 8.079 with the standard deviation (SD) of 0.212. Regarding the banking performance variables, deposit was in increasing trend due to structural changes of Nepalese economy and effort of financial sector reform program. Increase in RGDP leads to an increase of money supply that helps the banking sector to have more deposits and consequently lead to higher lending. In addition, that results in more investments and hence leads to fast economic growth. The overall mean value of the deposit was 10.38 with SD of 1.87. Growing loan and advances indicated that the role of loan was expanding fast as a source of funding for economic activities in the country. The mean value of the loan and advances was 10.11 with SD of 1.90. Similarly, increasing assets would provide stronger, resilient to shocks and capable of funding the real sector and, therefore, enhancing economic growth. The mean value of assets was 10.69 with SD of 1.88.

3.2 Augmented Dickey Fuller Test

The results of the ADF test are reported in Table 2. The Schwartz Bayesian Criterion (SBC) is used to determine the optimal number of lags included in the test. The ADF test results suggest that at the 1 and 5 % significance level could not

Table 1 Descriptive statistics of the employed variables

Variable	N	Minimum	Maximum	Mean	SD
RGDP	36	6.95	7.60	7.26	0.23
DEPOSIT	36	7.07	13.33	10.40	1.86
LOAN and ADV	36	7.08	13.13	10.11	1.90
ASSET	36	7.51	13.58	10.69	1.88

The Impact of Commercial Banking Performance on Economic Growth

Table 2 Augmented dickey fuller unit root tests for level and 1st difference variables

Level			1st difference		
Variable	Intercept	Trend and intercept	Variable	Intercept	Trend and intercept
RGDP	−0.292 (0.916)	−1.917 (0.624)	RGDP	−6.124(0.0000)	−5.957(0.0001)*
Deposit	−2.193(0.212)	−1.645(0.753)	Deposit	−4.195(0.0024)	−4.31 (0.0087)*
Loan	−0.086 (0.943)	−1.92 (0.620)	Loan	−4.947(0.0003)	−4.88 (0.0021)*
Assets	−1.376 (0.582)	−0.80 (0.955)	Assets	−2.470(0.1315)	−2.57 (0.2946)

*Indicate that the variable is significant at 1, 5 and 10 % respectively

reject the null hypothesis for any variables, which meant that the unit root problem existed and the series were non-stationary. Almost all the variables including deposit, loan and advances, and assets have non-stationary both when include intercept and when include intercept and trend at level while all the variables were stationary at first difference except assets. This result gives support to the use of ordinary least square to determine the relationships of banking performance and economic growth.

3.3 Ordinary Least Square Results

The Unit Root test resulted in the time series data were non-stationary. Thus, the analysis was performed using the regression analysis and the model was estimated by ordinary least square. The value of R-Square, the coefficient of determination, represented the correlation between the observed values with the predicted values of the dependent variable and provided the adequacy of the model. Here the value of the R-Square was 0.963 that meant the independent variable in the model could predict 96.3 % of the variance in the dependent variable (Table 3). The p value was given by 0.000 that was less than 0.05, resulting in the significance of the model. The regression models assume that the error deviations are uncorrelated. The constant parameter was 6.908.

Deposit and economic growth were negatively but significantly related with the economic growth. Deposit coefficient with a value of −0.268 implied that if

Table 3 Regressions between commercial banking performance and real GDP

Variable	Coefficient	t-statistic	Probability
C	6.908	90.427	0.000
Deposit	−0.268**	−2.248	0.0316
Loan	0.120	1.617	0.1157
Assets	0.256***	1.763	0.0874
R-squared	0.96		
F-statistic (probability)	282.52 (0.000000)		

**, ***Indicate that the variable is significant at 1, 5 and 10 % respectively

deposit was increased by a unit, economic growth was threatened as it made RGDP to decline by 0.268. Deposit is expected to positively impact on economic growth as suggested by Mckinnon (1973) and Shaw (1973). This can be attributed to the fact that commercial banking in Nepal has not strategically positioned banks to adequately mobilize enough deposit that would positively affect the economy. Apart from that, one major cause for this is the Nepalese banking system is dominated by the public sector banks in terms of shares of deposits and their mobilization (Jha and Hui 2012). Joint venture and domestic private banks have very few branch networks and are concentrated in urban centers. This showed little capacity to absorb small urban or rural savings and serve the credit needs of the small borrowers, rural or urban. The commercial banks limited their operations to large amounts- both in acceptance of deposits and lending. The minimum balance requirements in some Kathmandu Valley banks make them unapproachable even to the middle class families in the Valley. Bulk investors also have ability to negotiate on interest rates, which small investors lack. More evidences are available that average households in rural areas can save smaller amounts. But the commercial banks have shown little interest in this kind of savings. Given the oligopolistic cartel ling, they are over liquid and have no need to innovate for deposit mobilization. The negative but significant result also indicates that the commercial banks deposit in a country is crucial to the economic development.

A positive but insignificant relationship is established between loan and advances and economic growth. This result is inconsistent with the finding of Aurangzeb (2012) for the Pakistan banking sector. The loan and advances coefficient was 0.120, implying that a unit increase in loan and advances leads to an increase in RGDP by 0.120 units. The implication of this is that commercial banks were unable to find an appropriate client for ending their excess liquidity in the market and the funds directed to them by banks have not been optimally utilized to increase their productivity which inevitably improves economic growth. Besides, this might be due to the commercial banks are concentrated in urban areas. The excess of credit programs has made very little dent in the rural credit market. More than 80 % of the borrowing households have still to depend on non-formal sources for their credit needs. Majority of the credit programs have been unable to directly cater to the needs of the bottom 20 % households because the poor lack other resources and knowledge to benefit from the deposit-credit programs. Also, the negligence of commercial banks in the credit administration process increases the chances for non-performing loans and other classified assets and this has adverse effect on bank capital and the economy at large. This effect reduces the ability of commercial banks to further extend credit and may cause bank distress or failure since the profitability of a bank is directly related to the credit it grants. The lending rate charged on credit by banks is on the high side which deters investors from borrowing to embark on productive activities and give room for moral hazard. This runs contrary to the findings of Aurangzeb (2012) for the Pakistan banking sector. Although the relationship of loan and advances and RGDP found weak they did not mean that loan did not matter for the economy. Much to the contrary, they reinforce the argument that credit has a crucial role on overall productivity via a correct allocation of financial resources, a conclusion that has already found convincing empirical backing (Bebczuk and Garegnani 2007). The assets of the commercial banks plays significant and equal role in economic growth as signified by the positive and statistically significant coefficient, which meant that the assets of the Nepalese commercial banks relative to domestic economy did add value to its economy. Therefore, commercial banking sector is necessary to represent an added value for local economy. This result contradicts the finding of Awdeh (2012) which shows that the assets are insignificant with economic growth for the Lebanon banking sector. The asset coefficient was 0.256, implying that a unit increase in asset leads to an increase in RGDP by 0.256 units respectively. According to the Ordinary Least Square model, the result showed that, deposit, and assets were significant effect on the economic growth. For that reason, hypotheses 1 and 3 have been accepted and have a significant impact on economic growth of Nepal while loan and advances was not significant with economic growth. And hypothesis 2 has been rejected.

3.4 Tests for Causality

Table 4 shows that Granger Causality test between the commercial banking performance in terms of deposit, loan and advance and asset and economic growth in Nepal. The F-statistics and its corresponding values of the probability suggest that none of the commercial banking performance measures seemed to Granger cause the economic growth in Nepal. This suggests that neither the deposit, nor loan and advances, nor commercial banks assets Granger caused the economic growth in Nepal. Therefore, this study was accepting the null hypothesis. On the other hand, the results of Granger Causality models reveal that growth in the economic activities in Nepal (represented by RGDP) did not seem to Granger cause the deposit, loan and advances and assets. It might be either the size of market is too small. Most of the banks are highly concentrated in urban areas, particularly in Kathmandu valley and even more, particularly in city centers and commercial banks might not be efficiently transforming deposit into loan and that most of the loan they grant might not be for productive purposes.

Table 4 Granger causality test for commercial banking performance with economic growth in Nepal

Null hypothesis	F-Statistic	Probability
DEP does not Granger cause GDP	2.36771	0.1116
GDP does not Granger cause DEP	0.46194	0.6346
LOAN and ADV does not Granger cause GDP	1.59730	0.2197
GDP does not Granger cause ADV	0.15989	0.8530
ASSETS does not Granger cause GDP	2.23719	0.1249
GDP does not Granger cause ASSETS	0.39836	0.6750

4 Conclusions

Unit root test confirmed the non-stationary of all the variables at level whereas stationary at the first difference. Regression results indicate that deposits and assets have significant impact on the economic growth of Nepal while loan and advances was not significant impact on the economic growth. Although the commercial banks assets in the economy has increased but none of the expectations of liberalization, such as extension of the organized credit market to more rural areas, increasing access for smaller borrower or more efficient and productive use of financial resources seem to have been achieved. The loan and advances did not improve economic activities in Nepal. The Granger-Causality test confirmed the commercial banks performance did not Granger causes the economic progress. As a result it is required to grant an appropriate operating situation for the banking industry to execute its services in Nepal and should be encouraged to lend more to the productive sector.

Acknowledgments The authors would like to thank the National Science Foundation of China (71173060, 70773028 and 71031003) for their supports.

References

Acharya M (2003) Development of the financial system and its impact on poverty. Econ Rev 15, Nepal Rastra Bank

Anwar S, Nguyen LP (2011) Financial development and economic growth in vietnam. J Econ Financ 35:348–360

Aurangzeb KA (2012) Contributions of banking sector in economic growth: a case of Pakistan. Econ Financ Rev 1(2):45–54

Awdeh A (2012) Banking sector development and economic growth in Lebanon. Int Res J Financ Econ 100:53–62

Bebczuk R, Garegnani L (2007) Autofinanciamiento empresario y crecimiento económico. Ensayos Económicos 47, Central Bank of Argentina

Beck T, Levine R (2004) Stock markets, banks and growth: panel evidence. J Banking Financ 28:423–442

Goldsmith RW (1969) Financial structure and development. Yale University Press, New Haven, CT

Granger CWJ (1969) Investigating causal relations by econometric models and cross-spectral methods. Econometric 37:424–438

Hsu CM, Lin SM (2000) Financial development and endogenous growth model. Ind Free China 9:21–47

Jha S, Hui X (2012) A comparison of financial performance of commercial banks: a case study of Nepal. Afr J Bus Manage 27:7601–7611

Kharel RS, Pokhrel DR (2012) Does Nepal's financial structure matter for economic growth? Working paper serial number NRB/WP/10

McKinnon RI (1973) Money and capital in economic development. Brookings Institution, Washington, DC

Nepal Rastra Bank (2011) Banking supervision report 2010. Bank Supervision Department, Nepal Rastra Bank, Nepal. http://www.nrb.org.np/bsd/reports/Annual_Reports–Annual_Bank_Supervision_Report_2010.pdf

Poudel NP (2005) Financial system and economic development. Nepal Rastra Bank in 50 years. Nepal Rastra Bank, Kathmandu

Rajan R, Luigi Z (1999) Financial systems, industrial structure and growth. University of Chicago, Mimeo

Shaw ES (1973) Financial deepening in economic development. Oxford University Press, New York

Talliman Ellis, Moen (1998) Commercial banking, organization and decision theory. Federal Reserve Bank of Atlanta, U ms, pp 381–404

Data and Information Fusion for Bio-Medical Design and Bio-Manufacturing Systems

Yuan-Shin Lee, Xiaofeng Qin, Peter Prim and Yi Cai

Abstract This paper presents methods of data analysis and information fusion in bio-medical design and bio-manufacturing system performance improvement. Technical methods of data and information fusion and the new industrial engineering role in bio-medical design and bio-manufacturing systems development are discussed in this paper. A case study in regenerative medicine is presented to validate the product development lead time estimation, and to demonstrate that the total processing time can be significantly reduced through integrated approach. In overview, this research develops a method for transforming uncontrolled biological procedures into reproducible, controlled manufacturing processes.

Keywords Bio-manufacturing · Information fusion · Design and manufacturing · Bio-medical application

1 Introduction

Biomedical engineering is the application of the combination of engineering principles and design concepts with medicine and biology technologies for healthcare purposes. Bio-manufacturing was originally referred as the use of living organism to manufacture a product, which is considered as a higher discipline of biomedical engineering. Bio-manufacturing is referred to the emerging research areas that compass manufacturing methods that use a biological organism, or parts of a biological organism, in an unnatural manner to produce a product, as well as products designed to detect, modify, maintain and study biological organisms for use as new manufacturing agents (Couto et al. 2012).

Y.-S. Lee · X. Qin (✉) · P. Prim · Y. Cai
Department of Industrial and Systems Engineering, North Carolina State University, Raleigh, USA
e-mail: yslee@ncsu.edu

Y.-K. Lin et al. (eds.), *Proceedings of the Institute of Industrial Engineers Asian Conference 2013*, DOI: 10.1007/978-981-4451-98-7_177,
© Springer Science+Business Media Singapore 2013

Fig. 1 Bio-manufacturing and its related areas

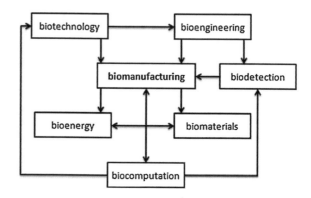

Figure 1 shows the new emerging biomanufacturing and regenerative medicine and its related areas. The new emerging research includes efforts from biotechonoloy, bioengineering, biodetection, biomaterials, biocomputation, and bioenergy. Within the region of bioengineering, Regenerative Medicine Manufacturing (RMM) has been thriving during the past two decades and gradually becoming one of the most value-added, competitive industries in the nation, or even over the world.

Regenerative medicine is always referred as the technologies creating living functional cells, tissues or organs to repair and replace damaged ones due to age, disease, damage or other defects. It is considered as an evolutional treatment offering the potential of lifetime cures for unmet medical needs by stimulating previously damaged organs to heal themselves or creating tissues and organs for implantation. In the last few years, we have been collaborating with Wake Forest University Institute of Regenerative Medicine (WFIRM) to work on regenerative medicine and bio-manufacturing systems development, which refers to viewing regenerative medicine from a systems engineering perspective by modeling, parameterizing, optimizing and controlling the process to achieve better results. In this paper, we will address the new industrial engineering role in the bio-manufacturing and biomedical engineering areas. A case study of regenerative medicine will be discussed.

2 Bio-Manufacturing and Regenerative Medicine

In regenerative medicine, a significant effort to increase tissue area and volume has been explored using mechanical strain methods. The major intent is to plastically deform the biological material so that the surface area increases. In some cases, the strain has also produced rapid cell and tissue growth but the results being reported seem inconsistent and effort to find effective methods to expedite the regeneration of tissue continues. Although a variety of tissue has been studied in the literature, our activity has focused primarily on skin tissue. The primary drive force is the high fatality of severe burned patients with large area of thermally or chemically burned skin tissues. To treat patients with severe burns, small skin grafts can be harvested from healthy places on the body, quickly expanded to much larger area compared to the original area outside of the body, and then transplanted back to a burned area of the body.

With the rapid growth of regenerative medicine research, a new set of products—and processes to create those products—are emerging. For the Industrial Engineer who will work with the commercial production of these products, it will be necessary to use engineering specifications to define both the products and the related processes. These terms include quantifying measurable attributes in the biological domain such as functionality, viability, and metabolic activity, as well as traditional attributes such as time, count, and size. To regenerate healthy skin for treating the severe burned patients, a small area of skin harvest from the patient needs to be stretched and grown into a larger area of healthy skin. After a successful clinical study, researchers began investigating the biaxial expansion of skin by testing human and pig skin from the back and upper legs. We are currently in the first phase of this study, and are attempting to establish an upper limit to the amount of force that can be applied to skin graft specimens without tearing them. A more effective and efficient skin expansion bioreactor design still remains a major challenge in regenerative medicine. In this paper, we present our efforts in exploring the promise of skin regeneration by investigating the skin growth factors and developing mechanical methods to expedite the regeneration of skin in the laboratory for safe implantation.

Figure 2 shows the preparation process of skin expansion. The objective of this research effort is to describe the expansion as a manufacturing process with sufficient specifications, such that the process is reproducible within given tolerances. The primary structural element in most mammalian soft tissue is collagen, which forms a mesh-like network to carry and deflect tensile and compressive stresses, which in turn act to deform this network. Induced stresses, which are large enough and/or occur over large enough periods of time, create permanent tissue deformation. Of particular interest, cells within these deforming tissues often respond with tissue growth, not only by increasing the collagen network strength, but also

Fig. 2 Procedure of skin expansion for regenerative medicine

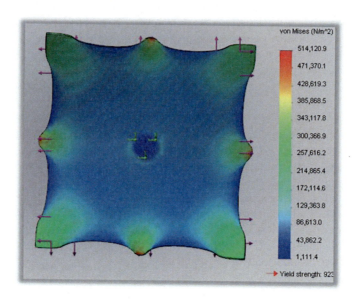

Fig. 3 Analytical modeling of soft tissue expansion for regenerative medicine

by proliferating, which increases the tissues' functional capacity. As such, tissue expansion can achieve different goals, each with a specific set of process parameters that seek to optimize those goals. Figure 3 shows how collagen-based, soft tissue responds to mechanical deformation and the effects of different process parameters based on the analytical modeling.

Many of the current skin stretching practices cause skin tissue failures during skin expansion. It is critical to understand and identify the causes of the failure in skin expansion and to develop a bioreactor that can effectively deliver the healthy regenerated skin tissues. In this paper, we have explored and examined how collagen-based, soft tissue responds to mechanical expansion and the effects of different skin expansion parameters. Figure 3 shows an example of skin stretching by biaxial skin stretching in both X- and Y-directions. Several clamping points were loaded with force to clamp the skin tissue during the skin expansion process.

3 Information Fusion in Bio-Manufacturing Systems

Information fusion (IF) is the concept of a group of data driven methods, which refers to a series of processes that integrate and synergize the information acquired from multiple sources to result in more accurate and reliable description of the situation, or even reveal hidden aspects of the situation (Dasarathy 2001). Such sources can be various types of sensors, databases, and other information gathered by human. The goal of IF is to help people make decisions that are better than those only based on individual source without synergy. Information fusion can be simply considered as the combination of sensor fusion and data fusion. Sensor fusion, as is easily understood, is to establish multi-dimensional sensor array to ensure a sufficient sampling information over individual source. Data fusion always includes data processing with various mathematical methods. Conventional data fusion methods include: Kalman Filter, Bayesian Decision Theory, Dempster-Shafer Evidential Reasoning, Neural Network, Expert System, Blackboard Architecture, Fussy logics, etc. In this paper, we look into the possible tools to be used in information fusion for bio-manufacturing systems development.

Information fusion plays an important role in the context of production and manufacturing process as discussed in our earlier work in (Lee et al. 2013). Figure 4 shows the information flow of conventional manufacturing system based on the part geometry, dimension and quality control data and information.

The creation of novel manufacturing technologies and engineering skills may support long-term presence that secure the entire value system. Based on that, we believe it is reasonable to introduce the concept of information fusion and apply the corresponding methods in the regenerative medicine manufacturing. Figure 5 shows the sequential steps involved in producing cells, which can be modeled based on a multi stage manufacturing process as presented in our earlier work in (Lee et al. 2013).

When information fusion is applied in conventional manufacturing, one of the most essential decisions to make is to determine which process is the best choice. Common constraints in process selection often include material attributes, geometry complexity, and properties requirement of the part, etc. Various criteria, including reduction of the total processing time, minimized total cost, and optimal product quality are always applied (Jin and Shi 2012). From the perspective of information fusion, such constraints and criteria, as well as the information acquired from multiple sensors, measured by all kinds of facilities are collected to serve as data flow to support the decision making.

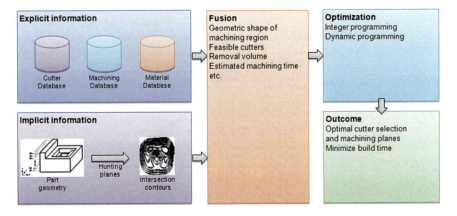

Fig. 4 Data flow and information fusion for process planning and milling operations

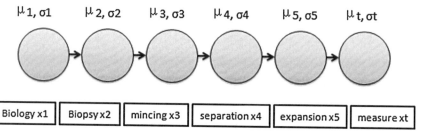

Fig. 5 Procedure of steps in bio-manufacturing processes

Table 1 Information flow for bio-manufacturing and regenerative medicine manufacturing

	Conventional manufacturing	Regenerative medicine manufacturing
Input	Discrete and continuous, can be loaded from each stage of the process	Discrete and continuous, can be loaded from each stage of the process
Constraints	Processing time, cost, quality, etc.	Cell survival rate, cost, function, etc.
Output	Optimal process selection	Optimal process selection
Noise	System error, white noise, etc. Accumulative, propagating along with data flow	System error, white noise, etc. Accumulative, propagating along with data flow
Flow direction	Forward with time flow direction, parallel or serial	Forward with time flow direction, parallel or serial

The following Table 1 presents the common points shared and differences between a conventional manufacturing process and the regenerative medicine manufacturing (RMM) process from the process structure point of view. In Table 1, it can be seen that differences exist between elements from the regenerative medicine manufacturing and the conventional manufacturing process. Concerning all the issues shown in Table 1, in this paper, a hybrid information fusion method is suggested to support the decision making process.

Figure 6 shows the data flow and information fusion in RMM process. Concerning the high randomness and the accumulation and propagation of the variances from the starting point to downstream steps, a data cleaning and classification process is needed before the fusion. Instead of directly inputting raw data into the model, raw data is cleaned and classified into qualitative data and quantitative data. Quantitative data is directly passed to the fusion step of the modeling; however, qualitative data need to be preprocessed in advance. After a series of preprocessing procedures, such as data mapping, noise removal, key factor identification, the remaining data is discretized and passed into the fusion step.

As shown in Table 1 as well as Figs. 4 and 6, there are many differences between the regenerative medicine manufacturing and the conventional manufacturing processes. This is due to the following reasons found in this research: First of all, the material varies from individual to individual, since the cells are

Data and Information Fusion

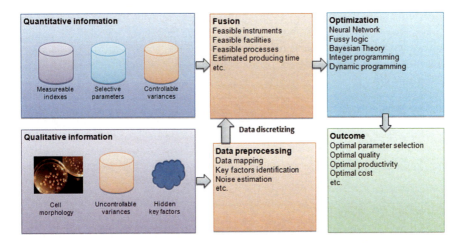

Fig. 6 Information fusion of regenerative medicine manufacturing systems

sampled from patients through biopsy process. As mentioned earlier, individual difference may severely impact the outcome of the entire process dependent on different age, gender, disease state etc. Secondly, manual operation introduces non-neglectable variances, which indicates inconsistency of the process. These variances vary from patches to patches, which yields lot-to-lot difference in ultimate outcome of the RMM process. Thirdly, unlike the conventional manufacturing, the discretizing of the data through the entire process is almost impossible. Thus, the traditional simple optimizing algorithms may not be directly applied. Furthermore, some of the RMM processes are still in state of being developed. The process itself is not completely well defined yet and some of the key factors remain unidentified.

Due to the high diversity of data, optimization algorithms may include soft computing methods, machine learning methods and other conventional statistical methods, the preference of which depends on the specific situation of the data. Optimal outcomes are designed to be generated throughout the model. However, optimization highly depends on each step of the data processing including the preprocessing, since the high randomness and huge amount of noises severely increase the risk of failing to identify the key factors. Nevertheless, the method is supposed to provide an acceptable optimal solution within a certain tolerance space, which is much better than the trial and error methods.

4 Conclusions

In this paper, an information fusion modeling technique is presented for modeling the bio-manufacturing systems for regenerative medicine manufacturing. The challenges related to bio-manufacturing with living cell therapies were presented.

Possible solution approaches to these challenges were investigated through both processing perspective and business perspective. This paper presents the possible future industrial engineering role in solving the challenges of the emerging new area of regenerative medicine manufacturing associated with information fusion and modeling.

Acknowledgments This work was partially supported by the National Science Foundation (NSF) Grants (CMMI-0800811, CMMI-1125872) to North Carolina State University. Their support is greatly appreciated. The authors also gratefully acknowledge the helpful comments and suggestions of Dr. Richard Wysk, Dr. Paul Cohen of North Carolina State University and Dr. Sang Jin Lee, Dr. James Yoo, Dr. Anthony Atala of Wake Forest University Institute of Regenerative Medicine, which have significantly improved this paper.

References

Couto DS, Perez-Breva L, Cooney CL (2012) Regenerative medicine: learning from past examples. Tissue Eng Part A 18(21–22):2386–2393. (120725085622009, Mary Ann Liebert, Inc. publishers)

Dasarathy BV (2001) Information fusion–what, where, why, when, and how? Inf Fusion 2(2):75–76

Jin R, Shi J (2012) Reconfigured piecewise linear regression tree for multistage manufacturing process control. IIE Trans 44(4):249–261

Lee YS, Qin X, Cai Y (2013) Data and information fusion for multi-stage manufacturing systems. In: Proceedings of 2013 industrial and systems engineering research conference (ISERC), San Juan, Puerto Rico, 18–22 May 2013, (CD) Paper Number: ISERC2013-1415

Evaluating the Profit Efficiency of Commercial Banks: Empirical Evidence from Nepal

Suvita Jha, Xiaofeng Hui and Baiqing Sun

Abstract This study offered an application of a non-parametric analytic technique, namely data envelopment analysis for measuring the performance of the Nepalese commercial banking sector. It explored the efficiency of the Nepalese commercial banks with the use of interest expenses and loan loss provision as inputs and net interest income, commission income and other operating income as outputs for the period of 2005–2010. It was also observed the effects of scale and of the mode of ownership (public sector, joint venture and domestic private sector) on bank behavior and therefore, on bank performance in the Nepalese banking industry. The public sector banks most recently in the analyzed period were observed to perform relatively more efficient than joint venture banks and domestic private banks with respect to their profit efficiency due to the large scale of branch networks.

Keywords Nepalese commercial banks · Data envelopment analysis · Profit efficiency · Ownership

1 Introduction

The banking sector is one of the most tightly regulated industries, especially in the developing countries. Although regulation varies in intensity from country to country, bank performance in emerging economies is often criticized due to lack of

S. Jha (✉) · X. Hui · B. Sun
School of Management, Harbin Institute of Technology, 150001 Harbin, People's Republic of China
e-mail: sovi1977@yahoo.com1

X. Hui
e-mail: xfhui@hit.edu.cn2

B. Sun
e-mail: wiseseaman@163.com3

competition and management skills. How the increased competitive pressure will affect banks depends partially on how efficiently they are running. Bank efficiency depends on the quality and associated risk on loans, and if they are not controlled, then the scores calculated may lead to erroneous conclusion (Mester 1996). A majority of studies focused on cost efficiency while research on the revenue and profit efficiency has been much scarcer (Ariff and Can 2008) despite the fact that analyzing profit efficiency constitutes a more important source of information for bank management than the partial assessment offered by the cost efficiency analyses. In fact, a profit-based approach is better in capturing the diversity of strategy responses by financial firms in the face of dynamic changes in competitive and environmental conditions. The profitability approach, which is a relatively newer approach, Drake et al. (2006) proposed the use of a profit-oriented approach in data evaluation analysis (DEA) context that is in line with the approach of Berger and Mester (2003). With reference to Nepal, Gajurel (2010) has used DEA to evaluate the cost efficiency of Nepalese banking sector whereas Jha et al. (2013) has reported technical efficiency of commercial banks under intermediation approach using deposits and interest expenses as inputs and loan and advances and interest income as outputs. There has been no any report about profit efficiency of the Nepalese commercial banks.

Due to the small amount of reliable work done on Nepalese commercial banking sector despite the unprecedented growth, Nepal witnessed several bank crashes one after another causing severe panic in the general population. This necessitated restructuring of the banking sector to increase its financial efficiency. This study is more relevant in the banking field as it includes loan loss provision as an input in computation of profit efficiency to verify the impact of credit risk on the efficiency score. The task of this paper is to examine profit efficiency of the Nepalese commercial banks during the period 2005–2010 using DEA approach.

2 Data and Methodology

2.1 Data and Variables

In this analysis included 18 commercial banks, which have been established before 2005 in Nepal. Many literatures suggest the use of homogeneity conditions for decision-making units in a model (Yeh 1996) and encourage the use of DEA for firms with similar resources and operations providing similar products and services (Oral and Yololan 1992). Therefore, commercial banks that started its operations after 2005 are not included in this investigation. In the present study, we adopted the profit-oriented approach to find out the profit efficiency using interest expenses and loan loss provision as inputs and three outputs—net interest income, commission income and other operating income. The required data for the input and output variables were mainly obtained from the Nepal Rastra Bank Bulletin published by the Central Bank of Nepal (Nepal Rastra Bank 2011), annual audited financial statements of the commercial banks published by the respective banks. Efficiency scores were estimated with MATLABR2010 a program whereas the ordinary calculations were done in Excel.

2.2 Details of the Model

We used DEA to estimate technical, pure technical and scale efficiencies of the Nepalese commercial banks for the period of 2005–2010. One of the well-known advantages of DEA, which is relevant to this study, is that it works particularly well with small samples. The research applied the DEA technique based on constant return on scale (CRS) and variable return on scale (VRS) assumptions. The detail of the model is explained in our previous work (Jha et al. 2013). The relative profit efficiency under CRS (PECRS) and profit efficiency under VRS (PEVRS) were measured as described by Jha et al. (2013). The scale efficiency was found by PECRS/PEVRS. The efficiencies among not only individual selected banks but also three types of the Nepalese commercial banks based on ownership—public sector banks (PSB), joint venture banks (JVB) and domestic private banks (DPB) were compared in this study.

3 Results and Discussion

3.1 Efficiency Estimates-Individual Commercial Banks

Table 1 shows the profit efficiency scores under CRS of individual banks. Among the public sector banks, Nepal Bank Limited (NBL) was fully efficient in year 2007, 2009 and 2010 while Rastriya Banijya Bank Ltd (RBBL) and Agricultural Development Bank Ltd (ADBL) were only fully efficient in year 2009 and 2010 because the share of non-performing loan (NPL) is very high from 2005 to 2008. Poor evaluation, insufficient follow up and supervision of loan distribution ultimately resulted in massive booking of poor quality assets, the level of which remains high. Hence, more provisions had to be made. In 2009 and 2010, this situation slightly improved and the growth in NPL was reduced. Among the joint venture banks, Nabil Bank Ltd (NABIL) bank was fully efficient from 2005 to 2009 and Standard Chartered Bank Ltd (SCBL) showed the consistency at its efficiency scores for all the years. Both NABIL and SCBL, which are holding highest share of deposits, have shown relatively better performance than rest while Nepal Bangladesh Bank Ltd (NBBL) from 2008 to 2009 and Everest Bank Ltd (EBL) in 2010 showed fully efficient. No any domestic private banks had 100 % efficiency from 2005 to 2010. Nepal Credit and Commerce Bank Ltd (NCCBL)

Table 1 Individual commercial banks technical efficiency under CRS

	Bank	Profit efficiency under CRS						Avg
		2005	2006	2007	2008	2009	2010	
PSB	NBL	0.487	0.780	1.000	0.981	1.000	1.000	0.875
	RBBL	0.410	0.689	0.811	0.843	1.000	1.000	0.792
	ADBL	0.698	0.620	0.743	0.882	1.000	1.000	0.824
	Avg	0.532	0.696	0.851	0.902	1.000	1.000	0.830
JVB	NABIL	1.000	1.000	1.000	1.000	1.000	0.978	0.996
	SCBL	1.000	1.000	1.000	1.000	1.000	1.000	1.000
	HBL	0.464	0.536	0.492	0.689	0.693	0.759	0.606
	NSBI	0.369	0.382	0.421	0.489	0.462	0.690	0.469
	NBBL	0.233	0.406	0.901	1.000	1.000	0.792	0.722
	EBL	0.456	0.622	0.788	0.993	0.770	1.000	0.771
	Avg	0.587	0.658	0.767	0.862	0.821	0.870	0.761
DPB	NIBL	0.716	0.595	0.617	0.634	0.563	0.996	0.687
	BOK	0.476	0.622	0.577	0.656	0.743	0.927	0.667
	NCCBL	0.213	0.290	0.397	1.000	1.000	1.000	0.650
	LBL	0.291	0.291	0.783	0.624	1.000	0.891	0.647
	NIC	0.457	0.357	0.471	1.000	0.618	0.869	0.629
	MPBL	0.998	1.000	0.867	0.572	0.527	0.442	0.734
	KBL	0.620	0.600	0.571	0.486	0.493	0.740	0.585
	LXBL	0.546	0.634	0.725	0.605	1.000	1.000	0.752
	SBL	0.587	0.665	0.763	0.709	0.692	0.678	0.682
	Avg	0.545	0.562	0.641	0.699	0.737	0.838	0.670
Avg		0.557	0.616	0.718	0.787	0.809	0.876	0.727

from 2008 to 2010, Lumbini Bank Ltd (LBL) in 2009, Nepal Industrial and Commercial Bank Ltd (NICBL) in 2008 and Laxmi Bank Limited (LXBL) from 2009 to 2010 were fully efficient. DPB have also shown a good efficiency trend. This shows that less slack was produced by these banks. Thus, the fully efficient banks armed with modern banking technology and business performance. It is remarkable here that the method of resource use in the efficient banks is functioning well, and featuring no misuse of resources. In the strength of DEA terminology, these commercial banks can be termed as internationally efficient banks. Other banks were deviating a lot from the optimal input–output mix. It shows that some large banks were efficient. Hence, it is using economies of scale to take advantage and able to maximize profits. However, a few other large banks are unable to achieve profit efficiency in the same way as their counterparts have taken. This is due to internal factors relating to non-interest income and to external factors like loan loss provisions.

Table 2 shows the profit efficiency scores under VRS of individual banks. RBBL and ADBL were found operating efficiently with the efficiency score of 1. Similarly, NABIL and SCBL were fully efficient from 2005 to 2010 while NBL from 2006 to 2010, NBBL from 2008 to 2010 and EBL in 2010 showed 100 %

Table 2 Individual selected commercial banks pure technical efficiency under VRS

	Bank	Profit efficiency under VRS						Avg
		2005	2006	2007	2008	2009	2010	
PSB	NBL	0.912	1.000	1.000	1.000	1.000	1.000	0.985
	RBBL	1.000	1.000	1.000	1.000	1.000	1.000	1.000
	ADBL	1.000	1.000	1.000	1.000	1.000	1.000	1.000
	Avg	0.971	1.000	1.000	1.000	1.000	1.000	0.995
JVB	NABIL	1.000	1.000	1.000	1.000	1.000	1.000	1.000
	SCBL	1.000	1.000	1.000	1.000	1.000	1.000	1.000
	HBL	0.521	0.570	0.522	0.690	1.000	0.786	0.681
	NSBI	0.530	0.594	0.738	0.798	0.545	0.723	0.655
	NBBL	0.309	0.505	0.985	1.000	1.000	1.000	0.800
	EBL	0.544	0.677	0.876	0.938	0.870	1.000	0.817
	Avg	0.651	0.724	0.854	0.904	0.903	0.918	0.826
DPB	NIBL	0.716	0.604	0.636	0.886	1.000	1.000	0.807
	BOK	0.627	0.675	0.953	0.999	0.833	0.944	0.838
	NCCBL	0.379	0.589	0.953	1.000	1.000	1.000	0.820
	LBL	0.568	0.775	1.000	1.000	1.000	1.000	0.891
	NIC	0.566	0.538	0.701	1.000	0.840	0.952	0.766
	MPBL	1.000	1.000	0.981	0.966	0.624	0.476	0.841
	KBL	0.684	0.692	0.867	0.840	0.859	0.922	0.811
	LXBL	1.000	1.000	1.000	1.000	1.000	1.000	1.000
	SBL	1.000	1.000	1.000	1.000	0.880	0.779	0.943
	Avg	0.727	0.764	0.899	0.966	0.893	0.897	0.858
Avg		0.742	0.790	0.901	0.951	0.914	0.921	0.870

efficient among JVBs. Among DPBs, LXBL was fully efficient while Siddhartha Bank Ltd (SBL) from 2005 to 2008, Nepal Investment Bank Ltd (NIBL) from 2009 and 2010, Nepal Credit and Commerce Bank Ltd (NCCBL) from 2008 to 2010, LBL from 2007 to 2010, NIC in 2008, Machhapuchhre Bank Ltd (MPBL) from 2005 to 2008 showed fully efficient. As per Table 3, when the scale efficiency of the individual bank was examined, all the PBS were fully efficient in 2009 and 2010. Among JVB, NABIL from 2005 to 2009 and SCBL from 2005 to 2010 were determined as competent banks while the scale efficiency score of NBBL from 2008 to 2009 and EBL 2010 showed 100 % efficient. NCCBL from 2008 to 2010, NIC in 2008, MPBL in 2006, LXBL from 2009 to 2010 showed 100 % efficient among DPB.

3.2 Efficiency Score: Ownership Classification

Table 4 demonstrates average profit efficiency classified on the basis of ownership. There was an increasing trend in their mean of yearly efficiency measures of the

Table 3 Individual commercial banks scale efficiency

	Bank	Profit efficiency under VRS						Avg
		2005	2006	2007	2008	2009	2010	
PSB	NBL	0.534	0.780	1.000	0.981	1.000	1.000	0.883
	RBBL	0.410	0.689	0.811	0.843	1.000	1.000	0.792
	ADBL	0.698	0.620	0.743	0.882	1.000	1.000	0.824
	Avg	0.547	0.696	0.851	0.902	1.000	1.000	0.833
JVB	NABIL	1.000	1.000	1.000	1.000	1.000	0.978	0.996
	SCBL	1.000	1.000	1.000	1.000	1.000	1.000	1.000
	HBL	0.891	0.941	0.941	0.998	0.693	0.965	0.905
	NSBI	0.695	0.643	0.570	0.613	0.848	0.955	0.721
	NBBL	0.752	0.803	0.915	1.000	1.000	0.792	0.877
	EBL	0.838	0.919	0.900	1.059	0.885	1.000	0.933
	Avg	0.863	0.884	0.888	0.945	0.904	0.948	0.905
DPB	NIBL	0.999	0.987	0.970	0.716	0.563	0.996	0.872
	BOK	0.760	0.921	0.605	0.657	0.892	0.982	0.803
	NCCBL	0.563	0.493	0.417	1.000	1.000	1.000	0.746
	LBL	0.513	0.375	0.783	0.624	1.000	0.891	0.698
	NIC	0.807	0.663	0.672	1.000	0.736	0.913	0.798
	MPBL	0.998	1.000	0.884	0.593	0.844	0.929	0.875
	KBL	0.907	0.868	0.658	0.579	0.574	0.803	0.731
	LXBL	0.546	0.634	0.725	0.605	1.000	1.000	0.752
	SBL	0.586	0.664	0.763	0.708	0.786	0.870	0.730
	Avg	0.742	0.734	0.720	0.720	0.822	0.932	0.778
Avg		0.750	0.778	0.798	0.826	0.879	0.949	0.830

commercial banks of Nepal. The average annual score based on CRS for PSB, JVB and DPB from 2005 to 2010 were determined 83, 76 and 67 % respectively. It indicated that PSB exhibited higher average efficiency compared to their counterparts. This result is contrary to Ariff and Can 2008 which showed that the JVB are profit-efficient relative to the PSB for Chinese banks. One major cause for this is the Nepalese banking system is dominated by PSB in terms of deposit mobilization and PSB hold larger share of total deposit of the banking system, thereby resulting in higher profits. It is often argued that larger banks possess more flexibility in financial markets and are better able to expand their credit risks (Cole and Gunther 1995). Due to having relatively very few branch networks with limited assets, JVB and DPB have little capacity to deposit mobilization in small urban or rural savings and make lower profit than PSB.

The level of managerial efficiency as revealed by PTE score under VRS is more in PBS relative to JVB and DPB. In VRS specification, the average score of the efficiency of the commercial banks of Nepal varied from 0.65 to 1. PSB showed the consistency in their performance for each year with the score 1 whereas DPB had higher profit efficiency under VRS (86 %) than that for JVB (82 %). The pure technical inefficiency is generally caused by inappropriate management practices

Table 4 Average efficiency of commercial banks

Years	Technical efficiency				Pure technical efficiency				Scale efficiency			
	PSB	JVB	DPB	Avg	PSB	JVB	DPB	Avg	PSB	JVB	DPB	Avg
2005	0.53	0.59	0.54	0.55	0.97	0.65	0.73	0.78	0.55	0.86	0.74	0.72
2006	0.70	0.66	0.56	0.64	1	0.72	0.76	0.83	0.70	0.88	0.73	0.77
2007	0.85	0.77	0.64	0.75	1	0.85	0.90	0.92	0.85	0.89	0.72	0.82
2008	0.90	0.86	0.70	0.82	1	0.90	0.97	0.96	0.90	0.95	0.72	0.86
2009	1	0.82	0.74	0.85	1	0.90	0.89	0.93	1	0.90	0.82	0.91
2010	1	0.87	0.84	0.90	1	0.92	0.90	0.94	1	0.95	0.93	0.96
Avg	0.83	0.76	0.67	0.75	1.00	0.82	0.86	0.89	0.83	0.91	0.78	0.84

but in case of Nepal, higher meant profit efficiency under VRS for PSB was consistent with the PSB have a large volume of operations and resources in Nepal. They would be able to get in benefits of economies of scale and earn more profit. On the other hand, JVB and DPB have relatively lower profit due to higher costs in their small branch networks. As stated before, scale efficiency score for each bank can be obtained by taking a ratio of technical efficiency score to pure technical efficiency score. The value of scale efficiency equal to 1 means that the bank is working at most dynamic scale size which corresponds to constant returns-to-scale. Furthermore, being scale efficiency less than 1 shows that the bank is not operating at its optimal scale. The scale efficiency of the commercial banks showed the mean efficiency each year by decomposing profit efficiency under CRS into profit efficiency under VRS and scale efficiency to gain insight into the main sources of inefficiencies. The scale efficiency varied in between 0.55 and 1. The average scale efficiency of the commercial banks was 84 %, indicating that the actual scale of production has differed from the most productive scale size by about 16 %. According to Hassan and Sanchez (2007) if scale inefficiency is greater than profit efficiency under VRS, it means wrong mix of inputs and outputs for the reasons beyond their control inefficiency. Here, scale inefficiency was 16 % whereas profit inefficiency under VRS was 11 %, suggesting wrong mix of inputs and outputs.

4 Conclusions

Concerning the ownership, the public sector banks were more efficient than joint venture banks and domestic private banks during the study period of 2005–2010 because public sector banks have higher the size of total assets. The commercial banks that have expanded their operations in large scale appeared to be more profit efficient than those operating only at a small scale. Not only this, increasing loan activity of the public sectors banks made them more efficient than joint venture banks and domestic private banks, which have limited networking and urban concentration. The individual commercial banks yearly analysis showed that the efficiency of majority of the banks in Nepal is still not fully efficient. Thus it is still needed for commercial banks in Nepal to improve their efficiency.

Acknowledgments The authors would like to thank the National Science Foundation of China (71173060, 70773028 and 71031003) for their supports.

References

Ariff M, Can L (2008) Cost and profit efficiency of Chinese banks: a non-parametric analysis. China Econ Rev 19:260–273

Berger AN, Mester LJ (2003) Explaining the dramatic changes in performance of U.S. Banks: technological change, deregulation and dynamic changes in competition. J Financ Intermed 12:57–95

Cole RA, Gunther JW (1995) Separating the likelihood and timing of bank failure. J Bank Financ 19:1073–1089

Drake LM, Hall MJB, Simper R (2006) The impact of macroeconomic and regulatory factors on bank efficiency: a non-parametric analysis of Hong Kong's banking system. J Bank Financ 30:1443–1466

Gajurel DP (2010) Cost efficiency of Nepalese commercial banks. Nepalese Manage Rev

Hassan K, Sanchez B (2007) Efficiency determinants and dynamic efficiency changes in Latin American banking industries. Network financial institute working paper no. 32, Indiana State University, USA

Jha S, Hui X, Sun B (2013) Commercial banking efficiency in Nepal: application of DEA and Tobit model. Inform Tech J 12:306–314

Mester L (1996) A study of bank efficiency taking into account risk-preferences. J Bank Financ 20:1025–1045

Nepal Rastra Bank (2011) Banking supervision report 2010. Bank Supervision Department, Nepal Rastra Bank, Nepal.http://www.nrb.org.np/bsd/reports/Annual_Reports–Annual_Bank_Supervision_Report_2010.pdf

Oral K, Yololan O (1992) An empirical study on analyzing the productivity of bank branches. Iie Trans 46:282–294

Yeh QJ (1996) The application of data envelopment analysis in conjunction with financial ratios for bank performance evaluation. J Oper Res Soc 47:980–988

Explore the Inventory Problem in a System Point of View: A Lot Sizing Policy

Tsung-Shin Hsu and Yu-Lun Su

Abstract This article presents a study on the make-to-order inventory management problem in an anonymous manufacturer which produces precision screw and bolt for European customers. Seeing that this company keeps about 800 active wide varieties of screw and bolt items, of which the manufacturing normally undergoes six processing steps each requires 1–4 h setup before production proceeds, the company is forced to keep inventory-item in their make-to-order production. In this clear and certain inventory problem setting, we can make a study from system point of view to effectively reduce the inventory and meet customers' requirements. Although the operations of business are preceded by the operation mechanism, it is easily found that we still can enhance inventory performance by some control factors such as lot-size policies and parameters. To facilitate the study we develop a system simulator according to the company's SOP as the tool to visualize the inventory problem then achieve effective management.

Keywords Make-to-order · Inventory management · Lot-size policies · WMS management systems · Inventory turnover · ROI

1 Introduction

Businesses have long recognized that good inventory management is valuable; it is especially essential for those the inventory is a large percentage of assets in its balance sheet. Having the necessary items just in time to be sold or be consumed is

---

T.-S. Hsu (✉) · Y.-L. Su
Department of Industrial Management, National Taiwan University of Science and Technology, Taipei, Taiwan, Republic of China
e-mail: tshsu@mail.ntust.edu.tw

Y.-L. Su
e-mail: D9801010@mail.ntust.edu.tw

obviously the best practices in maintaining high inventory turnover and satisfying customers as well. Stockless production and purchasing, lean manufacturing are well known practices for these purposes. However, uncertainty is unavoidable; reducing the lot size is a next choice. Seeing that amounts of inventory are resulted from triggering the needed replenishment orders, large lot sizes may result in excess amounts of inventory items in the stockroom. On the other hand, small lot sizes can lead to lost sales and higher production costs; we can not say it is a smart practice for the manufacturers especially for that setup costs are high.

This article presents a study on the make-to-order inventory problem in an anonymous manufacturer named as XYZ-company. XYZ-company is a precision screw and bolt producer which services European customers more than 30 years. It provides more than 800 product items, of which each piece of product undergoes 6 manufacturing processing steps including heading, thread rolling, drilling, coating and polishing normally. Since the setups before production precede take 1–4 h in every processing steps, and customer orders are mostly repeated and some in small quantities, order by order production is uneconomical. Therefore, XYZ-company is forced to keep inventory-item in their make-to-order production, and making up an inventory high as 58 % of its total asset, slowing down inventory turnover as 3.23 annually.

The devil of inventory seems in the details of the order-sizing. Since Harris (1913) proposed the EOQ, most of the inventory studies focused on determining the least cost inventory. Literature review articles (Chen et al 2012; Yu et al. 2012; Lin and Chung 2012; Duc et al. 2010; Wee et al. 2007; Abad 2008) showed that thousands studies based on the same theory foundations. These studies vastly extended to the assumptions of different demand patterns, covered single and multiple products, multiple warehouses or layers, existence of price discount and delayed payment, perishable products, and so on. Based on EOQ model, many studies focused on the effect of the parameter changing, Porteus (1985) discussed the effects of reducing setup cost, Jordan (2nd Quarter 1987) discussed the cost and profit affected by declining interest rates. Excepting least cost model, Schroeder and Krishnan (1976) viewed inventory as an financial investment and developed the lot size by optimizing the ROI.

2 The Problem

The conflict of management functions always exists in XYZ-company; finance wants to hold down inventory, manufacturing wants to maximize their production performance, while sales people emphasize on meeting customers' requirements. Providing a systematic solution to effectively managing the inventory is sensible. Since inventory related operation of XYZ-company is unchangeable and well known, and it has high inventory percentage of the asset and low inventory turnover, we can take this inventory as a whole and make a systematic study; instead of just obtain piecewise results of the inventory to this issue. Moreover, we can practically study to realize how effective the lot sizing rules XYZ-company's inventory is.

Although XYZ-company's inventory related operations are guided by the implemented standard operation procedures, managers still can take certain actions, such as override lot size decisions, reduce setup costs; eliminate sluggish inventories etc., to enhance inventory performances.

XYZ—company normally keep more than 800 product items in its stockroom, among them we pick AA series which contain 25 active product items, contribute 5 % of sale amount in average. Draw from its ERP database one year ago, we found XYZ-company received 53 incoming customer orders of AA product series in the planning horizon of 4 months. Among these 53 orders, 29 orders can be fulfilled and shipped to customers directly because inventories can cover the requirements; that is only 24 replenishment orders which cover 13 product items are to be produced. From these historical data, we can simulate since all these 53 orders are completed and shipped, measured by dividing sales by average inventory value.

The study design simulate of production and sales. To facilitate the study of the actually operated inventory system in XYZ we develop a system simulator which imitating the company's operation procedures according to the SOP in use. By means of this simulator we can understand and visualize the dynamics of the inventory related operation and then undergo the study.

3 Research Design

Figure 1 depicts a causation that simplifies the study of the inventory of XYZ-company, in which the causes are controllable factors; the effects are performance indicators of inventory; the rectangular box in-between expresses the system simulator that simulating XYZ-company's operation and inventory processes.

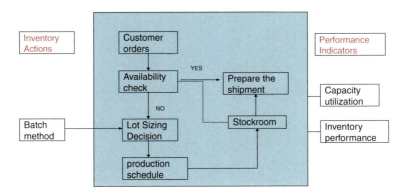

Fig. 1 System simulator

3.1 System Simulator

To facilitate a realistic inventory study such as which of XYZ-company, we develop a system simulator which imitating the company's inventory related SOP in use. By means of this, in coming customer orders can be processed automatically by the developed software; we can manipulate XYZ-company's inventory, visualize how the inventory related operations work, and measure the effects result from the inventory decisions. The simulator is developed according to the procedures and rules as follow:

1. Inventory checking: When receiving a customer order, warehouse people normally will check whether there are enough amounts of inventory for the required items. If the answer is yes shipment orders are issued accordingly, else replenishment orders will be issued instead.
2. Lot size decision: When receiving a replenishment order, operation people have to determine how large the production lot should be and then the order is scheduled to produce. Once this decision has been made, the quantity produced in excess of requirement becomes part of inventory in the stockroom. Lot sizing rules which compromising the costs of setup and inventory is essential.
3. Scheduling and production: Replenishment orders in XYZ-company normally are scheduled according to FCFS rule. The system kernel will use this rule to determine the sequence of replenishment order to be produced.
4. Order completing and reporting.
5. Order picking and preparing for shipment.

3.2 Inventory Actions

Using the system simulator as the kernel, we can test what kind of effects is resulted from the inventory actions we take. Similar to the people operated system guided by the SOP, the incoming customer orders are processed and inventory related operation generated by the software. We will emphasize our study on lot sizing; the comparing rules are traditional EOQ, ROI-based EOQ, and the experience based order lots.

3.3 Performance Indicators

Inventory turnover, in order to reduce costs to the minimum of the company will get more benefits than the expansion of the material reserves, lowest cost Order Quantity does not necessarily mean profitable. The performance of this inventory study covers compromising the performance of production and inventory. From production performance and output value, the amount shipped value/average

Explore the Inventory Problem in a System Point of View: A Lot Sizing Policy 1533

inventory, average inventory performance. These movements result in the creation of material and financial accounting documents.

4 Analyze the System

In this study, the XYZ–Company order data, a total of 53 orders, 25 kinds of products, input simulator, after simulation, 53 orders only 24 orders need to produce 29 orders inventory ready to ship directly; 13 kinds of products need to produce 12 kinds of products are in stock and ready to ship directly. Production order items the different bulk procurement policy, regardless of the number of orders size, the EOQ procurement batch policy, simulation results Table 1.

Inventory and out of storage shown in Fig. 2.

Average inventory amount of $3,689,626 by the end of the data can be seen, the output value of $4,644,236, the shipments total amount of $4,581,750. Under the same conditions, the use of the ROI procurement volume, the analog data shown in Table 2.

Inventory and out of storage shown in Fig. 3.

Average inventory amount of $3,584,198 by the end of the data can be seen, the output value of $3,153,563, the shipments total amount of $4,581,750.

Large orders in the Order 10,000, ROI order quantity policy, and orders less than 10,000 orders, the EOQ purchasing bulk policy, the analog data shown in Table 3.

Inventory and out of storage shown in Fig. 4.

Table 1 EOQ procurement batch policy simulation results

Working days	7	8	13	14	22	23	38	39	45
Amount of storage	0	368021.9599	0	440926.2	0	301575.3	0	809480.7	0
Amount of inventory	3499906	3867927.96	3690428	4131354	3588354	3889930	3712429.515	4521910	3976910

	46	52	53	59	60	68	69	94	95
	253170.7236	0	239931.6	0	372237	0	239931.6	0	1618961
	4230080.91	4230081	4470013	3379013	3751250	3443249.506	3683181	2793431	4412392

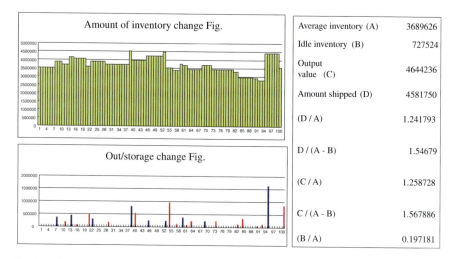

Fig. 2 EOQ inventory and out of storage

Table 2 ROI procurement batch policy simulation results

Working days	8	9	11	15	16	29	30	34	35	37
Amount of storage	0	425587.5	0	0	293212.5	0	679575	0	91942.5	0
Amount of inventory	3499906	3925494	3747994	3847009	4068221	3419721	4099296	4099296	4191239	4191239

38	40	41	46	47	50	51	67	68
80902.5	0	208312.5	0	135915	0	323610	0	815490
4272141	3727141	3935454	3935454	4071369	4071369	4394979	2995979	3811469

By the end of the data that the average inventory amount of $3,544,073, the output value of $3,104,128, shipped a total amount of $4,581,750.

Compare the three procurement batch policy, in the amount of the average inventory is, EOQ procurement batch >the ROI procurement batch> EOQ & ROI purchasing bulk; in the output value of the total amount is, EOQ procurement batch> the ROI procurement batch> EOQ & ROI procurement batch; while shipments total amount of three bulk Policy are the same, so the three bulk Policy inventory Turnover the EOQ & ROI procurement bulk > ROI procurement bulk > EOQ procurement bulk, comparative data shown in Table 4.

Explore the Inventory Problem in a System Point of View: A Lot Sizing Policy 1535

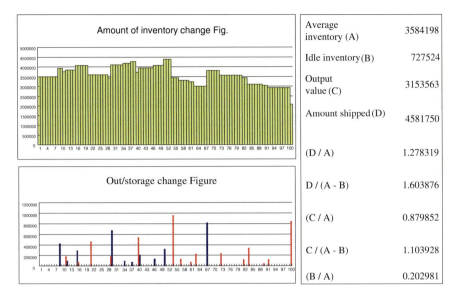

Fig. 3 ROI inventory and out of storage

Table 3 EOQ & ROI procurement batch policy simulation results

Working days	7	8	10	11	15	16	29	30	34	35
Amount of storage	0	367861.5	0	99015	0	301504	0	679575	0	91942.5
Amount of inventory	3499906	3867768	3867768	3789283	3789283	4018787	3370287	4049862	4049862	4141804

37	38	40	41	45	46	50	51	67	68
0	80902.5	0	208312.5	0	135915	0	323610	0	815490
4141804	4222707	3677707	3886019	3886019	4021934	4021934	4345544	2946544	3762034

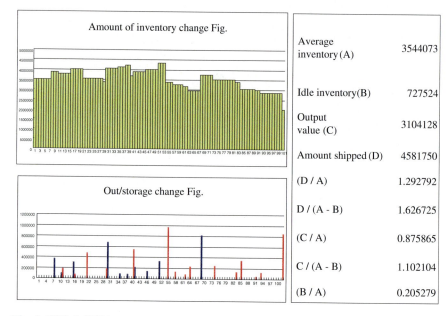

Fig. 4 EOQ & ROI inventory and out of storage

Table 4 Policy comparison

Batch strategy	Average inventory	Out value	Amount shipped	Output value/average inventory
EOQ model	3,689,626	4,677,236	4,581,750	1.241793
ROI model	3,584,198	3,153,563	4,581,750	1.278319
10,000 or above adopt ROI	3,544,073	3,104,128	4,581,750	1.292792

5 Discussion and Conclusion

The results appear from Table 4. EOQ model, the highest average inventory, the highest output value; ROI model, inventory costs are low compared to the EOQ model, the output value is lower than the EOQ model; mining EOQ bulk when demand is less than 10,000 procurement, higher than that the 10,000 mining ROI bulk purchases, both mixed-mode, the lowest cost of inventory, the output value of the lowest. Compare the three procurement batch policy, in the amount of the average inventory is, EOQ procurement batch> the ROI procurement batch> EOQ & ROI purchasing bulk; in the output value of the total amount is, EOQ procurement batch> the ROI procurement batch> EOQ & ROI procurement batch; while shipments total amount of three bulk Policy are the same, so the three bulk Policy inventory Turnover the EOQ & ROI procurement bulk > ROI procurement bulk > EOQ procurement bulk. In different batch Policy, the average inventory costs, output and output value/average inventory change scenarios.

Therefore, it is recommended that a small variety of small orders EOQ model is more suitable for large orders recommended ROI mode, this two mixed strategy inventory costs can be greatly reduced. Emergency inserted single need to increase capacity utilization, and not allowed out of stock recommended EOQ model, because of the EOQ the highest capacity and rush orders will not stock.

References

Abad PL (2008) Optimal price and order size under partial backordering incorporating shortage, backorder and lost sale costs. Int J Prod Econ 114(1):179–186

Chen X, Hao G, Li X, Yiu KFC (2012) The impact of demand variability and transshipment on vendor's distribution policies under vendor managed inventory strategy. Int J Prod Econ 139(1):42

Ton Hien Duc T, Luong HT, Kim YD (2010) Effect of the third-party warehouse on bullwhip effect and inventory cost in supply chains. Int J Prod Econ 124(2):395

Jordan PC (1987) The effects of declining interest rates on order sizing, inventory, and investment. Prod Inventory Manage 28(2):65

Lin SD, Chung KJ (2012) The optimal inventory policies for the economic order quantity (EOQ) model under conditions of two levels of trade credit and two warehouses in a supply chain system. Afr J Bus Manage 6(26):7669

Porteus EL (1985) Investing in reduced setups in the EOQ model. Manage Sci 31(8):998–1010

Schroeder RG, Krishnan R (1976) Return on investment as a criterion for inventory models. Decis Sci 7(4):697

Wee HM, Yu J, Chen MC (2007) Optimal inventory model for items with imperfect quality and shortage backordering. Omega 35(1):7

Yu Y, Wang Z, Liang L (2012) A vendor managed inventory supply chain with deteriorating raw materials and products. Int J Prod Econ 136(2):266

Harris FW (1913) How many parts to make at once, factory. Mag Manage 10(2):135–136

Global Industrial Teamwork Dynamics in China and Southeast Asia: Influence on Production Tact Time and Management Cumulative Effect to Teamwork Awareness-1/2

Masa-Hiro Nowatari

Abstract This research is international comparative survey on the teamwork awareness (TA) of China (four plants/Dalian, Shanghai: two, Guanzhou) and, Malaysia, Thailand and Vietnam in Southeast Asia. TA of the first line workers is compared between plants through country according to same category products. And those seven plants are surveyed from the production tact time (PTT) and the cumulative management effect (CME) from point of view. It clears TA are different by PTT and CME even if same category products. PTT effects on task orientation (TO) positively and CME effects on people orientation (PO) positively in the teamwork appraisal factors (TAF), respectively. And it confirms TA has stronger TO in Chinese workers, but has stronger PO in Southeast Asian workers.

Keywords Production tact time · Management cumulative effect · Teamwork awareness · Teamwork appraisal factors · China · Malaysia · Thailand · Vietnam

1 Introduction

The Industrial teamwork dynamics (ITD) has been developed to clarify relationship between line worker's teamwork awareness and team productivity through Japan domestic case studies in 1990s. From 2000s, it has been stepping up from ITD to global industrial teamwork dynamics (GITD), and studying various influences on TA by religion and national wealth through many overseas cases outside Japan. In this research, comparison study with China and countries in Southeast Asia bloc is carried out by GITD. Seven overseas plants belonging to the same Japanese parent company are surveyed, and workers in the front line are

M.-H. Nowatari (✉)
Department of Management Science, Faculty of Engineering, Tamagawa University, Tamagawa Gakuen, 6-1-1, Machida, Tokyo, 194-8610 Japan
e-mail: nowatari@eng.tamagawa.ac.jp; masahiro.nowatari@gmail.com

research subjects. To the Japanese manager who carries out local plant, understanding teamwork of local workers is important to localizing Japanese-style production management. However, TA of local line workers has a tacit level understanding yet, and staying Japanese manager has misunderstood no difference, local worker's TA is same as Japanese. Therefore, large opportunity loss has been revealed. It is important to make teamwork as practical knowledge from tacit knowledge into explicit knowledge. If it sets up it as significant management tool, it will be able to bring up huge effect in labor-intensive assembly line, especially. Also it contributes to mutual understanding between Japanese managers and local manager, workers in every country, each other positively. On the other hand, it needs to meet quickly with changing market demand, and PTT has to match with market speed. PTT is key index on production management. Moreover, construction on localization of Japanese-style production management requires many years, it is introduced MCE as glocalization (global-localization) index.

2 Research Subject

Research subject is first line worker in the plants of China, Malaysia, Thailand and Vietnam on Japanese-owned local electrical plants. TA is tested influences by PTT and MCE, and especially internal structure of TA is tested by same category product.

3 Production Tact Time and Teamwork Awareness

Production tact time (PTT) means production interval as the whole plant and is based by market demand speed. PTT must realize manufacturing system based on invest by evaluating return on investment (ROI) according to production strategy. It consists of 4 viewpoints, where (location), what (production item), how many (production scale) and how to produce. This paper focuses on same product category globally, and confirms relationship between PTT and Teamwork awareness (TA).

4 Survey Method

4.1 Teamwork Appraisal Factors

In order to measure TA, teamwork appraisal factor (TAF) is set up. TAF has two categories, "task orientation (workability): TO" and "people orientation (cohesiveability): PO". And TAF consists of twelve factors (sixty questionnaire items) of six factors on TO (thirty questionnaire items), and six factors on PO (thirty questionnaire items). Moreover, every questionnaire item is appraised on five stages evaluation (1, 2, 3, 4, 5), and more positive figure means favorable and strong awareness to teamwork. TO means capability for carrying out the task (job, work) given as a work team, and PO is capability for maintaining team as a work team (Table 1).

Questionnaire method had been carried out from 2007 until 2011 with factory tour, interview, collecting data and fact finding.

4.2 Surveyed Plants

Seven plants of four plants (Dalian, Shanghai: two, Guanzhou) in China, and three plants (Malaysia, Thailand, Vietnam) classified into Southeast Asia bloc, were surveyed. Number of respondents sum total is 5,668 persons, and questionnaire collect ratio is around eighty percentages. In addition, MCE has specified as operation years which is continuing period of each plant from starting operation to surveyed year. And, unit of PTT is "a second/a piece". On size of employee, plant of greater than two thousands are D and S, and other five plants are from four hundreds to eight hundreds. The product H are produced in plant D and V, the product P is only S, the products Y are in three plants which are R, B and T, and the products M are produced at G and T. Only the plant T produces two products Y and M, and the others plant produces a single product. The operation years of plant B is overwhelmingly thirty-five years long. Plant D, S, R and T are around ten years and G and V as short

Table 1 Teamwork appraisal factors (TAF)

Productivity orientation	Team work appraisal factors (P = 12, p = 60)	Number of questionnaire items
TO: Task orientation (workability) (P = 6, p = 30)	1. Level of work management	5
	2. Training and instruction to subordinates by the leader's superior	5
	3. Training and instruction to subordinates by the leader	5
	4. Care of subordinates by the leader	5
	5. Ability to accomplish work	5
	6. Conformance to job requirement (progress, quality and quantity)	5
PO: People orientation (cohesive ability) (P = 6, p = 30)	7. Cohesiveness	5
	8. Atmosphere of group	5
	9. Human relations	5
	10. Morale	5
	11. Mutual cooperativeness on working	5
	12. Satisfaction	5

as from two to four years. Even plant R, B and T produce same product Y category, PTT is slightly varying from two to four seconds according to product specification and automation level. So PTT can be divided into three levels roughly, ten seconds or less group for product Y and product M derived from product Y (product M's PTT is close to product Y's PTT) in plant T, sixty seconds group for product H and P, and three-hundreds seconds group for product M (Table 2).

4.3 Teamwork Process

Teamwork process in this research is based on Steiner in social psychology. ITD is an interdisciplinary approach by Group Dynamics (GD) and Industrial Engineering (IE), and is an evaluation system which analyzes quantitatively internal structure of a work team's teamwork awareness. The adaptability of resources (man, machine in a team) to given task (job, work into as a team) effects to team productivity, and it is considered as teamwork process which is team members' interaction of TAF (Fig. 1).

4.4 Hypothesis

To keep up PTT, production line is set up and worker assignment is organized, adequately. Work team is grouped together here, their work team's resources are adapted into daily given task, and teamwork process is formed. On the other hand, localization of Japanese-style production management needs many years to make up high productivity same as Japan domestic. So this glocalization needs indispensable many years to go beyond social culture of each country. To confirming it, it introduces MCE as glocalization index. And, following hypothesis will be tested.

Hypothesis Teamwork awareness is influenced by corporate culture even if producing same category product.

5 Result and Consideration

5.1 Social Background of Teamwork Awareness

Shorten PTT leads decreasing workload in a worker, but leads increasing repeat number of work. So it seems shorten PTT leads strengthen TO awareness, because learning effect is early got through decreasing work contents and increasing repetition, and it is brought up simple management, effective instruction and

Table 2 Researched plant

Country	City	Plant	Began operation	Surveyed year	Number of employee	Number of answerer All	Number of answerer Staff	Number of answerer Line	Collected ratio (%) All	Working year All Mean/SD	Age All Mean/SD
China	Dalian	D	1993	2007	2,300	2,104	231	1,873	91.5	4.07/4.17	22.96/5.41
	Shanghai (1)	S	2005[a]		2,400	1,550	470	1,080	64.6	3.61/3.48	19.75/5.41
	Shanghai (2)	R	1994		460	423	74	349	92.0	4.07/4.86	26.40/8.24
	Guangzhou	G	2005		470	424	155	269	90.2	1.14/0.80	22.05/4.68
Malaysia	Kuala Lumpur	B	1974	2009	789	419	55	364	53.1	8.08/8.08	29.52/8.42
Thailand	Ayutthaya	T	2001	2010	436	312	81	231	71.6	3.07/1.94	28.49/4.79
Vietnam	Ho Chi Minh	V	2007	2011	451	436	81	355	96.7	1.30/1.01	24.43/4.13

Country	City	Plant	Main religion	Management cumulative effect (MCE) (Year)[b]	Production Tact Time (PTT) (second/unit)[d]	Product	Production line system
China	Dalian	D	Buddhism	14	60	H	U-Line/I-Line
	Shanghai (1)	S		12	60–300	P	U-Line
	Shanghai (2)	R		13	2	Y	Full automatic line
	Guangzhou	G		2	300	M	U-Line/I-Line
Malaysia	Kuala Lumpur	B	Islam	35[c]	4	Y	Semi automatic line
Thailand	Ayutthaya	T	Buddhism	9	4–10	Y & M	U-Line/Semi automatic line
Vietnam	Ho Chi Minh	V	Buddhism	4	60	H	U-Line/I-Line

* Figures in Table 2 are on surveyed year, respectively
[a] Mergered year from three keiretsu companies into S
[b] Continuing year from beginning operation to surveyed year
[c] The first overseas plant from the parent company in Japan
[d] Average production interval time as whole plant

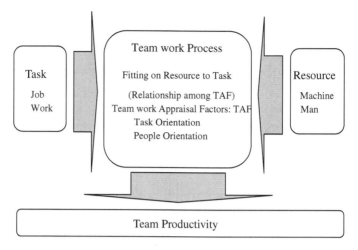

Fig. 1 Industrial teamwork dynamics: ITD

capability improving. More automated work seems TO become downward by less manual work. On the other hand, it seems PO is influenced by MCE. Daily management effort has accumulated knowhow from internal and external management activity, and this effort has made up strong PO. Although employee' continued working year may effect on TA, but local employee of Southeast Asia has shorter working year. Furthermore, it seems increasing number of kinds of product must lead complexity to daily operation and it may effect on TO.

5.1.1 TO Versus PO in China

In China, TO is higher than PO in plant D, R, G except S. In plant S, higher PO bases on reason management policy is "High wages than High efficient", it different from other plant's policy "High efficient than High wages". Therefore, plants of China have tendency stronger TO than PO.

5.1.2 TO Versus PO in Southeast Asia

Plant B in Malaysia, plant T in Thailand and plant V in Vietnam are belonging to the Southeast Asia bloc. Their TA is different from plants in China. PO is strongly recognized than TO (Table 3).

Table 3 Line/comparision between task orientation and people orientation

Country	City	Plant	Number of employee	Mean TAF	Mean T0	Mean P0	T0	Comparision (larger signs •) T0	Comparision (larger signs •) P0	SD TAF	SD T0	SD P0	F0	Comparision (larger signs •) T0	Comparision (larger signs •) P0
China	Dalian	D	1,873	3.99	4.05	3.98	**	•		0.722	0.751	0.689	**	•	
	Shanghai(1)	S	1,080	3.83	3.83	3.84	–			0.715	0.761	0.666	**	•	
	Shanghai(2)	R	349	4.29	4.30	4.28	–	•		0.634	0.643	0.625	–	•	
	Guangzhou	G	269	3.91	3.95	3.87	–	•		0.673	0.694	0.650	–	•	
Malaysia	Kuala Lumpur	B	364	3.70	3.63	3.76	*		•	0.818	0.808	0.823	–		
Thailand	Ayutthaya	T	231	3.35	3.31	3.38	–		•	0.940	0.972	0.907	–	•	
Vietnam	HoChiMinh	V	355	4.55	4.53	4.56	–		•	0.611	0.641	0.581	*	•	

* $p < 0.05$, ** $p < 0.01$

5.1.3 PTT Versus TA in China

Shorten ordering of PTT is set up plant R, D, G and S (G and S are same), and it confirms higher ordering of TAF mean, means of TO and PO are able to also set up plant R, D, G and S as same. Especially, plant R has highest PTT and highest TAF mean. When PTT is shorter, TAF mean has higher. That is remarkable feature in China (Tables 2, 3).

5.1.4 PTT Versus TA in Southeast Asia

Same as Sect. 5.1.3, shorten ordering of PTT is set up plant B in Malaysia, plant T in Thailand and next plant V in Vietnam, but it confirms higher ordering of TAF mean, means of TO and PO are able to set up plant V, plant B and next plant T, not same as PTT ordering. There is no clear relationship between PTT and TAF mean. This reason based on Plant B that has multi-ethnic country workers including terminable temporary worker from Indonesia, China, India, Nepal, Vietnam, Bangladesh and they have different religions, languages and customs (Tables 2, 3).

5.1.5 CME Versus TA in China

Here, MCE is defined as continuing operation years, so it seems if MCE is larger, TAF mean has larger. Operation years of four plants in China are divided into two groups, first group is plant D, S and R which is long group around ten years, and next group is plant G in shorter year. Plant S is unified by one with three plants located in Shanghai area, and after unification has passed only two years. Therefore, plant S remains influence of corporate culture of three plants respectively at the surveying time (Tables 2, 3).

5.1.6 CME Versus TA in Southeast Asia Bloc

Higher ordering of TAF mean is plant V, plant B and next plant T, but longer ordering of MCE is plant B, plant T and next plant V. There is no clear relationship between TAF mean and MCE (Tables 2, 3).

5.2 Teamwork Awareness Difference Between Plant Produced Same Category Product

5.2.1 Hypothesis Test

It tests TA differences among plants according to same product category. These comparison are plant D in China and plant V in Vietnam which produce product H, plant R in China and plant B in Malaysia which produce product Y, and plant G in China and plant T in Thailand which produces product M, respectively. Mean and SD of TAF are completely different between plants of China with plants in Vietnam, Malaysia and Thailand. Also TO mean of China is higher than PO, but each plant in Southeast Asia bloc has higher PO mean than TO with significant difference on t0 and F0 (Table 4).

Moreover, it tests differences between line workers by discriminant analysis. All correct ratio are higher more than eighty percentages, and it can be said TA between plants differs greatly even if produce same category product.

The weight of each TAF is reviewed as discriminant coefficients with statistical significance. Plant D and V which are producing same category product H, have statistical significant difference except factor 3 as between China and Vietnam. Plant R and B which are producing the same category product Y, have statistical significant difference except factor 3, 6, and 10 as between China and Malaysia. Furthermore, plant G and T which are producing the same category product M, have statistical significant difference except factor 2, 4, 5 and 10 as between China and Thailand. Also all comparisons are more influenced on PO by religion as social culture in each country. Furthermore, common important factors which discriminant coefficient is more than 1.0 are "7.Cohesiveness", "9.Human Relations" and "12.Satisfaction" through all analyzing. These three factors belong to PO and it seems strongly influence by MCE (Table 5).

Therefore,

Hypothesis Teamwork awareness is influenced by corporate culture even if producing same category product" is accepted.

5.2.2 Teamwork Unity

Factor loading values (FLV) on first and second principal component on principal component analysis based on mean value, is analyzed and unity as a work team is checked based on convergence of factor loadings values. Furthermore, relationship between these unity and PTT is considered. All FLV on first principal component are positive larger number, and can be understood all as basic teamwork factors. Second principal component can be understood to divide into two clusters, TO and PO. All cumulative contributed ratio (CCR) until second principal component is

Table 4 Line/comparison between countries

Product		H				Y				M			
Plant		D(n = 1,873)		V(n = 355)		R(n = 349)		B(n = 364)		G(n = 269)		T(n = 231)	
Country		China (Dalian)		Vietnam (Ho Chi Minh)		China (Shanghai(2))		Malaysia (kuala Lumper)		China (Guangzhou)		Thailand (Ayutthaya)	
TAF	Statistics	Mean	SD	Mean	SD	Meas	SD	Mean	SD	Mean	SD	Mean	SD
TO	1–6	4.02	0.767	4.51**	0.655**	4.28	0.656	3.65**	0.798**	3.93	0.685	3.28**	0.961**
PO	7–12	3.92	0.695	4.55**	0.598**	4.25	0.631	3.76**	0.824**	3.85	0.635	3.39**	0.895**
All	1–12	3.97	0.734	4.53**	0.628**	4.27	0.644	3.71**	0.813**	3.89	0.662	3.34**	0.930**
PTT (Production tact time) (second/unit)		60		60		2		4		300		4–10	

* $p < 0.05$, ** $p < 0.01$

Table 5 Line/comparision between Country

TAF	Product	H	Y	M
	Plant	D(n = 1,873)	R(n = 349)	G(n = 269)
		V(n = 355)	B(n = 364)	T(n = 231)
	Statistics	DC [a]	DC	DC
		12.476 [b]	−8.156	−6.695
TO	1	0.369**	2,180	1.358**
	2	−0.428**	0.556**	0.045
	3	0.036	−0.194	−0.774**
	4	−0.237**	0.535**	0.164
	5	−1.430**	−0.552**	0.282
	6	−0.592**	−0.194	1.396**
PO	7	−1.049**	−1.327**	−1.356**
	8	−0.282**	1.625**	1.325**
	9	3.895**	1.033**	1.381**
	10	0.612**	0.164	0.128
	11	−2.741**	−0.827**	−0.679**
	12	−1.257**	−1.082**	−1.548**
Correct ratio (%)		83.62	81.77	81.80
Discrimination function		**	**	**

* $p < 0.05$, ** $p < 0.01$
a Discriminant coefficient, b Constant

over sixty-five percentage, thus two clusters is able to fully explain. Some factors are mixed beyond own belonging cluster, but TAF are divides into two cluster TO and PO, generally.

It shows two-dimensional figure by FLV of first principal component is taken along a horizontal axis, and second principal component is taken along a vertical axis.

Each area set up as multiplied by distance between coordinate points on extreme value as the positivest and the negativest on first and second principal components axis, is defined as a unity sense of teamwork. There is difference in TA between China and Southeast Asia bloc, difference between plant D and V is especially great, plant D area is about four times weaker than plant V in TAF, about three times stronger in TO, and plant V has about around thirty times remarkably stronger than plan D in PO. Therefore, hypothesis is accepted also on this analysis. According to smaller area of TAF in order, it sets up plant T, V, R, B G and D. Plant T, R and B has 10 s or less PTT, they belong to smaller TAF area. Here, it means smaller area has strong sense of unity. According to smaller area on TO, it sets up ordering to plant D, T, R, G, V and B. It is able to confirm plant D, R, G in China has smaller area and sense of unity is stronger. And same as to smaller area on PO, it set up ordering to plant V, B, T, R, G and D. It can be said plant D, R, G in China has larger area, but plant V, B, T belong in the Southeast Asia bloc has stronger sense of unity as cohesive ability.

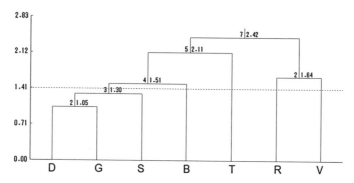

Fig. 2 Line/relationship among plants (r0 = 0.747)

5.2.3 Relationship Among Plants

Interrelationship among each plant is confirms through cluster analysis. Here, group average method (clustering method) and Euclidean distance (valuation scale) are adopted. Plant D, G, and S in China form a cluster as Chinese be made same social cultural sphere, and plant B and T in the Southeast Asia bloc add to this cluster with little later. But a cluster is formed by plant R and V, independently. Plant R is full automatic line and it perfectly different from others, and plant V has small MCE and complex management problems (Fig. 2).

6 Conclusion

This research certifies that Production Tact Time (PTT) and MCE (Management Cumulative Effect) affect into Teamwork Awareness (TA) of line worker's team from point of global. In test of hypothesis, it confirms that PTT shortening affects into Task Orientation (TO) positively.

Acknowledgments This research activity has been supported by OMRON Corporation as overseas social survey during 2007–2011. I appreciate very much for all of the people concerned in Japan and overseas plants, and especially thanks to all local peoples who cooperated directly answering to questionnaire. And graduation research student from 2008 to 2010 of my laboratory is summarizing fundamental investigation of this research. I introduce them and consider it as acknowledgement here.

Global Industrial Teamwork Dynamics in Malaysia—Effects of Social Culture and Corporate Culture to Teamwork Awareness—2/2

Masa-Hiro Nowatari

Abstract Research purpose is to verify how social culture and corporate culture effect into teamwork awareness (TA), respectively. Answerer is native employee on Japanese-owned local electrical plants in multi-racial nation Malaysia. Confirmation on TA difference caused by social culture bases on propagated religion in born country on each employee, and by corporate culture confirms as corporate attributes based on management cumulative effect (MCE). It clears that effect of corporate culture is stronger than social culture, and that corporate culture as social group process has been growing up glocalization (global-localization) while working time. This is second research related to Global Industrial Teamwork Dynamics (GITD) after first Chinese social survey. Propagated religions are Islam has Malaysia, Indonesia, Buddhism has China, Vietnam and Hinduism has India. Here, awareness concerning teamwork in daily production activity is caught as a group process in social psychology, and the internal structure is considered. Finally, the hypothesis is tested through statistical tests and discriminant analysis, principal component analysis.

Keywords Religion · Islam · Buddhism · Hinduism · Race · Malaysian · Chinese · Indian · Indonesian · Management cumulative effect · Teamwork awareness · Teamwork appraisal factors

1 Introduction

Teamwork on making product in the first line worker's team is able to pick up Group Process in Social Psychology, and this global research has been developing since 2000. It takes quarter century since Japanese manufacturing industry started

M.-H. Nowatari (✉)
Department of Management Science, Faculty of Engineering, Tamagawa University, Tamagawa Gakuen 6-1-1, Machida, Tokyo, 194-8610 Japan
e-mail: nowatari@eng.tamagawa.ac.jp; masahiro.nowatari@gmail.com

overseas development, and this trend will be strongly increasing more and more. On localizing of Japanese-style production management, it needs to unify with local social culture always. So, staying manager from home country must construct and create their plant's original culture by themselves according to glocalization (global-localization) from point of global view, and it must localize their original Japanese-style production system through local propagated religion and social culture. And it should include teamwork management in this process. Heaping up daily management activity makes growing up corporate culture beyond social culture while working hours, and strengthen original glocalized (global-localized) production management. This glocalization is based on management cumulative effect (MCE), and it means continuing year from beginning operation. MCE is able to accumulate management knowhow and wisdom through many trial and error related to daily management problem solving.

2 Research Objective

Research objective is to verify how social culture and corporate culture effect into teamwork awareness (TA) in Malaysia. Consideration on social culture bases on propagated religion in born country on each employee, and corporate culture confirms as corporate attributes, respectively. To certifying it, hypotheses are set and tested. First of all, t test (Welch method) and F-test is done as basic statistical test according to all, task orientation (TO) and people orientation (PO) of teamwork appraisal factors (TAF). Next, it considers TAF differences between religions and between plants by correct discriminant ratio on discriminant analysis. It certifies TAF differences between plants are more strongly than between religions. This is fruit of their efforts on daily localizing activity based on Japanese-style production system.

3 Research Plants

3.1 Surveyed Plants

Surveyed plants are in Malaysia on Japanese-owned local electrical plants. Plant A and C have strong MCE based on around forty operating years, and plant B has twenty over years. Mean working year of employee has seven to ten years in each plant, respectively. And mean age is around thirty years old. So, number of staying Japanese manager is very few, and local managers have been managed themselves plant. These plants have clear production management strategy and have acquired various certifications as ISO and so on. This survey was done from 2008 to 2010 and basic analyzing was done by graduation thesis of my senior students.

Their parent company, headquarter is in Japan and is listed company in the Tokyo stock market, and are typical company, having large capital, sales and employee in the world (Table 1).

3.2 Personal Attribute

Each plant has been employed many ethnic races, plant B has many Chinese and Indonesian than Malaysian. And plant C has most ethnic races and two-way employ system, permanent and contract. Working average year of key person has around ten years and average year is thirty years old, respectively (Table 2).

3.3 Propagated Religion

It shows propagated religion and national wealth as gross domestic product for each person, nominal (GDP) on every country. Islam area has been propagated in Malaysia (state religion), Indonesia and Bangladesh, and Hinduism area has India and Nepal. Moreover, Buddhism area has China and Vietnam but propagation rate is unknown. Malaysia has higher GDP, so it is able to employ person from neighbor countries (Table 3).

4 Survey Method

4.1 Teamwork Appraisal Factors

In order to measure TA, teamwork appraisal factor (TAF) is set up. TAF has two categories, "task orientation (workability): TO" and "people orientation (cohesiveability): PO". And TAF consists of twelve factors (sixty questionnaire items) of six factors on TO (thirty questionnaire items), and six factors on PO (thirty questionnaire items). Moreover, every questionnaire item is appraised on five stages evaluation (1, 2, 3, 4, 5), and more positive figure means favorable and strong awareness to teamwork. TO means capability for carrying out the task (job, work) given as a work team, and PO is capability for maintaining team as a work team. Questionnaire method had been carried out with factory tour, interview, collecting data and fact finding (Table 4).

Table 1 Researched plant

Plant	A	B	C
State	Selangor	Penang	Selangor
Surveyed year	2008	2008	2009
Established year	1973	1991	1973
Capital (billion yen)	3	1	0.5
Sales (billion yen)	30	10	3
Product items	Semiconductor, system LSI	Facsimile, MFP device	Electronic relay
Number of employee	1411	574	800
Number of answerer (Line + Staff)	615 (504 + 111)	476 (265 + 211)	419 (364 + 55)
General manager	Japanese	Japanese	Japanese
Department head	Local 100 %	Local 70 %	Local 100 %
Number of japanese staying staff	3	5	4
Production management strategy	First foothold of semiconductor manufacturing in Southeast Asia	100 % export global development	First overseas plant from the parent company in Japan Export main, having worldwide share
Characteristic of production management	a. Full automatic line and labor-intensive handwork in clean-room	a. Cell production system for full synchronized production (UA line, SMT line, U line with cart, U & I line with cart),	a. Continuous production system from making mold, parts making to assembly,
	b. Promotion on manpower rearing of next generation (IT, WTTP), employee participating	b. Added value management strategy.	b. 24 h full operation,
	c. Corporate culture beyond race difference and social culture, Objective achievement by all unison	c. Logistics (VMI, JIT, KANBAN),	c. Full automatic production 65 % and semi automatic production 35 %
	d. Line company, CSR activity,	d. Multi-skill operator	
	e. Japanese speaking ability person is 58 employees.	e. Line by mixed race/ethnic,	
		f. TVMS	

(continued)

Table 1 (continued)

Plant			A	B	C
Certification			ISO 9002, ISO 14001, ISO 9001, OHSAS 18001	ISO 9002, ISO 14001, ISO 9001, OHSAS 18001	ISO 9002, ISO 14001, ISO 9001, TS 16949
Personal attribute	Working year (year)	Mean	9.99	6.76	8.08
		SD	7.964	5.157	8.078
	Age (year)	Mean	31.48	30.44	29.52
		SD	8.774	8.505	8.421
Parent company (headquarter in Japan)[a]	State		Tokyo	Tokyo	Kyoto
	Established year		1875	1950	1933
	Capital (billion yen)[b]		440	40	60
	Sales (billion yen)[b]		6,000	350	620
	Number of employee[b]		210,000	20,000	36,000

[a] Figure is on 2011 financial year, [b] International

Table 2 Personal attribute in plant

Plant (number of answerer)	Nationality (race)	Number of answerer n	Personal attribute Working year (year) Mean	SD	Nationality unknown, n = 68 Age (year) Mean	SD	Pamanent/contract (Line/staff)
A(n = 562)	Malaysia	393	9.27	7.955	30.29	8.728	Par. (Line/staff)
	China	32	12.43	7.450	35.50	6.564	Par. (Staff)
	India	137	10.33	7.734	32.31	8.255	Par. (Line)
B (n = 476)	Malaysia	90	7.97	4.357	33.66	9.057	Par. (Line/staff)
	China	157	8.06	6.005	33.22	7.894	Par. (Line/staff)
	Indonesia	131	4.05	2.833	23.64	3.040	Par. (Line)
	India	41	10.71	5.217	36.92	9.096	Par. (Line/Staff)
	Nepal	28	2.79	0.357	25.64	4.931	Par. (Line)
	Vietnam	29	2.84	0.375	27.47	5.048	Par. (Line)
C = (404)	Malaysia	249	10.57	8.216	31.97	8.426	Par. (Line/staff)
	China	12	14.70	10.646	35.08	10.379	Par. (Line/staff)
	Indonesia	98	2.16	1.087	22.84	3.396	Con.(Line)
	India	15	14.20	10.051	34.53	7.936	Par. (Line)
	Nepal	3	3.00	1.732	23.00	2.828	Con.(Line)
	Vietnam	4	2.00	0	22.00	2.160	Con.(Line)
	Bangladesh	23	2.13	2.801	26.69	5.498	Con.(Line)

Table 3 Propagation and national wealth

Religion	Islam			
Country	Malaysia	Indonesia	Bangladesh	
Propagation rate (%)[a]	State religion	88.8	89.7	
GDP (US $)[b]	6,999	2,590	621	
Religion		Buddhism	Hinduism	
Country	China	Vietnam	India	Nepal
Propagation rate (%)[a]	–	–	80.5	80.6
G D P (US $)[b]	3,678	1,060	1,031	642[c]

[a] The ministry of foreign affairs (http://www.mofa.go.jp/mofaj), 2009 surveyed
[b] GDP/Gross domestic product for each person, nominal JETRO (http://www.jetro.go.jp/world/), 2009 surveyed
[c] The ministry of foreign affairs (http://www.mofago.jp/mofaj), 2010 surveyed

4.2 Hypotheses

At localizing process on Japanese-style production management, it is important understanding local employee's TA that different from Japanese, and certifying TA differences among religion from point of global view. In recent years, Japan

Table 4 Teamwork appraisal factors (TAF)

Productivity orientation	Team work appraisal factors (P = 12, p = 60)	Number of questionnaire items
TO : task orientation (workability) (P = 6, p = 30)	1. Level of work management	5
	2. Training and instruction to subordinates by the leader's superior	5
	3. Training and instruction to subordinates by the leader	5
	4. Care of subordinates by the leader	5
	5. Ability to accomplish work	5
	6. Conformance to job requirement (progress, quality and quantity)	5
PO : people orientation (cohesiveability) (P = 6, p = 30)	7. Cohesiveness	5
	8. Atmosphere of group	5
	9. Human relations	5
	10. Morale	5
	11. Mutual cooperativeness on working	5
	12. Satisfaction	5

parent companies have been deploying positively to make global standard into too diversity of Japanese-style production management in every local plant from point of a bird's-eye view. Global department called "Monozukuri-Innovation Head Office" supports this activity. And it begins proposing new management concept called "glocalization (global-localization)" as utilizing local-management based on global standard. This is second age after first manufacturing only age on localization. So first of all, it confirms TA difference among religions as social culture, and secondly it certifies difference among plants from corporate culture for employee according to same religion. Following hypotheses are set up and tested.

Hypothesis 1 Teamwork awareness is affected by religion as social culture.

After answerers on each plant are reclassified into their religion based on hometown country from Table 2, basic statistical method and discriminant analysis are used and tested.

Hypothesis 2 Teamwork awareness is different among plants even if same race.

After confirming their corporate culture on each plant, it certifies each plant has been particular cumulative wisdom on production management through repeated trial and error on daily management activity since established, and this wisdom is completely different each other even if they have same propagated religion and same race belonging to same mother country. This MCE means corporate culture, it seems stronger MCE brings up more unique TA. If it is true, it will be certified high correct discriminant ratio on discriminant analysis and many statistical significant on basic statistics among plants.

Table 5 Religion/reliability

Religion	Islam	(Number of factor : 12) Buddhism	Hinduism
Number of answerer n	984	234	224
Krombacher's α	0.966	0.964	0.915

Hypothesis 3 Teamwork awareness is stronger affected by corporate culture than social culture.

Here, it confirms how affect magnitude to corporate culture and social culture to TA according to results of Hypothesis 1 and 2. Each plant has been positively deployment by OJT, KAIZEN, 5S, QC Circle in localizing process of Japanese-style production system, and they have been made original corporate culture as unique MCE. It seems there is two cultures that corporate culture as employee activity during working time, and social culture as resident activity outside working time.

5 Result and Consideration

5.1 Social Culture and Teamwork Awareness

It considers religion as social culture how affect to TA according to Islam area including Malaysia, Indonesia and Bangladesh, and Buddhism area including China and Vietnam, and Hinduism including India and Nepal. Those three religions believer are extracted from each plant through Tables 2 and 3. Before hypothesis test, it confirms Krombacher's α reliability coefficient for certifying internal consistency of TAF measure. All Krombacher's α composed TAF have 0.9 over, it is no problem (Table 5).

On testing of Islam criterion, it is able to confirm high statistical significance with Buddhism mean, but there is no significance with variance. On the contrary, there is no significance with mean with Hinduism, but variances of TO and TAF are able to confirm statistical significance. Moreover, on testing of Buddhism criterion, it is able to confirm high statistical significance with Hinduism mean, and variances of TO and TAF are able to confirm statistical significance. In this way, it can confirm statistical significance on mean than variance among religions. TA is different among Islam and Buddhism, moreover Buddhism and Hinduism too, but there is no difference between Islam and Hinduism.

Confirmation on internal consistency of TAF based on principal component analysis, cumulative contributed ratio until second principal component is good as seventy percentages over. "5.Ability to Accomplish Work" and "6.Conformance to Job Requirement" on TO belong with PO in Islam and Buddhism, and

Table 6 Religion/TAF comparison (discriminant analysis)

TAF	Criterion	Islam (n = 984)		Buddhism (n = 234)
	Comparison	Buddhism n = 234 DC[a]	Hinduism n = 224 DC	Hinduism n = 224 DC
	Constant	−1.102	0.956	−2.792
TO	1	0.278**	−0.208**	0.717**
	2			
	3			
	4	−0.451**		−0.573**
	5	0.288**	0.638**	−0.298**
	6		−0.399**	0.445**
PO	7	0.722**	0.505**	
	8	−0.570**	−0.546**	
	9	−0.273**	−0.492**	0.386**
	10	0.268**		0.427**
	11		0.191*	−0.395**
	12			
C D R (%)[b]		64.53	59.77	65.07
Discriminant function		*	*	*
		Average CDR: 63.12 (%)		

$* \ p < 0.05$, $** \ p < 0.01$
[a] Discriminant coefficient, [b] Correct discriminant ratio

Table 7 Plant/reliability

	(Number of factor: 12)			
Plant	A			
Race	Malaysian	Chinese	Indian	
Number of answerer n	393	32	137	
Krombacher's α	0.932	0.974	0.952	
Plant	B			
Race	Malaysian	Chinese	Indonesian	Indian
Number of answerer n	90	157	131	41
Krombacher's α	0.934	0.969	0.968	0.944
Plant	C			
Race	Malaysian	Indonesian		
Number of answerer n	249	98		
Krombacher's α	0.952	0.929		

"12.Satisfaction" belongs into TO and "6.Conformance to Job Requirement" is into PO in Hinduism.

Moreover, TA difference among religions is analyzed by discriminant analysis, and discriminant function is got statistical significance, respectively. But average correct discriminant ratio is lower as around sixty percentage. Common confirmed factors are "1.Level of Work Management" and "5.Ability to Accomplish Work" of TO, and "9.Human Relations" of PO, those factors affects to TA difference among religions (Table 6).

Therefore, (Hypothesis 1) "Teamwork awareness is affected by religion as social culture" is accepted on between Islam and Buddhism, and Buddhism and Hinduism, but between Islam and Hinduism is rejected. So it seems Islam and Hinduism has similarity teamwork awareness, each other.

5.2 Corporate Culture and Teamwork Awareness

Through same procedure as above Sect. 5.1, all answerer are divided into each race in each plant except small answerer race from Table 2. Similarity of questionnaire content of TAF are confirmed, and all Krombacher's α have 0.9 over on race at each plant, it is no problem, too (Table 7).

On basic statistic test, Malaysian belongs to Islam area has statistical significance on differences of Mean and SD at TO, PO and TAF between plant A (criterion) and plant B, and between plant A and C, and between plant B and C, respectively. This tendency is same in between plant B and C on Indonesian. Again, it confirms Chinese belongs to Buddhism and Indian belongs to Hinduism have statistical significance on Mean at TO, PO and TAF between plant A (criterion) and plant B and SD has same tendency except TO. So statistical significance between means is able to confirm through all TO, PO and TAF, TA is completely different among plants even if they are same race.

Therefore, (Hypothesis 2) "Teamwork awareness is different among plants even if same race" is accepted on mean, but on SD is rejected by no statistical significance on TO in Malaysian and Chinese.

On principal component analysis on each plant, cumulative contributed ratio until second principal component has around seventy percentage over, it is good. Plant A has phenomenon "6.Conformance to Job Requirement" on TO is taken into PO. And plant B has "6.Conformance to Job Requirement" on TO is taken into PO and, "12.Satisfaction" belongs into PO is taken into TO. Moreover plant C has "5.Ability to Accomplish Work" and "6.Conformance to Job Requirement" on TO are taken into PO.

And TA difference between plants based on same race confirms according to discriminant analysis. Average correct discriminant ratio is seventy-three percentages though only Chinese has no statistical significance on discriminant function. This ratio has larger than religion value sixty-three percentages. Every plats have unique corporate culture as MCE, and it is large different from others.

Table 8 Plant/TAF comperison (discriminent analysis)

TAF	(Race) Criterion Comparison		Malaysian			Chinese		Indian		Indonesian	
		A (n = 393)			B (n = 90)	A (n = 32)		A (n = 137)		B (n = 131)	
		B	C		C	B		B		C	
		n = 90	n = 249		n = 249	n = 157		n = 41		n = 98	
		D C[a]	DC		DC	DC		DC		DC	
	Constant	−6.129	−3.302		2.324	−2.127		−7.395		7.210	
TO	1	1.045**	0.813**		−0.639**	−0.855**		1.223**		0.471**	
	2	−0.583**			0.493**	−0.826**					
	3							−0.946**		−1.148**	
	4	0.457**	−0.280**		−0.763**	0.659**				−0.640**	
	5							−0.752**		−0.411@	
	6									1.488**	
PO	7				−0.443**					−0.996**	
	8										
	9		−0.437**								
	10	0.860**	0.638**		−0.403**	0.869**		0.785**			
	11	−0.439**			0.628**	−0.962**					
	12	0.366@	0.241**					1.091**		−0.454*	
CDR (%)[b]		75.36	63.24		70.21	68.78		79.78		83.41	
Discriminant Function		**	*		@	—		*		**	
								Average CDR: 73.46(%)			

@ $p < 0.10$, * $p < 0.05$, ** $p < 0.01$
[a] Discriminant coefficient, [b] Correct discriminant ratio

Corporate culture affects more strong to TA than social culture as religion. Each plant has been particular cumulative wisdom through repeated trial and error on daily management activity since established plant, and this wisdom content is completely different from others even if same propagated religion, same race belonging to same mother country. Common confirmed factors are "1.Level of Work Management", "2.Training and Instruction to Subordinates by The Leader's Superior" and "4.Care of Subordinates by The Leader" of TO, and "10.Morale", "11.Mutual Cooperativeness on Working" and "12.Satisfaction" of PO. Those factors are large difference of TA among plants (Table 8).

5.3 Corporate Culture and Social Culture

As above mentioned, it confirms corporate culture affects more strong to TA than social culture according to statistical significant test on basic statistic and discriminant analysis.

Therefore, (Hypothesis 3) "Teamwork awareness is stronger affected by corporate culture than social culture" is accepted.

6 Conclusion

In multi-racial nation Malaysia, it confirms particular corporate culture as management cumulative effect make affect into teamwork awareness than social culture as propagated religion on each race in their mother country. On (Hypothesis 1) "Teamwork awareness is affected by religion as social culture", and (Hypothesis 2) "Teamwork awareness is different among plants even if same race", it certifies interesting knowledge. Both hypotheses are accepted partially not all, so continued research activity is needed. (Hypothesis 3) "Teamwork awareness is stronger affected by corporate culture than social culture" is accepted and interesting for future management localizing, and it seems Japanese style production management will make firm glocalization (global-localization) system through management cumulative effect based on daily production activity.

Acknowledgments This research activity has been supported by OMRON, TOSHIBA and TOSHIBA TEC corporations as overseas social survey during 2008–2009. I appreciate for all of the people concerned in Japan and overseas Malaysia plants, and especially thank to all local peoples who cooperated directly answering to questionnaire. And I have no words to express my gratitude on kindness supporting of staying Japanese staff, local Malaysian and Chinese staff within very busy during two years, respectively. And graduation research student from 2008 to 2009 of my laboratory is summarizing fundamental investigation of this research. I introduce and give an acknowledgement to them in here.

Relaxed Flexible Bay Structure in the Unequal Area Facility Layout Problem

Sadan Kulturel-Konak

Abstract In this paper, relaxed flexible bay structure (RFBS) is introduced to represent the facility layout problem (FLP) with unequal area departments, which is a very hard problem to be optimally solved. The flexible bay structure (FBS), which is a very common layout in many manufacturing and retail facilities, makes the problem restrictive as departments have to be placed in parallel bays with no empty spaces allowed. The RFBS, however, relaxes the FBS representation by allowing empty spaces in bays, and therefore, it results in more flexibility while assigning departments in bays. In addition, departments are allowed to be located more freely within the bays, and they can have different side lengths as long as they are within the bay boundaries and do not overlap. The effectiveness of the new representation is shown using it in a hybrid probabilistic tabu search-linear programming (PTS-LP) approach. The comparative results show that improvements have been achieved by allowing partially filled bays using the RFBS.

Keywords Facilities planning and design · Unequal area facility layout · Relaxed flexible bay structure · Probabilistic tabu search

1 Introduction

The Facility Layout Problem (FLP) in the continuous plane is generally defined as to partition of a planar area into a number of departments with known area and interdepartmental material flow requirements. The goal is minimizing the total material handling cost which, in general, is a function of the distance and the amount of material flow between departments. The constraints of the problem

S. Kulturel-Konak (✉)
Penn State Berks, Management Information Systems, Tulpehocken Road, 7009 Reading, PA 19610, USA
e-mail: sadan@psu.edu

include satisfying the area requirements of the departments, the boundaries of the layout and restrictions on the departments' shapes, such as maximum aspect ratio and minimum side length. The output of the FLP is called block layout, which specifies relative location and shape of each department in the area. Therefore, a block layout should be easily transferable to the actual facility design.

Excellent review papers (Kusiak and Heragu 1987; Meller and Gau 1996; Singh and Sharma 2006; Drira et al. 2007) present an overview of research on block layout design. There are exact algorithms developed for different versions of the FLP in the continuous plane, such as Montreuil (1990), Meller et al. (1998), Sherali et al. (2003), Castillo et al. (2005), and Konak et al. (2006). The exact approaches to the FLP have a major limitation of being restrictive with the size of the problem that can be optimally solved due to computational intractability of the problem. Therefore, the majority of work on the FLP has focused on heuristic approaches promising to find good solutions in relatively short amounts of time. Meta-heuristic approaches such as Simulated Annealing (SA) (Meller and Bozer 1996), Genetic Algorithms (GA) (Banerjee et al. 1997; Gau and Meller 1999) and Tabu Search (TS) (Kulturel-Konak et al. 2007; Kulturel-Konak et al. 2004; Scholz et al. 2009), Ant Colony Optimization (ACO) (Komarudin and Wong 2010; Wong and Komarudin 2010), have been previously applied to the problem.

The FBS, which was first proposed by Tong as a continuous layout representation, is a very common layout in many manufacturing and retail facilities. In the FBS, departments are located only in parallel-bays with varying widths, and bays are bounded by straight lines on both sides. Departments are restricted to be located only in one bay, and they are not allowed to expand over multiple bays. In the original description of the FBS, the total area of the departments in a bay defines the width of the bay because bays are expected to be completely filled by departments, and all departments in a bay have the same width. Therefore, each bay must include a feasible combination of departments with respect to the shape constraints, which could be difficult to satisfy if departments have different size and shape constraints. These restrictions significantly limit the number of feasible layout configurations in the FBS as reported in Konak et al. (2006). As a result, the FBS may not yield block layouts with low material handling costs in highly constrained problems. Another drawback of the FBS arises if the facility area is bigger than the total area of the departments. Any extra space in the facility is handled in various ways in the literature, including using dummy departments, allocating empty space at one side of the facility (Konak et al. 2006), or allowing empty space within the bays and adjusting departments accordingly as in (Wong and Komarudin 2010; Kulturel-Konak and Konak 2011). One of the drawbacks of using dummy departments is the increased computational efforts to solve the problem.

Despite the disadvantages summarized above, the FBS also has some desirable features that make it a practical layout representation scheme. Therefore, the main motivation of this paper is to improve the FBS, particularly in the cases where the facility has an empty space, so that it would be more practical. First, the RFBS, allowing departments to be located more freely within the bays, was proposed by Kulturel-Konak and Konak (2011). Then, Kulturel-Konak (2012) improved the representation also allowing departments in a same bay having different side-lengths as long as they are within the bay boundaries and do not overlap. In this paper, the PTS-LP approach previously developed by Kulturel-Konak (2012) is presented as the solution approach. Then, results from two approaches incorporating slightly different RFBS approaches are compared using a set of test problems from the literature.

2 Methodology

2.1 Solution Representation

In a regular TS, the objective function under consideration should be evaluated for each and every element of the neighborhood of the current solution, which can be very costly from the computational standpoint. Therefore, PTS considers only evaluating a random sample of the neighborhood in creating solution candidates while solving complex problems to reduce computational effort. The PTS-LP uses a permutation encoding where a solution π is represented as an array of 2 N elements such that numbers from 1 to N represent corresponding departments, and zeros represent bay breaks. The solution array encodes the order in which the departments are lined up in the facility and the positions of the bay breaks in this department order. The actual FBS layout is constructed by assigning department s to bays from the top to the bottom based on the order given in the encoding and starting a new bay at each bay break position. For example, the layout given in Fig. 1 is represented as array (4, 0, 5, 3, 0, 1, 2, 0, 0, 0) which is decoded as follows. Department 4 is the first department to be assigned to the facility, and "0" succeeding department 4 indicates a bay break. The last two "0"s in the array indicate empty bays. Because the maximum possible number of bays is equal to the number of departments, the encoding array has 2 N elements (N zeros for bay breaks and N department indexes). To prevent a degenerate encoding, empty bays are always stored at the end of the array.

Fig. 1 The encoding scheme of the PTS-LP and RFBS representation

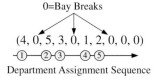
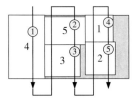

2.2 Local Move Operators

The PTS-LP uses two move operators, namely Insert(i, j) and Swap(i, j), to generate neighborhood solutions from the current solution. Operator Insert(i, j) removes the element in the ith position of the solution array and insert it to the jth position. Operator Insert(i, j) is a versatile move as it is possible to add or remove a bay, or to change the number of departments within bays. Operator Swap(i, j) swaps the elements at the ith and jth positions of the solution array. Swap(i, j) is performed only if both elements at the ith and jth positions are departments. Swap move does not change either the relative locations of the bay breaks in the current layout or the number of the departments in bays.

2.3 PTS-LP Approach

Given a candidate solution π, the actual locations of the departments are determined using an LP formulation, called Problem $RFBSLP(\pi)$ (Kulturel-Konak 2012). Problem $RFBSLP(\pi)$ uses the tangential support approximation for non-linear department area constraints from Sherali et al. (2003). The notation and variables used in Problem $RFBSLP(\pi)$ and the PTS-LP approach are as follows:

N, NB	Number of departments and bays, respectively
$NB(\pi)$	Number of bays in π
$b(i)$	Index of the bay where department i is located
$k(i)$	Position of department i (from the bottom) in bay $b(i)$
$\pi(b, k)$	Kth department (from the bottom) located in bay b
N_b	Number of departments in bay b
(x_i, y_i)	Coordinates of the centroid of department i
d_{ij}^x, d_{ij}^y	Distances between the centroids of departments i and j in the x and y axes directions
l_i^x, l_i^y	Side lengths of department i in the x and y axes directions
ub_i^x, ub_i^y	Maximum side lengths of department i in the x and y axes directions
lb_i^x, lb_i^y	Minimum side lengths of department i in the x and y axes directions
X_b	Location of bay break b in the direction of the x axis
h_i	The facility boundary violation of department i in the y axis direction
w_i	The facility boundary violation of department i in the x axis direction
f_{ij}	Amount of material flow between departments i and j
Δ	Number of tangential support points
π^*	The best solution in $N(\pi)$
π^{**}	The best-feasible-solution-so-far
t_{NI}	Number of iterations in which π^{**} has not been updated
P	Department pairs with positive flows, $P = \{(i, j): i < j, f_{ij} > 0\}$
C	Penalty term due to infeasible departments, $C = \sum_{(i,j) \in P} f_{ij}(W + H)$.

Problem $RFBSLP(\pi)$.

$$\min z = \sum_{(i,j) \in P} f_{ij}(d_{ij}^x + d_{ij}^y) + C(\sum_i h_i + \sum_i w_i)$$

Subject

$d_{ij}^x \geq \|x_i - x_j\|$	$\forall (i,j) \in P$
$d_{ij}^y \geq \|y_i - y_j\|$	$\forall (i,j) \in P$
$lb_i^x \leq l_i^x i^x$	$\forall i$
$lb_i^y \leq l_i^y i^y$	$\forall i$
$X_{b(i)-1} \leq x_i - 0.5 l_i^x$	$\forall i : b(i) > 1$
$X_{b(i)} i + 0.5 l_i^x$	$\forall i$
$x_i + 0.5 l_i^x - w_i \leq W$	$\forall i : b(i) = NB(\pi)$
$y_i + 0.5 l_i^y - h_i \leq H$	$\forall i : k(i) = N_{b(i)}$
$x_i - 0.5 l_i^x \geq 0$	$\forall i : b(i) = 1$
$y_i - 0.5 l_i^y \geq 0$	$\forall i : k(i) = 1$
$y_j - 0.5 l_j^y \geq y_i + 0.5 l_i^y$	$\forall i,j : b(i) = b(j)$ and $k(j) = k(i) + 1$
$a_i l_i^x + 4 \bar{x}_{ip}^2 l_i^y \geq 4 a_i \bar{x}_{ip}$,p	
$w_i, h_i, x_i, y_i, l_i^x, l_i^y \geq 0$	$\forall i$
$X_b \geq 0$	$\forall b$
$d_{ij}^x, d_{ij}^y \geq 0$	$\forall (i,j) \in P$

Problem $RFBSLP(\pi)$ determines the shape and locations of the departments within the bays while it ensures that departments don't overlap and department area constraints are satisfied. Problem $RFBSLP(\pi)$ can also be solved effectively because the FBS provides a tight LP lower bound.

Procedure PTS-LP

Step 1. Set $t = 1$. Randomly generate an initial solution.

Step 2. Apply Insert(i, j) on current solution $\pi\pi$ with probability λ_t for $i = 1,\ldots, N + NB, j = 1,\ldots,N + NB$ to create new solutions. Add new solutions to $N(\pi)$.

Step 3. Apply Swap(i, j) on current solution s with probability λ_t for $i = 1,\ldots, N + NB-1, j = i+2,\ldots,N + NB-1$ to create new solutions. Add new solutions to $N(\pi)$.

Step 4. Evaluate non-tabu solutions in $N(\pi)$ by solving Problem $FBSLP(\pi)$ of Kulturel-Konak (2012) optimally. Determine the best solution in $N(\pi)$, π^*, and add π^* to the tabu list. Set $\pi = \pi^*$

Step 5. Set $t = t+1$ and $t_{NI} = t_{NI} + 1$. If π^* improves upon π^{**}, set $\pi^{**} = \pi^*$ and $t_{NI} = 0$.
Step 6. Set $\lambda_t = 1 - t_{NI}(1 - \lambda_{min})/100$, and if $t < t_{max}$ go to Step 2.
Step 7. Increase Δ and evaluate π^{**} one more time.

3 Comparative Results

One of the main motivations of this paper is to remedy drawbacks of the FBS when the facility has empty space as in the case of problems F10, BA12, BA13, BA14, Tam20, Tam30, SC30, and SC35. In Table 1, the previous best-known FBS solutions of these problems are given. In Table 2, the best results found by the PSO-LS (Kulturel-Konak and Konak 2011) and the PTS-LP (Kulturel-Konak 2012) are compared. In this paper, the PTS-LP parameters are: $\lambda_{min} = 0.1$, $t_{max} = 2{,}000$, $\Delta = 20$ during search, $\Delta = 50$ for the best-feasible solution). Both approaches using RFBS improved their previous best-known FBS solutions. The PTS-LP was able to further improve the solution quality in each problem and the improvement is up to 5.61 %. However, the PSO-LS is much faster than the PTS-LP in finding the solutions.

Table 1 The best-known FBS solutions of the test problems with empty spaces

Problem	Best-known FBS	Reference
F10	9,020.75	
BA12	8,299.50	Wong and Komarudin (2010)
BA14	4,904.67	
Tam20	9,003.82	
Tam30	19,667.45	
SC30	3,679.85	Wong and Komarudin (2010)
SC35	3962.72	Wong and Komarudin (2010)

Table 2 Comparative results of approaches using RFBS

Problem	PSO-LS (Kulturel-Konak and Konak 2011)	CPU sec PSO-LS	PTS-LP (Kulturel-Konak 2012)	CPU sec PTS-LP	%Improve over PSO-LS
F10	9,020.75	2	8,583.53	689	4.85
BA12	8,129.00	10	8,021.00	1,988	1.33
BA14	4,780.91	19	4,696.37	3,725	1.77
Tam20	8,753.57	104	8,454.38	13,321	3.42
Tam30	19,462.41	924	19,322.98	43,167	0.72
SC30	3,443.34	873	3,362.17	14,120	2.36
SC35	3,700.75	1,842	3,493.18	18,847	5.61

Relaxed Flexible Bay Structure

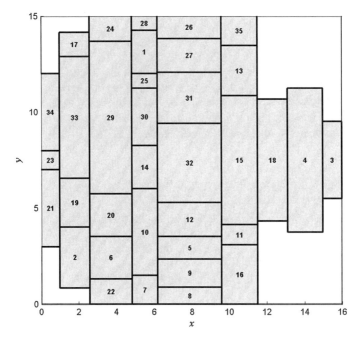

Fig. 2 The best RFBS solutions found for problem SC35 by the PSO-LS

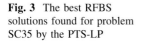

Fig. 3 The best RFBS solutions found for problem SC35 by the PTS-LP

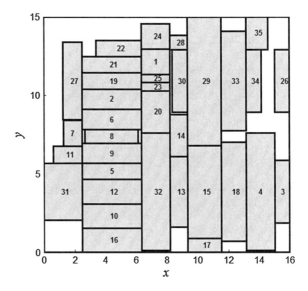

The detailed layouts of the new best-solutions found for SC35 using both approaches are given in Figs. 2 and 3. As seen from best layouts found, the PTS-LP can use the empty space in the facility much more efficiently than the PSO-LS; therefore, it is able to fine tune the layout.

4 Conclusions

This new approach, called RFBS, provides significant reduction in material handling cost since a larger number of department combinations can be considered in bays without violating department shape constraints. However, the results have shown that not only lower costs, but also more practical layouts can be obtained by the new relaxed FBS approach. It should be noted that the RFBS representation is still more restrictive compared to the general representation used in many exact formulations.

References

Banerjee P, Zhou Y, Montreuil B (1997) Genetically assisted optimization of cell layout and material flow path skeleton. IIE Trans 29:277–291

Castillo I, Westerlund T (2005) An e-accurate model for optimal unequal-area block layout design. Comp and Oper Res 32:429–447

Drira A, Pierreval H, Hajri-Gabouj S (2007) Facility layout problems: a survey. Annu Rev Control 31:255–267

Gau KY, Meller RD (1999) An iterative facility layout algorithm. Int J Prod Res 37:3739–3758

Komarudin, Wong KY (2010). Applying ant system for solving unequal area facility layout problems. Eur J Oper Res 202 (3):730–746

Konak A, Kulturel-Konak S, Norman BA, Smith AE (2006) A new mixed integer programming formulation for facility layout design using flexible bays. Oper Res Lett 34:660–672

Kulturel-Konak S (2012) A linear programming embedded probabilistic tabu search for the unequal-area facility layout problem with flexible bays. Eur J Oper Res 223:614–625

Kulturel-Konak S, Konak A (2011) A new relaxed flexible bay structure representation and particle swarm optimization for the unequal area facility layout problem. Eng Optimiz 43(12):1263–1287

Kulturel-Konak S, Norman BA, Coit DW, Smith AE (2004) Exploiting tabu search memory in constrained problems. INFORMS J Comput 16:241–254

Kulturel-Konak S, Smith AE, Norman BA (2007) Bi-objective facility expansion and relayout considering monuments. IIE Trans 39:747–761

Kusiak A, Heragu SS (1987) The facility layout problem. Eur J Oper Res 29:229–251

Meller RD, Bozer YA (1996) A new simulated annealing algorithm for the facility layout problem. Int J Prod Res 34:1675–1692

Meller RD, Gau KY (1996) Facility layout objective functions and robust layouts. Int J Prod Res 34:2727–2742

Meller RD, Narayanan V, Vance PH (1998) Optimal facility layout design. Oper Res Lett 23:117–127

Montreuil B (1990) A modeling framework for integrating layout design and flow network design. In: Proceedings of the 8th international material handling research colloquium, 19–21 June, Hebron, KY. Charlotte, NC, 43–58

Scholz D, Petrick A, Domschke W (2009) STaTS: a slicing tree and tabu search based heuristic for the unequal area facility layout problem. Eur J Oper Res 197:166–178

Sherali HD, Fraticelli BMP, Meller RD (2003) Enhanced model formulations for optimal facility layout. Oper Res 51:629–644

Singh S, Sharma R (2006) A review of different approaches to the facility layout problems. Int J Adv Manuf Technol 30:425–433

Wong KY, Komarudin (2010) Solving facility layout problems using flexible bay structure representation and ant system algorithm. Expert Syst Appl 37:5523–5527

Comparisons of Different Mutation and Recombination Processes of the DEA for SALB-1

Rapeepan Pitakaso, Panupan Parawech and Ganokgarn Jirasirierd

Abstract This paper aims to propose comparisons of mutation and recombination processes of differential evolutionary algorithm (DEA) to solve a simple assembly line balancing problem in which the cycle time is given, in order to minimize a number of workstations (SALBP-1). Firstly, we apply general DEA to solve SALBP-1 which has following steps: (1) generating an initial set of solutions, (2) applying mutation, (3) recombination, and (4) selection process. To extend our general DEA, we present 5 different types of mutations and 3 recombination processes to enhance the search capability of DEA. From the computational results we can conclude that the proposed heuristics can find 100 % optimal solution out of 64 test instances which is better than the algorithm proposed in the literature.

Keywords Simple assembly line balancing · Differential evolution algorithm · Binomial recombination · Exponential recombination

1 Introduction

An assembly line balancing problem (ALBP) is a problem that tries to assign a certain number of tasks to some workstations that lay along the conveyor belt or other logistic tools that perform similar functions. The assignment process is done

---

R. Pitakaso (✉) · P. Parawech · G. Jirasirierd
Department of Industrial Engineering, Ubonratchathani University, Ubonratchathani, Thailand
e-mail: enrapepi@ubu.ac.th

P. Parawech
e-mail: dekkop@hotmail.com

G. Jirasirierd
e-mail: ganokgarn@hotmail.com

when we want to keep some conditions such as predecessor constraints, cycle time constraints etc., in order to provide the best pre-defined objective functions such as a minimal number of workstations, minimal cycle time, maximal assembly line efficiency etc. These objectives can be considered individually or simultaneously (Scholl 1999; Baybars 1986).

Assembly line balancing problems (ALBPs) are normally solved and considered as precedent graphs. The graph consists of some numbers of nodes that correspond to a number of tasks. Each node in the graph represents a specific task, while an edge connecting two nodes represents the precedent relation between the corresponding tasks. Figure 1 illustrates an example of a precedent graph for an 8-tasks ALBP having processing times between 3 and 9 time units. The precedent constraints for example, task 6 must proceed after the completion of task 4, task 1, and task 2.

The simple assembly line balancing problem (SALBP) has been being distinguished into 2 types which are (1) SALBP-1 to minimize the number of stations for a given fixed cycle time, and (2) SALBP-2 to minimize the cycle time for a given number of stations. SALBP-1 is normally used when to set the new assembly line, while the SALBP-2 is used when to change the production process in the existing assembly line. In this paper, we focus only on SALBP-1 because it is one of the most used in textile industry which is now the blossoming industry in Thailand. SALBP-1 has a huge number of exact and heuristic techniques that are available in the literature. Exact algorithms are mostly based on the branch and bound method (Johnson 1988; Hoffmann 1992; Scholl and Klein 1997). Recently, some researchers turn their directions toward meta-heuristics techniques such as genetic algorithm, particle swam optimization and differential evolution algorithm (Kim et al. 1996; Andreas and Nearchou 2005; Andreas 2007). DEA presented in this paper, uses 5 different mutations and 3 recombination processes and compare their performance based on computational time used to find best solution with the solutions presented in the literature. This paper is organized as followed: Sect. 2 presents the proposed heuristic (DEA), Sect. 3 presents the computational results of the test instances, and Sect. 4 is the conclusion of the whole paper.

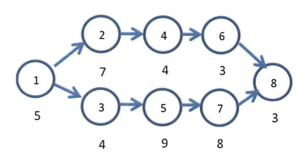

Fig. 1 Example of precedent diagram of ALBPs

2 Proposed Heuristic

In this paper, we present the differential evolution algorithm (DEA) to solve SALBP-1. We will now discuss about the general procedure for DEA which consists of several steps which are (1) construct a set of initial solution, (2) perform the mutation, (3) obtain new generation of solutions by recombination and selection process. The general procedure of DEA is shown in Fig. 2.

Figure 2 represents the general DE algorithm to solve general problems. The notations used in Fig. 2 are as followed: *NP* is predefined number of population. Ω is a vector representing the solution of the problem. Detail of applying DEA to solve SALBP-1 is to explain in the following sections.

2.1 Randomly Generated Vector Ω (Target Vector)

First step of the proposed algorithm is to generate a set of vector Ω, a number of vector Ω that will be randomly generated equals to NP (number of population which is the predefined parameter). Each vector composes of D components while D equals to number of tasks that will be assigned to a certain number of work stations. For example, in Fig. 1, the number of tasks shown is 8 thus each vector Ω will has dimension of 8 as shown in Fig. 3. If we set *NP* to 3 thus three vectors which has dimensions of 8 components will be generated as shown in Fig. 3.

The numbers shown in Fig. 3 represent values of target vector $X_{i,j,G}$, while *i* is population label. *j* is a component label which corresponds to vector dimension (D). In this case, vector dimension is 8 or D = 8 and *j* is running from 1 to 8 which has value of 0.92, 0.08,...,0.44 consecutively. The values of each component in a vector are randomly picked between predefined minimum value and maximum value. In our algorithm, we set minimum value as 0 and maximum value as 1.

Procedure of Differential Evolution algorithm (DEA)
Begin randomly generate a set of target vectors •;
    (Number of vectors that are generated equals to NP)
    while termination condition not satisfied **do**
        for j=1 to NP **do**
        Perform mutation process using formula (6)-(10)
        Perform crossover process using formula (11),(13)
        Perform selection process using formula (14)
        end for
    end while
End Begin

Fig. 2 The general procedure of DE algorithm

popu-lation	1	2	3	4	5	6	7	8
1	0.92	0.08	0.65	0.05	0.99	0.02	0.68	0.44
2	0.75	0.88	0.45	0.52	0.12	0.98	0.54	0.34
3	0.79	0.01	0.56	0.39	0.76	0.82	0.94	0.03

Fig. 3 A chromosome represented SALBP-1

The vector that is shown in Fig. 3, is the vector that is used in DEA, obtaining the result for assembly line balancing, we need to transform the target vector to illustrate the SALBP-1 solution. The transforming procedure shows in Fig. 4.

If the cycle time of a production line in Fig. 1 is set to 12 time units. The first assignment is done by assigning task 1 to workstation 1. Currently, workstation 1 has $U = 0$ or remaining processing time is 12 units (full cycle time). Then, task 1 is assigned to workstation 1. After the first assignment, i is set to 2. Now, we have to find the candidate tasks to enter workstation 1. After task 1 is assigned to station 1, task 2 and task 3 can now enter station 1 thus list $\theta = \{2, 3\}$ and they have component values in list $\theta = \{0.08, 0.65\}$. Then task 2 will be assigned to station 1 because it has the least value of the component in the vector. Currently, station 1 now has $U = 5$ thus the remaining processing time in station 1 equals to $12 - 5 = 7$. Processing time of task 2 equals to 7 thus task 2 can be assigned to station 1. Due to station 1 has no remaining time to put a single task in then work station 2 will be opened and re-done all processes mentioned above until all ask is assigned to a work station then the transforming process is terminated. The result of the transforming procedure is shown in Fig. 5.

Start a procedure of vector transforming
Begin with Set $K=1$ and $i=1$; $U=0$;
    While $i<= N$ **do**
        Find a candidate list that can be assigned to work station K
        considered precedent graph and name it as list •
        Select the task that correspond to the component in a vector that has
        the least value of a vector component in •and name it as task •
        Check processing time of • (P •) and set $U=U+P$ •
    if $U<=$cycle time **do** Assign •to station K and set $i=i+1$;
        else do $K=K+1$; $U=0$;
    end-if
    end-while

Fig. 4 A general procedure of the vector transforming

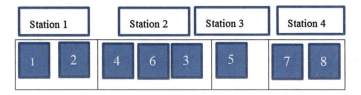

Fig. 5 A result of transforming process

2.2 Perform Mutation Process of Vector Ω

In the mutation process the mutant vector ($V_{i,j,G+1}$) will be generated from the target vector ($X_{i,j,G}$) by process mentioned in Sect. 2.1. Traditionally, the mutation process of DEA is performed using formula (6) in our algorithms we apply 5 mutation formulas drawn from Qin et al. (2009) which shown in formulas (6)–(10).

Denote r1, r2, r3, r4, r5 is the random vector which randomly select from a set of vector Ω. $X_{best,G}$ represents best vector found so far in the algorithm. F is predefined integer parameter, in the proposed heuristics is set equals to 2. i is vector number which run from 1 to NP and j is component number which run from 1 to D.

2.3 Perform Recombination Process on Mutant Vector $V_{ij,G+1}$

The result of mutation process is a set of mutant vectors $V_{i,j,G+1}$ which will apply the recombination process and get the trial vector ($U_{i,j,G+1}$) as the product of the recombination process. In the traditional DEA the binomial recombination formula (11) is applied additionally. In our algorithm, we use 1 exponential recombination formula (12) which drawn from Qin et al. (2009) and develop 1 more exponential recombination formula (13)

$$U_{i,j,G+1} = \begin{cases} V_{i,j,G+1} & \text{if } rand_{i,j} \leq CR \text{ or } j = Irand \\ X_{i,j,G} & \text{if } rand_{i,j} > CR \text{ or } j \neq Irand \end{cases} \quad (11)$$

$$U_{i,j,G+1} = \begin{cases} V_{i,j,G+1} & \text{when } randb_{i,j} \\ X_{i,j,G} & \text{if } randb_{i,j} \end{cases} > j \quad (12)$$

$$U_{i,j,G+1} = \begin{cases} V_{i,j,G+1} & \text{when } j \leq rand\, b_{i,j,1} \text{ and } j \geq rand\, b_{i,j,2} \\ X_{i,j,G} & \text{when } rand\, b_{i,j} < j < rand\, b_{i,j,2} \end{cases} \quad (13)$$

Let $rand_{i,j}$ be a random number run from 0 to 1 and CR is the recombination probability which is the predefined parameter in the proposed heuristics which is set to 0.8. $Irand$ is random integer number. Denote $randb_{i,j}$, $randb_{i,j,1}$ and $randb_{i,j,2}$

as random integer number which used to represent the component position of the mutant vector and they can run from 1 to D. Formula (12) shows that component 1 to component $randb_{i,j}$ in trial vector will equal to mutant vector ($V_{i,j,G+1}$) the remaining components will equal to target vector ($X_{i,j,G}$). Formula (13), the trail vector will equal to mutant vector only the component that has j less than or equals to $randb_{i,j,1}$ and is greater than or equals to $randb_{i,j,2}$. The remaining components will copy value from the target vector.

2.4 Perform Selection Process on the Trial Vector $V_{i,j,G+1}$ and the Target Vector $X_{i,j,G}$

The result of selection process will be the target vector in the next generation which will be used as the starting point of mutation process of the next iteration. The selection process is applied using formula (14)

$$X_{i,j,G+1} = \begin{cases} U_{i,j,G+1} & \text{if } f(U_{i,j,G+1}) \leq f(X_{i,j,G}) \\ X_{i,j,G} & \text{otherwise} \end{cases} \quad (14)$$

Let $f(U_{i,j,G+1})$ and $f(X_{i,j,G})$ be the objective function of the trial vector ($U_{i,j,G+1}$) and the target vector ($X_{i,j,G}$). The objective function of the trial and target vector can be transformed using the procedure that is mentioned in Sect. 3.1.

Table 1 Computation result of the proposed heuristics

Instance No-N-CT	Op.	An.	DE 1	2	3	4	5	6	7	8	9	10	11	12	13	14	15
1-7-6	6	6	6	6	6	6	6	6	6	6	6	6	6	6	6	6	6
2-7-7	5	5	5	5	5	5	5	5	5	5	5	5	5	5	5	5	5
3-8-20	5	5	5	5	5	5	5	5	5	5	5	5	5	5	5	5	5
4-11-48	4	4	4	4	4	4	4	4	4	4	4	4	4	4	4	4	4
5-11-62	3	3	3	3	3	3	3	3	3	3	3	3	3	3	3	3	3
6-9-6	8	8	8	8	8	8	8	8	8	8	8	8	8	8	8	8	8
7-9-7	7	7	7	7	7	7	7	7	7	7	7	7	7	7	7	7	7
8-11-10	5	5	5	5	5	5	5	5	5	5	5	5	5	5	5	5	5
9-21-14	8	8	8	8	8	8	8	8	8	8	8	8	8	8	8	8	8
10-28-342	3	3	3	3	3	3	3	3	3	3	3	3	3	3	3	3	3
11-30-25	14	14	14	14	14	14	14	14	14	14	14	14	14	14	14	14	14
12-30-30	12	12	12	12	12	12	12	12	12	12	12	12	12	12	12	12	12
13-70-176	21	22	21	21	21	21	21	21	21	21	21	21	21	21	21	21	21
14-70-364	10	10	10	10	10	10	10	10	10	10	10	10	10	10	10	10	10
15-83-8412	10	10	10	10	10	10	10	10	10	10	10	10	10	10	10	10	10
16-111-5755	27	28	27	27	27	27	27	27	27	27	27	27	27	27	27	27	27
17-111-8847	18	19	18	18	18	18	18	18	18	18	18	18	18	18	18	18	18
18-111-10027	16	16	16	16	16	16	16	16	16	16	16	16	16	16	16	16	16

Table 2 Summary of the computational result

	DE															An.
	1	2	3	4	5	6	7	8	9	10	11	12	13	14	15	
N-opt	64	64	64	64	64	64	64	64	64	64	64	64	64	64	64	61
AvC	0.65	0.54	0.51	0.65	0.55	0.54	0.49	0.45	0.35	0.68	0.54	0.54	0.72	0.71	0.71	2.0

3 Computational Framework and Result

We test our proposed heuristics with 64 test instances drawn from website: http://alb.mansci.de/index.php?content=search&content2=search&content3=classify-add2&content4=classify-add&searchstring=data%20set&sdl=true and compare our algorithm with Andreas and Nearchou (2005). In Table 1 we show computational result only for 18 out of 64 test instances but we will conclude the computational result of all test instances in Table 2. Our proposed algorithms compose of 15 sub-algorithms which are the combinations of 5 different mutation process and 3 recombination processes (DE-1 uses mutation formula (6) and recombination formula (11) while DE-2 uses mutation formula (6) and recombination formula (12) etc.). We test our algorithms on a computer that is run by Intel CPU core i5-2476 M at 1.6 GHz. We run our algorithm 5 times and each time we run 100 iterations. Let we define following abbreviations used in both table as following : *No* is a number of test instances, *N* is number of tasks a test instances, *CT* is the given cycle time of a test instances, *Op* is the optimal solution found in the literature, *An* is the best result published by Andreas and Nearchou (2005), *Nopt* is number of optimal solution found by an algorithm and *AvC* is an average computational time used by the algorithm.

From the computational results, the proposed heuristics, DE-1 to DE-15, can find 100 % optimal solution out of 64 test instances which are better than Andreas and Nearchou (2005) which can find optimal solution 61 cases. Andreas and Nearchou (2005) use average computational time 2.00 s while our algorithm use 0.35–0.72 s to execute our algorithm.

4 Conclusion

In this paper, we present the differential evolution algorithm (DEA) to solve SALBP-1. We present 15 DEA algorithms which use different mutations and recombination formulas. From the computational results, we find that our algorithms can find 100 % optimal solution while the best algorithm published so far can find 91.05 % optimal solution. We also use 64–84 % less computational time than that of Andreas and Nearchou (2005). Moreover, we reveal that the best combination of mutation and recombination process judged from the least computational time is using mutation formula (9) and recombination formula (13) which use 84 % computational time less than that of Andreas and Nearchou (2005).

References

Andreas CN (2007) Balancing large assembly lines by a new heuristic based on differential evolution method. Int J Adv Manuf Technol 34:1016–1029

Andreas CN, Nearchou (2005) A differential evolution algorithm for simple assembly line balancing. Paper present in the 16th International federation of autimatic control (IFAC) world congress, Prague, 4–8 July 2005

Baybars I (1986) A survey of exact algorithms for the simple assembly line balancing problem. Manage Sci 32:909–932

Hoffmann TR (1992) EUREKA: a hybrid system for assembly line balancing. Manage Sci 38:39–47

Johnson RV (1988) Optimally balancing large assembly lines with "FABLE". Manage Sci 34:240–253

Kim YK, Kim Y JU, Kim Y (1996) Genetic algorithms for assembly line balancing with various objectives. Comput Ind Eng 30:397–409

Qin AK, Huang VL, Suganthan PN (2009) Differential evolution algorithm with strategy adaptation for global numerical optimization. IEEE Trans Evol Comput 13:398–417

Scholl A (1999) Balancing and sequencing of assembly lines. Physica-Verlag publication, Germany

Scholl A, Klein R (1997) SALOME: a biderectional branch and bound procedure for assembly line balancing. INFORMS J Comput 9:319–334

An Exploration of GA, DE and PSO in Assignment Problems

Tassin Srivarapongse and Rapeepan Pitakaso

Abstract The optimum solution of assignment problem might not be found by using exact method for a large size assignment problem. The heuristic methods that will take acceptable time should be introduced to solve the problem. The common and well-known methods that are used in this study are genetic algorithm (GA), differential evolution (DE), and Particle Swarm Optimization (PSO). The goal of these three algorithms is the best solutions with appropriate time. We believe one of these three algorithms stands out from the others. We will focus on the best solution and shortest time consuming. In this study, we will classify our study into 3 parts. Firstly, we investigate similarities and differences of each heuristic. Secondly, test our theory by using the same assignment cost matrix. Finally, we compare the results and make conclusions. However, different problem might get different results and conclusions.

Keywords Genetic algorithm · Differential evolution · Particle swarm optimization · Assignment problem

1 Introduction

The assignment problem is the problem that needs to find an exact match between agent and job. One agent comes with his experiences which make him specialization. When an agent with specialization does a job, he will do it with his full

T. Srivarapongse (✉) · R. Pitakaso
Department of Industrial Engineering, Faculty of Engineering, Ubon Ratchathani University, Ubonratchathini, Thailand
e-mail: tassin66@hotmail.com

R. Pitakaso
e-mail: enrapepi@ubu.ac.th

experiences. If we compare this agent with the other agents that do not have a specialization, the agent without a specialization will produce more output compare to an agent with less specialization. We can say this is "Give a right man on a right job". The cost of giving a right agent to a right job will be low. If we can match all pairs of agent and jobs, the overall cost will be the best solution.

There are two main directions to solve the assignment problem. Firstly, the exact method is a technique that can find the best solution. These are some examples of the exact method: Hungarian method, Branch and bound, simplex method and so on. We also have a computer program such as LINGO is used to find an exact answer. Secondly, heuristic or metaheuristic are alternative method to find an answer that is close to the best answer within acceptable time. Heuristic method is tailored made for a specific problem. Metaheuristic method can solve general problems or can be applied to many problems.

One of the most important different between those two ways is the time that are used in order to find answer. The answer from exact method is the best answer that is time consuming because of it concerns all possible answers. The heuristic or metaheuristic can find the acceptable answer within satisfied time. For small size problem, exact method and heuristic or metaheuristic tend to be no different in time consuming. On the other hand, large size problem must take longer time for exact method but shorter time for heuristic or metaheuristic.

Heuristic is tailored for specific problem. It cannot use to solve general problem. For general problem, we have to switch to metaheuristic. We select some of metaheuristic methods which are more general methods. They can adopt to fit in many problems. Some of them can be suitable to solve in one problem but might not be fit in another problems. The selected problems are GA, DE and particle swarm optimization (PSO). GA and DE use some common techniques such as crossover and mutation which come from an ordinary reproduction of any life on earth. PSO adopt techniques of animal instinct to find their food like a swarm of bees or school of fish.

2 Genetic Algorithm

GA is an algorithm that uses a genetic of life form to create a new generation. GA is a consist of

1. Chromosome encoding
2. Initial population
3. Fitness function
4. Genetic operator (Selection, Crossover, and Mutation)

Chromosome encoding

In this study, we use value encoding for both of agents and jobs. The method that we use in chromosome encoding will cause the efficiency of algorithm. Time consumption will be more or less depends on this encoding (Table 1).

Table 1 Assignment cost table

Agents	Jobs										Total	Rank
	1	2	3	4	5	6	7	8	9	10		
1	89	70	38	94	4	70	18	15	79	6	483	2
2	57	41	4	47	88	83	46	43	18	75	502	3
3	23	53	87	12	45	97	8	65	68	82	540	5
4	33	73	66	63	86	3	4	12	85	80	505	4
5	86	49	33	21	6	50	9	37	30	2	323	1
6	74	11	31	98	76	94	82	43	83	52	644	10
7	67	6	75	62	51	2	97	65	40	89	554	6
8	82	34	5	88	45	27	76	63	49	89	558	7
9	33	19	89	100	82	87	3	82	83	42	620	9
10	26	76	87	94	2	83	24	69	85	54	600	8

The agent that will be the first assignment has the minimum sum of overall cost of each agent. This example, agent 5 is selected to be the first assigned agent. Thus, job that is assigned to agent 5 is job 10 because assignment cost of agent 5 to job 10 has lowest cost. After we assign a job to an agent, we will cut it off. Next, the second agent is agent 1 and he is assigned to job 5. It continues this process until the last agent is selected. However, if the minimum cost of assigning an agent to a job is already selected for that job, we will choose a next minimum cost and assign this job for this agent. For instances, the minimum cost of rank 7 or agent 8 is 5 for job 3 but job 3 is already assigned for agent 2. Therefore, we will assign job 2 for agent 8 instead of job 3. These are chromosome encoding that we use for all algorithms.

Initial population

For the first generation, we will use greedy random as a method to select initial population. The minimum cost of assignment should have more opportunity to be selected. We will convert assignment cost by using one over cost. Thus, the lowest cost will become new value that have more chance to be select.

Fitness function

After we have initial population, we have to define the fitness function that minimizes objective function under some constraints. The important constraints are the one to one assignment problem. Only one agent must be assign to a specific job. At the same time, only one job must be assign to a specific agent. The fitness function is the objective function and constraints are specified here.

The objective function

$$\min \sum_{i=1}^{n} \sum_{j=1}^{n} a_{ij} x_{ij}$$

Constraints

$$\sum_{j=1}^{n} x_{ij} = 1$$

$$\sum_{j=1}^{n} x_{ij} = 1$$

$$x_{ij} \in \{0, 1\}$$

Decision variable

$$x_{ij} \begin{cases} 1 & \text{if agent i is assigned to job j} \\ 0 & \text{otherwise} \end{cases}$$

a_{ij} is an assignment cost for assigning agent i to job j

This fitness function can be used again in other algorithms. Therefore, the fitness function in others will refer to this function.

Genetic operators

GA starts from selecting first generation parents from initial solutions. After selection, we will make them breeding and we called this is crossover. The crossover in this study, we use 2 points crossover method. For second generation, we make a decision that the new generation will mutate for how much depends on mutation probability which is 10 %. If we have 100 genes in each chromosome, we will have 10 % mutation or 10 genes. For this research, we will use insertion mutation as a technique for mutation. First, we select gene that we want to insert then select the insertion point. When it is done, we compute fitness function. We will use this new generation to be a parent for the next one. GA process will repeat over and over again until satisfied end condition.

GA Pseudo Code

1. Chromosome encoding

Value encoding
2. Generate initial solutions
3. Compute fitness functions
4. Select parents to be first generation
5. Crossover

Two point crossover
6. Mutation

Insertion mutation
7. Repeat 4, 5, and 6 until meet end requirement

3 Differential Evolution

DE is an evolutionary algorithm. Some parts of DE use the same name as GA which are mutation and selection. For every algorithm, they have to begin with chromosome encoding. This encoding will have an effect to all algorithms thus we will use the same encoding which is value encoding.

The final solutions should not be difference. The process of DE is as follow.

1. Initialization
2. Mutation
3. Recombination
4. Selection

Initialization

Firstly, define number of parameter and populations. Then specify upper and lower bound of each parameter. The samples that are in preferred range will be randomly selected. This process is called initialization and will be used in next step.

Mutation

All of selected sample have to go through all processes which are mutation, recombination and selection over and over again until meet end condition. Mutation will increase in the search area. It tends to move away from local area to another area. It might be close to the global optimization. Mutation vector is a vector that use three vectors combine together. The new mutated vector comes from one vector plus difference between another two vectors. This will move from one place to another place. It might be find a global optimization.

$$v_{i,G+1} = x_{r1,G} + F(x_{r2,G} - x_{r3,G})$$

where F is mutation factor which is between 0 and 2 $v_{i,G+1}$ is called the donor vector.

Recombination

The recombination is a selection between the donor vector ($v_{i,G+1}$) and the target vector ($x_{i,g}$). The chosen vector is called the trial vector ($u_{i,G+1}$). The donor vector will be the trial vector if the random number is less than CR. CR is the specified probability.

$$u_{j,i,G} = \begin{cases} v_{j,i,G} \text{ if } rand_{j,i} \leq CR \text{ AND } j = I_{rand} \\ x_{j,i,G} \text{ if } rand_{j,i} > CR \text{ } j \neq I_{rand} \end{cases}$$

where $i = 1, 2, 3, \ldots, n$
$j = 1, 2, 3, \ldots, d$

$$rand_{j,i} \leq CRj = I_{rand}$$

Selection

$$X_{i,G+1} = \begin{cases} u_{i,G} & \text{if } f(u_{i,G}) \leq f(x_{i,G}) \\ x_{i,G} & \text{otherwise} \end{cases}$$

$$i = 1, 2, \ldots, N$$

DE pseudo code

1. Chromosome encoding

Value encoding
2. Generate initial solution
3. Compute fitness function
4. Mutation
5. Recombination
6. Selection

4 Particle Swarm Optimization

Particle Swarm Optimization (PSO) is an algorithm that applies the perception of social interaction to the problem. PSO adopts social interaction of animals but not just one individual animal such as snake, eagle, and so on. PSO adopts social interaction of animals that depend heavily on their group for instances; swarm of bees, school of fish, or pack of dogs. They live together and also seeking for their food together and stored in their warehouse. One life depends on another one within their group. Hence, this algorithm is called Particle Swarm Optimization (PSO).

Each particle maintains its' path to find best solution. This is called personal best. As mention before, it is not alone thus it also depend on the neighbors. The best solution from the neighbors is called global best. Each particle finds its best solution within its area and tends to move and stick with their groups. Though, they will find the global best solution together.

Pseudo code of PSO

1. Initialize population
2. Evaluate fitness function
3. Find personal best fitness

4. Identify neighbors and find global best fitness
5. Change velocity and positions

Mathematical model for PSO

$$v_i^{k+1} = v_i^k + c_1 rand_1 \left(pBest_i - current_i^k\right) + c_2 rand_2 \left(gBest - current_i^k\right)$$

where v_i^{k+1} = velocity of agent i at iteration k
c_j = weighted factor
$rand_j$ = random number between 0 and 1
$current_i^k$ = current position of agent i at iteration k
$pBest_i$ = personal best of agent i
$gBest$ = global best of the group

5 Results and Conclusions

To prevent unexpected error from chromosome encoding, we will use value encoding for all algorithm in this study. Hence, the results show consequences of each algorithm. To test all algorithms, we generate three different size samples. The small and big samples are 15 × 15 and 100 × 100 respectively. Small size problem runtime is 15, 30 and 60 s. Big size problem runtime is 1:30, 3:00, and 6:00 min. The results that show in these tables are the percentage difference from optimal solutions of each sample size (Tables 2 and 3).

For big size problem, PSO has a better algorithm than GA and DE. It is close to the optimal solution than the others. The result in small size problem is different from bid size problem. DE's results are close to the optimal solution than the others. GA tends to be no different neither big size problem nor small size problem. For the future research, researcher should increase run time for DE on big size problem. Because the results from small size problem are closer to the optimal for both mean and minimum value. Another future topic for researcher is the test

Table 2 Results for big sample size (100 × 100)

Algorithms	Runtime (min)	Mean	Minimum
DE	1:30	26.36	25.15
	3:00	27.88	23.64
	6:00	21.82	19.70
GA	1:30	40.30	30.61
	3:00	25.15	18.48
	6:00	34.55	31.82
PSO	1:30	6.36	0.00
	3:00	1.52	0.91
	6:00	6.36	3.33

Unit percentage of the differences from optimal solution

Table 3 Results for small sample size (15 × 15)

Algorithms	Runtime (s)	Mean	Minimum
DE	15	20.51	19.66
	30	0.00	0.00
	60	0.00	0.00
GA	15	43.59	40.17
	30	44.44	31.62
	60	24.79	0.00
PSO	15	33.33	26.50
	30	19.66	19.66
	60	7.69	0.00

Unit percentage of the differences from optimal solution

of problem on more advance assignment problem such as location-allocation, p-median problem, quadratic assignment problem, generalize assignment problem or the others related problem.

References

Burkard RE (2000) Selected topics on assignment problems. Discrete Appl Math 123(2002):257–302

Das S, Abraham A, Konar A (2008) Particle swarm optimization and differential evolution algorithms: technical analysis, applications and hybridization perspectives. Stud Comput Intell (SCI) 116:1–38

Frieze AM, Yadegar J (1983) On the quadratic assignment problem. Discrete Appl Math 5(1983):89–98

Khan K, Sahai A (2012) A comparison of BA, GA, PSO, BP and LM for training feed forward neural networks in e-learning context. Int J Intell Syst Appl. doi:10.5815/ijisa.2012.07.03

Storn Rainer, Price Kenneth (1997) Differential evolution—a simple and efficient heuristic for global optimization over continuous spaces. J Global Optim 11:341–359

Taillard ED (1995) Comparison of iterative searches for the quadratic assignment problem. Location Sci 3(2):87–105

Author Index

A
Alfian, N., 1413
Ai, J., 1065
Ai, T. J., 1101
AlSaati, A., 1315
Amelia, P., 133
Amnal, A., 1081
Apichotwasurat, P., 861
Arai, Y., 165
Arvitrida, N. I., 133
Aryani, W. D., 1461, 1471
Astanti, R. D., 1065, 1101

B
Babbar, S. K., 1371
Bawono, B., 1101
Bintoro, A. G., 1065

C
Cai, Y., 1513
Chaiwichian, A., 1237
Chaklang, A., 311
Chan, F. T. S., 729, 739
Chan, H., 205
Chang, C., 505
Chang, J., 239, 381, 389, 479
Chang, K., 441
Chang, M., 9
Chang, P., 249, 257, 513
Chen, A., 371
Chen, A. C., 619, 681
Chen, A. H, 633, 665
Chen, A. P., 345
Chen, C., 219, 1191, 1307
Chen, F., 1217
Chen, G., 1001

Chen, G. Y., 907
Chen, H., 871, 975
Chen, K. Y., 521
Chen, L., 1279
Chen, M., 157, 879, 1225
Chen, P., 173
Chen, S., 45, 149
Chen, S. M., 521
Chen, T., 149, 1279
Chen, X., 381, 389
Chen, Y., 149, 265, 271, 569, 783, 871, 1183, 1405
Cheng, C., 271, 819, 1009
Cheng, M. S., 423
Chi, C., 819, 1335
Chiang, C., 665
Chiang, T., 1225
Chiang, Y., 1279
Chien, C., 1157
Chien, Y., 1343, 1421
Chihara, T., 197
Chiou, J., 505
Chiu, C., 219, 1199
Chiu, C. C., 363
Chiu, J., 1137
Chiu, W., 713
Choo, S. L. S., 433
Chou, C., 1279
Chou, S., 1071, 1081
Chow, J. C. L., 729
Chu, C., 73, 287
Chu, T., 1495
Chu, W., 985
Chuang, P., 783
Chuang, W., 531
Chung, C. Y., 423
Chung, S. H., 729, 739
Chung, T., 1377

D
Dai, J., 793
Dewabharata, A., 1071, 1081
de Weck, O., 1315
Dioquino, M. L. A., 549

E
Erawan, M., 1081

F
Fan, S. S., 457, 465
Fang, L., 611
Fang, Y., 1297
Farkhani, F., 889
Fitriani, K., 687
Fu, C., 801
Fu, K. E., 861

G
Gao, S., 775
Gradiz, O., 915
Gutierrezmil, H. Y. L., 633

H
Hao-Huai, C., 487
Haridinuto, 1353
Hartono, B., 1413
Herani, I. R., 1451
Hida, T., 229
Hong, I., 173
Hong, Y. S., 1047
Hsiao, M., 539
Hsieh, C., 371
Hsu, C., 765, 957, 1225
Hsu, M., 695
Hsu, P. Y., 423
Hsu, S., 1217
Hsu, T., 397, 1529
Hsu, Y., 957
Hsueh, K., 257
Hu, C. H., 29
Hu, J., 757
Hu, S., 327
Huang, C., 111, 915, 1441, 1451, 1495
Huang, H., 1025
Huang, K., 1009
Huang, L., 265
Huang, Q., 993
Huang, S., 465
Huang, Y., 577, 705, 1343

Huang, Y., 1137
Huarng, S., 925
Hui, X., 1503, 1521
Hung, T., 879

I
Inayati, I., 1461, 1471
Ishii, N., 181
Iwata, K., 157
Izumi, T., 197

J
Jang, R., 681, 953
Jeng, S., 127
Jewpanya, P., 89
Jha, S., 1503, 1521
Jhang, J., 1173
Jia, W., 559
Jiang, B. C., 73
Jiang, X., 381, 389
Jiang, Z., 559
Jirasirierd, G., 1571
Jou, Y., 593
Juan, Y., 303, 1441, 1451

K
Kachitvichyanukul, V., 89, 103, 749, 1117
Kaji, K., 1165
Kajiyama, T., 165
Kan, C., 345
Kang, S., 907
Kao, C., 899, 1183
Kao, H., 1031
Kao, M., 239
Kao, Y., 1479
Kasemset, C., 353
Ker, C. W., 1109
Ker, J., 1109
Khasanah, A. U., 649
Kisworo, D., 127
Koesrindartoto, D. P., 809
Ku, H., 1199
Kulturel-Konak, S., 1563
Kunnapapdeelert, S., 749
Kuo, C., 189, 287
Kuo, I., 1199
Kuo, P., 189, 319
Kuo, R., 649
Kuo, R. J., 1207
Kuo, Y., 319
Kurniati, N., 1353

Author Index

L
Laio, J., 327
Lee, B., 925
Lee, C., 695, 925, 945
Lee, K., 295, 695, 1039
Lee, M., 29
Lee, P., 415
Lee, T., 457, 899, 1137, 1513
Li, R., 79
Li, S., 249
Li, W., 53
Li, X., 497
Li, Y., 559, 845
Liang, S. M., 837
Liang, Y., 633, 665
Liang, Z., 319
Liao, C., 405, 1025, 1377
Liao, Y., 1405
Lin, C., 381, 389, 569, 775, 1137, 1157
Lin, C. J., 363
Lin, G., 531
Lin, H., 9, 319
Lin, J., 239, 1279
Lin, J. T., 219
Lin, S., 149, 205, 211, 1001, 1017, 1461, 1471
Lin, T., 1183
Lin, W., 641, 649, 1217
Lin, Y., 211, 249, 257
Line, S., 449
Liou, J. J. H., 295
Liu, C., 1289
Liu, J., 103, 1271, 1385
Liu, T., 899, 1183
Liu, X., 1271, 1385, 1395
Liu, Y., 271
Liu, Z., 611
Lo, C., 505
Lo, S., 381, 389
Logrono, D. O., 471
Long, H., 487
Low, C., 79
Lu, B., 845
Lu, K. Y., 423
Lu, S., 935
Lu, Y., 119, 471

M
Maftuhah, D. L., 1055
Maghfiroh, M. F. N., 279, 853
Martinez, I. A. G., 1147
May, M., 673

Menozzi, M., 577, 705
Miao, H., 149
Miller, L., 945
Mori, S., 809
Mustakim, A., 1431

N
Nadlifatin, R., 1017
Nakagawa, T., 157
Nakamura, S., 157
Nakamura, Y., 63
Nanthasamroeng, N., 965
Niu, B., 739
Nomokonova, E., 1001
Nowatari, M., 1539, 1551
Nurtam, M. R., 827

O
Oh, G., 1047
Ouchi, N., 165
Ou-Yang, C., 1431, 1441, 1451

P
Pai, K., 713
Pai, P., 993
Palacios, C., 915
Pan, Y., 975
Parawech, P., 1571
Pei, H., 899
Peng, L., 265
Persada, S. F., 1017
Perwira Redi, A. A. N., 279, 853
Pitakaso, R., 965, 1237, 1571, 1581
Poh, K. L., 1091
Pradhito, N. H., 1071
Praharsi, Y., 141, 549
Prasetio, M. D., 1017
Prim, P., 1513
Pujawan, I. N., 889
Purnomo, H. D., 141

Q
Qin, X., 1513

R
Rajamony, B., 335
Rau, H., 1289, 1297, 1421

Revilla, J. A. D., 1147
Rieza, M., 1441
Robielos, R. A. C., 1361
Rogers, J., 907

S
Saithong, C., 311
Sakhrani, V., 1315
Santosa, B., 1431
Sasiopars, S., 353
Sato, S., 809
Seo, A., 197, 229
Shi. L., 265
Shiang, W., 363
Shih, J., 935
Shih, P., 1199
Shih, Y., 819, 1335
Shimamoto, T., 1251
Shu, Y. S., 441
Siang, S., 1297
Srisom, A., 311
Srivarapongse, T., 1581
Su, L., 327, 539, 1405
Su, Y., 1529
Sulistyo, S. R., 1413
Sun, B., 1521
Sun, I, 1335
Supakdee, K., 965
Suryani, E., 1207, 1451
Suwiphat, S., 353
Syu, J, 303

T
Taguchi, T., 1127
Tanaka, K., 1127, 1165, 1251
Tanaka, R., 1251
Tang, L. C., 1385
Tien, F., 793
Ting, C., 449
Togashi, T., 181
Trappey, A. J. C., 1025
Tsai, C., 111, 985, 993
Tsai, D., 53
Tsai, H., 925, 945
Tsai, P., 757
Tsai, Y., 53
Tseng, Y., 1405
Tsui, K., 657
Tu, C., 985, 1217
Tu, N., 721, 845

V
von Helfenstein, S. B., 1325

W
Wang, C., 73
Wang, F, 119, 1495
Wang, H., 1441
Wang, H. C., 1451
Wang, J. Y., 1343
Wang, K., 889, 1039
Wang, L., 149
Wang, S., 793, 1031
Wang, Y., 1091, 1109
Wee, H. M., 127, 371
Wee, H., 141, 549
Wee, P., 371
Wee, P. K. P., 127
Wen, C., 619
Wen, C. L., 127
Wen, S. Y., 319
Weng, W., 641, 1031, 1487
Wibowo, T. I., 1065, 1101
Widodo, E., 1055
Wirdjodirdjo, B., 133, 1055, 1071
Wisittipanich, W., 1117
Wong, C. S., 729, 739
Wu, C., 1479
Wu, C. H., 801
Wu, G., 79
Wu, K., 19, 37, 1421
Wu, M. X., 521
Wu, S., 53
Wu, W., 471, 957
Wu, Y., 593
Wu, Y. C., 319

X
Xing, Y., 657

Y
Yang, C., 827, 899, 1183
Yang, D., 559
Yang, K., 611
Yang, L., 1233
Yang, M., 757
Yang, S. L., 879
Yap, P. A. B., 433
Yasid, A., 1207
Yasin, M., 1261

Yee, T. S., 335, 433
Yeh, R., 1353
Yeh, W., 295
Yen, C., 1031
Yoshikawa, S., 1
Young, M. N., 441
Yu, C., 265, 585, 603, 625
Yu, C. K., 397, 405, 415
Yu, V. F., 279, 853, 1261
Yuan, Y., 845
Yu-Chien, L., 487

Z

Zain, Z. M., 335
Zhang, C., 1395
Zhang, J., 1165
Zhang, Q., 497
Zhao, S., 1271, 1385
Zhao, X., 157
Zhou, J., 721, 845